Atomic and Molecular Processes in Controlled Thermonuclear Fusion

NATO ADVANCED STUDY INSTITUTES SERIES

A series of edited volumes comprising multifaceted studies of contemporary scientific issues by some of the best scientific minds in the world, assembled in cooperation with NATO Scientific Affairs Division.

Series B: Physics

RECENT VOLUMES IN THIS SERIES

Volume 46 – Nondestructive Evaluation of Semiconductor Materials and Devices
edited by Jay N. Zemel

Volume 47 – Site Characterization and Aggregation of Implanted Atoms in Materials
edited by A. Perez and R. Coussement

Volume 48 – Electron and Magnetization Densities in Molecules and Crystals
edited by P. Becker

Volume 49 – New Phenomena in Lepton-Hadron Physics
edited by Dietrich E. C. Fries and Julius Wess

Volume 50 – Ordering in Strongly Fluctuating Condensed Matter Systems
edited by Tormod Riste

Volume 51 – Phase Transitions in Surface Films
edited by J. G. Dash and J. Ruvalds

Volume 52 – Physics of Nonlinear Transport in Semiconductors
edited by David K. Ferry, J. R. Barker, and C. Jacoboni

Volume 53 – Atomic and Molecular Processes in Controlled Thermonuclear Fusion
edited by M. R. C. McDowell and A. M. Ferendeci

Volume 54 – Quantum Flavordynamics, Quantum Chromodynamics, and Unified Theories
edited by K. T. Mahanthappa and James Randa

Volume 55 – Field Theoretical Methods in Particle Physics
edited by Werner Rühl

Volume 56 – Vibrational Spectroscopy of Molecular Liquids and Solids
edited by S. Bratos and R. M. Pick

This series is published by an international board of publishers in conjunction with NATO Scientific Affairs Division

A Life Sciences B Physics	Plenum Publishing Corporation London and New York
C Mathematical and Physical Sciences	D. Reidel Publishing Company Dordrecht, Boston and London
D Behavioral and Social Sciences	Sijthoff & Noordhoff International Publishers
E Applied Sciences	Alphen aan den Rijn and Germantown U.S.A.

Atomic and Molecular Processes in Controlled Thermonuclear Fusion

Edited by
M.R.C. McDowell
Royal Holloway College, University of London
Surrey, United Kingdom

and
A.M. Ferendeci
Boğaziçi University
Istanbul, Turkey
and
Case Western Reserve University
Cleveland, Ohio

PLENUM PRESS • NEW YORK AND LONDON
Published in cooperation with NATO Scientific Affairs Division

Library of Congress Cataloging in Publication Data

Nato Advanced Study Institute on Atomic and Molecular Processes in Controlled Thermonuclear Fusion, Castéra-Verduzan, France, 1979.
Atomic and molecular processes in controlled thermonuclear fusion.

(NATO advanced study institutes series: Series B, Physics; v. 53)
"Proceedings of the NATO Advanced Study Institute on Atomic and Molecular Processes in Controlled Thermonuclear Fusion, held at Chateau de Bonas, Castéra-Verduzan, Gers, France, August 13–24, 1979."
Includes indexes.
1. Controlled fusion–Congresses. 2. Fusion reactors–Congresses. 3. Collisions (Nuclear physics)–Congresses. I. McDowell, M. R. C. II. Ferendeci, A. M. III. North Atlantic Treaty Organization. IV. Title. V. Title: Molecular processes. VI. Series.
QC791.7.N37 1979 621.48'4 80-238
ISBN 0-306-40424-9

Lectures presented at the NATO Advanced Study Institute on
Atomic and Molecular Processes in Controlled Thermonuclear Fusion,
held at Chateau de Bonas, Castéra-Verduzan, Gers, France,
August 13 – 24, 1979.

A Division of Plenum Publishing Corporation
227 West 17th Street, New York, N.Y. 10011

Printed in the United States of America

PREFACE

The NATO Advanced Study Institute on "Atomic and Molecular Processes in Controlled Thermonuclear Fusion" was held at Château de Bonas, Castera-Verduzan, Gers, France, from 13th to 24th August 1979, and this volume contains the text of the invited lectures. The Institute was supported by the Scientific Affairs Division of NATO, and additional support was received from EURATOM and the United States National Science Foundation. The Institute was attended by 88 scientists, all of whom were active research workers in control of thermonuclear plasmas, or atomic and molecular physics, or both. In addition to the formal lectures, printed in this volume, which were intended to be pedagogic, more than twenty research seminars were given by participants.

The first half of the Institute was directed to introducing atomic and molecular theoretical and experimental physicists to the physics of controlled thermonuclear fusion. Most attention was paid to magnetic confinement, and within that field, to tokamaks. Mr. M. F. A. Harrison of U K A E A Culham Laboratory and the JET project, gave a comprehensive introduction to the many ways in which atomic and molecular physics contributes to the understanding, design and operation of current and proposed devices, in particular drawing attention to processes that are necessary for the operational control of proposed reactors, and the disposal ("exhaust") of burnt fuel. Dr. J. T. Hogan (Oak Ridge National Laboratory, U.S.A.) followed these lectures with a more detailed discussion of the general principles of magnetic fusion confinement in tokamaks, leading into a discussion of short-term trends. Particular emphasis is placed on beam deposition, plasma fueling and the effects of charge exchange.

Dr. Erol Oktay (U.S. Department of Energy) lectured on the atomic and molecular physics of power balance, reviewing recent work on maximum permissible impurity concentrations in various operating regimes. The physics of the plasma-wall interaction was discussed by Dr. K. J. Dietz (JET Project), concentrating on chemical and metallurgical effects due to the large concentration of hydrogen in the walls, the effects of oxygen and of unipolar arcs.

The remainder of the lectures were directed to a survey of the relevant atomic and molecular physics, especially collision processes, and its applications in diagnostics. Professor C. J. Joachain (Brussels) surveyed the present state of knowledge in atomic collision theory, after which topics of special relevance were discussed in more detail by others. Theoretical models for charge exchange (Professor B. H. Bransden (Durham)), ionization by electron impact (Professor I. E. McCarthy (Flinders)), electron impact excitation of impurity ions (Dr. D. Robb (Los Alamos Scientific Laboratory)) and recombination (Dr. C. Bottcher (Oak Ridge National Laboratory)) were chosen for detailed discussion.

The experimental techniques used in measuring cross sections were described by Professor K. T. Dolder (Newcastle) and Dr. F. J. de Heer (F.O.M. Institute for Atomic and Molecular Physics, Amsterdam). Professor Dolder dealt with many aspects of the measurement of electron impact ionization of atoms, excitation by electron impact, and with collisions between charged particles. Dr. de Heer then dealt in more detail with electron capture and ionization by positive ions.

Bound state properties, spectroscopy, and diagnostics were considered in turn in groups of lectures by Professor M. Klapisch (Racah Institute, Jerusalem), Professor I. Martinson (Lund) and Professor E. Hinnov (Princeton). Attention was directed primarily to complex systems in high states of ionization, with on the theoretical side, consideration of configuration interaction and relativistic effects on oscillator strengths, and on the experimental, measurement techniques and problems of interpretation and identification. Professor Hinnov's lectures illustrated the use of spectroscopic diagnostic techniques in experiments on the Princeton Large Torus with special reference to ions of iron.

We hope that in presenting these lectures to a wider public, scientists in many disciplines will find here a comprehensive introduction to the atomic and molecular physics of controlled thermonuclear fusion, which will assist them in helping to formulate and solve the problems posed by the next generation of machines.

We would particularly like to thank Dr. H. Drawin, Professor H. S. Taylor, Mr. M. F. A. Harrison and Professor C. J. Joachain for their advice on the content of the scientific programme. Dr. L. A. Morgan took much of the responsibility for the finances of the conference, and we all relied on Mrs. Margaret Dixon's constant willingness to take on any administrative problem, as well as providing the necessary secretarial assistance. The arrangements for the conference owe much to the forethought and continuous care of Professor J. C. Simon, Director of the Association Scientifique Culturel et Educative de Bonas, and his wife, Françoise. Mr. Paul Allan of Ian Allan (Travel) Ltd. and M. Benhamou of the Syndicat d'Initiative d'Auch helped solve our travel problems, where other organisations failed.

We would also like to thank the lecturers, who not only performed at exactly the right level, but produced their manuscripts on time, and participated fully in the Institute, particularly in that they enabled students to discuss scientific problems with them at all times.

We are deeply indebted to the NATO Scientific Affairs Division, especially to its Director, Dr. M. di Lullo, and the Accountant, Mr. K. Hey, for their encouragement and support throughout, and we very much welcomed the presence throughout the Institute, of Professor A. Boever of the NATO Science Committee, and his wife.

M. R. C. McDowell
A. M. Ferendeci

November 1979

CONTENTS

LIST OF LECTURERS

C. Bottcher
Physics Department, Oak Ridge National Laboratory, Post Office Box X, Oak Ridge, Tennessee 37830, U.S.A.

B. H. Bransden
Department of Physics, University of Durham, South Road, Durham, England.

F. J. de Heer
FOM Institute for Atomic & Molecular Physics, Kruislaan 407, 1098 SJ Amsterdam-Watergraafsmeer, The Netherlands.

K. J. Dietz
Plasma Systems Division, JET Joint Undertaking, Abingdon, Oxfordshire OX14 3EA, England.

K. T. Dolder
Department of Atomic Physics, The University, Newcastle Upon Tyne, NE1 7RU, England.

M. F. A. Harrison
UKAEA/Euratom Fusion Association, Culham Laboratory, Abingdon, Oxon OX14 3DB, England.

E. Hinnov
Plasma Physics Laboratory, Princeton University, James Forrestal Campus, P.O. Box 451, Princeton, New Jersey 08540, U.S.A.

J. T. Hogan
Fusion Energy Division, Oak Ridge National Laboratory, P.O. Box Y, Oak Ridge, Tennessee 37830, U.S.A.

C. J. Joachain
Physique Théorique, Faculté des Sciences, Université Libre de Bruxelles, Code Postal 227, Campus Plaine U.L.B., Boulevand du Triomphe, B-1050 Bruxelles, Belgium.

M. Klapisch
Physics Department, Racah Institute, Hebrew University, Jerusalem, Israel.

I. E. McCarthy
School of Physical Sciences, The Flinders University of South Australia, Bedford Park, South Australia 5042, Australia.

M. R. C. McDowell
Department of Mathematics, Royal Holloway College, (Univer (University of London), Egham Hill, Egham, Surrey TW20 OEX, England.

I. Martinson
Department of Physics, University of Lund, S-223 62 Lund, Sweden.

E. Oktay
Office of Fusion Energy, Department of Energy, Washington, D.C. 20545, U.S.A.

ATOMIC COLLISION PROCESSES IN MAGNETIC CONFINEMENT CONTROLLED THERMONUCLEAR FUSION RESEARCH

M. R. C. McDowell

Mathematics Department, Royal Holloway College

Egham Hill, Egham, Surrey, England

§1. INTRODUCTION

This NATO Advanced Study Institute is concerned with atomic and molecular processes in controlled thermonuclear fusion research. Most of our attention will be given to processes likely to be of importance in magnetic confinement devices, in particular in existing and planned tokamaks, since this approach to fusion is currently most advanced. Magnetic confinement devices other than tokamaks also show considerable promise; progress in stellerator research has recently been summarised by Shohet(1); there are advantageous features, but severe design problems for large scale machines. Long linear systems may avoid some macroscopic instability problems, but require very high particle densities ($n_e \simeq 10^{17}\,cm^{-3}$) and high magnetic fields ($>25T$) if laser heating is to be used(2). High energy neutral beam heating may be preferable, or heating by relativistic electron beams(3) but beam penetration and control may pose severe problems. End losses in current designs are unacceptably high and need to be reduced by end-plugs, either in the form of multiple magnetic mirrors, or possibly by solid frozen DT.

Various schemes have been advanced for obtaining controlled power output from fusion in inertially confined plasmas, whether heated by multiple lasers, relativistic electrons, or high energy heavy ion beams. Atomic processes play a fundamental role in determining the structure of, and energy transfer in such micro-plasmas, but the specific reactions involved, and the critical problem areas, are not yet clear.

For the remainder of this paper I will therefore confine my attention to some aspects of the atomic physics involved in tokamaks.

More detailed accounts of each of the problems involved will be found elsewhere in this volume.

It is well known that as the ion temperature increases above a few keV, the resistivity drops and ohmic heating no longer suffices. There are also difficulties associated with the onset of macro-instabilities at high current densities, which again limit the effectiveness of ohmic heating. Supplementary heating to ion temperatures necessary to achieve ignition must be provided, and adiabatic compression, neutral beam injection and R.F. heating are all being used alone or in combination.

Neutral beam injection, initially with H (and H_2) but eventually with D (and D_2), has the great advantage of contributing directly to the thermonuclear power output through beam-plasma collisions. If P_b is the injected beam power density, an important figure of merit in achieving break even, and eventually ignition, is the fusion power multiplication factor

$$Q = P_f/P_b \tag{1}$$

where P_f the fusion power density due to deposition of fusion products (α-particles for the D-T scheme) is given by

$$P_f = P_t + P_{bp} \tag{2}$$

the sum of the thermonuclear power density P_t, and the beam-plasma power density P_{bp}[4]. Since of the $E_F = 17.6$ MeV released per reaction only the 3.5 MeV from the α-particle contribute to the plasma energy balance, one needs $Q = 5$ before the total fusion α-particle power is equal to the beam power. Many authors, but most recently Jensen et al[5] have considered the maximum allowable fraction of various impurities to achieve $Q = 5$ operation, ignition and break even at various T_i (typically 10 and 20 keV). As the impurity fraction rises above a critical fraction, the required $n_e\tau$ values increase dramatically (Fig. 1). Jensen et al show that to achieve a given Q in a plasma with $n_D = n_T = \frac{1}{2}n_0$ one requires

$$n_e\tau = \frac{1.5[T_e(1 + fZ_o) + T_i(1 + f)]}{\dfrac{A \langle\sigma v\rangle E_F}{4(1 + fZ_o)} - (fL_Z + L_H)} \text{ s cm}^{-3} \tag{3}$$

which is of course a generalisation of the Lawson criterion ignoring the energy recycling efficiency. Here the simplification has been made of considering a single impurity of fractional concentration f and the mean charge Z_0, L_Z and L_H are the impurity and hydrogen emission rates per ion per free electron; T_e, T_i the electron and ion temperatures in keV, and

$$A = \frac{(1 + 0.2Q)}{Q - P_{bp}/P_b} \tag{4}$$

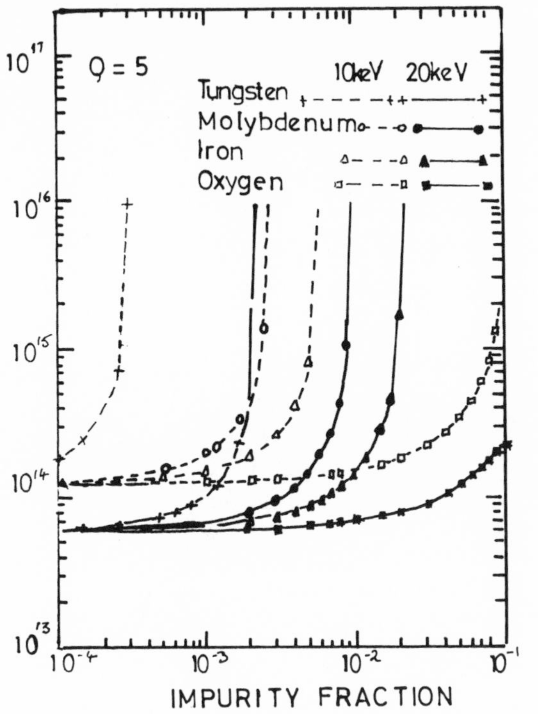

Fig. 1: $n_e\tau$ requirements as functions of impurity fraction f for Q = 5 beam driven D-T plasma at temperatures of 10 and 20 keV from (5).

§2. BEAM DEPOSITION AND INSTABILITY PROBLEMS

The uncertainties in the atomic data used in deriving these results are often substantial, though the results of Jensen et al[(5)] for $Z \leq 42$ agree well with those of other workers, and above that differ mainly in the enhanced values due to including dielectronic recombination. Coronal equilibrium is assumed, and in view of the uncertainty of the atomic data, is probably sufficiently accurate at this stage, provided diffusion velocities or temperature and density gradients are not too high. Change of ion state with charge transfer does not appear to have been included in any radiative calculations, and for highly stripped impurity ions may lead to selective excitation, and enhanced radiation losses. The best known example is[6,7]

$$C^{6+} + H(1s) \rightarrow C^{5+}(n = 4) + H^{+}$$

which proceeds via two avoided crossings, and has a maximum cross section of order 5×10^{-15} cm^2 at $v = 0.5$ au ($\simeq$ 4 keV/amu). This is further discussed by Bransden in Chapter .

However, beam heating (which with D or D/T beams also acts to replenish the fuel) depends on beam penetration and requires the energy to be deposited in the hot core of the plasma. The incident neutrals are introduced by pairs of anti-parallel beams (to avoid imparting appreciable momentum in the toroidal direction), aligned tangential to I_ϕ. The neutrals are rapidly ionized by proton impact ionization, electron impact ionization and charge exchange, while in the case of the molecular component dissociative reactions are also important. If there were no impurities, deposition would follow the ion density profile. However, the charge-exchanged neutrals from the plasma strike the limiter and provided they have more than a certain minimum energy cause sputtering. The sputtered and out-gassed impurities increase the density close to the limiter, and hence the deposition of beam energy there, and in turn the impurities contribute to further ionization of the beam[8], and there may also be contributions from charge exchange. It has often been assumed on the basis of a simple binary encounter model that the ionization cross section varied as q_{eff}^2, so that a rapid and disastrous build-up of beam deposition near the limiter might occur (the so-called beam-deposition instability). Recent measurements by Berkner et al[9] on target ionization of (H, H_2) by highly stripped ions of Fe ($11 \leq q \leq 22$) suggest a scaling more closely of $q^{1.43 \pm 0.05}$ from 0.28 to 1.1 MeV/amu and this is in good accord with classical Monte-Carlo calculations. The physically important quantity is the total loss cross section (ionization plus capture) and this has been calculated by Olson and Salop[10] in a three body classical model (electron, hydrogen atom, ion) in which the ion is represented by a hydrogenic ion with an effective charge q_{eff}, fixed to fit the ionization potential from the most important levels. They find

$$\sigma_{LOSS} \simeq 2.5\ 10^{-16}\ q^{1.25}\ \text{cm}^2$$

for fully stripped ions on H. Quantal distorted wave calculations which are in good agreement with the Monte-Carlo results in the energy range 1-10 keV/amu have recently been presented by Ryufuku and Watanabe[10]. They propose a scaling $\sigma(E/q^{0.45}) = \sigma(E)/q^{1.12}$ where $\sigma(E)$ is the cross section for p,H capture. More recently Crandall et al[11] and Meyer et al[12] have measured the capture cross section component for a wide variety of charge states of ions of O,Fe and other materials of interest on H and H_2 at velocities which span the range of injection velocities for the proposed TFTR. In particular they give results for all states of ionization of Q,O^{q+} ($1 \leq q \leq 8$) which we show in Fig. 2, and these are in remarkable agreement with the Monte-Carlo calculations.

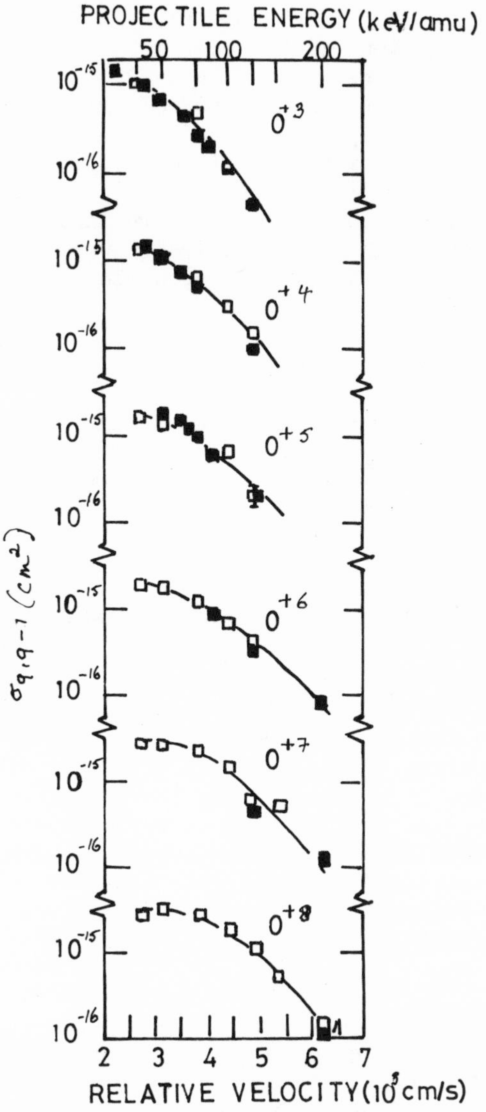

Fig. 2: Electron-capture cross section $\sigma_{q,q-1}$ for O projectiles as a function of projectile velocity. Open squares - Monte-Carlo calculations by Olson and Salop (10); solid squares - experimental results (from (12)).

They find $\sigma_{cx} \simeq \sigma_o q^p$, with $2 < p < 3$, but ionization being more important at higher beam energies, say > 60 keV for H.

A detailed model of radial impurity transport in tokamaks has been considered by Duchs et al(13) who show that edge deposition is more important for higher energy beams. For TFTR parameters useful injection pulse times are unlikely to exceed 200 ms. They suggest a scheme in which the beam pulse is followed immediately by adiabatic compression, with a compression ration $C \simeq 1.5$ achievable (in the ATC device a C of $\simeq 2.3$ was reached) which increases the temperature by a factor of about $C^{4/3}$. This drives the plasma from the outer limiter. A pulse of cold neutral gas is then injected at the limiter, until the density near the edge is a maximum for any radial distance. The cool blanket depresses the energy of the charge exchange neutrals in it below the sputtering threshold, and since impurities diffuse towards the density maximum, those in the outer region stay there, while those in the central hot plasma diffuse outwards $(\nabla T_i \cdot \nabla n < 0)$.

§3. RADIATION LOSSES

A completely pure D-T plasma is an unrealistic aim. There will always be some outgassing of the walls and limiters, and of course the end products of the reaction, He^{++} will capture free electrons by radiative recombination, and eventually by dielectronic. The distribution of He^{++} in the plasma is not well understood and further work is required on its transport properties(14). As it diffuses outwards towards the walls, charge exchange with neutral H may become important, and in particular in low density regions capture from highly excited (Rydberg) states of H^o should be considered(15). Post will discuss methods of measuring the He^{++} distribution function during the Institute.

There is no need to spell out in detail the various atomic processes which contribute to radiation loss. Most studies assume an optically thin plasma, and a local coronal equilibrium model. That is, the distribution of charge states of any impurity ion is governed by the equilibrium between electron impact ionization and recombination. It is also a good assumption, in most cases, that only the ground state population of any charge state need be considered. Bremsstrahlung is normally assumed to be due to hydrogenic ions of charge q_{eff}, though a recent study by Summers and McWhirter(32) shows that corrections due to ion structure may be significant and radiation from non-electric-dipole transitions may also be an important correction, especially for He-like ions, leading to an overall increase of a factor of two in the temperature range 1.5×10^7 to 4×10^7 K.

§3a. IONIZATION CROSS SECTIONS

Our knowledge of electron impact ionization cross sections is very limited. The experimental data up to 1977 have been reviewed by Dolder(16) and were mainly restricted to low stages of ionization of the alkalis and alkaline earths, of H and the rare gases. There has been no recent measurement of electron impact ionization of atomic hydrogen, the results below 200 eV being uncertain by at least 10%. It seems likely (Harrison, private communication) that the gas cell measurements for ionization of He may also be unreliable at low energies (< 50 eV)(33). This is not serious in itself, but the semi-empirical formulae(17) used in many coronal equilibrium studies, rely heavily on fits to this data. The most recent tabulations by Post et al(17) combine the various semi-empirical formulae to yield an ionization rate from level n of charge state q at T keV

$$C_{qn} = 1.86\ 10^{-7}\ n_e n_q\ T^{\frac{1}{2}}\ e^{-x_n}\ (1 - e^{-x_n})\left(\frac{I_H}{I_{qn}}\right)^2 \bar{G}_{qn}\ \text{cm}^{-3}\ \text{sec}^{-1}$$

where $x_n = (I_{qn}/kT_e)$ and the effective Gaunt factor is

$$\bar{G}_{qn} = 12.18\ e^{-n/(n+5)}(1 + 0.0335n)\left(1 - \frac{0.622}{q} - \frac{0.0745}{q^2}\right)|f(y)|$$

with

$$f(y) = 0.23 + 0.046\,y + 0.1074y^2 - 0.0459\,y^3 - 0.01505\,y^4$$

and $y = \log_{10}(x_n/4)$.

An alternative set of rates based on the Exchange-Classical Impact Parameter formulation of Burgess has been given by Summers(17), and the accuracy of this formulation has been compared with the data by Burgess et al(17). Recent measurements(27) of ionization rates for a number of ions of C, N, O and Ne in a theta pinch plasma at 80 eV give results generally within a factor of two of the ECIP rates, except for O^{5+} and Ne^{7+}.

Recent experimental work by Crandall et al(18) on ionization of C^{3+}, N^{4+} and O^{5+} from threshold to 1500 eV shows that Coulomb-Born calculations give an adequate account of the cross section near the lowest energy maximum, but that inner shell excitation ($1s^2 2s \rightarrow 1s2sn\ell$) followed by autoionization produces a second higher maximum at higher impact energies, and suggests that at least for these Li-like ions, this effect increases in importance with increasing Z, in almost linear fashion.

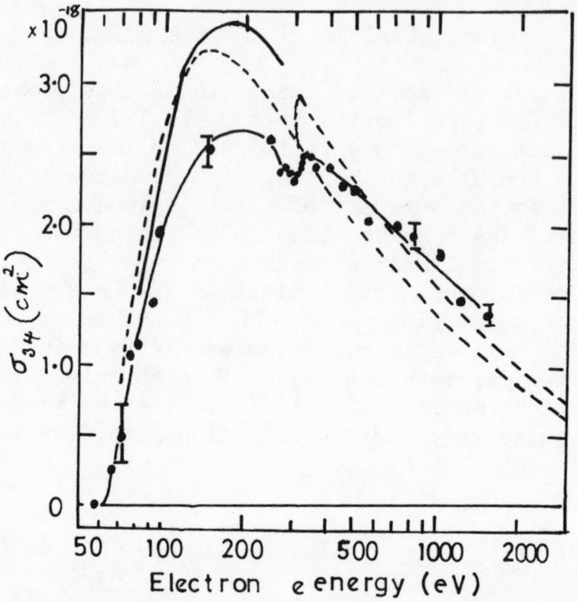

Fig. 3: Cross section for electron impact ionization of C^{3+} from (18). The connected full circules are present data; the broken curve is scaled Coulomb-Born of Golden and Sampson (28); the full bold curve is Coulomb-Born by Moores (19); the short-dashed curve is the summed excitation cross sections for $1s^22s \rightarrow 1s2s2\ell$ given by Magee et al (29) added to the scaled Coulomb-Born beginning at the 294 eV excitation threshold. The error bars shown are typical statistical uncertainties at 90% confidence level.

Theoretical studies have been almost entirely restricted to the Coulomb-Born approximation and variants of it to include exchange effects, and the latest results will be reported by Jakubowicz, for Li-like and Be-like ions(19). Together with a Coulomb-Born treatment of inner shell excitation in the autoionization region, and the assumption that all such excitation leads to ionization, they provide reasonably accurate estimates of the cross section at and above the first maximum. However, such approaches are of very uncertain validity close to threshold, and for highly stripped high Z impurities it is only the threshold region which is important in the tokamak regime; for Fe^{25+} the ionization threshold is at 8.5 keV. Relativistic atomic structure effects may be important for $Z > 20$ and certainly by $Z = 45$. In any event, more information is urgently required, particularly for all stages of ionization of C^{q+}, O^{q+}, N^{q+} and Fe^{q+}. There are suggestions from the experimental work(18) that the high energy asymptotic behaviour of the cross section

($Q(E) \to AE^{-1}$ (log E + B) as $E \to \infty$) does not set in to rather higher energies than predicted by theory. Standard theoretical treatments are costly, and for ions of Fe and heavier elements likely to be time consuming and expensive. Alternative approaches must be sought. McCarthy and McDowell[20] have shown that accurate ionization cross sections may be obtained from a knowledge of the absorptive part of the optical potential due to continuum states, including exchange, and preliminary results by Stelbovics and by Demetriades will be reported here on methods of obtaining the exchange part of the second order optical potential.

§3b. EXCITATION CROSS SECTIONS

Theoretical studies of electron impact excitation of positive ions have been reviewed by Seaton[21] and a full review of the currently available theoretical and experimental data is being prepared by Henry[22]. There are still very few experimental studies[16]. Where comparison is possible, there appears to be agreement between theory and experiment to within ± 20% near threshold, and better at higher energies, in most cases[23]. However, there remains a discrepancy of almost a factor of two at threshold between theory and experiment for the very simple case $e + He^+ (1s) \to e + He^+ (2s)$. Further, the Coulomb-Born approximation and its variants fail badly at impact energies $X (= E_i/I_z) < 4$, and many state close-coupling calculations are required in the region of interest just above threshold. The effects of resonances are important in some cases, for although the resonances themselves are narrow compared to the spread of electron energies in keV plasmas, when averaged over maxwellian temperature distributions, they may change the average cross section by as much as a factor of two for forbidden transitions (Fig. 4). It is important to note that positive ion excitation cross sections do not go to zero at threshold. For low Z impurities the situation is more promising, in particular for ions of C, N and O. For $q > 4$, the Coulomb interaction dominates, and even quite close to threshold distorted wave calculations for excitation of low lying excited states in optically allowed transitions give close agreement with more detailed close coupling calculations. Berrington et al[28] show that for electron impact excitation of O^{4+} resonances converging to the $n = 3$ thresholds increase the $2s^2\ ^1S^0 \to 2s\ 2p\ ^3P^0$ collision strength by about a factor of two below these thresholds, though there is only a 10% effect for the singlet-singlet transition. While the individual resonances are narrow, the range of impact energies over which the effect is important is of order 20 eV. Effects on the rate coefficient are less significant at high electron temperatures, being 15% at $T_e = 3.5 \times 10^5$K, for the singlet-triplet collision. Contributions from levels $n \geq 4$ and the continuum have not been included. There are, however, much more severe problems in treating intercombination transitions, for example O^{6+} ($1^1S \to 2^3S$, 2^3P).

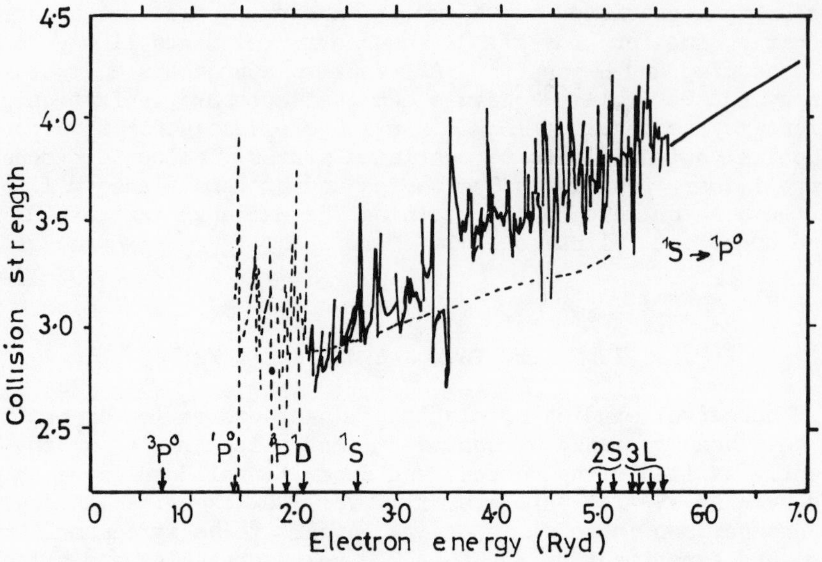

Fig. 4: The collision strength for the $2s^2\ ^1S^0 \rightarrow 2s2p\ ^1P^0$ transition in O v. The broken line is a six state calculation, the solid line a twelve state. (From (30)).

For higher Z ions, relativistic effects on the atomic structure begin to become important by $Z = 20$, and must be taken into account for $Z \geq 45$. Some studies have been carried out by Burke and Scott(24) and results will be presented here by Scott for Be-like ions. Relativistic effects on electron impact excitation of some lithium-like ions have been examined by Callaway et al(26) who find the main effect arises from the spin-orbit splitting of the 2p states. For C^{3+} the cross section for excitation of the $2p_{\frac{3}{2}}$ state is almost exactly twice that for $2p_{\frac{1}{2}}$ at all energies: above the $2p_{\frac{3}{2}}$ threshold the non-relativistic results agree to within 5% with the sum of $\sigma_{\frac{1}{2}}$, $\sigma_{\frac{3}{2}}$, but are a factor of 3 too large below $E_{2p_{\frac{3}{2}}}$, and in addition a non-relativistic calculation gives the wrong threshold, the discrepancy being as much as 4Ry for W^{71+}.

Most tokamak radiation loss studies have employed simple semi-empirical formulaes(25) in terms of oscillator strengths and Gaunt factors to estimate excitation cross sections, of the form

$$\sigma_{ij}(E) = \frac{8\pi}{\sqrt{3}} \frac{1}{E} \frac{1}{\Delta E_{ij}} f_{ij} \overline{G} \pi a_o^2$$

with

$$\overline{G} = \frac{6\sqrt{3}}{8\pi} A [1 + B \ln\{1 + \beta(\frac{E}{\Delta E_{ij}} - 1)\}]$$

where A, B and β are scaled to known results for light elements (17). A reassessment of the reliability or otherwise of such formulae and those for ionization is urgently required: uncertainties of a factor of two in ionization rates can lead to changes of factors of four in the estimated radiative losses for high Z impurities (17), so that it would be desirable to have results accurate to ± 20%.

§4. SUMMARY

I have indicated some aspects of the physics of magnetically confined hot plasmas where a detailed understanding of atomic processes is of vital importance. The spatial and temporal deposition of energy by neutral beam injection is controlled by the charge exchange and ionization cross sections of atomic hydrogen on protons and impurity ions at energies from a few to a few hundred keV. We now have much more confidence in our knowledge of these cross sections at least for low Z impurities, with theory and experiment in good agreement at energies above 40 keV/amu. This suggests that the beam deposition instability, while a serious problem, is not so dramatic as was feared, but that it may be necessary to go to schemes involving pulsed beam injection followed by adiabatic compression and cool gas boundary layer formation.

Present studies of radiative losses provide sharp limits of the maximum tolerable level for a given impurity for operation at a given Q, and suggest that the efficient use of carbon-based limiters, divertors, scrape-off and cool gas blankets may all be necessary. However, the atomic data used in the radiative loss calculations is very uncertain, and may be unreliable by a factor of two or more in either direction, though comparison of theoretical predictions with available diagnostic data for several systems suggests the uncertainties are not much more than this (5), even at the plasma edge, and for the PLT studies are in general less than 20%. Nevertheless, detailed studies will be necessary to confirm the reliability of the coronal model, and for efficient operation.

In particular, our knowledge of the electron impact ionization cross sections of most impurities is so poor that agreement of current calculations on mean Z and radiative losses with observation may be entirely accidental, so that extrapolations to the next generation of machines may be misleading.

All present calculations of coronal equilibrium ignore any effects of the magnetic field on the ionization cross sections and recombination rates. At fields of 5×10^4 gauss, atomic hydrogen levels above about $n = 14$ become unbound, though the principal quantum number at which these effects occur increases with Z^2 for other hydrogenic ions. Oscillator strengths change dramatically at high n in strong magnetic fields, while sharp resonances occur in photoionization cross sections at energies corresponding to the Landau levels. Detailed investigations are now underway(31), but the overall significance of such effects remains uncertain: Summers and McWhirter(32) argue that the effects of magnetic fields on recombination rates are likely to be unimportant, but this has not been convincingly demonstrated.

REFERENCES

(1) J. L. Shohet "State of Stellerator Research", Comm. Plas. Phys. Cont. Fus. 3, 25, (1977).
(2) F. F. Chen "Alternate concepts in magnetic fusion", Phys. Today 32, 36, (1979).
(3) V. Bailey, J. Benford, R. Cooper, D. Dakin, B. Ecker, O. Lopez, S. Putnam and T. S. T. Young, Proc. 2nd Intl. Topical Cont. on High Power Electron and Ion Beam Research and Technology, (Lab. Plasma Studies, Cornell, Ithaca N.Y. 1978).
(4) D. L. Jassky, Nucl. Fusion, 17, 309, (1977).
(5) R. V. Jensen, D. E. Post, W. H. Grasberger, C. B. Tarter and W. A. Lokke, Nucl. Fusion, 17, 1187, (1977).
(6) J. Vaaben and J. S. Briggs, J. Phys. B. 10, L521, (1977).
(7) A. Salop and R. E. Olson, Phys. Rev. 16, 1811, (1977).
(8) J. P. Girard, D. A. Marty and P. Moriette, Plasma Physics and Controlled Nuclear Fusion Research (Proc. 5th Int. Conf. Tokyo 1974) 1, 681, (1975) [IAEA, Vienna].
(9) K. H. Berkner, W. G. Graham, R. V. Pyle, A. S. Schlachter, J. W. Steams and R. E. Olson, J. Phys. B. 11, 875, (1978).
(10) R. E. Olson and A. Salop, Phys. Rev. A16, 531, (1977). H. Ryufuku and T. Watanabe, Phys. Rev. A19, 1538, (1979).
(11) D. H. Crandall, R. A. Paneuf and F. W. Meyer, Phys. Rev. A19, 504, (1979).
(12) F. W. Meyer, R. A. Paneuf, J. H. Kim, P. Hvelplund and P. H. Stelson, Phys. Rev. A19, 515, (1979).
(13) D. F. Düchs, D. E. Post and P. H. Rutherford, Nucl. Fusion, 17, 565, (1977).
(14) H. W. Drawin, Atomkemenergie, 33, 182, (1979). Physikal Blätter, 35, (1979).

(15) M. Burniaux, F. Brouillard, A. Jognaux, T. R. Govers and S. Szucs, J. Phys. B. 10, 2421, (1977).
(16) K. Dolder and B. Peart, Rep. Proc. Phys. 39, 693, (1976).
(17) H. W. Drawin, Report EUR-CEA-FC-387, Fontenay-aux-Roses, (1967); W. Lotz, Astrophys. J. Supp. 14, 207, (1967); T. Kato, Report IPPJ - AM2, Nagoya University, Japan, (1977); D. E. Post, R. V. Jensen, C. B. Tarter, W. H. Grassberger and W. A. Lokke, Atom. Data and Nuc. Tables, 20, 397, (1977); H. P. Summers, Mon. Not. R. Astron. Soc. 169, 663, (1974); A. Burgess, H. P. Summers, D. M. Cochrane and R. W. P. McWhirter, Mon. Not. R. Astron. Soc. 179, 275, (1977).
(18) D. H. Crandall, R. A. Paneuf, D. E. Hasslequist and D. C. Gregory, J. Phys. B. 12, L249, (1979).
(19) D. L. Moores, J. Phys. B. 11, L403, (1978); H. Jakubowicz and D. L. Moores "Total cross sections for electron impact ionization of Li-like and Be-like positive ions", to be submitted to J. Phys. B.
(20) I. E. McCarthy and M. R. C. McDowell, J. Phys. B. 13, (1979) in press.
(21) M. J. Seaton, Adv. Atom. Mol. Phys. 11, 83, (1975).
(22) R. J. W. Henry, Phys. Reports, in preparation.
(23) M. A. Hayes, D. W. Norcross, J. B. Mann and W. D. Robb, J. Phys. B. 11, L429, (1977).
(24) P. G. Burke and N. S. Scott, to be submitted to J. Phys. B.
(25) H. W. Drawin, Invited papers, International Symposium and Summer School on the physics of ionized gases (SPIG), I.O.P. Beograd (1979), Ed. R. K. Janev, p.633.
(26) J. Callaway, R. J. W. Henry and A. P. Msezane, Phys. Rev. 19, 1416, (1979).
(27) W. L. Rowan and J. R. Roberts, Phys. Rev. A19, 90, (1979).
(28) L. B. Golden and D. H. Sampson, J. Phys. B. 10, 2229, (1977).
(29) N. H. Magee Jr., J. B. Mann, A. L. Merts and W. D. Robb, LASL Report LA6691MS, (1977).
(30) K. A. Berrington, P. G. Burke, P. L. Dufton, A. E. Kingston and A. L. Sinfailam, J. Phys. B. 12, L275, (1979).
(31) H. S. Brandi, B. Koiller, H. G. P. Lins de Barros and L. C. M. Miranda, Phys. Rev. A18, 1415, (1978); K. Onda, J. Phys. Soc. (Japan) 45, 216, (1978). S. M. Kara and M. R. C. McDowell, (1979) submitted to J. Phys. B. W. A. M. Blumberg, W. M. Itano and D. J. Larson, Phys. Rev. A19, 139, (1979).
(32) H. P. Summers and R. W. P. McWhirter, J. Phys. B. 12, 2387, (1979).
(33) See however, T. D. Märk and F. de Heer, J. Phys. B. 12, L429, (1979) and M. F. A. Harrison, A. C. H. Smith and E. Brook, J. Phys. B. 12, L433, (1979).

THE RELEVANCE OF ATOMIC PROCESSES TO MAGNETIC CONFINEMENT AND THE CONCEPT OF A TOKAMAK REACTOR

M F A HARRISON

UKAEA/Euratom Fusion Association

Culham Laboratory, Abingdon, Oxon, OX14 3DB, UK

1. INTRODUCTION

The origins of research into controlled nuclear fusion may possibly be traced to the discovery of deuterium, whose existence was postulated (albeit on the basis of incorrect measurements) by Birge and Menzel in 1931, and to the discovery by Oliphant, Harteck and Rutherford in 1934 that new isotopes, T and He^3, were produced in disintegration experiments involving D-D reactions. The possibility of using nuclear fusion as a source of energy was explored in many laboratories but the first statement regarding the scale of research into its controlled application appeared in 1958 in the form of concurrent publication in Nature[1] of several papers describing the US and UK programmes. This was shortly followed by publication of comparable work in the USSR. The world-wide effort now amounts to over 3000 scientists and engineers but so complex are the scientific and technological problems that it is still not possible to predict with absolute certainty the route that will eventually lead to an economic and reliable fusion reactor.

The most crucial requirement of any nuclear fusion power generator must be a reactor vessel wherein fusion reactions proceed in such a manner that an adequate gain in usable energy can be attained. The nature of the fusion reactions requires that the energy of the colliding nuclei be substantial so that the fuel must be hot ($kT \sim 10$ to 100 keV) and therefore thermally insulated from the walls of its containing vessel. The basic source of energy is provided by the collision of nuclei but the detailed mechanisms by which the fuel is heated and by which it loses energy to the vessel depend upon a wide range of atomic collision processes.

To achieve the necessary degree of thermal insulation it is necessary that the hot fuel be physically removed from contact with the walls of its container and two approaches are presently envisaged. In the first, the density of the fuel is modest ($\sim 10^{14}$ cm^{-3}) but the fuel is ionized and the kinetic pressure of the resultant plasma is balanced by the confining forces of an external magnetic field so that the diffusion of ions and electrons and the conduction of energy to the walls of the container are very substantially reduced; the concept is called magnetic confinement. The second approach, which in contrast can only operate impulsively, is to produce the fuel in the form of a solid pellet and then to launch this pellet into a large vacuum chamber. Whilst in free flight and well removed from the walls, the pellet is subjected to an intense but uniformly distributed pulse of energy either from a number of laser beams or from beams of relativistic energy particles. The pulse is sufficiently rapid and intense to both heat and inertially compress the pellet before ablation and radiation losses become significant. Fusion reactions can thus take place within the extremely dense but thermally insulated region of the pellet.

During the last two decades many books on fusion research have been written. Early examples such as Glasstone and Lovberg[2] and Rose and Clark[3] still provide excellent backgrounds to the problems of magnetic confinement and its associated plasma physics but these cannot include the more recent developments. An example of an up to date and simple survey is that of Hagler and Kristiansen[4]. Current progress can be followed in the various proceedings of the "International Conferences on Plasma Physics and Controlled Nuclear Fusion Research", the last of which took place in Innsbruck[5], and within Europe there is also a series of conferences on "Controlled Fusion and Plasma Physics[6]. Up to date accounts of selected topics can be found in the lectures presented at the Plasma Physics Summer School (Culham 1978)[7] and there are many related "topical meetings" held from time to time in the various countries that are active in fusion research.

To date there is no single reference work that deals comprehensively with the relevant areas of atomic and molecular physics although a valuable compendium of papers can be found in the publication of the IAEA Advisory Meeting on Atomic and Molecular Data for Controlled Thermonuclear Fusion[8] and these may be supplemented by recent papers of Drawin[9,10] and by Summers and McWhirter[11]. Aspects of the subject receive detailed consideration from time to time in papers dealing with plasma modelling, plasma heating, plasma surface interactions and plasma diagnostics. There are also relevant publications in the closely related field of astrophysics.

The aim of the present paper is to consider the basic concept of a fusion reactor in such a manner that the associated atomic and molecular aspects of the problem can be viewed in a balanced perspective. The discussion is limited to the method of magnetic confinement and particular emphasis is given to the tokamak device. It is hoped that the presentation will provide the reader with some feel for the practical problems of reactor design and also provide sufficient indication of the nuclear, plasma, surface and atomic physics problems so that he will be stimulated to seek out the detail which of necessity must be omitted from this brief account. Unless stated otherwise Gaussian units are used throughout in order to provide ready cross reference with atomic data sources and with useful references in astrophysics. Energies and temperatures (kT) are quoted in eV and the symbols c and e are used for the velocity of light and the electronic charge.

Atomic processes relevant to the concept of magnetic confinement fall into the following broad categories:

a. Processes that are fundamental to the loss of energy from the hot plasma of nuclear fuel and to the equipartition of energy amongst the electrons and ions that form this macroscopically neutral collection of particles.

b. Processes associated with the addition of energy and particles to the confined plasma, ie, heating and fuelling.

c. Processes associated with the boundary of the confined plasma, ie, the detailed mechanisms of plasma-surface interactions.

d. Processes associated with impurity elements within the plasma and which can cause unacceptable losses of energy.

e. Processes that are necessary for the operational control of the reactor, ie, the maintenance of correct temperature and density and also the exhaust of "burnt" fuel.

f. Processes that are invoked as the basis for instruments used to monitor the conditions within the plasma, ie, diagnostic techniques.

The emphasis placed upon these roles and on the absolute need for knowledge of atomic processes has changed from time to time, dependent upon the particular needs of the moment, but in the long term these aspects of atomic physics are irrevocably involved in the evolution of a fusion power reactor.

2. NUCLEAR REACTIONS AND THE CONCEPT OF A POWER REACTOR

The release of energy in any nuclear reaction is dependent upon a reduction in mass (ie 931 MeV/amu) and it is fortunate that substantial reductions occur in fusion collisions between the light nuclei. This fact has favoured the choice of the stable isotope deuterium as fuel in a fusion reactor. Deuterium, whose natural abundance is about 1 part in 6500, can be readily extracted from water and it has been conservatively estimated that sea water could provide sufficient fuel to meet the predicted energy needs of the world for at least 10^9 years (see for example Ref.3). The most significant reactions involving deuterium are:

$$D + D \begin{cases} \nearrow (T + 1.01\ \text{MeV}) + (p + 3.03\ \text{MeV}) \\ \searrow (He^3 + 0.82\ \text{MeV}) + (n + 2.45\ \text{MeV}) \end{cases}$$

$$D + He^3 \rightarrow (He^4 + 3.67\ \text{MeV}) + (p + 14.67\ \text{MeV})$$

$$D + T \rightarrow (He^4 + 3.52\ \text{MeV}) + (n + 14.06\ \text{MeV})$$

where the two DD reactions occur with almost equal probability. Despite the obvious convenience of an input of non-radioactive fuel, the DD reactions are not favoured, firstly because the energy released is on average only 3.66 MeV compared with 17.58 MeV for DT, and secondly the technical difficulties of increasing this low

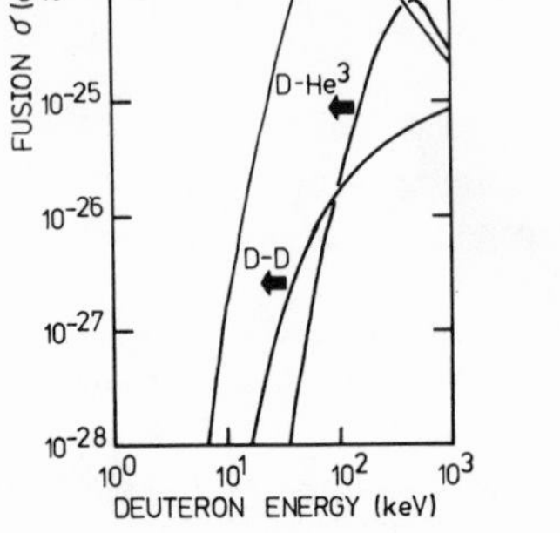

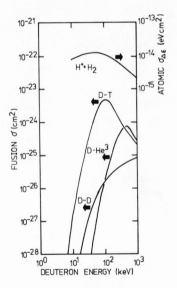

Fig.1 Cross sections for the fusion reactions D-T and D-D (total) taken from Ref.12 together with D-He^3 from Ref.3. Also shown is the energy loss cross section $\sigma_{\Delta E}$ for H^+ in H_2 gas[12].

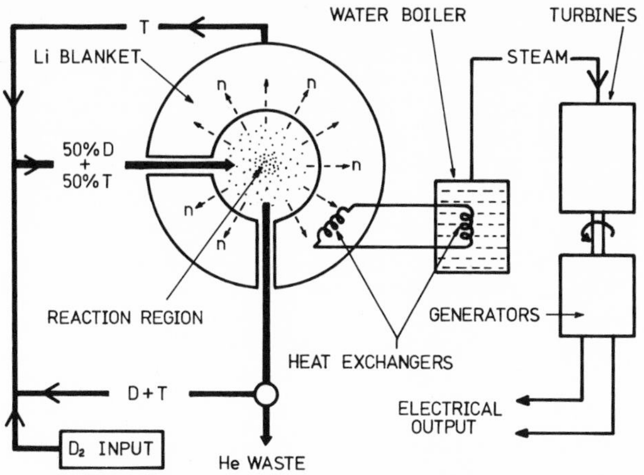

Fig.2 A simplified cycle for fuel and reaction products in a DT fusion power reactor.

yield by invoking the DHe^3 reaction (from He^3 bred by DD reactions) is too great to contemplate at present.

The technical advantages of selecting the DT reaction can be appreciated by considering the reaction cross sections which are shown in Fig.1. Not only is the peak magnitude of the DT cross section some 50 times greater than that of DD (total) but it occurs at about 100 keV, which is very much lower than the corresponding energies for both DD and DHe^3. The powerful electrostatic repulsion between the two charged nuclei results in low reaction probabilities at low collision energies and, in addition, the energetic product of a DT reaction is a neutron. This is in marked contrast to fission reactions which are initiated by neutrons and yield energetic charged nuclei and so the DT process presents technical problems which are almost the reverse of those encountered in a fission reactor.

Not surprisingly, a conceptual fusion reactor differs in most respects from a fission reactor; their one point of similarity is that they both rely on steam as a medium to provide their power output. A highly simplified diagram of the flow cycle of fuel and reaction products for a continuously operated reactor is shown in Fig.2. A mixture of 50% D + 50% T gas is fed into a reaction vessel where sufficient energy is introduced for fusion reaction to take place. The 14.06 MeV neutrons released in the reaction readily pass through the walls of the vessel and enter a surrounding blanket that contains lithium. This blanket serves the dual purpose of neutron absorber and tritium breeder. Breeding occurs predominantly through the exothermic reaction

$$n + Li^6 \rightarrow He^4(2.1\ \text{MeV}) + T(2.7\ \text{MeV})$$

which breeds a balanced supply of tritium and also raises the total energy produced per fusion event to 22.38 MeV. In practice some of the neutron energy is taken up in the mechanical structure of the blanket and a more reasonable practicable yield of energy is $Q_F \approx 20.1$ MeV per fusion event of which $Q_\alpha = 3.52$ MeV remains (with the α-particles) within the reaction vessel and the remainder is available for extraction from the blanket. The deposition of energy heats the blanket from which energy is extracted by circulation of a heat exchange fluid. This fluid transfers heat to a steam boiler which is coupled to steam turbines and electrical generators. Any unburnt fuel together with He atoms arising from DT fusion is exhausted from the reaction vessel and recycled after the He has been separated and exhausted as waste. The correct composition of input fuel is maintained by adding tritium bred in the blanket and deuterium provided from an external supply. A useful general survey of fusion reactor concepts is given by Ribe[13] and up to date information can be obtained from Ref.14.

3. THE NEED FOR PLASMA CONFINEMENT

An obvious condition for any reactor is that the mean free path for fusion collisions, $\lambda_{DT} = [n_T \sigma_{DT}(E)]^{-1}$, must be smaller than the dimensions of the reaction vessel. Accepting for the moment that other constraints restrict the target density (assumed here to be tritium) to $n_T \sim 10^{14}$ cm^{-3}, then the minimum value of λ_{DT} is about 2×10^9 cm and this occurs for deuterium particles with an energy of 100 keV so that the average reaction time is about 6.5s. Clearly, any device that must accommodate such large dimensions would be an impracticable proposition.

If the target were neutral gas then a second fundamental problem arises due to the competing effects of atomic collisions. The amount of energy involved in any atomic event is always much smaller than in a fusion event, but the atomic cross sections are much larger so that the rate at which energy is interchanged by atomic processes can dominate the energy balance. To illustrate the fundamental significance of this problem let us consider a hypothetical system in which a gaseous target of tritium is bombarded by a beam of 100 keV D^+ ions. Although the atomic data for these particular partners is not known, it is reasonable for the present argument to apply directly the data for H^+ ions in H_2 gas[12]. The stopping cross section,

$$\sigma_{\Delta E}(E) = \frac{1}{n}\frac{dE}{d\ell} \qquad [\text{eV cm}^{-2}],$$

which is plotted in Fig.1, is a measure of the energy lost by atomic processes such as ionization, excitation and charge exchange as the H^+ passes through the target gas. Thus for a target density of 10^{14} cm^{-3}, the ions in a 100 keV beam of D^+ would on average lose ΔE = 50 keV energy in a distance $\ell \approx 4 \times 10^4$ cm and this energy would eventually be transferred to the walls of the container by radiation and particle impact. Not only is the energy lost from the system but it would cause serious problems at the vessel wall. If the intensity of the deuterium beam entering the target is I_0(ions s^{-1}) then the average fusion power produced in the path length ℓ can be expressed as

$$P_\ell = I_0[1 - \exp(- \ell/\overline{\lambda}_{DT})]Q_F$$

where $\overline{\lambda}_{DT}$ is averaged over the energy range 100 to 50 keV. The energy efficiency over this path length is thus

$$\varepsilon_\ell = P_\ell/I_0\Delta E \approx 4.9 \times 10^{-3}$$

which is obviously useless. An actual situation would be much worse because $\varepsilon(\ell)$ must be integrated over the complete path length and the fusion cross section at distances greater than ℓ decreases rapidly due to the decreasing energy of the beam.

Both of these particular problems are substantially reduced if the target consists of fully ionized hydrogen isotopes. Firstly there are then no bound electrons so that ionization, excitation and charge exchange do not cause losses of energy; even so, Coulomb scattering of the incident particles by the electrons and ions of the target plasma can cause serious energy losses if the target is not very hot (see Ref.2 and Section 6). Secondly, a collection of charged particles can be so confined by a magnetic field that even the Coulomb scattered particles can then travel for distances comparable to λ_{DT} without colliding with the walls of a container of practical dimensions. A third advantage is that the ionized gas is a good conductor of electricity and so it can be heated by passing a current through it. In principle, the whole collection of particles can become sufficiently hot for fusion reactions to occur so that the concept of a beam and a target is not required. The magnetic confinement approach to a fusion reactor is therefore based upon the following cycle of operations.

a. Introduce neutral D and T gas into the reaction vessel.

b. Ionize this gas mixture and therefore form a plasma within a confining magnetic field.

c. Heat this plasma by passing a current through it; this process is called "ohmic heating".

d. Use ohmic heating and probably additional energy

from external sources to heat the plasma to a sufficient temperature for fusion reactions to take place.

e. Contain the α-particles from these fusion reactions by means of the magnetic field and so utilise their energy (3.52 MeV) to maintain the temperature of the hot plasma. When the α-particle power is equal to the energy losses from the plasma the system has become "ignited" and external sources of heating are no longer required.

f. This condition is maintained for a time compatible with the technical constraints of the particular device (the "burn-time" ranges from a few seconds to continuous operation). If necessary, more DT fuel is added and the waste products He and unburnt DT are exhausted.

Present-day experimental techniques extend to about mid way through cycle (d); laboratory plasmas have been heated to high temperatures (∿ 5 keV) but ignition has not yet been achieved.

4. PHYSICAL CONDITIONS FOR A FUSION PLASMA

The rate at which fusion collisions occur within a homogeneous plasma where the particle velocities correspond to a Maxwellian distribution is given by,

$$R_F = n_D n_T <\sigma_{DT} v> \qquad [cm^{-3}\ s^{-1}] .$$

Here $<\sigma_{DT}v>$ is the rate coefficient (ie the product of (cross section x velocity) averaged over the Maxwellian distribution corresponding to an ion temperature T_i and shown here in Fig.3). If n_i is the total density of ions in the plasma and $n_i/2 = n_D = n_T$, then the useful fusion power density $P_F(T_i)$ is given by,

$$P_F(T_i) = \frac{n_i^2}{4} <\sigma_{DT}v>Q_F = 8.05 \times 10^{-13} n_i^2 <\sigma_{DT}v> \qquad [W\ cm^{-3}], \quad (1)$$

where Q_F = 20.1 MeV. Each α-particle has an energy Q_α = 3.52 MeV and so a fraction of the fusion power, namely

$$P_\alpha(T_i) = \frac{n_i^2}{4} <\sigma_{DT}v>Q_\alpha = 1.41 \times 10^{-13} n_i^2 <\sigma_{DT}v> \qquad [W\ cm^{-3}] \quad (2)$$

can be contained and used to maintain the temperature of the reacting plasma.

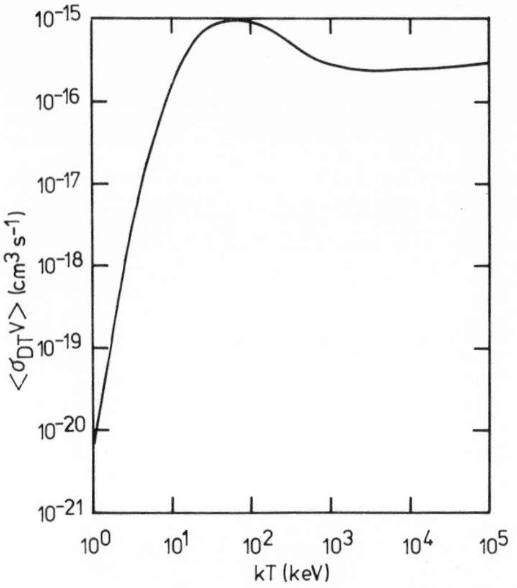

Fig.3 Rate coefficient $<\sigma_{DT}v>$ for the DT fusion reaction as a function of kinetic temperature kT.

If the plasma ions are hydrogen-like (ie protons or fully stripped ions) then there are a negligible number of bound electrons and so energy cannot be lost by line radiation. However, free-free collisions do occur and their effects are significant. Acceleration of a free electron in the attractive Coulomb field of a free ion gives rise to bremsstrahlung radiation and, at the temperatures of interest to fusion, the energy is carried away by photons in a spectral region ($\lambda < 10$ Å) over which the plasma is transparent. The power lost by bremsstrahlung radiation over all wavelengths from a hydrogen-like plasma can be expressed as

$$P_{br}(T_e) = 1.69 \times 10^{-32}\, n_e\, T_e^{\frac{1}{2}} \sum (n_z\, Z^2) \qquad [\mathrm{W\ cm^{-3}}] \qquad (3)$$

where n_z is the ion density in each fully stripped charge state Z. Here it should be noted that the presence of even a small concentration of fully stripped ions of high Z will cause a major enhancement in the bremsstrahlung power loss. If the ions are not fully stripped, account must be taken of the distortion of their asymptotic Coulomb field by the bound electrons (see for example Ref.11) and so expressions such as (3) must be applied with caution.

A magnetically confined plasma will, in idealised conditions, sustain its temperature (ie become ignited) when $P_\alpha(T_i) > P_{br}(T_e)$ and it is important to consider the relevant conditions for a pure

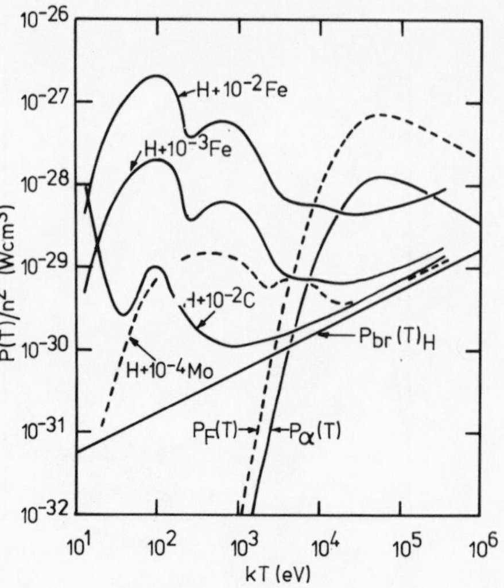

Fig.4 Power density functions $P(T)/n^2(kT)$ for total usable fusion power $[P_F(T)]$, α-particle power $[P_\alpha(T)]$ and bremsstrahlung radiation $[P_{br}(T)]$ from a DT plasma. Also shown are the composite power loss functions $[P_r(T)]$ corresponding to the addition of $C_x n_i$ impurity ions of the elements $C[C_x = 10^{-2}]$, $Fe[C_x = 10^{-2}$ and $10^{-3}]$ and $Mo[C_x = 10^{-4}]$; data are taken from[11].

hydrogenous plasma (ie $Z = 1$, $n_i = n_e = n$ and $T_i = T_e = T$). The power functions, $P_\alpha(T)/n^2$, $P_F(T)/n^2$ and $P_{br}(T)/n^2$, are plotted as functions of T in Fig.4 and it can be seen that ignition occurs at $T \approx 5$ keV and thereafter fusion reactions render the plasma exothermic. If confinement can be maintained, the plasma would continue to heat itself until a second balance between $P_\alpha(T)/n^2$ and $P_{br}(T)/n^2$ occurs at a temperature in excess of 1 MeV. The kinetic pressure of the plasma is $p = (n_i + n_e)\,kT = 2nkT$ and, if we assume a particle density of $n = 5 \times 10^{14}$ cm^{-3}, then the plasma pressure at a temperature of 1 MeV is about 780 (ATM) and this would probably exceed the limits of the containing magnetic field (from Section 5 it can be seen that this pressure would require a field of at least 140 kG, which is excessive for a practical magnet system). Thus a reactor must be controlled to work in a temperature regime between the ignition point and that dictated by a reasonable intensity of confining magnetic field. At densities of about 5×10^{14} cm^{-3} the acceptable total fusion power density is of order 10 W cm^{-3} which, although low, is acceptable for fusion reactor designs and implies, when allowance is made for gradients in density and temperature, that the volume of a power reactor (of say 5000 MW thermal output) is of order 1000 m^3.

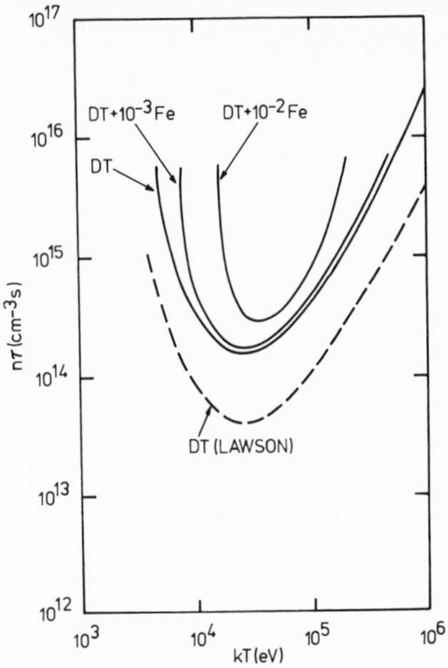

Fig.5 The criterion for ignition defined by equation (4), $n\tau_E(T)$, is shown for a pure DT plasma and for DT contaminated with iron, ($C_x = 10^{-2}$ and 10^{-3}). Also shown is the Lawson criterion for "break-even" power production when an energy conversion efficiency of 0.4 is assumed.

Present-day studies are limited to plasma regimes below the ignition conditions. The maximum temperature so far achieved is about 5.5 keV[15] but to attain ignition it is also necessary to simultaneously attain appropriate conditions of density and confinement time for both particles (τ_p) and energy (τ_E).

The average energy per particle in a Maxwellian distribution is $\frac{3}{2}$ kT and so the energy required to raise the cold field to a temperature T within a fully ionized plasma is $3(n_e kT_e + n_i kT_i)/2 = 3$ nkT. This can also be regarded as the energy density within the hot plasma. Suppose that this energy is lost (by radiation, conduction and particle convection, etc) in a time τ_E and it is further assumed that the plasma can be heated to T within a time much smaller than τ_E, then the energy balance for ignition is given by

$$\tau_E P_\alpha(T) = 3\ nkT + \tau_E\ P_{br}(T)$$

so that

$$n\tau_E = \frac{3\ kT}{[P_\alpha(T) - P_{br}(T)]n^{-2}} \qquad [cm^{-3}\ s]\ . \qquad (4)$$

This function is plotted in Fig.5 and it can now be seen that for $T \approx 5$ keV the requirement for ignition is that $n\tau_E \approx 4 \times 10^{15}$ $cm^{-3}s$ but that these constraints drop significantly with increasing T.

This type of analysis was first introduced by Lawson[16] to determine the "break-even" conditions of a power reactor operating for a burn-cycle of duration τ_L. He considered that all the energy delivered to the walls of the vessel in a time τ_L could be converted into usable energy with an efficiency ε. If the plasma can be heated to a temperature T in a negligible time, then the energy produced per unit volume by fusion reactions during the burn-time is $E_R = \tau_L P_F(T)$ and the energy lost by bremsstrahlung and by cooling of the plasma is $E_S = \tau_L P_{br}(T) + 3\ nkT$. All of this energy $(E_R + E_S)$ will be deposited on the walls of the reactor by the end of the burn-time but the fraction recovered must, after conversion, be at least equal to the expenditure of energy (E_S) during the burn-time. Thus conditions for a nett gain in energy can be expressed as

$$\varepsilon(\tau_L\ P_F(T)\ +\ \tau_L P_{br}(T)\ +\ 3\ nkT) > (\tau_L P_{br}(T)\ +\ 3\ nkT)$$

or more simply

$$\varepsilon(R + 1) > 1\ ,$$

where $R = E_F/E_S$. Rearrangement yields

$$n\tau_L \geqslant \frac{3\ kT}{\left[P_F(T)\frac{\varepsilon}{(1-\varepsilon)} - P_{br}(T)\right]n^{-2}} = F(T)\quad , \tag{5}$$

so that $n\tau_L$ can be determined as a function of T. The function has a very similar shape to the ignition curve and is shown for a pure DT plasma where $\varepsilon = 0.4$ in Fig.5. The minimum value is $n\tau_L \approx 4 \times 10^{13}$ $cm^{-3}s$ at $T \approx 25$ keV; however, it is more customary to consider lower temperature conditions such as 10^{14} $cm^{-3}s$ at about 9 keV.

The preceding discussion has been restricted to a pure (ie $Z = 1$) plasma. In practice impurities with low Z such as C and O are frequently present due to surface desorption and other allied boundary effects; also high Z elements such as Fe and Mo are introduced by sputtering of the walls of the vacuum vessel and other structural members of the containment device. The presence of impurity ions which have bound electrons introduces sources of line radiation which, since fusion plasmas are optically thin to high energy photons, results in substantial power losses. Radiation is emitted due to electron collisions with impurity ions that lead to excitation, recombination and bremsstrahlung radiation. The power lost can be determined on the basis of coronal modelling (see for example Ref.11 and the brief discussion in Section 8 of this paper). The total radiated power density $P_{tx}(T_e)$ for an impurity element, X, with density $n_x = C_x n_i$ must be added to the bremsstrahlung losses from the DT ions so that the power density $P_r(T_e)$ radiated from the plasma can be determined, ie

$$P_r(T_e) = P_{br}(T_e) + C_x n_i n_e F_{tx}(T_e)$$

where the radiated power loss function $P_{tx} = n_e C_x n_x F_{tx}(T_e)$ is discussed in Section 8.

Such data have been plotted in Fig.4 in the form

$$P_r(T_e)/n^2 = P_{br}(T_e)/n^2 + C_x F_{tx}(T_e) \tag{6}$$

for the impurity elements, carbon ($C_x = 10^{-2}$), iron ($C_x = 10^{-2}$ and 10^{-3}) and molybdenum ($C_x = 10^{-4}$) using the assumption that $n_e = n_i$ and $T_e = T_i = T$. The charge state of the impurity ion increases with increasing electron temperatures (see Fig.13) and so bremsstrahlung losses become predominant at high temperatures. However, in this regime there is an appreciable surplus of heating power available from the contained α-particles so that modest concentrations ($C_x \sim 10^{-3}$) of high Z impurities can be accepted in an ignited reactor. Indeed such losses have been invoked as a mechanism to maintain the plasma temperature and hence its pressure within a viable operating regime[17]. In the lower temperature regimes the power lost by line radiation from very small concentrations of impurity elements greatly exceeds that available from fusion processes so that ignition conditions are extremely sensitive to the presence of impurities.

The effect of impurities upon ignition can be assessed bv modifying equation (4) to the form

$$n\tau_E = \frac{3\ kT}{[P_\alpha(T) - P_r(T)]n^{-2}} \tag{7}$$

and the consequent dependence of $n\tau_E$ upon T can be seen from the curves for plasmas containing Fe impurities presented in Fig.5. The constraints imposed upon the plasma by impurities can preclude ignition in a practical device because energy must be fed into the plasma until ignition is attained and the wastage of energy by radiation can escalate the cost and strain the technical feasibility of the external power sources to an intolerable extent. In conclusion it must be stressed that, although impurity elements enhance bremsstrahlung losses at high temperatures, their major impact lies in the pre-ignition regime and it is not surprising that present-day preoccupation with research in this plasma regime has led to a somewhat pessimistic long term view of the effect of impurities within the plasma of an ignited reactor. However, a contradictory view must be taken when considering the materials problems of power loading on the walls of the reactor[18]. From this particular viewpoint it is desirable to minimise the amount of photon radiation falling on the wall by minimising the impurity content of the ignited plasma.

In a magnetically confined plasma the centripetal acceleration of the charged particles gyrating around the magnetic field lines gives rise to emission of radiation called cyclotron (or synchrotron) radiation. Ion velocities are relatively low and so their emission can be neglected but that associated with the electrons can be significant. The problem is complex and related to the operating parameters of a particular device. The power emitted in a magnetic field of intensity B[G] is

$$P_{cy} \approx 6.21 \times 10^{-21} B^2 n_e T_e \qquad [\mathrm{W\ cm^{-3}}] \tag{8}$$

and in an isothermal plasma $n_e kT_e = n_i kT_i = B^2/16\pi$, see equation (12), so that this expression is often presented in the form

$$P_{cy}(T_e) \approx 5.00 \times 10^{-31} n_e{}^2 T_e{}^2 \qquad [\mathrm{W\ cm^{-3}}]. \tag{9}$$

However, the radiation is emitted mainly in the infrared and μ-wave region so that the plasma is not completely optically thin and, moreover, the walls of the containing vessel act in some measure as reflectors. The problem is not discussed further other than to point out that cyclotron radiation power losses can limit the exothermal temperature rise of an ignited reactor at a temperature somewhat lower than the second crossing of the bremsstrahlung curve and that this effect can be observed in Fig.9 of Section 6.

The conditions considered in this section are based upon the assumption that the plasma is homogeneous in both temperature and density and that losses by diffusion and other plasma processes are negligible. In practice these assumptions are not true and to obtain some idea of the consequences it is necessary to consider the physical properties of plasma confinement by a magnetic field.

5. MAGNETIC CONFINEMENT AND THE TOKAMAK DEVICE

A particle of mass m and unit electronic charge e moving in a uniform magnetic field of strength B is constrained to perform a

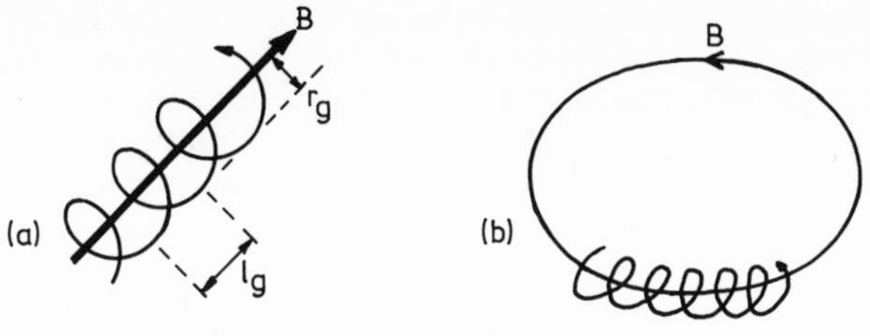

Fig.6 (a) Trajectory of a charged particle gyrating around a magnetic field line; the gyroradius is r_g and the pitch ℓ_g (see text).

(b) Gyration around a closed field line.

circular trajectory whose radius r_g (ie gyromagnetic radius) is

$$r_g = \frac{mv_\perp c}{eB} .$$

Here $v_\perp$ is the particle velocity perpendicular to the field. Charged particles within a plasma have a random distribution of velocities so that, whilst movement perpendicular to the field is constrained, particle velocities in the direction parallel to the field, $v_{||}$, remain unchanged. Thus the trajectories form spirals of radius r_g (and pitch $\ell_g = v_{||}\ 2\pi/\omega_g$) that gyrate around the field lines as shown in Fig.6(a). Here ω_g is the gyromagnetic frequency

$$\omega_g = \frac{eB}{mc} .$$

To achieve plasma confinement it is therefore necessary either to impede this free flow parallel to the field or else to bend the field lines back upon themselves so that they become "closed" and thereby allow particles to gyrate around the resultant ring of magnetic field in a manner illustrated in Fig.6(b). An experimental approach to confinement based on the linear concept is that of magnetic mirrors. Here regions of increased magnetic field are located at each end of the device and the localised increase in magnetic field reflects both ions and electrons by the principle of conservation of the magnetic moment of the gyrating particles. However, some losses of particles with $v_{||} >> v_\perp$ are unavoidable and this problem has formed a serious obstacle to the evolution of a mirror fusion reactor. Consequently, the present paper is limited to considerations of confinement systems that are based upon closed, or effectively closed, lines of force.

An assessment of the magnitude and topography of the field needed to confine charged particles can be obtained in the following simple manner. The plasma is considered as a collection of singly charged positive and negative particles moving randomly (ie collisions between the particles are neglected) and $n_i = n_e$ is assumed. Then the force exerted on a particle by any electric field that is present is $e\underline{E}$ and the force exerted by the magnetic field is $e(\underline{v} \times \underline{B})$. In a steady state the force on all the particles in a unit volume (ie the force density) is just balanced by the rate of momentum transfer which is equal to the gradient in pressure. It therefore follows that

$$n_i e[\underline{E} + \frac{1}{c}(\underline{v_i} \times \underline{B})] = \nabla p_i$$

and

$$- n_e e[\underline{E} + \frac{1}{c}(\underline{v_e} \times \underline{B})] = \nabla p_e$$

The difference in particle velocities gives rise to a nett rate of movement of charge, $n_e(\underline{v_i} - \underline{v_e})$ which is equal to a current density $\underline{j}$ so that the nett gradient in pressure is given by

$$\frac{1}{c}(\underline{j} \times \underline{B}) = \nabla p \ . \tag{10}$$

If the electric field does not vary with time then Ampere's law (as expressed by Maxwell's equations) is

$$\nabla \times \underline{B} = \frac{1}{c}(4\pi\underline{j}),$$

which in combination with equation (10) yields) the result that

$$\nabla(p + \frac{B^2}{8\pi}) = 0$$

for a magnetic field in which the lines are straight and parallel. Thus,

$$p + \frac{B^2}{8\pi} = C \ ,$$

where C is a constant. The quantity $B^2/8\pi$ is the energy density of the magnetic field and it can also be regarded as the magnetic pressure of the field. If the plasma is completely confined by an external field of strength B_o then the pressure at the outside of the system must fall to zero so that $C = B_o^2/8\pi$ and

$$p + \frac{B^2}{8\pi} = \frac{B_o^2}{8\pi} \quad . \tag{11}$$

Thus the maximum pressure that can be confined is $B_o^2/8\pi$ and in the simple case, where $B = 0$, the balance can be expressed as

$$(n_i kT_i + n_e kT_e) = 2\ nkT = \frac{B_o^2}{8\pi} \tag{12}$$

and the parameter,

$$\beta = \frac{2\ nkT}{B_o^2/8\pi} , \tag{13}$$

is a measure of the efficiency of utilisation of the external field and it is of particular importance in the economics of both experiments and conceptual reactors. If we assume that there is no magnetic field within the plasma of a fusion reactor operating in the regime $n \approx 5 \times 10^{14}\ cm^{-3}$ and $T \approx 15$ keV, then the plasma pressure is $2\ nkT = 2.4 \times 10^7$ dynes cm^{-2} and the intensity of the confining field must be $B_o = 25$ kG.

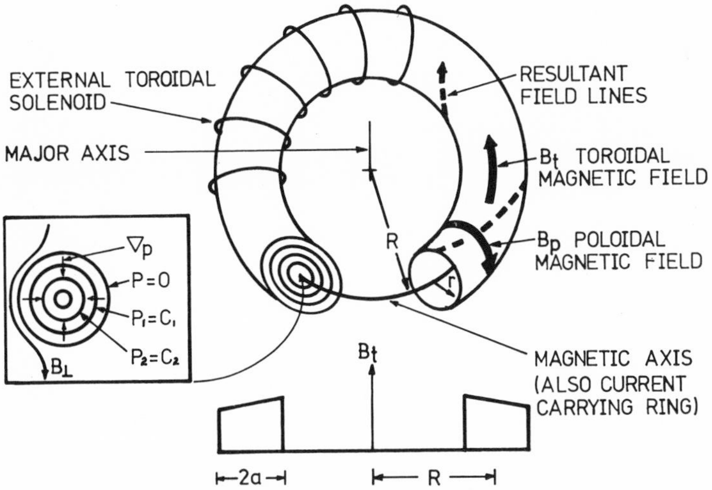

Fig.7 Basic aspects of a magnetic confinement system based upon toroidal geometry. Inset are details of the minor section showing nested magnetic surfaces and the field $B_\perp$. Radial characteristics of the toroidal field $B_t(R)$ are also indicated.

A more rigorous analysis shows that the pressure gradient in equation (11) should really be $\nabla p_\perp$ which means that the magnetic field can support a plasma pressure only in the direction perpendicular to the field lines and so a confined plasma must be completely surrounded by magnetic field. The field lines must therefore form a closed system and in this context p = C defines a closed surface with $\underline{\nabla p}$ normal to the surface which is not crossed by any field line. Such a system must be toroidal and of the form illustrated schematically in Fig.7, where the major radius of the torus is defined by R and the minor radius by r.

The simplest closed line system in which an approach to equilibrium can be obtained is that generated by an isolated current carrying conductor located on the magnetic axis. This generates a poloidal field of intensity $B_p = 2I/10a$[G] where r = a is the minor radius of the plasma and I [in units of A] is the current flowing in the toroidal direction around the ring. If a = 200 cm, then for the preceding reactor conditions, I ≈ 25 MA. In this configuration the drift surfaces of electrons and ions form nested surfaces within the torus as shown inset in Fig.7. An alternative method of providing a toroidally shaped magnetic field is to pass a current through an external toroidal solenoid and thereby generate a field B_t whose lines lie parallel to the magnetic axis. However, in practice both B_t and B_p are required to maintain equilibrium conditions. The geometry of a torus is such that the field must be more intense on the inner side of the torus (ie B_t decreases as a function of the major radius R as shown in Fig.7). Thus if only the B_t field is present, there is a gradient in magnetic field pressure that drives the plasma outwards towards the weaker regions of B_t. A more detailed treatment shows that the particle trajectories (which are cycloidal within an inhomogeneous magnetic field) cause a separation of positive and negative charges[2] in the vertical direction (parallel to the major axis) and this separation is the cause of the outward drift of the bulk plasma. However, the toroidal field lines can be twisted so that they rise above and then fall below the mid plane (ie the plane of the magnetic axis) and this motion does not reverse the direction of the vertical separation of charge so that it is sometimes towards and sometimes away from the mid plane. The nett effect of charge separation can thereby be annulled. To achieve this requirement a poloidal field must be present so that the field lines resulting from the combination of B_t and B_p spiral around the magnetic axis. Tamm[19] has shown for a system with axial symmetry that the radius of gyration of particles around the lines of force is dependent upon B_p(ie $r_g = mv_\perp c/eB_p$) so that the poloidal field is the most important magnetic component for equilibrium, stability and for any process such as diffusion that arises due to the scattering of the guiding centres of particle trajectories from field line to field line.

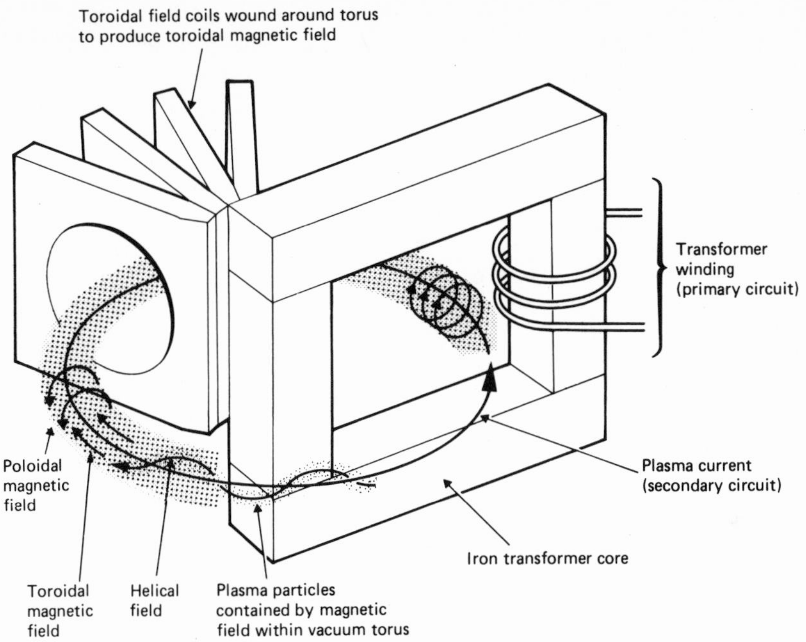

Fig.8 Basic aspects of a tokamak system.

It must be noted that in toroidal geometry the poloidal field is also more intense towards the inner side of the torus. Thus there is a gradient in B_t parallel to the major axis which also results in an expansion of the plasma ring. To compensate for this effect an external vertical field $B_\perp$ (see inset in Fig.7) must be introduced to maintain equilibrium. The efficiency of the effective confining field is measured by the parameter $\beta_p = 8\pi(n_i kT_i + n_e kT_e)B_p^{-2}$ and in a steady state condition there is probably a limiting condition $\beta_{p(max)} = R/a$ beyond which the closed system cannot be maintained.

There are several variants of the toroidal field concept classified by the ratio B_t/B_p but in the tokamak B_t is some 6 to 9 B_p and the B_p field is generated by inducing a current to flow within the plasma itself. This paper is restricted to a discussion of the tokamak but it should be noted that other devices, particularly the stellarator, have many aspects in common. The tokamak is illustrated schematically in Fig.8 where it can be seen that the toroidal body of the plasma is confined by a strong B_t field generated by an external system of toroidal solenoids. The plasma is linked with the core of a transformer and transformer action is used to induce a current to flow in the toroidal direction within the electrically conducting plasma. This current, I_p, generates the B_p field and also heats the plasma by the process of Coulomb collisions between the driven electrons and the slower moving ions. It should be noted that the basic tokamak is essentially a pulsed device because of its transformer action. At present the pulses last for less than 1s but pulses of 1000 s duration or so are feasible (see for example Ref.13) and it is not inconceivable that the tokamak might eventually be operated in a virtually continuous mode. A simple review of the physics of tokamak devices is presented in Ref.20.

To enhance the ability to both heat and to confine the plasma it is obviously desirable to maximise the plasma current. However, an upper limit must be imposed because the equilibrium conditions of the plasma can be grossly disturbed by many sources of magneto-hydrodynamic instabilities which are sensitive to the magnitude of the plasma current. The twisting of the field lines around the magnetic axis generates a rotational transform, ι, where ι is the angle in the poloidal (θ) direction. A "safety factor"

$$q \equiv \frac{2\pi}{\iota} = \frac{a}{R}\,\frac{B_t}{B_p} = \frac{5a^2 B_t}{R I_p} \tag{14}$$

can be defined. Thus each magnetic field line makes q transits of the torus before being twisted back to its original poloidal angle (θ) and the effective length of the line is $2\pi Rq$. Assessments of instabilities in present-day tokamaks indicate that $q_a \gtrsim 2.5$ where a is the value of r at the plasma boundary; values of q(r) where $r < a$ are smaller than q_a. In present-day devices $\beta_p \lesssim R/a$ so that $B_t \gtrsim B_p$ (2.5 R/a) and an upper limit of B_t = 30 to 50 kG is imposed by technological and economic aspects of the external toroidal solenoids. The use of superconducting magnets in future devices will raise B_t to values in excess of 100 kG but only at considerable financial expense.

It is apparent that the tokamak does not make effective use of its very expensive toroidal field windings and when the pressure balance β_B of the total magnetic field is considered it would appear[20] that $\beta_B \lesssim a/6R$ is a reasonable value for a tokamak with a circular minor cross section, so that $\beta_B \approx 0.05$. This factor might be improved by using a plasma of non-circular geometry (in the poloidal plane). Despite these disadvantages the tokamak still offers the most promising route to providing the plasma conditions needed for a fusion reactor although its toroidal shape poses many problems to the engineers who must both build and maintain a fusion power station (see for example Ref.21). Basic parameters of some representative tokamak devices are presented in Table 1.

Table 1 Basic Parameters of Representative Tokamak Devices.

DEVICE	COUNTRY	R (cm)	a (cm)	B_t(kG)	I_p(MA)	q
			EXISTING			
TFR	FRANCE	98	20	50	0.30	4.1
PULSATOR	FRG	70	12	28	0.08	3.6
ALCATOR	USA	54	9.5	75	0.10	6
DITE	UK	112	23	28	0.20	3.5
PLT	USA	130	40	35(46)	0.60	3.6
T-10	USSR	150	37	35(50)	0.40	
			PROPOSED OR UNDER CONSTRUCTION			
ASDEX	FRG	154	40	30	0.50	
PDX	USA	140	45	50	0.50	
JT60	JAPAN	300	100	50	3.00	
TFTR	USA	248	85	52	2.50	
JET	EEC	296	125 (a)† 210 (b)	28	3.80	2.8
			CONCEPTUAL REACTOR			
CCTR MkIIA*	UK	740	210 (a)† 370 (b)	41	11.70	2.5

*Culham Conceptual Tokamak Reactor MkIIA (5830 MW thermal; 2500 MW electrical) Ref.22.
†(a)(b) Elliptical minor cross section.

A comprehensive survey of experimental data for tokamak performance has been published by Pfeiffer and Waltz[23].

6. THE EFFECTS OF ELASTIC COLLISIONS UPON PLASMA TRANSPORT

In the preceding discussion elastic collisions between the electrons and the ions have largely been neglected. These are essentially classical in nature but the effects of ion charge Z must be taken into account so that those atomic processes which control the ion charge state distribution within the plasma are of some significance. It can be readily shown (for example see Ref.2) that Coulomb scattering is not dominated by single events but by the accumulation of a large number of collisions that eventually cause scattering through an angle equal to 90° where it can be considered that the initial energy of the incident particle has been shared with its numerous collision partners. Scattering through

90° in the presence of a magnetic field also implies that the incident particle has moved transversely across the field by a distance $\Delta r \sim r_g$. Within a plasma the Coulomb field of each charged particle is partially screened by the presence of the other charged particles. The mobile electrons screen the field of the more slowly moving ions so that the Coulomb field of the ion tends to zero at a distance λ_D defined by,

$$\lambda_D = \left(\frac{kT}{4\pi n_e e^2}\right)^{\frac{1}{2}} = 7.43 \times 10^2\left(\frac{T}{n}\right)^{\frac{1}{2}} \qquad [\text{cm}], \tag{15}$$

where λ_D is usually referred to as the Debye length. Thus in Coulomb scattering it is necessary to introduce a parameter $\Lambda = \lambda_D/b_o$ where b_o is generally assumed to be the smallest impact parameter at which small angle scattering can occur.

The electrical resistivity of a plasma, η, is a parameter of considerable importance in the tokamak device. Firstly the induced plasma current is $I_p = \pi a^2 j$ (where j is the current density [A cm^{-2}] and this must generate the poloidal magnetic field $B_p = \pi aj/5$. Thus the induced loop voltage V_ℓ must be adequate to drive the current I_p through the resistive plasma, ie,

$$V_\ell = 2\pi Rj\eta = 10\eta B_p\, R/a \qquad [\text{V}] .$$

Secondly, the driven current is used to heat the plasma electrons which then transfer their energy to the ions. The ohmic heating power density is given by,

$$P_\Omega = \eta\, j^2 \qquad [\text{W cm}^{-3}] .$$

The effects of the magnetic field can be neglected when considering these problems because the current flow is parallel to B and the resistivity can be determined on the assumption of a uniform plasma in the manner described by Spitzer[24]. The resistivity is defined as

$$\eta = \frac{c}{en_e}\,\frac{P_{ei}}{j}$$

where P_{ei} is the total momentum transferred per unit volume and time to the slow moving ions by collisions with the more mobile electrons. For a fully ionized hydrogen plasma

$$P_{ei} \approx m_e(v_i - v_e)n_e n_i \nu_{ei}$$

where $v_i - v_e$ is the drift velocity of the electrons through the ions and ν_{ei} is the frequency at which electrons transfer momentum to the ions and is assumed here to be the frequency for 90° scattering. Since $j = n_e e(v_i - v_e)/c$ it follows that,

$$\eta \approx \frac{c^2 m_e n_i \nu_{ei}}{e^2 n_e}$$

and for Maxwellian distribution of particle velocities,

$$\nu_{ei} = \frac{4(2)^{\frac{1}{2}}\pi\, n_i\, Z^2 e^4\, \ell n\Lambda}{m_e^{\frac{1}{2}}(3\, kT_e)^{3/2}} = \frac{3.0 \times 10^{-6} n_i\, Z^2\, \ell n\Lambda}{T_e^{3/2}} \quad [s^{-1}]\,. \tag{16}$$

The parameter $\ell n\Lambda$ lies in the range of 10 to 20 for most plasmas of interest here.

Consider now a hydrogen plasma with a small concentration of impurities such that an impurity ion of charge Z has a population density $C_z n_i$ where n_i is the density of the hydrogenous ions. The momentum transfer may be calculated by assuming that the electron drift velocity with respect to all ions is the same and noting that $\nu_{ei} \propto z^2$ so that,

$$P_{ei} = n_e\, m_e\, \nu_{ei}^{\ 1}(v_i - v_e)\ [n_i + n_i \sum_z C_z\, Z^2],$$

where $\nu_{ei}^{\ 1}$ refers to hydrogenous ions with Z = 1.

Then,

$$\eta = \frac{c^2 m_e}{e^2}\ \frac{\nu_{ei}^{\ 1}\, n_i}{n_e}\ [1 + \sum_z C_z\, Z^2]$$

and, because $n_e = n_i(1 + \sum_z C_z Z)$ for an electrically neutral plasma, the mean effective value of the ionic charge can be expressed as

$$\langle Z\rangle = \frac{1 + \sum_z C_z\, Z^2}{1 + \sum_z C_z\, Z}\ .$$

Spitzer and Harm[25] quote a more accurately derived expression for resistivity, namely,

$$\eta = \frac{\pi^{3/2} m_e^{\frac{1}{2}} e^2 c^2 \langle Z\rangle \ \ell n\Lambda}{\gamma_E \ 2(2\ kT_e)^{3/2}} = \frac{3.04 \times 10^{-3} \langle Z\rangle \ \ell n\Lambda}{\gamma_E T_e^{3/2}} \ [\text{ohm cm}] \qquad (17)$$

where,

$\gamma_E = 0.58(Z = 1),\ 0.68(Z = 2),\ 0.78(Z = 4),\ 0.92(Z = 16)\ 1.0(Z = \infty)$.

The significance of atomic impurities in the hydrogen plasma can now be appreciated by the fact that 1% of O^{8+} increases η by 1.4 times and 1% of Fe^{26+} causes η to be 4.3 times that of a pure hydrogen plasma.

It must be strongly emphasised that η decreases with increasing temperature (ie $\eta \propto T^{-3/2}$) and, since I_p should be maintained approximately constant in a tokamak, the ohmic heating power decreases as $T^{-3/2}$ and it is unlikely that ohmic heating alone will raise the plasma to its ignition temperature. Thus the introduction of a small amount of impurities can be beneficial because of the consequent increase in $\langle Z\rangle$ but this improvement demands an increase in the induced loop voltage V_ℓ. Present-day experiments tend to be limited by the stored energy (volts seconds) available to drive the primary of the transformer so that an increase in plasma resistivity due to impurities generally causes a reduction in the available plasma current so that the overall performance of the tokamak is degraded.

The transport of energy parallel to the magnetic field is dominated by collisions between the mobile electrons so that the significant collision frequency is ν_{ee}. This frequency is comparable in magnitude to ν_{ei}[24] and a coefficient of thermal conductivity for a plasma where Z = 1 has been quoted by Spitzer, namely,

$$\kappa_{||} = 4.5\left(\frac{2}{\pi}\right)^{3/2} \varepsilon \ \frac{(kT)^{5/2}\ k}{m_e^{\frac{1}{2}} e^4 \ \ell n\Lambda} = \frac{315\ T_e^{5/2}}{\ell n\Lambda} \ [\text{WeV}^{-1}\ \text{cm}^{-1}] , \qquad (18)$$

where the parameter ε allows for the thermoelectric effect. It is evident that even a small difference in electron temperature ΔT_e can cause a substantial transport of energy (proportional to $\Delta T_e (T_e)^{5/2}$) along the field lines. However, direct losses of energy by this process occur only if the magnetic field lines intercept the mechanical structure of the torus.

If the electron and ion temperatures are not equal then energy can be transferred by collisions between the mobile electrons and the slower ions. Spitzer has derived an expression for the equipartition time for a group of "test" particles moving through a group of "field" particles denoted by the subscript, f, namely,

$$t_{eq} = \frac{3\ m\ m_f\ k^{3/2}}{8(2\pi)^{\frac{1}{2}} n_f Z^2 Z_f^2 e^4\ \ell n\Lambda} \left(\frac{T}{m} + \frac{T_f}{m_f}\right)^{3/2} , \qquad (19)$$

which leads to an energy transfer term of the type

$$P_{ei} = 7.54 \times 10^{-28} \frac{n_e^2 Z\ \ell n\Lambda}{A_p} \frac{(T_e - T_i)}{T_e^{3/2}} \quad [\mathrm{W\ cm^{-3}}] \qquad (20)$$

where A_p is the atomic weight of the plasma ions (ie, 2.5 for DT).

The present treatment of the plasma has so far assumed that both temperature and particle densities are homogeneous. This is far from true because particles and energy are continuously lost from the plasma boundary and so the maximum values of both temperature and density will, in general, coincide with the magnetic axis. Axial symmetry is maintained throughout the torus but radial gradients - dn/dr and - dT/dr must exist throughout the plasma; in present-day devices the magnitudes of these gradients are of order $5 \times 10^{12}\ \mathrm{cm^{-3}/cm}$ and 100 eV/cm respectively but in reactors these values could be an order of magnitude greater.

The outward flux of particles across the magnetic field $\Gamma_\perp$ can be defined by

$$\Gamma_\perp = D_\perp dn/dr .$$

Here $D_\perp$ is the diffusion coefficient for radial transport across the plasma and can be expressed as

$$D_\perp \sim (\Delta r^2)/\Delta t$$

where Δr is the mean radial step caused by scattering collisions in a mean time step Δt. In a pure hydrogen plasma electrons and ions must be lost at equal rates in order that the overall electrical neutrality of the plasma be maintained and, in addition, the minimum radial step can be approximated to the gyroradius r_g of the charged particle. Thus it is reasonable to make the approximation,

$$D_\perp \sim r_g^2\ \nu_{ei} \qquad [\mathrm{cm^2\ s^{-1}}] . \qquad (21)$$

Since r_g is greatest for the ions and $\nu_{ei} \approx \nu_{ee}$ it follows that ions tend to diffuse at a faster rate than the electrons. However, when allowance is made for the need to maintain ambipolarity it is found[2], for the case where $T_i \sim T_e$, that

$$D_{amb\perp} \approx 2D_{e\perp},$$

ie, twice that of the slower diffusing electrons. The loss of charged particles from the confined plasma also means that the amount of energy transported by each ion pair is also lost. The energy density transported by convection can therefore be expressed as

$$P_{D\perp} = \Gamma_\perp(3\ kT_e + 3\ kT_i)/2,$$

and whilst this may be substantial in the hot plasma near to the magnetic axis it decreases towards the plasma boundary where the temperature is much lower.

The temperature gradient - dT/dr causes losses of energy due to thermal conduction radially across the magnetic field. This can be treated in a somewhat similar way to particle diffusion but, because no nett movement of charge is involved, the constraints imposed by ambipolar transport are not relevant to this problem. The power flux density Ψ can be expressed as

$$\Psi = -\ \kappa_\perp\ dT/dr$$

where $\kappa_\perp$ is the coefficient for thermal conductivity perpendicular to the magnetic field. It is convenient to make use of the relationship

$$\kappa_\perp \approx \frac{\kappa_{||}\nu_c^2}{\omega_g^2} \qquad (22)$$

where ν_c is the collision frequency, ω_g the gyrofrequency of the conducting particles and $\kappa_{||}$ is the coefficient parallel to the field lines (ie in conditions of zero magnetic field). Since $\omega_g = eB/mc$ it is evident that the radial thermal flux is very substantially reduced by the magnetic field and it is this property coupled to the comparable effect on the convective losses that provides the thermal insulation between the core of hot plasma and the walls of the torus. It should also be noted that $\omega_g \propto m^{-1}$ so that the ions tend to be the more effective conductors of energy across the field.

It must be strongly emphasised that the "classical"

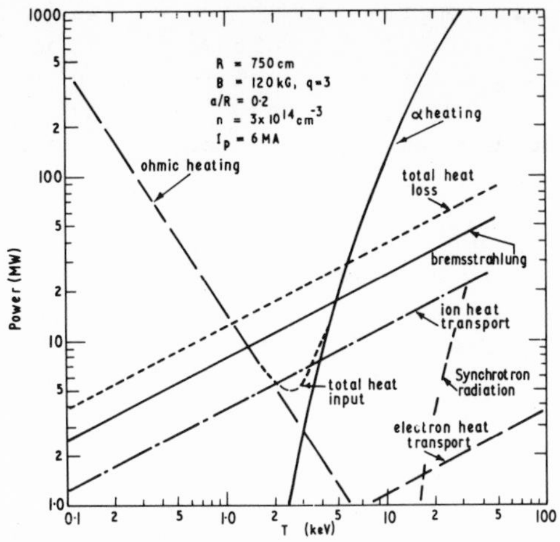

Fig.9 Power balance for a pure DT reactor taken from Sweetman[26], neo-classical heat losses were assumed (see Ref.20).

coefficients for diffusion and thermal conduction considered in the preceding simple discussion are appreciably modified in a toroidal environment. The topography of the magnetic field and the collisionality of the particles moving along the field lines of effective length $2\pi Rq$ introduce a large number of effects that enhance the losses of particles and energy (for example see Ref.20) but these effects do not alter the overall concepts of particle and energy transport.

A convenient example of the combined effects of these various processes in a DT reactor plasma has been given by Sweetman[26]. He assumed that no impurities were present and that the radial profiles of density and temperature could be expressed as $n(r) = (1 - r^4/a^4)$ and $T_i(r) = T_e(r) = (1 - r^4/a^4)$. His results, together with the other relevant reactor parameters are shown in Fig.9 where it is evident that the balance between the ohmic heating power and the α-particle heating power occurs at a temperature of about 3 keV. However, at this temperature, the total energy losses from bremsstrahlung and radial transport are about 10 times greater than the energy input from ohmic heating. Clearly, such a device will not ignite and some form of additional heating is necessary.

7. NEUTRAL BEAM INJECTION HEATING AND SOME EFFECTS OF CHARGE EXCHANGE

A variety of mechanisms for providing external sources of plasma heating have been studied; for example, injection of μ-wave radiation, adiabatic compression of the plasma by a rapid increase in the effective magnetic field intensity and, more extensively, injection of intense beams of energetic particles. Although each route has its protagonists, the most promising approach appears to be heating by neutral beam injection and this subject has recently been reviewed by Green[27].

To heat the plasma using an external source of energetic particles it is necessary to meet the following requirements:

a. Each injected particle must be appreciably more energetic than the electrons and ions of the plasma.

b. Particles must be injected at such a rate that the bulk plasma can gain an adequate amount of energy during the confinement time of the injected particle.

c. The particles must be able to pass through the strong magnetic field that confines the plasma and yet be trapped in the plasma.

d. When inside the plasma, the particles must travel for a distance sufficient for there to be an efficient equipartition of energy.

The technique of neutral beam injection can go some way towards meeting all of these requirements. Firstly, fast atoms or molecules are formed by neutralisation of ions that have themselves been extracted from an ion source with energies in the region of 100 keV (the process of neutralisation is either charge exchange of positive ions in a gas target or possibly, in the future, stripping of negative ions). The neutral species pass unimpeded through the confining magnetic field but, during penetration of the plasma, they become ionized and thereby trapped within the field. Trajectories of the trapped ions can be so long that several mean free paths for energy transfer can be accommodated within the torus.

The instantaneous rate of transfer of energy to the plasma ions can be derived from equation (19) and is given by

$$\frac{1}{E_{in}} \frac{dE_{in}}{dt} = 1\ 1.8 \times 10^{-7} \frac{A_i^{\frac{1}{2}} Z^2 n \ \ell n\Lambda}{A_p E_{in}^{3/2}} \qquad [s^{-1}] \ , \tag{22}$$

where E_{in}, A_i, Z are respectively the energy (in eV), the atomic weight and the charge of the trapped ions and A_p is the atomic weight of the plasma ions ($\equiv$ 2.5 for DT). Assuming that $v_e \gg v_{in}$, then the instantaneous transfer rate to the electrons is,

$$\frac{1}{E_{in}} \frac{dE_{in}}{dt} = - 3.3 \times 10^{-9} \frac{Z^2 \, n \, \ell n \Lambda}{A_i T_e^{3/2}} \quad [s^{-1}] , \tag{23}$$

so that the ratio,

$$\frac{\text{transfer rate to ions}}{\text{transfer rate to electrons}} = \frac{56}{A_p} \left(\frac{A_i T_e}{E_{in}} \right)^{3/2} .$$

Fusion reaction rates depend upon plasma ion temperature and so it is beneficial to maximise the transfer of beam energy to the DT ions. The obvious choice for the injected beam species is D atoms. This element is not radioactive, it is naturally abundant and it will not contaminate the DT plasma. Moreover, energetic D atoms can possess sufficient velocity to penetrate to the central regions of the plasma before being ionized so that the beam power can be deposited in the hot core of the plasma where DT fusion reactions are most likely to occur especially during the ignition phase. Since $A_i = 2$ and $A_p = 2.5$, equality in the energy transfer rates between the electrons and the ions is given by,

$$E_{in} \approx 16 \; T_e ,$$

so that beam energies in the range 100 to 200 keV are desirable.

These energies are also compatible with the requirements for beam trapping and the principal atomic processes that cause the beam to become ionized and thereby confined within the magnetic field are:

Reaction	Process	Cross Section
$D + {D^+ \atop T^+} \rightarrow D^+ + {D \atop T}$	charge exchange	$\sigma_x \, (v_i)$
$D + {D^+ \atop T^+} \rightarrow D^+ + e + {D \atop T}$	ion impact ionization	$\sigma_{ip} (v_i)$
$D + e \rightarrow D^+ + e + e$	electron impact ionization	$\sigma_i \, (v_e)$

where v_c is the atom-ion collision velocity. The depth of plasma penetrated by D atoms is dependent upon the effective mean free path. For the heavy particle collisions $v_{in} >> \langle v_i \rangle$ (ie, the beam velocity is greater than the average velocity of the DT ions) so that $\lambda_x = [n_i \sigma_x (v_{in})]^{-1}$ and $\lambda_{ip} = [n_i \sigma_{ip} (v_{in})]^{-1}$, but for electron collisions $\langle v_e \rangle >> v_{in}$, so that $\lambda_i = v_{in} [n_e \langle \sigma_i (v_e) v_e \rangle]^{-1}$. The

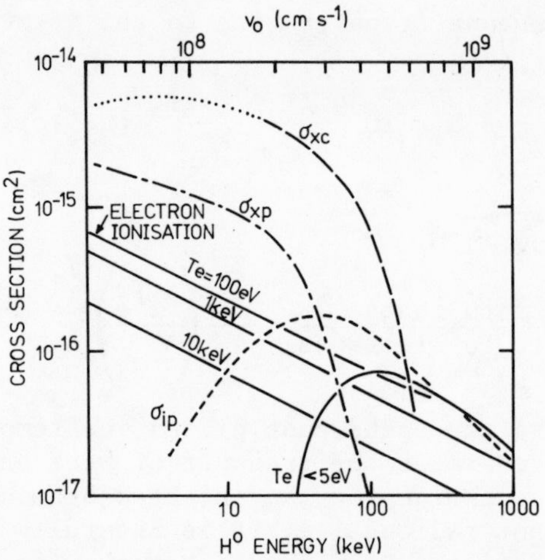

Fig.10 Cross sections of the atom-proton collision processes (σ_{xp},σ_{ip}) responsible for beam trapping plotted as a function of collision velocity v_0. Recent experimental data (σ_{xc}) for $H + C^{5+}$ are taken from[28] and extrapolated to lower energies using[29]. Electron collision data are plotted in the form $<\sigma_i v_e>/v_0$.

beam attenuation over a path length ℓ can be expressed as

$$I = I_o \exp(- n\sigma'\ell) = I_o \exp(- n\ell/D)$$

where $n_e = n_i = n$ is assumed and D is a characteristic plasma "thickness" for $\exp^{-1}$ attenuation of the beam; D can be expressed as

$$D = \frac{1}{\sigma'} = \left[\sigma_x(v_{in}) + \sigma_{ip}(v_{in}) + \frac{<\sigma_i(v_e)v_e>}{v_{in}}\right]^{-1} \ [\mathrm{cm}^2]. \qquad (24)$$

The relevant cross sections together with the electron ionization term are plotted as functions of beam energy in Fig.10 (actually the data are for H^0 and H^+ but they are applicable to D^0 and D^+ or T^+ at values of $v_{in} = v_0$). It is evident that the beam is trapped mainly by charge exchange with the plasma ions when v_{in} is less than about 2×10^8 cm s^{-1} and also that, for hot plasmas of interest to fusion, electron impact ionization provides a relatively minor contribution. Trapping by ionization in collisions with the plasma ions becomes increasingly more important as v_{in} increases.

High velocity injection is beneficial because it ensures penetration to the hot plasma core and favours energy transport to

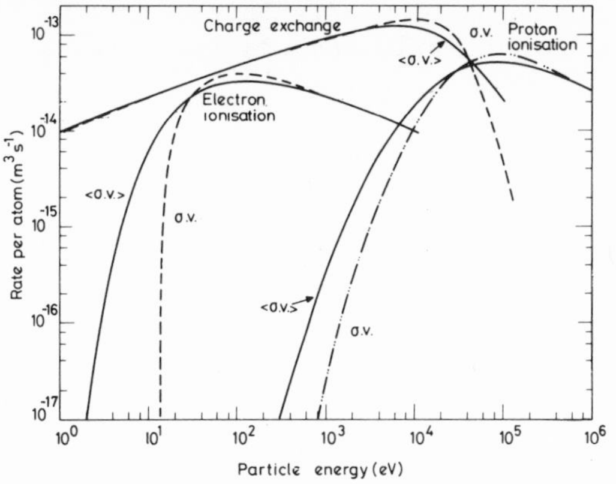

Fig.11 Rate coefficients and cross sections for charge exchange and ionization in atomic hydrogen, taken from[30].

the DT ions but, equally important, it minimises the probability of neutralisation of plasma ions by charge exchange collisions. A plasma ion that is neutralised is no longer confined within the magnetic field and this hot atom will be lost to the walls of the torus unless it is re-ionized during its passage through the plasma. Since charge exchange is an adiabatic process the neutralised ions have the same average energy as the plasma ions* and their mean free path for re-ionization by electron impact can be expressed as $\lambda_i = \langle v_i \rangle (n_e \langle \sigma_i v_e \rangle)^{-1}$, but during their passage through the plasma such atoms will also experience other charge exchange collisions with plasma ions and this process has a mean free path $\lambda_x = \langle v_i \rangle (n_i \langle \sigma_x v_i \rangle)^{-1}$. Collisions occur as random events so that the probability $\pi_o(\ell)$ that an atom (but not the same atom) can be transported for a distance ℓ from an initial neutralisation site can be expressed as

$$\pi_o(\ell) \sim \left[1 - \left(\exp - \frac{\lambda_i}{\lambda_x}\right)\right]^{\frac{\ell}{\lambda_x}}$$

$$\sim \left[1 - \left(\exp - \frac{\langle \sigma_i v_e \rangle}{\langle \sigma_x v_i \rangle}\right)\right]^{\frac{n \langle \sigma_x v_i \rangle \ell}{\langle v_i \rangle}} . \qquad (25)$$

The rate coefficients $\langle \sigma_x v_i \rangle$ and $\langle \sigma_i v_e \rangle$ are shown for H^+ and H in Fig.11 and the coefficient for charge exchange is greater in magnitude at all temperatures. There is thus a high probability that neutralised plasma ions can escape. The plasma becomes opaque only

*The ability for unconfined atoms to retain the imprint of the plasma ion temperature is used as a diagnostic technique to measure T_i.

when values of $n\ell$ are particularly large or when the ratio $\langle\sigma_i v_e\rangle/\langle\sigma_x v_i\rangle$ is particularly favourable to ionization, ie at temperatures near to 50 eV.

Charge exchange can also occur with impurity ions that are present in the plasma and the cross sections for such processes are both large in magnitude and much less sensitive to increases in v_{in} than are the comparable reactions of H with H^+, D^+ and T^+. As an example, data for,

$$H + C^{5+} \rightarrow H^+ + C^{4+},$$

taken from[28,29] are plotted in Fig.10. It is evident that small concentrations of highly charged impurity elements can affect beam penetration so that the spatial distribution and concentration of such impurities (discussed in Section 8) are of considerable significance to neutral beam heating.

Efficient application of the processes by which the fast atom beam is formed is an essential prerequisite for the success of neutral beam heating. Hydrogen isotopes exist as diatomic molecules under normal atmospheric conditions and when ionized in an ion source molecular gases yield an ion beam that contains a mixture of species, eg, D^+, D_2^+, D_3^+. The ion beam must then be converted into a neutral beam by charge exchange (or dissociative charge exchange) in a neutral gas target, ie, (D^+ in D_2), (D_2^+ in D_2) etc. The product atom beam has almost the same velocity as the incident parent ions and so, to attain the requisite values of v_{in}, beams of molecular parent ions must be accelerated from the ion source by means of a very high potential difference and severe technological problems are encountered[27]. It is not practicable to produce a completely pure beam of D^+ atoms and, since all species in the beam must be accelerated through the same potential, then the presence of molecular components introduces a distribution in v_{in} and hence a spread in beam penetration.

A particularly difficult problem arises because the probability of ion neutralisation by charge exchange in the neutraliser gas target is small in the range of velocities of interest and so the efficiency of conversion to atoms is low. One solution to this problem is to use D^- as the parent species and invoke the stripping reaction

$$D^- + D_2 \rightarrow D + e + D_2$$

as a route to atom beam formation. This reaction has a large cross section at the requisite velocity and the beneficial effect of D^- ions upon the power efficiency

$$\eta_o = \frac{\text{power in atom beam leaving neutraliser}}{\text{power in ion beam entering}}$$

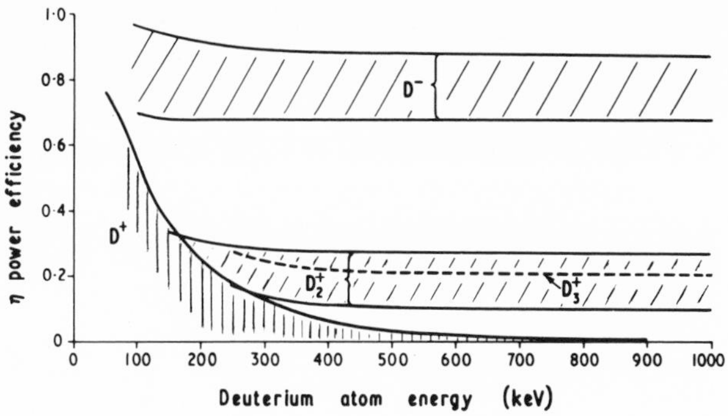

Fig.12 Power efficiency η_0 for atom production from D_1^+, D_2^+, D_3^+ and D^- beams taken from[31]. Hatched area indicates variation with degree of ionization of target. Ionization reduces η_0 for D^+ and increases η_0 for D_2^+ and D^- ions.

has been determined by Riviere[31] and is shown in Fig.12. The requirement for an intense source of D^- ions poses severe problems. One possible solution is to convert D^+ to D^- by reactions of the type

$$D^+ + Na \rightarrow D^- + Na^{++} \ ;$$

The equilibrium yield of D^- is high, but unfortunately only at energies in the range of a few keV, so that additional acceleration of the D^- beam is necessary prior to its entry into the gas stripping cell. An alternative approach is to use direct acceleration of D^- ions formed in an ion source wherein the D^- yield has been enhanced by plasma interaction at a caesiated surface. An up to date appraisal of H^- (or D^-) technology may be attained by scrutiny of[32]. It must be stressed that the concept of D^- injection is much less advanced than that of D^+ ions where injection systems yielding 2.4 MW for $\sim$ 0.3s are currently in operation.

Such powerful beams must produce substantial ionization of the gas within the neutralising target and the effects can be seen from Fig.12 where the hatched areas show changes in power efficiency due to collisions of the beam with electrons in a fully ionized target. Processes such as,

$$D_2^+ + e \rightarrow D^+ + D + e + e$$

and

$$D^- + e \rightarrow D + e + e$$

enhance the efficiency but, for D^+ ions where neutralisation is strongly dependent upon charge exchange, ionization of the target decreases the efficiency of atom beam formation.

8. POWER LOSSES DUE TO INELASTIC COLLISIONS OF ELECTRONS WITH IMPURITY IONS

Inelastic atomic collisions occur between the electrons and the ions within the confined plasma. Any unstripped ion of element X with a charge state z+ is likely to be further ionized by processes of the type

$$e + X^z \rightarrow e + e + X^{z+1} \quad \text{(outer and inner shell ionization)}$$

and

$$e + X^z \rightarrow e + X^{z*} \rightarrow e + e + X^{z+1} \quad \text{(auto-ionization)}$$

and the rate coefficient for these combined processes can be conveniently expressed as $S_i(T_e) = \langle\sigma_i v_e\rangle$. Ions are neutralised by two body radiative recombination,

$$e + X^{z+1} \rightarrow X^z + h\nu$$

and, if sufficient electrons are present, three body recombination,

$$e + e + X^{z+1} \rightarrow e + X^z,$$

takes place. Electron impact excitation of unstripped ions occurs,

$$e + X^z \rightarrow e + X^{z*} \rightarrow e + X^z + h\nu,$$

and the rate coefficient for excitation can be expressed as $S_{mn}(T_e) = \langle\sigma_{mn} v_e\rangle$ where the transition is from level m to n. Electron collisions at energies very close to the excitation threshold result in slow, inelastically scattered electrons that can be trapped within the Coulomb field of their target ion. A doubly excited ion is thus formed and the process leads to dielectronic recombination in the following manner,

$$e + X^{z+1} \rightarrow X^{z**} \begin{cases} \nearrow X^{z*} + h\nu & \text{(dielectronic recombination)} \\ \searrow X^{z+1} + e & \text{(auto-ionization).} \end{cases}$$

The characteristic time for any inelastic electron collision process can be expressed as $\tau_e = (n_e\langle\sigma v_e\rangle)^{-1}$ and in a magnetically confined fusion plasma this time is appreciably greater than the radiative lifetimes ($\tau_n \propto z^{-4}$) of most of the excited states of the highly charged ions. Furthermore, the plasma is virtually transparent to the X-ray photons that are characteristic of $n \rightarrow m$ transitions in highly charged ions. Thus excitation occurs at a rate that is governed only by electron collisions and the environment is somewhat comparable to that of the solar corona (see for example Ref.33). The lifetimes of the excited states are generally so short that an ion resides mostly in its ground state m = o (or, where applicable, in a metastable state) and excurses for only brief periods into an excited state. The equilibrium density population of excited state n of an ion in charge state z can therefore be expressed as

$$\frac{n_{zn}}{n_{zo}} = \frac{n_e S_{on}(T_e)}{A_{no}} \tag{26}$$

where A_{no} is the rate of spontaneous decay from $n \to o$ (ie the Einstein A coefficient). The power radiated is given by

$$P_{zn}(T_e) = 1.6 \times 10^{-19} A_{no} n_{zn} \chi_{on} \qquad [\mathrm{W\ cm^{-3}}]$$

where χ_{no} is the threshold energy of the collision induced transition $o \to n$. Analysis of the losses of energy caused by line radiation shows that the major components are carried by but a few resonant transitions from the ground or metastable states so that the total line radiation power lost, $P_{z1}(T_e)$, for a particular ion can be determined from

$$P_{z1}(T_e) = \sum_n P_{zn}(T_e)$$

where the summation is of limited extent.

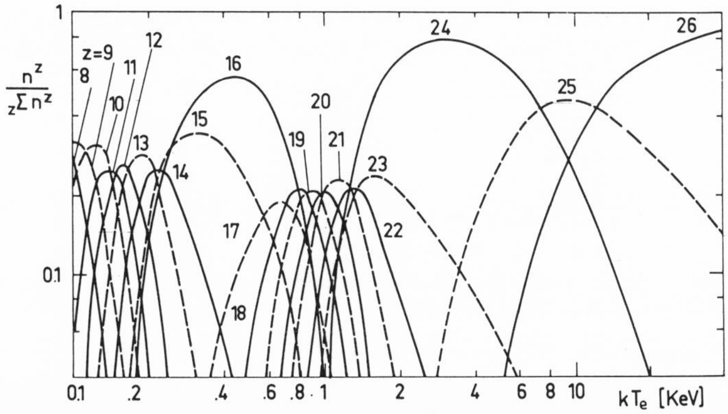

Fig.13 Charge state distribution for iron plotted as a function of electron temperature. Data are taken from[34].

The charge state balance $n_{(z+1)}/n_z$ can be determined on the assumption that ionization takes place only from the ground state and the time dependent situation can be stated in the form,

$$\frac{dn_z}{dt} = n_e\left\{- n_z S_{iz}(T_e) + n_{(z-1)} S_{i(z-1)}(T_e) - n_z\alpha_z(T_e) + n_{(z+1)}\alpha_{(z+1)}(T_e)\right\}. \quad (27)$$

Here z can range from 0 to the fully stripped condition z = Z and $\alpha_{(z+1)}$ is the coefficient for all forms of recombination. The steady state solution is of the form

$$\frac{n_z}{n_{(z+1)}} = \frac{\alpha_{(z+1)}(T_e)}{S_{iz}(T_e)}$$

and the steady state charge balance within a plasma can be determined as a function of temperature T_e provided that the total number density of particles of a particular element $N_x = \sum_z n_z$ is prescribed. Results for iron taken from[34] are shown in Fig.13.

The power lost due to recombination is of the general form $P_{zr}(T_e) = n_e n_z \alpha_z(T_e)\Delta E_e$, where ΔE_e is the total energy lost by the colliding electrons, and it is convenient to consider the sum of all the power losses associated with a charge state z in the form of a total power loss coefficient,

$$P_{tz}(T_e) = \left[P_1(T_e) + P_r(T_e) + P_d(T_e) + P_{br}(T_e)\right]_z = n_e n_z F_{tz}(T_e),$$

where the subscripts r and d refer to two and three body recombination and to dielectronic recombination respectively. For a particular element X the radiated power loss is $P_{tx}(T_e) = \sum_z P_{tz}(T_e)$ where the summation is dependent upon the number of charge states populated at a particular temperature T_e. The radiated power function[35] is shown for iron in Fig.14 in the form,

$$F_{tx}(T_e) = P_{tx}(T_e)/n_e n_x \qquad [\mathrm{W\ m^3}],$$

and the contributions from the various processes that make up this total loss are also shown.

The accuracy of this type of analysis is dependent upon the precision with which the parameters $S_i(T_e)$, $S_{mn}(T_e)$, $\alpha_{(z+1)}(T_e)$ etc can be calculated. The plasma environment (ie, its relatively high electron density and its large magnitude of B and the electric field, $E = e[\underline{B} \times \underline{v}]$) perturbs the atomic characteristics so that data determined for use in the solar corona must be applied with caution when considering fusion plasmas. The effects of the plasma environment are particularly dominant upon the higher excited states because the rate of inelastic electron collision tends to

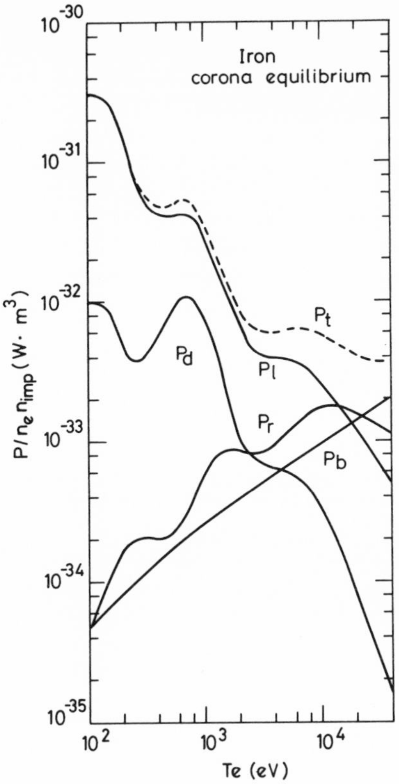

Fig.14 Radiated power function $F_{tx}(T_e) = P_{tx}(T_e)/n_e n_x$ for iron; also shown are the separate losses due to line radiation (P_l), radiative recombination (P_r), dielectronic recombination (P_d) and bremsstrahlung radiation (P_b). Data are taken from[35]. Note that 10^{-6} Wm^3 = W cm^3.

increase as n^4 (ie $\sigma(n) \propto n^4$) and the coronal conditions break down at some limiting value of n; further, the ionization thresholds are low ($\chi_i(n) \propto n^{-4}$) and field ionization effects can take place; indeed, within a plasma the ionization continuum is lowered somewhat. Substantial contributions to dielectronic recombination occur from states where $n > 50$ and so this process is especially sensitive to plasma effects.

Application of the steady state coronal model requires that equilibrium conditions for collisions between electrons and ions be attained and the longest collision time is usually that for recombination (as a typical example, let $n_e = 10^{14}$ cm^{-3} and $\alpha = 10^{-12}$ cm^3 s^{-1}, so that $\tau_e \approx 10^{-2}$s). Thus coronal conditions apply only to a plasma that is quiescent for a period comparable to τ_e and where the ions do not drift during the time τ_e into regions where either n_e or T_e are significantly changed. Further, the validity of assuming Maxwellian velocity distributions must always be questioned in laboratory plasmas. Finally, a magnetically confined plasma must contain some concentration of neutral atoms and, because charge exchange cross sections for collisions of H(or D,T) with highly charged ions are so large (see Fig.10), even small concentrations of atoms can significantly perturb the charge state balance predicted by the coronal model.

Axisymmetric gradients in density and temperature exist in the torus and the density of power lost due to impurity radiation in a radial element lying between r and r + dr is given by,

$$\frac{P(T_e,r)}{2\pi r dr} = n_e(r)\ n_x(r)\ F_{tx}(T_e,r) \qquad [\text{W cm}^{-3}], \tag{29}$$

and can be determined if the radial distributions of n_e, n_x and T_e are known. Impurity ions are unlikely to be uniformly distributed throughout the plasma but, if it is assumed that there is no gradient in temperature, then classical diffusion theory[36] indicates that the radial flux of an impurity element X can be expressed as

$$\Gamma_x = K(z n_x \nabla n_i - n_i \nabla n_x), \tag{30}$$

where z is the charge state of the impurity and the parameter K allows for the effects that toroidal geometry imposes upon classical diffusion. The first term in equation (30) is dominant because of the factor z and so the flux of highly charged impurities is largely dependent upon the gradient in hydrogen ion density, ∇n_i, so that impurities diffuse inward if n_i peaks on the magnetic axis. The steady state solution yields $n_x(r) \propto [n_i(r)]^z$ which could cause unacceptably high radiation losses within the hot plasma core. This simple concept is substantially modified when temperature gradients are taken into account and also when more realistic diffusion coefficients are included (a brief survey of recent literature is given in Ref.40) and the most recent models of plasma behaviour indicate that the inward diffusion of impurities need not be excessive.

Some indication of the radial distribution of impurities can be obtained by assuming that a prescribed flux of impurity atoms is released from the torus wall with a velocity v_0 and that this velocity is retained by each charge state as the element moves

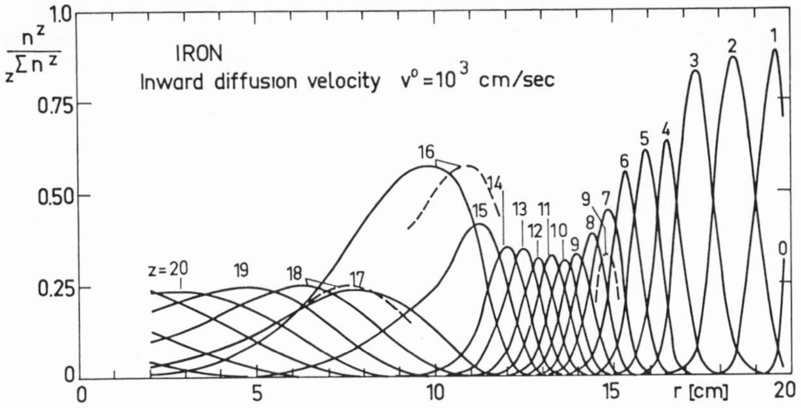

Fig.15 Radial distribution of the charge state of iron calculated for the TFR device[35] assuming a constant inward diffusion velocity of 10^3 cm s^{-1}. The dashed curves show results obtained assuming a static, steady state coronal distribution.

inward through the plasma. An analysis based upon this concept[35] is shown in Fig.15 for the density and temperature gradients relevant to the TFR experiment (very approximately these are 3×10^{12} cm^{-3}/cm and 80 eV/cm extending for some 10 cm inward from the wall). It should be noted that a drift velocity of 10^3 cm s^{-1} causes some violations of the τ_e requirement of a steady state coronal model and so due allowance has been made. In effect, ions have insufficient time to equilibrate with the local electron temperature and so a particular charge state z experiences a higher electron temperature than would pertain in an equilibrium condition. Thus, both the depth of penetration of a charge state z and the total radiated power are increased. By contrast an opposite effect can occur near to the boundary of the plasma because here outward diffusion may cause an ion to leave the plasma in a shorter time than that required for recombination. Therefore the ion leaves in a higher charge state than would be predicted by the steady state model.

It is evident that the charge states z of impurities lie in radial bands throughout the torus (ie "onion skins" of 'width' Δr) and it should be noted that the power radiated is dependent upon both the volume of the band ($4\pi^2 r$ R Δr) and the magnitude of $P_{tx}(T_e)$ within each band and that the latter increases with decreasing temperature (see Fig.14). Both aspects provide an environment in which a modest concentration of impurities in the cooler, outer layers of the plasma can radiate all of the energy released by the α-particles in the hot core of the plasma but only if this energy can be transported outwards in the form of convection and conduction. The effect of impurity radiation is to cool the electrons and thereby introduce a "short circuit" for the transport of energy through the insulating layer provided by magnetic confinement. The process can be tolerated provided that the increase in temperature gradient consequent upon edge cooling does not cause an excessive transport of power from the plasma core to the outer radiating bands and also provided that inward diffusion of impurities does not spread this "short circuit" effect too deeply into the plasma.

9. PLASMA MODELLING

A confined plasma must maintain continuity of particle transport and also a balance in both momentum and energy. For a toroidal system it is convenient to consider transport in one dimension only, namely radially from the magnetic axis to the wall of the torus. The analysis is complex and the degree of sophistication is often limited by the extent of local computing facilities.

A simplified expression adapted from[37] is given below for a DT plasma (ie, Z = 1) that is heated by both a neutral beam and by α-particle power.

$$\frac{\delta n_i}{\delta t} = \frac{1}{r}\left(rD_\perp \frac{\delta n_i}{\delta r}\right); \qquad n_e = n_i + 2n_\alpha + \sum_x n_x Z_x; \tag{31-32}$$

$$\frac{\delta n_i T_i}{\delta t} = -4.28 \times 10^{-11} n_i n_e \frac{(T_i - T_e)}{T^{3/2}} + \frac{2}{3r}\frac{\delta}{\delta r}\left(rn_i \kappa_{i\perp} \frac{\delta T_i}{\delta r}\right) - \frac{1}{r}\frac{\delta}{\delta r}(rn_i v_i T_i)$$

|← ion-electron transfer →| |← conduction →| |← convection →|

$$+ 4.17 \times 10^{15} \{n_D n_T \langle\sigma_{DT} v\rangle U_{\alpha i} + P_{in}(U_{bi} + f(Q_\alpha/E_{in})U_{\alpha i})\} - P_{cx} \tag{33}$$

$$\frac{\delta n_e T_e}{\delta t} = 4.28 \times 10^{11} n_e n_i \frac{(T_i - T_e)}{T^{3/2}} + \frac{2}{3r}\frac{\delta}{\delta r}\left(rn_e \kappa_{e\perp} \frac{\delta T_e}{\delta r}\right) - \frac{1}{r}\frac{\delta}{\delta r}(rn_e v_e T_e)$$

$$+4.17 \times 10^{15}\{n_D n_T \langle\sigma_{DT} v\rangle U_{\alpha e} + E_j + P_{in}(U_{be} + f(Q_\alpha/E_{in})U_{\alpha e}) - P_{br} - P_{cy} - \sum_x P_{tx}(T_e)\} \tag{34}$$

$$\frac{\delta B_p}{\delta t} = 10^5 \times \frac{\delta E}{\delta r}; \qquad \frac{\delta j}{\delta t} = 7.96 \times 10^{11} \frac{1}{r}\left(r\frac{\delta E}{\delta r}\right); \qquad E = \eta j \tag{35-37}$$

Here E is the toroidal electric field (V cm^{-1}) so that $P_\Omega = Ej$ and $U_{\alpha i}$, $U_{\alpha e}$ are the fractions of the α-particle energy going to the ions and to electrons respectively. U_{bi} and U_{be} are equivalent fractions of the injected beam power, P_{in}, and a fraction, f, of this beam is assumed to collide with the plasma ions and so produce fusion reactions that give rise to additional heating by α-particles. Power losses by charge exchange are accounted for by the parameter P_{cx}. In these expressions time t is in 10^{-3}s and power in W cm^{-3}.

Most of the parameters are strongly interrelated, for example the diffusion coefficients $D_\perp$ and the thermal conduction coefficients $\kappa_\perp$ together with the resistivity η vary across the plasma depending upon local conditions of n, B, T and q(r) (see for example Ref.20); the power deposited by the neutral beam at radius r depends upon beam penetration. The impurity radiation loss, $P_{tx}(T_e)$, is also dependent upon the inward diffusion of impurities which must be treated differently from that of DT (see Section 8). Complications also arise from the high initial energy of the α-particles (ie, 3.52 MeV) and the fact that if α-particles are not exhausted and replaced by fresh DT then the fusion reaction rate decreases with time. The model plasma must also be bounded by conditions that describe adequately the interface between the plasma and the torus wall and both the physics and chemistry of this region are highly complex.

The general method used to obtain a simultaneous solution of the model equations is to prescribe an initial set of radial profiles for n, T and j (usually parabolic) together with accepted forms for $\kappa_\perp$ and η. Particles are allowed to diffuse to the boundary where conditions for n and T_e and T_i are prescribed. Particles that reach the boundary are then returned to the plasma by means of a neutral particle source term that makes some allowance for the velocity of the desorbed and reflected particles. Release of impurities by sputtering of the torus wall may also be included. The initial spatial distributions are evolved with time and sets of the radial characteristics of density and temperature together with the radial flow of power are determined. Beam heating can be switched on and off as required.

An example of this method is provided by Hughes[38] who has used his HERMES code to model the JET plasma. The basic parameters of JET are given in Table 1 and Hughes prescribed the following initial conditions. Time t = 0; plasma DT; $n_D = n_T = n_i/2$.

$$n(r) = (\hat{n} - n_a)\,(1 - r^2/a^2) + n_a,$$
$$T(r) = (\hat{T} - T_a)\,(1 - r^2/a^2) + T_a,$$

where $\hat{T}_e = \hat{T}_i = 100$ eV on the magnetic axis and $T_{ea} = T_{ia} = 10$ eV at the boundary (ie, pedestal) point. The comparable density parameters are

$$\hat{n}_e = \hat{n}_i = 5 \times 10^{13}\ cm^{-3} \text{ and } n_{ea} = n_{ia} = 5 \times 10^{12}\ cm^{-3}.$$

The plasma current flowing at t = 0 is 4.8 MA and neutral beam heating is provided by 120 keV D atoms starting at 7s with a beam power of 4.5 MW. Thereafter the injection power rises linearly to 45 MW at 8s and then remains constant.

The coefficient for the recycling of atoms from the torus wall is taken as unity up to 7s. Thereafter it is reduced to 0.9 to allow for the increased intensity of the outward flux of energetic charge exchange atoms produced by the 45 MW neutral injection beam (see[40] and Section 10). All hydrogenous atoms leave the wall with

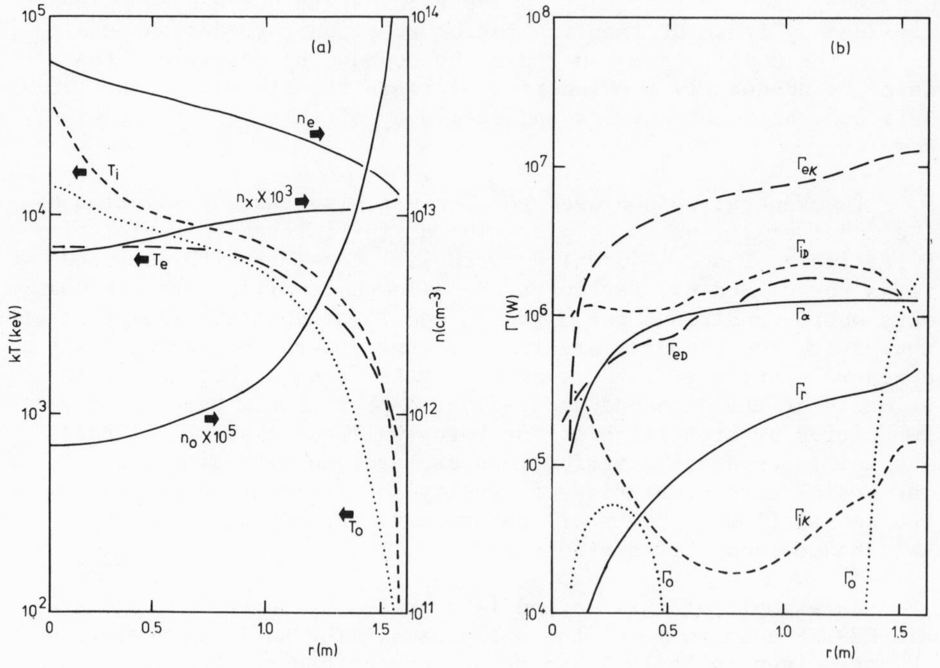

Fig.16 A model of the JET plasma computed by Hughes using the HERMES code[38].

(a) Shows radial profiles of density (n_e, n_x, n_o) and temperature (T_e, T_i, T_o).

(b) Shows profiles of the outward flow of power Γ associated with conduction ($\Gamma_{i(e)K}$), convection ($\Gamma_{i(e)D}$), neutral particle flow (Γ_o), α-particle heating (Γ_α) and radiative losses (Γ_r).

an energy E_o = 3 eV (which is probably rather low for real conditions). Contamination of the plasma by iron impurities is accounted for by ascribing an inward flux (5 x 10^{12} cm^2 s^{-1}) during the time t = 0 to 7s and increasing this to 2 x 10^{13} cm^2 s^{-1} for t > 7s. This is equivalent to sputter yield of ≈ 3.7 x 10^{-3} (see Section 10). The distribution of iron charge states and their radiative power loss is determined using the unified ion model[39]. Data for the plasma at t = 8s are presented in Fig.16.

Radial profiles of density and temperature are shown in Fig.16(a). Outward diffusion coupled with recycling at the wall has caused the initial parabolic density profile to become almost square. There is very appreciable attenuation of the DT neutral density (n_o) towards the magnetic axis but it should be remembered that the atoms in the centre of the plasma are mainly those of the high velocity neutral beam and this fact is evident from the high temperature T_o of the atoms in this region. Neutral injection has raised the ion temperature above that of the electrons but the concentration of impurity ions has not become excessive, ie, $C_{x(max)} \sim 10^{-3}$ near to the wall.

Radial losses of power from the plasma are shown in Fig.16(b) in for form of an outward loss flux,

$$\Gamma = 4\pi^2 R\int_o^r P(r)rdr = 2\pi^2 Rr^2\overline{P} \qquad [W],$$

where P(r) is the power density at r for each particular loss process and $\overline{P}$ is the average over o→r. The plasma has not yet reached ignition because Γ_α does not exceed the total power losses. In fact α-particle heating occurs only within r ≈ 60 cm so that the volume of plasma in which fusion reactions take place is only about 0.15 of the total plasma volume. The major loss process is electron conduction (Γ_{eK}) and this greatly exceeds ion conduction (Γ_{iK}). It should be noted that this is in contradiction to the classical description presented in Section 6 but is a consequence of applying more realistic coefficients for electron thermal conduction (for example see Ref.20). Losses from electron convection (Γ_{eD}) and ion convection (Γ_{iD}) are very comparable except close to the axis where the condition $T_i >> T_e$ enhances the ion transport term. Neutral particle transport Γ_o appears in two branches because the injected beam causes a nett inward flow of energy in the regime r ≈ 40 to 130 cm but it should be noted that charge exchange losses to the wall are substantial. The total radiated power loss Γ_r is relatively modest because of the low concentration of iron but it increases substantially towards the wall because of the enhancement in $P_{xr}(T_e)$ at lower temperatures (see Fig.14). Data close to the wall are uncertain because the mesh steps used in this particular calculation are 8.75 cm (ie greater than most atomic mean free paths) and also the proximity of the wall introduces a wide range of atomic and surface processes that are not presently incorporated into the code.

10. THE PLASMA BOUNDARY

The torus wall is bombarded by ions and electrons that diffuse outward from the confined plasma and their flux is defined by $\Gamma_{ip} = D_{\perp}dn/dr$ (ion pairs $cm^{-2}\ s^{-1}$). The energy carried within the plasma by each ion pair is $3k(T_i + T_e)_b/2$ where the temperatures correspond to the local conditions at the boundary but this global partition of particle energy is distorted within a distance $\sim \lambda_D$ of a surface. Here a region of substantial potential gradient (called a plasma sheath) is set up in order to retard the more mobile particles and maintain equality in the loss of positive and negative charge. When $T_e \sim T_i$, the sheath potential is electron repelling for particles moving parallel to the field lines (ie $U_s \sim 3\ kT_e$), but the sheath is ion repelling for motion radially across field lines. If the sheath potential is electron repelling then plasma ions are accelerated and strike the surface with enhanced energy and so the impact energy of multiply charged impurity ions can be substantially enhanced.

The surface is also subjected to neutral atom bombardment arising from charge exchange processes within the plasma. Neutral atoms arriving at the wall have a distribution in energy that relates to the ion temperature within the plasma (see Section 7) but their energy flux is also strongly dependent upon the depth of plasma penetrated by the parent atom that has initiated the charge exchange reaction. Thus fast neutral beam injection causes the loss of very energetic atoms from the hot core of the plasma whereas the return of slow neutralised plasma ions from the torus wall (a process called recycling) gives rise to a much less energetic flux of charge exchange atoms.

The torus is also subjected to fluxes of X-ray photons that penetrate only a short distance into the bulk material of the wall and therefore give rise to localised heating. Moreover, the photons interact with adsorbed atoms on the surface and so add to the mechanisms by which gas is released from the wall.

The wall responds to these physical and physico-chemical insults by a variety of mechanisms which are discussed in a recent review by McCracken and Stott[40]. The incident ions and atoms are either reflected with a fraction of their incident energy or else adsorbed within the bulk material. The adsorbed atoms diffuse within the bulk material in a manner dictated by both the local temperature and the local gradient of their density, but eventually atoms reappear on the surface as adsorbed gas and are once more released by the fluxes of particles and energy that the plasma presents to the surface. A degassed wall can act for a short time as a particle pump but after receiving a limiting fluence of bombarding particles (the magnitude of which is dependent upon wall temperature) the wall becomes saturated and the incident and outgoing fluxes reach equilibrium. The probability of release of an energetic (ie, reflected) or a slow (ie, desorbed) particle is dependent on the velocity of incident particles because the faster species tend to bury themselves within the bulk material whereas the slower have a greater chance of reflection. These mechanisms lead to the recycling of hydrogenic elements and to the desorption of light impurity elements such as carbon and oxygen.

Energetic impact of ions and atoms causes sputtering of the wall material and this mechanism is a serious source of high Z impurities. The sputter yields for DT on stainless steel reach a maximum of $\approx 2 \times 10^{-2}$ at about $E_0 = 1$ keV and drop to about 10^{-3} in the region of $E_0 = 50$ eV; indeed there is a threshold for sputtering predicted for iron to be $E_0 = 32$ eV for D^+ and 22 eV for T^+ (note E_0 relates to mono-energetic incident particles and not to the plasma temperature). The self-sputtering yields for high Z elements are much greater, often exceeding unity at their maxima.

Since a potential difference exists between the plasma and its boundary surface, it is possible for an arc to strike whenever a localised spot on the surface becomes sufficiently hot to emit electrons and to vaporise the surface material. These arcs are called unipolar (ie, the plasma acts as the anode) and their highly localised and intense outward flow of current is maintained by a much more diffuse flow of plasma electron current that returns to the area of the wall surrounding the arc spot. The arc spot is fed by conduction through the bulk material. Arcs can be a powerful source of high Z impurity species and are potentially more troublesome that sputtering.

The problems of surface interactions are compounded in a reactor by the additional presence of the neutron flux that must pass through the torus wall adjacent to the plasma and enter the surrounding blanket. Thus damage can occur throughout the bulk material of the structure which is also subjected to thermal and mechanical shocks by the pulsed nature of the tokamak device (for example see Ref.18). The processes have profound effects upon the design and cost of the structure and the techniques needed for the maintenance of a fusion power station. Estimates of acceptable levels of power loading at the wall range between 200 to 600 W cm^2 and it is likely that the useful lifetime of the device can be enhanced if energy and particle losses due to inelastic atomic processes can be minimised. An assessment of the current status of wall problems can be obtained from the proceedings of the conferences on Plasma Surface Interactions in Controlled Fusion Devices (the most recent was held at

Culham in 1978[41]) and also from the various meetings dealing with reactor concepts, eg Ref.14.

The detrimental effects of contamination upon an unignited plasma (see Section 4 and 5) have provided a strong motivation to maintain clean (ie hydrogenous) conditions within the plasmas of present-day experiments. It is customary to shield the wall by means of a sacrificial orifice plate or rod called a limiter and this can be made of low Z materials such as carbon. It is also customary to prepare the surface of the torus wall by discharge cleaning followed by the deposition of a coating of titanium metal that serves as a getter. However, no matter how successful these techniques may be for short pulses of plasma (eg duration < 0.5 s), they have little relevance to the elimination of impurities throughout the much longer burn-times (200-2000 s) of a reactor. It is therefore essential in the long term to provide some positive method of preventing (or at least reducing) the influx of impurities from the torus wall.

The most promising approach for a long-burn reactor appears to be the use of a divertor. By introducing a local perturbation in the outer magnetic surfaces within the torus it is possible to "divert" some field lines and direct these through a small aperture in the torus wall and into a separate divertor chamber attached to the torus. A fraction of the area of the outer magnetic surfaces is thereby rendered open. Plasma particles and energy flow readily along the open field lines and enter the divertor chamber but radial flow to the torus wall is still impeded by the unperturbed area of the closed magnetic surface. Plasma that has entered the divertor chamber must be neutralised and pumped away so that some form of target is required. This may present a gas, liquid or solid face to the plasma but the neutralised plasma atoms together with any impurities released at the target or walls of the chamber need not flow back into the torus because both the aperture in the torus wall and the open area of the magnetic surfaces can be made quite small. The rotational transform imposed upon the field lines in a tokamak means that each of the spiralling field lines must eventually pass through the localised perturbation so that particles gyrating around the field line will then enter the divertor rather than diffuse to the torus wall.

It should be noted that a divertor also provides a means by which outward diffusing helium ions produced in DT fusion reactions can be removed from the plasma. It is therefore likely to be an essential component of a long burn-time reactor which must obviously be provided with facilities for introducing DT fuel and exhausting He and "unburnt" DT from the hot plasma.

Perturbation of the magnetic surfaces may be introduced in a

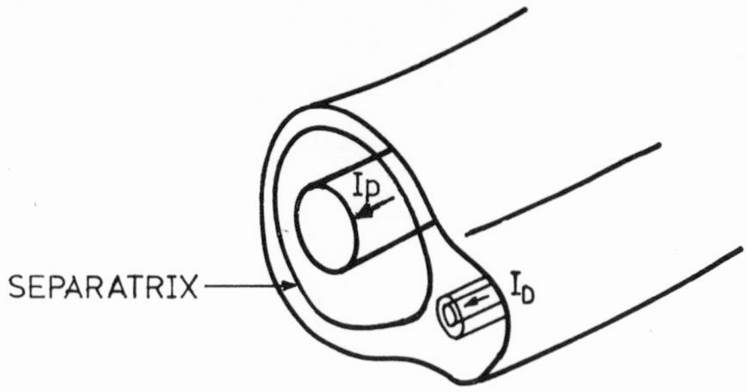

Fig.17 The concept of a poloidal divertor showing the conductor that carries current I_D in the toroidal direction and the approximate location of the separatrix. (Note that in practice a more complex system of conductors is required[42].)

number of ways; for example, since the B_p field in a tokamak is produced by a current I_p flowing through the plasma then an external ring conductor carrying a current I_D that flows in the same direction can be used to generate an external magnetic field that opposes B_p over a small region of poloidal angle θ. The open area is thus axisymmetric with the major radius R, so that the concept is called a poloidal divertor[42] and it is illustrated in Fig.17. An alternative approach called a bundle divertor[43] restricts the region of perturbation in both the poloidal and toroidal directions by locating two small coils close to the wall of the torus as shown in Fig.18. In both systems the field perturbation extends for only a short radial distance X into the torus and so the magnetic surfaces that do not experience the perturbation (ie, within $r = 0$ to $(a - X)$) remain closed and thereby confine the plasma. The outermost closed surface is called the separatrix. Plasma diffusing across the separatrix will enter the divertor chamber after travelling for a time $\tau_{||}$ along the open field line and during this time it continues to diffuse radially outwards towards the wall. For each particular divertor geometry and plasma condition it is possible to characterise this diffusion by a scale length Δ so that the plasma density in the region outside the separatrix (called the scrape-off layer) can be expressed as

$$n(X) = n_s \exp(-X/\Delta),$$

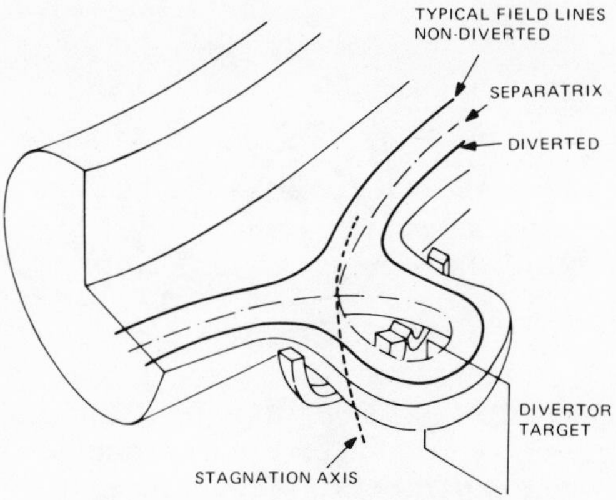

Fig.18 The bundle divertor as used on DITE[43].

where n_s is the density at the separatrix. Thus the density near to the wall is greatly attenuated if $\chi > \Delta$ and the efficiency with which plasma particles are exhausted can be defined as

$$\varepsilon_\chi = 1 - \exp\,(-\,\chi/\Delta). \tag{38}$$

Fluxes of particles (eg atoms from charge exchange and energetic α-particles) and also fluxes of photons will impact upon the torus wall and thereby release impurity elements. Nevertheless, these impurities can be ionized within the scrape-off layer and transported along the open field lines into the divertor before inward diffusion carries them over the separatrix and into the confined plasma. The efficiency of this shielding action can be expressed as

$$\varepsilon_s = 1 - \exp\left(1 - \int_a^{a-\chi} n(x)\,\frac{<\sigma_i.v_e>}{v_o}\right) = 1 - \exp\left(-\,\varepsilon_\chi \Delta n_s\,\frac{<\sigma_i.v_e>}{v_o}\right). \tag{39}$$

It is evident that the ability for a divertor to shield the plasma from impurities is strongly dependent upon the rate coefficient for ionization of neutral atoms of wall materials such as iron. It is unlikely that steady state coronal conditions for many of the higher charge states can be established during the short time ($\sim 10^{-3}$s) that the ionized impurities spend in the scrape-off plasma, but even so it is important to know how much energy can be radiated within the torus because the amount of power that passes into the divertor chamber has a strong influence upon the design of the target and chamber and also upon its associated vacuum system. This is especially important for a power generating reactor where the exhaust power could be ∿ 500 MW.

The divertor concept is elegant but it is also costly and difficult to install because of the complex structure needed for its magnetic system. An alternative solution has therefore been sought especially for tokamak devices with burn-times of the order of a few seconds. The most deleterious impurities are those with high Z and, if arcing is neglected, these arise from sputtering of the torus wall by both energetic ions and charge exchange atoms. However, sputtering yields are low if the bombarding particles are not energetic and incident ion energy can be made low by ensuring that the temperature of the plasma in contact with the torus wall is also low. Recycling of neutral species in a low temperature boundary plasma also results in low energy atoms returning to the wall but charge exchange in the hot core of the plasma will maintain an outward flux of energetic atoms unless it is possible to interpose a barrier of plasma that is opaque to fast atoms. Equation (25) indicates that the plasma becomes opaque (ie, the probability of re-ionization exceeds the probability of atom transport over a distance ℓ) when $n\ell >> \langle\sigma_x\rangle$ and when $\langle\sigma_x v_i\rangle/\langle\sigma_i v_e\rangle$ tends to a minimum, ie for $T_e \sim 10$ to 100 eV (see Fig.11). It is therefore reasonable to assume that the wall is shielded from fast atom impact if a boundary plasma, of thickness ℓ, and density $n = 10^{15}/\ell$ can be maintained at $T_e \approx 20$ eV. This concept has been called a cool mantle[44]. In practice the high density in the mantle might be obtained by an input of cold hydrogenous gas or (as is considered in[44]) by a low energy neutral beam that penetrates a short distance and thereby generates an optimum density profile. Elements such as iron are still sputtered but at a much reduced rate and the radiation from these impurities helps to maintain the low temperature of the mantle. It has been argued that the peak in plasma density near to the torus wall helps to concentrate impurities in this region (ie, in a manner analogous to equation (30)) and thereby reduces inward diffusion. The success of this cool mantle concept which will be tested in JET is strongly sensitive to atomic collisions. To produce an adequate model of the boundary plasma it is necessary to consider inelastic collision processes involving metals atoms such as iron at electron temperatures as low as 1 eV. The validity of steady state, coronal modelling is questionable and it is also necessary to assess the effect of charge exchange collisions of the type

$$H^+ + Fe^+ \rightarrow H + Fe^{2+}.$$

The cross section for this reaction is large and somewhat similar to that of $H^+ + H$, at least at H^+ energies as low as 1.5 keV[45].

11. SURVEY OF RELATED SUBJECTS

Preceding sections have emphasised the physical conditions of a magnetically confined plasma in which fusion reactions take place but the present paper does not extend to equally detailed discussions of additional facilities that are needed to produce, maintain and monitor the conditions of the plasma environment. In present-day experiments the plasma is first formed by electrical breakdown of neutral molecular hydrogen (or deuterium) gas that has been fed into an evacuated torus. In principle, the mechanism of breakdown is not important provided that the electrons are supplied with sufficient energy to initiate a cascade process; the subject is discussed in Refs.2 and 3. However, it should be noted that the neutral gas is initially in contact with the torus wall so that impurities (particularly C and O) may be introduced into the plasma during the breakdown process; thus methods of gas filling and pre-ionization in which wall contact is minimised are likely to be beneficial.

In a reactor (or even a long pulse experiment) it will be necessary to refuel the hot plasma with atoms of D and T. These atoms must penetrate to a substantial depth within the plasma in order to enter the hot reacting core but they must not perturb excessively the conditions for plasma equilibrium. Fast neutral beam injection can meet this requirement but probably at an unacceptable cost in both energy and additional hardware. Present-day opinions differ as to the most effective route but they can be summarised as follows:

a. Feed a brief "puff" of low energy gaseous fuel into the edge of the plasma and thereby raise the edge density. The ionized fuel is then transported inward by diffusion. However, it should be noted that the high edge density also enhances outward diffusion and so the process is likely to be wasteful in fuel.

b. Produce the fuel in the form of solid pellets and fire these pellets into the plasma by means of a mechanical pellet launcher. The surface of the pellet is vaporised by plasma bombardment and thereby forms a protective atmosphere around the solid core which can thus penetrate to an appreciable depth. Pellet velocities of 10^5 cm s^{-1} are predicted for acceleration methods based upon high speed centrifuge techniques.

c. Produce the fuel in the form of charged "clusters" (ie, agglomerates which on average comprise $\sim 10^3$ atoms and a single charge). These clusters are electrostatically accelerated to velocities equivalent to 1 to 10 keV/amu. The charged clusters pass readily through the confining magnetic field and are then progressively broken down into molecules by collisions with the plasma. High acceleration potentials are required to attain reasonable penetration but unfortunately the acceleration field must be kept low because the charged clusters are readily fragmented by the impulse that they experience during acceleration.

Information on the current status of refuelling can be found in[46].

It is axiomatic that neutral beam injection systems require intense sources of ions. All sources presently regarded as suitable for injection comprise a hot filament emitting electrons into a discharge chamber filled with molecular gas. A simplified concept of the main channels of the ionization cycle is as follows:

$$e + D_2 \rightarrow e + e + D_2^+$$

$$e + D_2^+ \rightarrow e + D_2^{+*} \rightarrow e + e + D^+ + D$$

$$e + D_2^+ \rightarrow D + D \text{ (dissociative recombination)}$$

and to a much lesser extent

$$e + D \rightarrow e + e + D^+.$$

Most of the D atoms strike the interior walls of the discharge chamber and thereby recombine to form D_2 which thus completes the cycle. The need to optimise the yield of D^+ has already been noted in Section 7. Comparable ionization cycles involving surface recombination are experienced by molecular species such as CO that are released from the torus walls. It should be noted that the balance between charge exchange and electron impact ionization processes (already discussed for H and H^+ in Section 7) swings heavily in favour of ionization for molecules desorbed into an hydrogenous plasma because charge exchange is not now a resonant symmetrical process.

Atomic processes also influence the optical properties of ion beams which expand under the action of their own space charge unless compensated by the presence of slow electrons. High energy beams ionize residual gas within the beam transport system and thereby provide some measure of space charge compensation. By contrast, the presence of residual gas causes scattering of fast atom beams (mainly through charge stripping collisions) which introduces problems due to the consequent bombardment of the walls of the beam line by energetic scattered particles. Recent references to ion and atom beam formation may be found in [27,47].

The α-particles formed by DT fusion events take $\sim$ 0.1s to lose

$\exp^{-1}$ of their initial energy[2] and travel $\sim 10^9$ cm during this time. Atomic processes can cause losses of high energy (E > 1 MeV) α-particles and in particular charge exchange reactions of the type

$$He^{++} + X^{z} \rightarrow He + Z^{z+2}$$

have relatively large cross sections and will be significant whenever the density of X^z is substantial.

Plasma diagnostic techniques are an essential component of fusion research because measurement of parameters such as n,T and the local values of B and E within the plasma are greatly complicated by the large fluxes of energetic particles and by the intense magnetic fields. In addition, the dynamic nature of the plasma calls for good resolution in both time and space. Mechanical devices cannot be placed in regions of high n and T and so external observations are made of the fluxes of photons and neutral particles emitted from the torus. Conversely, electromagnetic radiation or neutral particles can be beamed through the plasma whose properties are inferred from both scattering and attenuation of the beams. It is evident that atomic processes form the basis of several diagnostic techniques and that data for the particular mechanisms invoked must be both precise and detailed. Information about diagnostic techniques can be found in several reference works, for example[48].

12. CONCLUDING REMARKS

It is abundantly evident that atomic and molecular processes play a major role in fusion research but it is also clear that an assessment of their true significance must take account of many other interrelated areas such as plasma transport, the physics of surfaces and materials and also engineering requirements. At the present stage of fusion research it would be presumptuous to state precisely what additional knowledge of atomic physics is essential to the development of a fusion reactor but some general requirements are obvious.

There is a continuing need for more accurate diagnostic measurements and also for new diagnostic techniques. Improvements in plasma modelling are crucial, not only for the interpretation of existing experiments but also for the reliable prediction of the performance of future devices. More sophisticated treatments of the plasma boundary must be included in the models and there is a sparsity of atomic data relating to the metallic elements which dominate in this region. Powerful neutral beam injection is likely to be a major aspect of future devices and improved techniques for the production of fast atom beams will be of major significance.

At the present time uncertainties in plasma transport processes probably exceed uncertainties in the accuracy of existing atomic data but this would not be true if some significant atomic process were overlooked. Such an example would be the fact that a small concentration of fast Mo or W in a neutral injection beam would probably preclude ignition. Thus, in the author's opinion, one of the most important requirements at present is an overall awareness of the delicate balance that must be maintained between these conflicting fields of physics, technology and engineering.

REFERENCES

1 Nature. 181 (1958).

2 S. Glasstone and R.H. Lovberg, Controlled Thermonuclear Reactions (D. van Nostrand Co.Inc., New York, 1960).

3 D.J. Rose and M. Clark, Plasmas and Controlled Fusion (The M.I.T. Press, Cambridge, 1961).

4 M.O. Hagler and M. Kristiansen, An Introduction to Controlled Thermonuclear Fusion (Lexington Books, Lexington, 1977).

5 Plasma Physics and Controlled Nuclear Fusion Research in 1978 (Proc. 7th Conf. Innsbruck, 1978) IAEA, Vienna (1979).

6 Controlled Fusion and Plasma Physics (Proc. 8th Europ. Conf., Prague, 1977).

7 Lectures from Culham Plasma Physics Summer School, 1978 ed. J.J. Field and P. Reynolds (in course of publication). See also Plasma Physics, Lectures from the Culham Plasma Physics Summer School ed. B.E. Keen (Institute of Physics, Conf.Ser.No.20, 1974).

8 Physics Reports (Section C of Physics Letters) 37, 1978, "Atomic and Molecular Data for Controlled Thermonuclear Fusion" invited papers at IAEA Advisory Meeting, Culham, November 1976.

9 H.W. Drawin, Journal de Physique, Suppl.2, 40, C1-73 (1979).

10 H.W. Drawin, in "Plasma Physics 1978, Lectures given at the IXth Summer School on Phenomena in Ionized Gases, Dubrovnik, 1978, ed. R.K. Janer.

11 H.P. Summers and R.W.P. McWhirter, J.Phys.B: Atom. & Molec. Phys., 12, 2387 (1979).

12 Atomic Data for Controlled Fusion Research, Oak Ridge National Laboratory, ORNL-5206 and 5207 (1977) ed. C.F. Barnett.

13 F.L. Ribe, Rev.Mod.Phys., 47, 7, 1975.

14 Fusion Reactor Design Concepts (Proc.Tech.Comm. Meeting and Workshop, Madison 1979) IAEA, Vienna (1978).

15 H. Eubank et al. (P.L.T. Group), 7th Int.Conf. on Plasma Physics and Controlled Nuclear Fusion Research, Innsbruck (1978) Paper IAEA-CN-37-C-3.

16 J.D. Lawson, Proc.Phys.Soc. (London) 106A, Suppl.2, 173 (1959).

17 A Fusion Reactor Power Plant (Princeton University Plasma Physics Laboratory Report MATT-1050, August 1974) ed. R.G. Mills.

18 R.W. Conn, Proc. 3rd Int.Conf. on Plasma Surface Interactions, Culham 1978, p.103.

19 I.E. Tamm, in Plasma Physics and the Problem of Controlled Thermonuclear Reactions, ed. M.A. Leontovich (Pergamon Press, 1961) Vol.1, p.35.

20 Status and Objectives of Tokamak Systems for Fusion Research, edited by a Review Panel, S.O. Dean et al., WASH-1295 UC-20 (1973).

21 J.T.D. Mitchell, Am.Nucl.Soc. 3rd Topical Meeting on Technology of Controlled Nuclear Fusion, Santa Fe (1978) p.954.

22 R. Hancox and J.T.D. Mitchell, Plasma Physics and Controlled Nuclear Fusion Research, Berchesgaden, 1976, III. IAEA, Vienna (1977) p.193.

23 W. Pfeiffer and R.E. Waltz, Nuclear Fusion, 19, 51 (1979).

24 L. Spitzer, Physics of Fully Ionized Gases (John Wiley & Sons Inc, New York, 1962).

25 L. Spitzer and R. Harm, Phys.Rev. 89, 977 (1953).

26 D.R. Sweetman, Nucl.Fusion 13, 157 (1973).

27 T.S. Green, 10th Symp. on Fusion Technology, Padova, Italy (1978).

28 T.V. Goffe, M.B. Shah and H.B. Gilbody, J.Phys.B: Atom.Molec. Phys., in course of publication (1979).

29 A. Salop and R.E. Olson, Phys.Rev.A 16, 1811 (1977).

30 R.L. Freeman and E.M. Jones, Culham Laboratory Report CLM-R137 (1974).

31 A.C. Riviere, Neutral Injection Heating of Toroidal Reactors, Appendix 3 (D.R. Sweetman ed.) Culham Laboratory Report CLM-R112 (1971).

32 Proceedings of the Symp. on the Production and Neutralisation of Negative Hydrogen Ions and Beams, Brookhaven National Laboratory (1977) BNL 50727 (1977).

33 R.W.P. McWhirter, Physics Reports (Section C of Physics Letts.) 37, 165 (1978) Atomic and Molecular Data for Controlled Thermonuclear Fusion, invited papers at IAEA Advisory Meeting, Culham, November 1976.

34 Equipe TFR, Nuclear Fusion, 17, 1297 (1977).

35 C. Breton, C. de Michelis and M. Mattioli, Nuclear Fusion, 16, 891 (1976).

36 J.B. Taylor, Proc. IAEA Workshop on Fusion Reactor Design Problems, Culham 1974. Nuclear Fusion Special Supplement (1974) p.403.

37 R.W. Conn and J. Kesner, Proc. 2nd Conf. on Surface Effects in Controlled Fusion Devices, San Francisco (1976) p.1.

38 M.H. Hughes, private communication (1979).

39 R.V. Jenson, D.E. Post, W.H. Grasberger, C.B. Tarter and W.A. Lokke, Nuclear Fusion, 17, 1187 (1977).

40 G.M. McCracken and P.E. Stott, Culham Laboratory Preprint CLM-P573 (to be published in Nuclear Fusion).

41 Proc. 3rd Int.Conf. on Plasma Surface Interactions in Controlled Fusion Devices, Culham (1978).

42 M. Keilhacker, reprinted from Tokamak Reactors for Breakeven: A Critical Study of the Near-Term Fusion Reactor Programme. International School of Fusion Reactor Technology, Erice (1976) p.171.

43 P.E. Stott, C.M. Wilson and A. Gibson, Nuclear Fusion, 18, 475 (1978).

44 A. Gibson and M.L. Watkins, Proc. of 8th Europ. Conf. on Controlled Fusion and Plasma Physics, Prague (1977) Vol.1, p.31.

45 R.G. Montague, D.A. Hobbis and M.F.A. Harrison, to be published (1979).

46 Proc. of the Fusion Fuelling Workshop, Princeton 1977, CONF-771129 (1978).

47 Proc. 2nd Symp. on Ion Sources and Formation of Ion Beams, Berkeley (1974), LBL-3399 (1974).

48 Plasma Diagnostics, ed. R.H. Huddleston and S.L. Leonard (Academic Press, New York, 1965).

GENERAL PRINCIPLES OF MAGNETIC FUSION CONFINEMENT

JOHN T. HOGAN

OAK RIDGE NATIONAL LABORATORY FUSION ENERGY DIVISION

P. O. BOX Y, OAK RIDGE, TENNESSEE 37830 USA

1. INTRODUCTION

Recent results with neutral beam-heated tokamaks have led to increasing optimism about the prospects for a successful magnetic fusion reactor. The conceptual scheme for providing a fusion driver for a fission-fussion hybrid system has been validated, and a demonstration of 'break-even' ('D-T fusion power out = injector power in') seems quite likely with the TFTR (Tokamak Fusion Test Reactor) device, now scheduled to begin operation in Princeton in 1981. While breakeven, or $Q = 1$, performance should be satisfactory for hybrid systems, in which the fusion driver provides fuel makeup for fission reactors, an economically attractive pure fusion scheme will require $Q \sim 30$-40. To achieve this level, the plasma will have to be maintained in a high-β state for many energy and particle replacement times. This more stringent requirement imposes rigorous demands on our understanding of fusion systems. In particular, it means that we must obtain better understanding of the plasma particle and energy balance, must optimize neutral beam deposition, and must control the plasma energy and particle flux to the exterior through improved limiters or with magnetic divertors. In addition, our diagnostic evaluation of the scaling of particle and energy balance must be improved.

It seems clear that the joint work of fusion and atomic/molecular physicists must intensify if these goals are to be attained. Hence, it is the purpose of these lectures to introduce the concepts and trends in present magnetic confinement experiments, especially as they affect the joint work of atomic/molecular and fusion physicists. Processes which have a significant effect on the goal of producing a clean, high-β discharge for long pulse lengths are described, with emphasis on the fusion physics side. This should serve as a background for the more specialized discussions in the Institute, and should be helpful in following the literature in areas where there is a significant interaction of atomic/molecular and fusion physicists.

We will first introduce some basic fusion concepts; or rather, we will recommend sources for further study by those interested. Next we will briefly review recent experiments and discuss current trends and near-term experiments. We will then describe, in more detail, some specific plasma physics processes which enter in many atomic/molecular physics-related problems: neoclassical and MHD transport processes, the divertor/scrape-off plasma, neutral beam deposition and thermalization, and plasma fueling processes. Finally, we examine a case study in the interaction between atomic/molecular and fusion physics: the way in which calculation and measurement of the impurity-hydrogen charge exchange trapping cross-section has affected the understanding of fusion devices.

2. FUSION CONCEPTS

These lectures will dwell entirely on processes and problems arising with tokamak magnetic confinement devices. Of course, there are many others, and they are quite successful, but they are viewed to be somewhat further removed from the reactor stage at present. In any event, the tokamak is now the subject of the most active collaboration between A&M and fusion scientists.

Useful introductions to tokamak physics have been prepared by Artsimovich, Coppi and Furth.[1] More extensive discussions of recent neutral beam systems have been given by Murakami and Eubank.[2] A survey of fission-fusion hybrid ideas has been presented by Bethe.[3] For more detailed study, two extensive reviews of tokamak experimental results have been prepared: by Artsimovich,[4] and Furth[5] in Nuclear Fusion.

We will consider only the essential concepts which are needed for the detailed lectures at the Institute.

2.1 Confinement Properties: Theory

Neoclassical theory. Neoclassical theory is based on some of the earliest (or classical)[6] ideas about plasma transport. Essentially, while particles execute closed orbits in the magnetic confinement geometry (Fig. 1), distant collisions, resulting in small

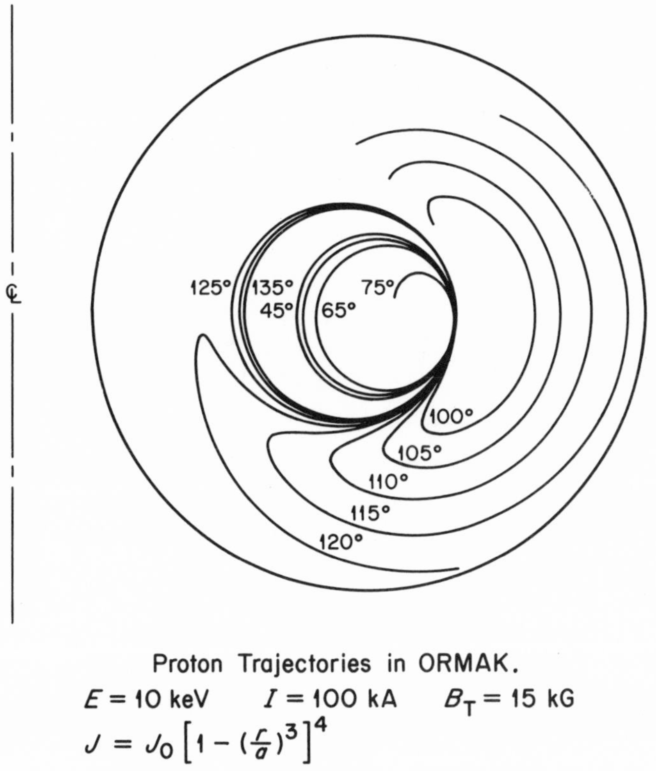

Fig. 1. Cross-section of the tokamak plasma showing drift orbits for various pitch angles ($\theta = \cos^{-1} v_{\parallel}/v$). Small changes in this angle due to weak collisions can lead to large spatial excursions.

momentum transfer, can cause large changes in spatial diffusion by alteration of the particles' orientation with respect to the magnetic field. In this theory, the controlling rate for diffusion is the rate at which particles collide: the electron-ion scattering rate (time for a 90° deflection of an electron by scattering from ions) governs electron processes, while the same rate for ion-ion collisions determines the ion loss processes. Because the electron-ion scattering time is shorter and leads to much faster diffusion of electrons than ions, a strong electric field is set up due to the separation of charges, and this electric field pulls the ions after the electrons at a faster rate than the simple collisional processes would allow. The requirement of ambi-polarity of fluxes

(1) $\Gamma_e = \Gamma_p + \sum_k k\Gamma_k$

thus constrains the transport rates in a plasma with electrons, protrons, and more highly charged ions (α particle fusion products or impurities.)

Neo-classical theory[7] predicts that the ratio of electron/ion thermal conductivities should be small ($\chi_e/\chi_i \lesssim \sqrt{m_e/m_i}$). While both χ_e and χ_i are ill-measured in tokamak experiments, the observed upper bound to χ_e actually exceeds the observed upper bound to χ_i, making neo-classical theory an unlikely model for electron behavior. Nonetheless, values for χ_i are in tolerable agreement with neo-classical predictions. Further, results from controlled experiments with impurity ion injection confirm some detailed features of the neo-classical model. Equation 1, when applied to the case of protons and a single impurity with charge Z, predicts that $n_z = (n_p)^Z$ in equilibrium. Thus, impurities injected from the outside should settle in the core and have a spatial profile more sharply peaked than that of the protons. It is found that externally introduced impurities do migrate to the core, however, the profile is not necessarily more sharply peaked.

This discrepancy may be resolved by consideration of a class of processes which have only recently been calculated.

MHD oscillations. The static magnetic field used as a model in Fig. 1, and which is used for neo-classical transport calculations, is only an approximation. The strong longitudinal field B_T is required to stabilize extremely violent motion which would destroy the plasma in an Alfvén time (a/v_a) if this field were absent. However, it is nonlinear stability which is so provided. This means that small stable oscillations in the field are present which can have a significant effect on transport rates. The attainment of high-β in future reactors also requires the stabilization of these oscillations, as we shall discuss.

Sawteeth and saturated islands. Stability to these fast modes is described entirely by the Kruskal-Shafranov 'Safety factor' q, which typically varies from a value near unity in the core, to 2-4 on the exterior. For values near unity, oscillations with $e^{i(m\theta - m\phi)}$ (m=n=1), can occur. These show a periodic behavior, with intermittent growth followed by expulsion of plasma from the core. This 'sawtooth' behavior is one of the more striking experimental features of its' behavior, and a theory by Kadomtsev has been successful in describing the resulting transport.

When q=m/n (m-2,3,4.... n=1), perturbations can also grow in the intermediate plasma region between the core and the edge. These oscillations stabilize nonlinearly, and create finite amplitude helically symmetric field structures super-imposed on the axially symmetric tokamak geometry. Thus, particles can move rapidly across this spatial region ($\Delta/a \lesssim 20\%$) much more quickly than by neo-classical processes. There is a strong effect on τ_E scaling when these 'islands' are produced.[8] There is an effect on impurity transport to be expected as well, since there may be a 'shorting-out' of the radial electric field required for ambipolarity. This 'screening' effect has been invoked to explain the results of experiments in which Argon was deliberately injected into a tokamak discharge.

For high values of $\beta(\gtrsim 1.5\%)$ a new class of non-symmetric instabilities with $m \approx n$ is expected to occur. Localized in the outer weak field regions of the torus, these 'ballooning' modes are thought to set the upper limit on attainable β. When the energy available for plasma expansion exceeds that available for stabilization by field line bending, the plasma could be lost catastrophically at the sound speed. In this case, MHD effects would not simply enhance the transport rate, but would terminate the experiment.

2.2 Confinement Properties: Experiments

Particle and energy confinement. The measured confinement properties in tokamak experiments are described by two empirical formulae: the first, popularly known as 'Alcator scaling,'[11] predicts that

$$(2) \quad \tau_E \equiv 1.9\ 10^{-6}\ a^2 \left(\frac{\bar{n}_e}{10^{13}}\right) q^{\frac{1}{2}}$$

The second, proposed by Mirnov,[8] accounts for the degradation due to MHD oscillations and predicts

$$(3) \quad \tau_E = 3 \cdot 10^{-9}\, a l \sqrt{\bar{n}_e/10^{13}}$$

Each of these must be used in conjunction with an empirical description of the density scaling proposed by Murakami et al.[12]

$$(4) \quad \bar{n}_e = 2 \cdot 10^{11}\, B_T/R$$

These experimental rules for τ_E must be understood in context. They incorporate all the loss processes: line radiation, ionization losses, as well as conduction and convection. Moreover, even the effective minor or cross-section radius may vary strongly from discharge to discharge in a device with a given nominal value of a.

Neutral beam heating experiments in PLT have indicated an improvement of electron confinement similar to Alcator scaling, but with χ_e decreasing with increasing T_e. The scaling $\chi_e \sim T_e^{-\frac{1}{2}}$ is known as 'PLT scaling.'[10]

As noted earlier, ion confinement processes appear to be adequately described by the neo-classical model.

Explicit determination of τ_p seems to be most difficult. Neutral beam heating provides a direct power input, which can be estimated; however, the particle input from the limiter and walls is almost unknown. It is conventionally assumed that $D/\chi_e = 1/5$ since this model leads to reasonable behavior in simulation codes.

2.3 Magnetic Geometry

The basic configuration shown in Fig. 1 has been generalized in recent experiments. As seen in Fig. 2, the cross-section has been altered to produce Dee and Doublet shaped plasmas. Larger currents can be passed through these configurations, while retaining the same value of the Kruskal-Shafranov safety factor. Also by maximizing the effect of the high-field side of the torus against the 1/R toroidal falloff of B_T, this shaping may serve to stabilize the plasma at higher β values than those thought possible with a circular cross section.

Further changes in the geometry are necessary to provide channelling of the plasma efflux to regions which are specially fitted for handling high heat loads and pumping. These magnetic divertor configurations are being tested at present in the DIVA and PDX experiments.

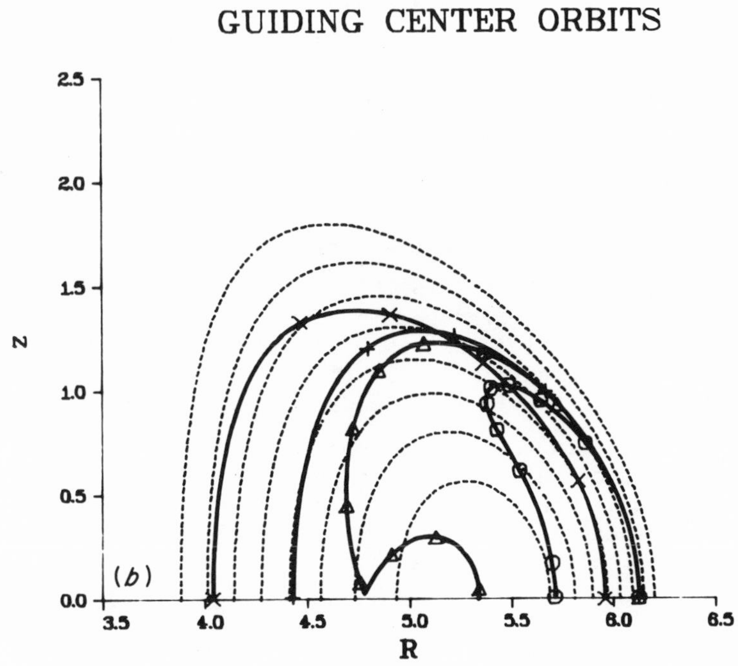

Fig. 2. By elongating the plasma cross-section, while maintaining the same value of toroidal magnetic-field, the total stable current can be increased.

An even more radical change has been tried successfully in the DITE experiment. Toroidally localized coils have been employed to lead a bundle of magnetic field lines out to a plate, and thus shown the effectiveness of this form of divertor. Both bundle and axi-symmetric (or poloidal) divertors will be the subject of important trials in the next few years. A double toroidal, or Spitzer, divertor has begun operation at the Kurchatov Institute in Moscow. In this variant, the field is disturbed locally at two toroidal locations.

3. RECENT EXPERIMENTS AND NEAR-TERM TRENDS

3.1 Neutral Beam Heating Experiments

The most dramatic results in recent fusion work have occurred in a series of neutral beam injection experiments with increasing levels of beam power. The early 'high power' injection experiments ($P_b \gtrsim P_{OH}$) in Ormak, ATC and TFR produced ion temperatures in the 2 keV range, with $T_i > T_e$ for the first time in tokamaks. Last year's experiments with PLT, in which up to 2.2 MW of 40 keV neutrals were injected, produced in temperatures in excess of 6 keV, D-D neutron output in excess of 100 W and an equivalent D-T Q value of 5%. Simple extrapolation of the <u>same parameters</u> to the case of 120 keV D^o injection in TFTR leads to a $Q_{DT} \approx 1$, since the over-whelming contribution to the neutron output comes from beam-thermal and beam-beam reactions. The PLT experiment has been devoted to ion cyclotron wave heating experiments recently, and good results have been obtained here also. Heating efficiencies of $\sim$2 eV/kW injected, comparable with those in neutral beam heating, have been obtained.

The PDX experiment has produced 460 kA plasma discharges, and in the next year will begin a joint PPPL/ORNL 6 MW neutral beam heating project. Equivalent $Q_{DT} \sim 10\%$ are expected.

The ISX-B experiment at ORNL has a circular/dee shaped plasma which allows an elongation up to 1.6. The low toroidal field facilitates studies of limits to β, and values of β in excess of those theoretically predicted to be unstable to ballooning modes have recently been achieved. ($\beta \sim 2.2\% > \beta_{CRIT} = 1.5\%$.)

Doublet III plasmas with currents in excess of 2 MA have been produced, and injection with up to 7 MW of 80 keV H^o will be under-taken in the next few years.

DITE tokamak has carried out 1 MW injection experiments, and T-II in Moscow is equipped with 0.5 MW of power. The Asdex divertor experiment is under construction at Garching. TFR 600, a continuation of the successful TFR device at Fontenay-aux-Roses, is conducting a joint study of neutral beam and wave heating.

The IAEA's World Survey of Fusion Facilities may be consulted for a more detailed accounting.

3.2 Trends in the Experiments

Confinement scaling. The central task of the present experiments is to study the parameter dependence of the transport processes. It is desireable, in this respect, to minimize the effects of any process other than plasma transport. However, it is inevitable that there will be some impurity constituents and that the role of recycling of the working gas between the plasma and the limiter or wall will be significant.

The important result of PLT neutral beam heating experiments was to show that the saturation level of possible collisionless instabilities was quite low. It had earlier been feared that transport rates would be greatly enhanced over present values in the high temperature regime needed for reactors.

The Alcator series of experiments is the prototype for a class of "high-field" ($B_T \gtrsim 10T$) reactor designs. In these experiments the highest absolute values of $n_e\tau_E$ have been achieved, along with the highest densities ($\geq 10^{15}$ cm^{-3}). The provision of wave heating (needed because neutral beams require impractically high energy to penetrate these dense plasmas) is expected to increase the temperatures above the presently observed 1 keV level.

Shaping studies. Scaling with temperature will result as varying amounts of power are applied. Scaling with size, now conventionally accepted as size),^{2} will be tested in detail. By varying the cross-section shape, it is hoped that the higher currents which can be stably produced will reduce neo-classical conduction losses. Elongation should produce a higher β_{CRIT}, and this supposition will be tested in high-power beam injection experiments.

Impurity production. At present, the detailed nature of impurity production and transport is unknown. High power experiments are limited in duration to $\sim$150 ms because of constraints on existing neutral beam technology. The gross measure of impurity content, a weighted average of the ionic charge which enters calculations of momentum transfer, is

$$(5)\quad Z_{eff} = \sum_{\substack{\text{all positive}\\ \text{species}}} \frac{n_k Z_k^2}{n_e}$$

An accumulation rate of Z_{eff} = 0.05/sec could not be detected in present experiments, and yet would be fatal for long pulse reactors.

It has been proposed that impurities can be introduced by

- sputtering of the surface by escaping charge exchange neutrals
- desorption by charged particles (ions or neutrals leaving the plasma
- evaporation or melting of the limiter
- arcing
- self-sputtering of the wall or limiter by escaping impurity ions.

The radiation losses produced by these impurities have reached catastrophic levels in some experiments. Tungsten impurities, in particular, have foiled electron heating experiments on the Ormak device and produced hollow electron temperature profiles in PLT. After replacement of tungsten limiters with water-cooled graphite, neutral beam heating experiments on PLT registered unprecedented temperature rises. A specific study of the effects of heavy metals was made on the ISX-A tokamak, which employed SS limiters from earliest operation. Energy confinement times exceeding "Alcator scaling" were produced. In addition, tungsten was deliberately introduced in a laser blow-off experiment, and characteristic degradation of τ_E and concomitant hollow T_e profiles were observed.[13]

There is a trend in present experiments away from heavy-metal limiters (such as M_o and W) and toward carbon or stainless steel. This trend may, however, be only temporary. In the PLT device the effective plasma radius was decreased by ∿10% when the graphite limiter was used, and the power input levels planned for PDX (with similar size) are three times these injected into PLT. The ISX-A device was relatively impurity-free with stainless steel limiters, but only ∿500 KW of ohmic heating power was produced. The ISX-B experiment, with 3 MW of injected power planned, may have different results. The exceeding difficult atomic physics problems regarding ionization and recombination of heavy metals (such as W and M_o)

will at least temporarily, be accorded a low priority by many fusion researchers.

Plasma fueling. The Alcator-A tokamak has produced the highest values of $n_e\tau_E$ among fusion devices, attaining high densities ($\sim 10^{15}$ cm^{-3}), by controlled injection of cold ($\sim$few eV) gas. A key paradox for confinement theory is the rate at which the spatial distribution of the density relaxes. It seems to be anomalous in comparison with these transport models which perform reasonably well in predicting steady state values of parameters. The edge interaction region is complex, however, and it may be that oxygen is admitted with the injected gas. If so, and if the $O^{n+} + H_0 \rightarrow O^{(n-1)} + H^+$, charge exchange rates significantly exceed those for $H^0 + H^+$, then an important anomaly would be removed.

Plasma fueling by injection of frozen hydrogen pellets has recently been given successful trials on the ISX-A and ISX-B experiments. The pellets are injected at a speed $\sim$1 km/s, and can be deposited at will across the plasma cross section. Interferometric and holographic studies have been used to follow the pellet in flight, and local densities during ablation of $\sim 10^{19}$ cm^{-3} have been recorded. The interaction of pellets with stored fast ions from an injected neutral beam ($\sim$40 keV) shows beneficial effects: the stored fast ion energy favorably accelerated the pellet ablation and subsequent heating of the cold plasma which is produced.

High-β experiments. If the trends we have discussed in confinement scaling continue, then the production of the requisite $\sim$10-20 keV reactor plasma temperatures seem assured. Reactor studies have shown, however, that high temperature must be accompanied by high density, so that the reactor's fusion power density is sufficient to economically repay the overhead costs in an acceptably small power plan.[14] This implies that the β values must lie in the range of 4-6%. Typical β values in the PLT neutral beam heating experiments were $\lesssim$1%, because of the high (3.2T) toroidal fields. In ISX-B however, the combination of low toroidal field ($\lesssim$1.2T) and high injected power ($\sim$1.5 MW) lead to a test of the properties of high β plasmas.

Preliminary tests, in a circular plasma configuration, have now been completed. They show (Fig. 3) that the β values attained exceed those previously calculated to be unstable to ballooning modes. As the beam power is increased from 0.3 to 1 MW a steadily increasing rise in MHD activity is observed, which may signal the onset of ballooning instability at a higher β. Nonetheless, the theoretical guidelines for attainable β values have been achieved for the circular configuration and attention is now turning to produce the dee-shaped plasmas which will yield the high β_{CRIT} needed for reactors.

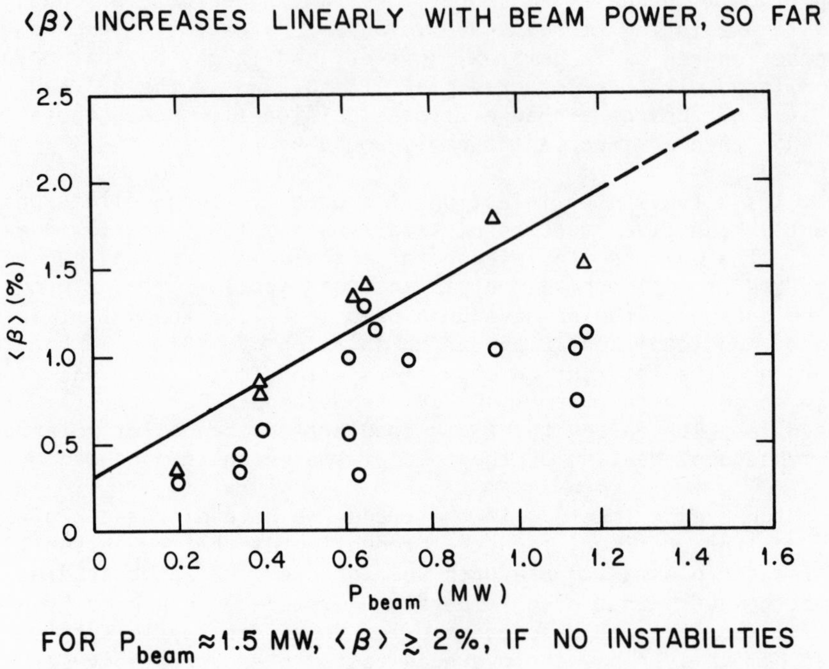

Fig. 3. The values of $\beta^* \left(\equiv \frac{2\mu_o [\int dV p^2]^{\frac{1}{2}}}{B_T^2}\right)$ attained by the ISX-B tokamak for different levels of injected power. Previously, it was thought that $\beta^* \sim 2\%$ was a limiting value.

4. FUSION PHYSICS PROCESSES

4.1 Neoclassical Transport of Impurities

The bulk of atomic physics work in the fusion field relates to the behavior of impurities in the plasma. The impurities, as an ion species, behave, in a first approximation, according to neoclassical theory.

There are some important paradoxes with respect to the plasma physics aspects of these models, however, which have cast some doubt on the utility of attempting detailed measurements of impurity behavior. These relate to puzzling asymmetries in the impurity light emission seen in tokamak experiments which can reverse polarity even during the course of a discharge. Burrell and Wong[15] have recently analyzed this situation, and we shall describe their resolution of the paradox.

Another contradiction concerns the apparent lack of a continual accumulation of impurities in the core of the discharge. Argon injection experiments on the T_4 tokamak may have resolved this issue, since sawtooth oscillations (in reality, minor disruptions expelling plasma from the core) were present. In addition, the important effect of possible charge exchange reactions as a recombination mechanism was demonstrated. We pursue the subject in more detail, since it is ubiquitous in atomic physics studies, and we shall describe the neoclassical effects here.

<u>General features of the theory</u>. The mathematical aspects of the calculations are thoroughly described by Hazeltine and Hinton. Hawryluk, Suckewer, and Hirshman have recently given a review of the status of the transport of low Z impurities in tokamaks.[16]

The evolution of impurity ions in a tokamak discharge is described by

$$(6)\quad \frac{\partial n_j}{\partial t} = -\frac{1}{r}\frac{\partial}{\partial r}(r\Gamma_j) + n_e(n_{j-1}S_{j-1} - n_jS_j + n_{j+1}R_{j+1} - n_jR_j)$$

$$j = 1 \rightarrow Z$$

These equations, complemented by the requirement of charge neutrality

$$(7)\quad \sum_j n_jZ_j = n_e$$

and ambipolarity

(8) $\Sigma Z_j \Gamma_j = \Gamma_e$

will determine the temporal and spatial evolution, if the relevant rate coefficients are known. The rate coefficients are, in many cases, not known, and this hampers detailed study. What is known are the general properties of the neoclassical theory, and we shall describe these.

The moment of the kinetic equation describing the ion evolution gives, for momentum:

(9) $$n\underline{u}_\perp = \frac{1}{m\Omega} \hat{n} \times (\nabla \cdot \underline{\underline{P}} - \underline{F} - en\,\underline{E} + m\frac{\partial}{\partial t} n\underline{u})$$

Considering an expansion of the equations in a small parameter (scale length/gyroradius), the lowest order result is that the distribution functions are Maxwellian. For the component of $\underline{u}$ perpendicular to $\underline{B}$

(10) $$\frac{\partial}{\partial t} mn\underline{u} + \nabla \cdot \underline{\underline{P}} - en(\underline{E} + \frac{1}{c}\underline{u} \times \underline{B}) = \underline{F}$$

To next order, the requirement that the densities depend only on magnetic flux arises, and to second order, deviations in the poloidal direction can be maintained. In second order the cross-field diffusion flux is

(11) $$n\underline{u}_\perp = \frac{1}{m\Omega} \hat{n} \times [\nabla \cdot (\underline{\underline{P}} - \overline{p}\underline{\underline{I}}) - e(n\underline{E} + \hat{n}\nabla\phi]$$

The latter term introduces the changes due to moving flux surfaces, and is not a true diffusion effect. The first term expresses the key point: the source of enhanced diffusion in toroidal geometry arises from anisotropy in the plasma stress tensor. For high collision frequencies, the tensor is diagonal, but there are variations within a flux surface. For low collision frequencies, $\underline{\underline{P}}$ is non-diagonal and greatly enhanced anisotropy results.

For most applications, the impurities will have a high collision frequency, and so the essential ingredient in the theory is the asymmetry of the plasma pressure around a flux surface. To recapitulate, to lowest order (zero gyroradius) there is no deviation from a field line; to next order there is no deviation from a flux surface and to second order, where the theory makes its first non-trivial contribution, it is the asymmetry on a flux surface which provides the essential driving term.

Momentum balance in the direction parallel to the field leads to the flux

$$(12)\quad \Gamma_{\parallel}^{a} = -\frac{1}{Z}\frac{\partial P}{\partial\psi}\frac{1}{B}\left(1 - \frac{B^2}{\langle B^2\rangle}\right) + \frac{\langle \Gamma_{\parallel}^{a} B\rangle}{\langle B^2\rangle}$$

The variation over a flux surface is present in the first term. The force balance equation

$$(13)\quad \underset{\sim}{B} \cdot \nabla P_Z = - n_Z I_Z \; e\underset{\sim}{B} \cdot \nabla\psi + BR_u$$

can be integrated to give

$$(14)\quad \frac{n_Z(\theta)}{\langle n_Z\rangle} = -\alpha \sin\theta\left[c_o\left(\frac{Z_z}{Z}\frac{1}{n_i}\frac{\partial n_i}{\partial r} - \frac{1}{n_z}\frac{\partial n_z}{\partial r}\right) + \left\{c_o\left(1 - \frac{5}{2}c_i\right) - \alpha_2\right\}\right] + \gamma\frac{\partial T_i}{dr}$$

Depending on the relative strengths of ion density and temperature gradients, then, the symmetry (or lack of it) in the impurity spectral emission can vary quite widely. The characteristics can even fluctuate during the course of a single discharge. This effect is inherently present in neo-classical theory, and should be pronounced even for moderate ionization states. These (or neutral) stages will of course reflect the lack of symmetry imposed by the external sources.

<u>Argon injection experiments</u>. The deliberate injection of a trace amount of Argon into a T_4 discharge showed one neo-classically expected effect very clearly: within 10 msec the Argon appeared in the core. However, thereafter the rate of accumulation stopped, and the Argon reached a steady state level. It has been speculated by the authors of [17] that charge exchange between the ambient hydrogen neutrals (∿200 eV in energy) and Ar^{16+} may have been responsible for their observation. Since these rates are unknown in this energy range, however, no conclusions could be drawn. Another mechanism, arising from plasma physics, should be borne in mind.

Kadomtsev[18] has proposed a model for the sawtooth oscillations present in tokamaks. The current density on axis increases, the Kruskal-Shafranov safety factor drops below unity and an energy surplus is created in the magnetic field which can be released as plasma expansion energy. A reconnection of field lines occurs, with the resulting final state a uniformized image of the initial state. The plasma parameters are carried with the flux as it is redistributed. The reconnection criterion is

$$(15)\quad \psi^*(\equiv 0) = \psi^*(r_o)$$

and for a parabolic profile this implies flattening out to a radius r_o $r_o \sim \sqrt{2}\, r_s$ $[q(r_s) = 1]$. In the T_4 experiments $r_o/a \sim 0.3$. Thus, a substantial part of the plasma was subject to a periodic expulsion (outflux) mechanism and this can resolve the paradox with neoclassical theory.

A different explanation has been advanced by Mirnov.[19] He points out that large scale magnetic islands are typically produced in tokamaks. These magnetic structures are the result of nonlinearly saturated tearing instabilities. However, they effectively couple two previously disjoint flux surfaces ψ_1 and ψ_2, having electron temperature $T_e(\psi_1) \neq T_e(\psi_2)$. Thus an electrostatic barrier is set up to prevent instantaneous equilibration. The protons are attracted and impurity ions flow against the proton gradient. Thus there is an electrostatic sheath prohibiting further penetration of the Argon injected from the outside.

At present, although a detailed model is lacking, it appears that neoclassical theory with MHD effects, can explain the features of impurity transport in tokamaks. In each experimental situation, however, there are key unknowns in the atomic physics areas: ionization rates for electron impact, and charge exchange of hydrogen in the 0-1 keV energy range with multiply-charged impurity ions.

4.2 Divertor/Scrapeoff Plasma

Serious experimentation with divertor plasmas is now underway. The DITE bundle divertor and DIVA divertor showed beneficial effects on divertors on plasma properties.[20] The major results of the DITE experiment illustrate the role which divertors will play in a reactor: by reducing the ion-induced desorption of cold gas trapped on the wall, the plasma particle balance could be controlled. By ionizing metallic impurities and sweeping them into the divertor chamber, the scrapeoff layer could serve as a shield for the plasma.

Plasma energy was, in fact, coupled to the divertor plates, and this raises a concern with divertors: the energy flux may be too severe. Besides the multiple roles of recycle control, shield for impurities, and energy absorber, the divertor must pump the helium produced by D-T reaction products in a reactor.

The atomic physics role in determining divertor properties is similar to that in analyzing the conventional limiter.[21] Essentially one is balancing the parallel flow along field lines to the divertor box (or limiter) against the cross-field transport. The plasma models for parallel flow are still being worked out. The usual assumption is that ions flow at the sound speed into the divertor chamber if charge exchange with neutrals is unimportant

$$(16)\quad u_{\parallel} \sim \sqrt{2(T_e + T_i)/m_i} = c_s$$

Under these circumstances, however, the product $\bar{n}_e\Delta$ of the scrape-off layer may be insufficiently large to ionize the impurities produced at the first wall. In that case, neutral gas puffing will be introduced to build up the scrape-off density. However, charge exchange will balance the parallel pressure gradient, and the flow to the collector will be reduced

$$(17)\quad u_{\parallel} \sim -\frac{D}{n}\Delta P;\; D = \frac{1}{m_i\, n_o \langle\sigma v\rangle_{cx}}$$

The optimum scrape-off design will have three major physical features: ion parallel flow into the divertor, hydrogen gas puffing at a large enough rate to keep up the $\bar{n}_e\Delta$, and impurity production. The ionization rates will be determined by T_e, and it, in turn, will be determined by flow and radiative losses. Further, the radiative losses will be strongly influenced by the $H_o + A^{nt}$ charge exchange reactions. Thus, charge exchange cross-sections for the 0-100 eV range, for H_o with low ionization stages of Fe, Cr, Ni, Ti, are most crucially needed, since even crude estimates are lacking. Moreover, the electron impact ionization rates require better calculation measurement.

There is an important surface-related factor as well. As ions flow to the collector plate an electrostatic sheath is built up. The magnitude of this potential barrier depends in a critical way on the secondary electron emission which occurs in the divertor box. Assuming no secondary emission, present models give a parallel heat flux.

$Q_{\|} = 6\, T_e\, \Gamma_{\|}$, with $\sim 7\, T_e\, \Gamma_{\|}$ recorded on the DIVA experiment. However, 100 times this value is needed to explain the DITE experiments. Thus, there is some room for improvement in the models.

4.3 Neutral Beam Deposition

The neutral beam heating technique has been successfully proven. In view of the large role it will play in future tokamak experiments, we will describe the major physical processes.

The neutral atoms emerge from the beamline and are trapped in the plasma by charge transfer with plasma ions, by electron impact ionization, or by impact ionization on the plasma ions. Once charged, these energetic injected particles are confined by the Tokamak magnetic field. For the case of injection nearly tangential to the magnetic field shown in Fig. 4, the particles follow trajectories which also lie nearly along the fields. However, a projection of their orbits onto a plane perpendicular to the toroidal axis shows that there are important drift motions; their orbits are displaced outward from the axis of symmetry for particles injected nearly parallel to the plasma current, inward for those particles which are nearly anti-parallel. By calculating the resulting drift after trapping, and averaging the densities of the fast ions over the Tokamak radial flux surfaces, the deposition of fast ions is finally characterized by the function $H(r)$ shown in Fig. 4. It is the normalized density of deposited fast ions as a function of plasma minor radius. Various radial deposition profiles are shown as a function of a/λ. (a is the plasma minor radius, λ the mean free path against trapping by all the processes mentioned). That is,

$$(18) \quad \frac{a}{\lambda} = \int \Sigma n_e \langle \sigma_e v_e \rangle + n_H \sigma_x v_o + n_Z \sigma_{xz} v_o + n_H \sigma_i v_o + n_Z \sigma_{iz} v_o) ds$$

The integral is taken along the path of the beam from entrance to a point on the vacuum chamber wall opposite the injector.

Once deposited, the energetic ions interact with the background confined plasma. While many possible collective thermalization processes have been considered, the model of successive distant binary encounters, as described by the Fokker-Planck drift-kinetic equation, is thought to be accurate. The injected fast neutrals, now fast ions, thus transfer momentum and energy to the plasma and finally join the background, increasing the plasma number density. Calculations with the Fokker-Planck equation yield estimates for these processes which are consistent with the rates observed in a number of present experiments.

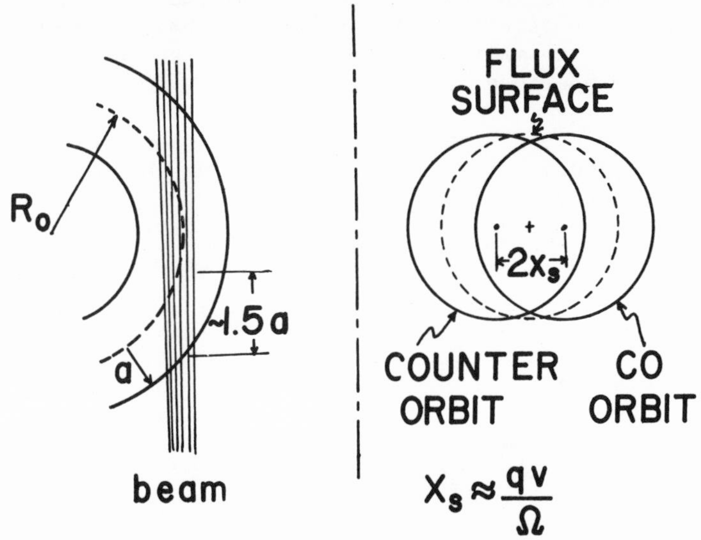

$$\dot{n}_f(r) = \frac{I_B/e}{(2\pi R_0)(\pi a^2)} H(r) = \frac{d\dot{N}/dr}{\text{DRIFT SURFACE AREA}}$$

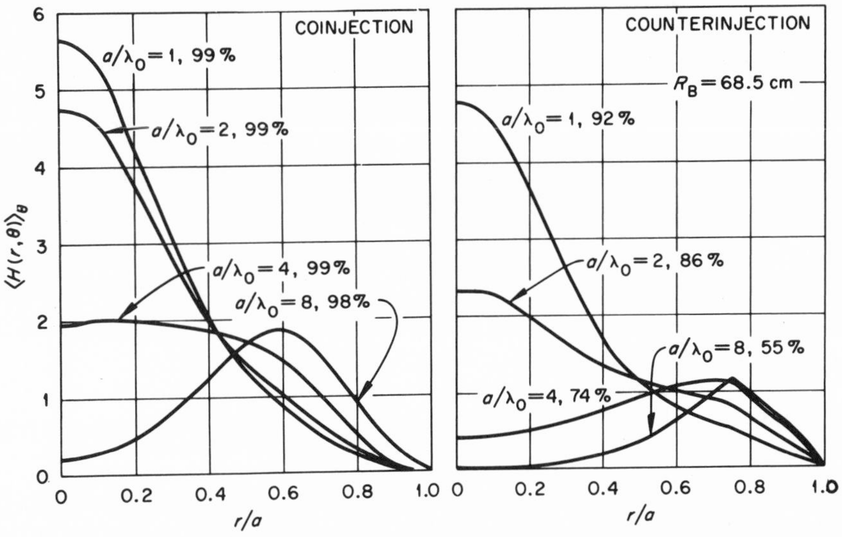

Fig. 4. Neutral beam deposition processes. (Top left) The beam enters the plasma chamber and neutrals are ionized. (Top right) Seen in cross-section, particles travelling in the direction of the plasma current are displaced outward, and those travelling in the opposite sense are shifted in. (Bottom) Resulting deposition profiles for co- and counter-injection.

The chief effect of contemporary injection experiments is to transfer energy to the background plasma, and thus to heat it to temperatures impossible to obtain by ohmic heating alone. The beam particle input is small by comparison with the particle sources from the walls and aperture limiters.

In some present experiments the beam energy transfer to the plasma is degraded by charge exchange of the thermalizing ions with residual low-energy hydrogenic neutrals in the plasma. While the ratio of hydrogenic neutral/hydrogenic ion densities is typically 10^{-5}, the resonance charge exchange cross section is large enough to produce a significant loss of beam ions during the time (of order 10 milliseconds) required for thermalization.

Figure 5 shows a typical computed distribution of fast ions in velocity space. (This is a solution of the Fokker-Planck equation.) There is an empty region in the space of particles traveling counter to the current. This so-called "loss region" is characterized by particle drift orbits whose spatial trajectories cause them to leave the machine and strike the wall. Perpendicular injection, such as will be done on the PDX experiment, also produces such a loss region, especially at low plasma current. The rate at which the beam ions are lost through this gap in velocity space is characterized by the rate of diffusion in the angle made by the ion velocity with the magnetic field. This 'pitch angle' diffusion rate is proportional to the rate of momentum exchange with the background electrons, and is directly proportional to the plasma Z_{eff}.

The cross sections and the rate coefficient needed for Eq. (18) are obvious examples of atomic data needs. The charge transfer and impact ionization cross sections are needed for quite a large number of species, and over a fairly wide range of energies. These quantities are also needed for plasma diagnostics and for estimates of plasma cooling due to impurity ions. One wishes to ascertain, largely from spectroscopy, the chemical composition (i.e., the n_k) and charge state distribution (i.e., Z_k) for the positive ion species in the plasma. Estimates of the plasma cooling processes will require knowledge of the dielectronic and radiative recombination rates for the impurity ions. These topics will be discussed elsewhere in the Institute.

Since these latter processes are important to the overall plasma energy balance, and are not unique to the neutral injection process, we will focus on the deposition cross sections in our discussion.

Making reference once more to Fig. 4, we note that beam depositiontion profile ceases to be peaked in the center and becomes flat for $a/\lambda \sim 4$. In present tokamaks, the energy confinement is better in the center of the device than at the edge; thus, the peaked

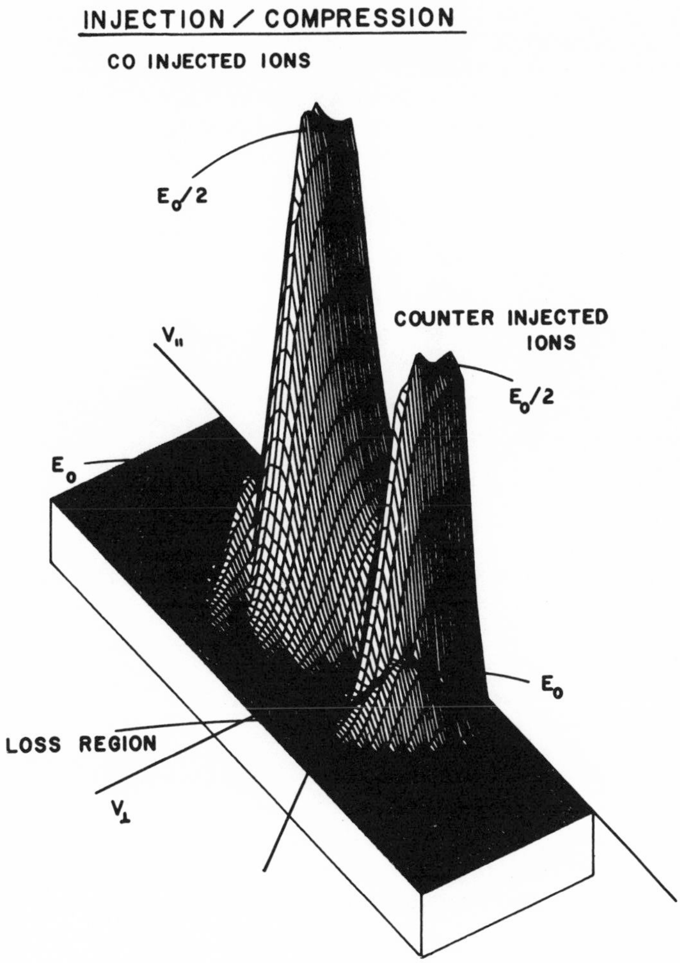

Fig. 5. A typical distribution of fast-ions in velocity space. ($v_{\|} \equiv \underset{\sim}{v} \cdot \underline{B}/B$). The loss region occurs for those values of pitch angle producing orbits which intersect the chamber wall.

profiles in Fig. 4 will produce an efficient plasma heating, while those peaked outwardly will produce poor heating. Not only that, but such outwardly peaked profiles will also produce a stronger interaction with the wall. Thus $a/\lambda \sim 4$ represents the lower limit of adequate beam penetration; larger values are deemed unsuitable.

A very effective cooperative atomic physics/fusion effort has succeeded in the past few years in finding the charge exchange cross section for interaction of impurities and injected ions in the range 25 keV < E_o/nucleon < 60 keV. This is the range of interest for experiments up to TFTR. It is found that the electron loss cross-section scales as $Z^{1.4}$, and thus $\lambda^{-1} \sim \bar{n}_e \sigma_H$. Hence, the increased capture rate when impurities are present just balances the deficit of scattering centers. Thus, previous estimates of neutral beam heating in impure plasmas turn out to be fortuitously correct! Moreover, a previously feared coupling between impurity production and beam trapping is diminished. If the cross-section scaling had been worse, say $\sim Z^2$, then $\lambda^{-1} \approx n_e Z_{eff}$. As more impurities are created the beam would be trapped closer to the wall, creating even more impurities, and so on.

While this instability will not occur when charge exchange is the dominant trapping reaction, it should be noted that work is still needed for impact ionization. The INTOR reactor design study jointly conducted by the US, USSR, CEC and Japan through IAEA, presently considers the use of 400 keV beams to provide auxiliary core heating. Recent measurements of H_2 impact ionization on multi-charged impurities suggests that scenarios requiring long pulse operation of 400 keV beams may well revive concern about the beam deposition instability.

4.4 Plasma Fueling

As noted previously, the puffing of neutral gas is used to increase the density. The density evolution is determined by

$$\text{(19)} \quad \frac{\partial n_e(r,t)}{\partial t} = \frac{1}{r}\frac{\partial}{\partial r}\left(rD\frac{\partial n_e}{\partial r}\right) + n_o n_e \langle\sigma v\rangle_{ion} + \Sigma_{impurities}$$

Given the gas inlet rate, and the rate at which gas is recycled from the walls, it follows that the net ionization input will depend on the neutral density inside the plasma. Fig. 6 shows the dependence of calculated neutral density profiles on the scaling of the $O^{n+} + H^0 \rightarrow O^{(n-1)} + H^+$ charge exchange reaction in the 0 - 1 keV range. Order of magnitude changes are seen in the central density which is calculated to result from a given edge density.

Isler and Crume[23] have recently reported an experiment in which certain spectral lines of O^{5+} and O^{6+} appear too anomalously intense to have been produced solely by electrons. Estimates from the calculated $C^{4+} + H^0 \rightarrow C^{3+} + H^+$ reaction allowed these authors to speculate that the emission from n = 4 states of O^{4+} and O^{5+} and the n = 5 (and possibly 4) states of O^{6+} were enhanced by charge exchange. The anomalous oxygen lines (115 Å for O^{5+} and 81.9 Å for O^{6+}) originate from levels which could be populated by charge exchange.

The fusion significance of this result is that radiative losses from oxygen are much larger than those which would be calculated neglecting charge exchange. As a result, the 'anomalous' transport needed to explain the observed behavior is reduced.

5. CASE HISTORY: CHARGE EXCHANGE CROSS-SECTIONS

As an example of the way in which specific new atomic physics information is used in fusion, we retrace the influence which calculation and measurement of beam trapping cross-sections have had on the understanding of neutral beam heating experiments.

Originally, several studies showed that impact ionization rate coefficient enhancement due to impurities in the 60 keV/AMU energy range could prove fatal for neutral beam experiments such as TFTR. By curtailing the effectiveness of heating after ∿300 msec, much of the research value of these experiments would be lost. This conceptual issue led to the identification of the need for knowledge of this cross-section as a high priority atomic physics issue. Several theoretical and experimental groups contributed to the measurement of the cross-sections[2,2] and it was found that the earlier predictions of a beam-dominated instability were too pessimistic. However, the effort has just begun to pay off in scientific terms. Previous to these measurements, a diagnostic beam experiment was underway on the TFR tokamak.[24] By measuring the attenuation of a low current, 10-20 keV beam, it was hoped to be able to extract the ratio n_p/n_e in a plasma with $Z_{eff} \sim 4$-6, which was rich in oxygen. Without detailed knowledge of the trapping cross-section, however, this experiment would have yielded n_p/n_e within too coarse limits. Once σ_{cx} was known, however, the experiment produced the first solid measurement of this ratio.

DEPENDENCE OF N_0 (0) ON OXYGEN CHARGE EXCHANGE

ASSUME CROSS-SECTION FOR $O^{n+} + H^0 \rightarrow O^{(n-1)^+} + H^+$ IS $\sigma = z^\alpha \sigma_{H^0 + H^+}$

MINOR RADIUS 20 cm

$T_e(r) = 1880 \, (1 - r^2/a^2)^{2.5}$ eV

$T_p(r) = 1000 \, (1 - r^2/a^2)$ eV

$n_e(r) = 8.10^{13} \, (1 - r^2/a^2)$ cm^{-3}

$n_p(r) = n_e(r) - \sum_{k=2}^{9} (k-1) \, n_{oxy}^{(k)}(r)$

$n_{oxy}^{(k)}(r) = C^{(k)}(r) n_{oxy}^T(r)$

$C^{(k)}(r) \rightarrow$ CORONAL EQUILIBRIUM

$n_{oxy}^T(r) = n_{oxy}(0)(1 - r^2/a^2) + 10^9$ cm^{-3}

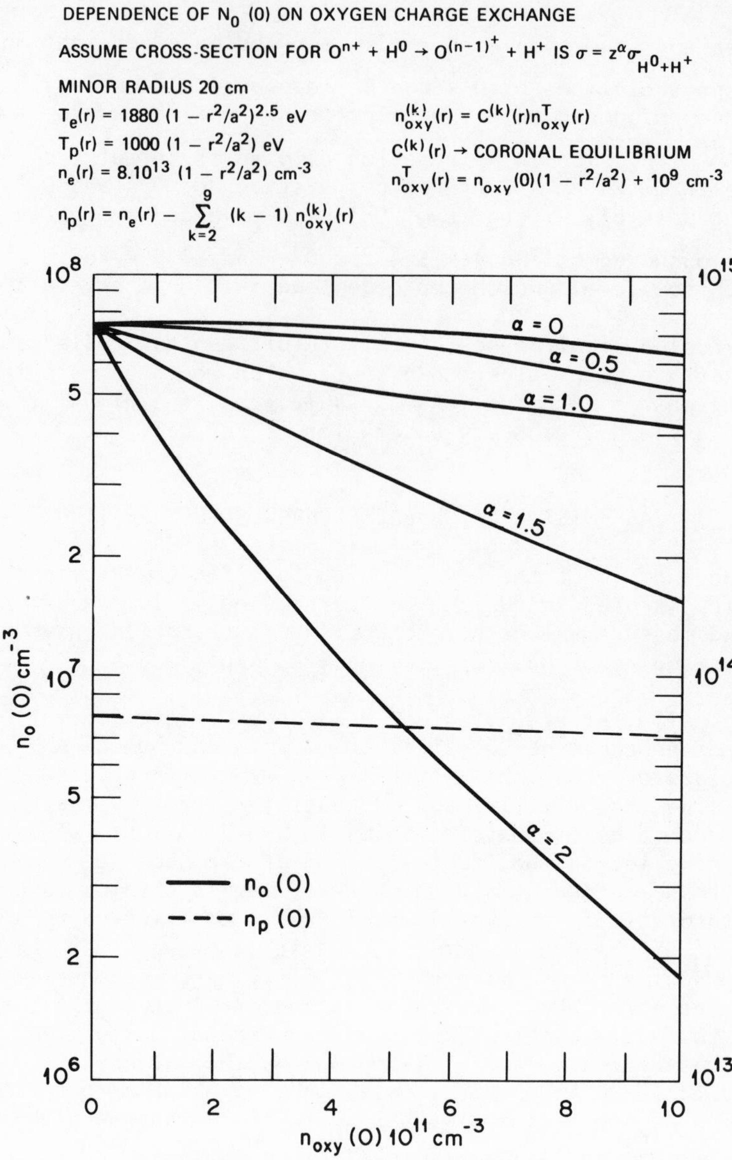

Fig. 6. Effects of the hydrogen-impurity cross-section on calculated neutral densities. For typical TFR parameters, a discrepancy of nearly two orders of magnitude can be produced by varying the scaling of the cross-section.

Direct observation of charge exchange trapping on impurities (O^{8+}) was reported on the ORMAK device.[25] Here there was some discrepancy with the detailed calculations, as to whether the charge states which should be populated by this process were, in fact, those observed.

However, the most important result of this collaboration has been found by Abramov et al.[26] They point out that in intense neutral beam injection experiments, the recombination of highly stripped Fe ions by charge exchange with H^o is orders of magnitude higher than radiative or dielectronic recombination rates. Hence, the charge state is reduced and radiative output enhanced. Goldston, et al. have shown that the neutral H produced by charge exchange of beam neutrals with thermal protons produces a neutral density $\sim 10^8$ cm^{-3} in PLT.[27] This neutral population is calculated to depress the ionization stage of Fe by several states, and may explain the large measured power loss from the center of PLT. The charge exchange of H^o and multiply charged ions thus has an important effect, not originally considered: the recombination due to this process enhances the line radiation cooling of the plasma core. As beam power is raised, the neutral atom density is raised with it. Thus higher beam power leads to a higher recombination rate, as well as a higher ionization rate, and not to a simple complete stripping of the impurity ion, and the consequent diminution of radiative losses. The, now known, charge exchange cross sections have led to a more accurate picture for the energy loss processes, in addition to refining our knowledge of the power input deposition.

DEFINITIONS OF SYMBOLS

α, α_2 coefficients in expressions for neo-classical fluxes[15]

β ratio of plasma to magnetic field pressure $2\mu_o P/B^2$

Γ flux of particles (number/cm^2/sec)

γ neoclassical coefficient[15]

Δ width (cm) of radial extent of a magnetic island, width of scrapeoff

ϕ toroidal angle

λ mean free path

ψ magnetic flux produced by the field due to the plasma current (poloidal flux)

ψ^* helical poloidal flux[18]

σ cross-sections: following are subscripts

- x charge exchange of injected H with plasma hydrogen
- xz charge exchange of injected H with impurities
- i impact ionization of injected H with plasma protons

- iz impact ionization of injected H with impurities
- e impact ionization of injected H with electrons

Σ source of electrons from ionization of impurities

$\tau_{E,P}$ energy and particle replacement times

r gyrofrequency $\frac{eB}{m}$

χ thermal conductivity

a plasma minor radius

$\underline{B}$ magnetic field

B_T magnitude of the toroidal component of the magnetic field

c_o neo-classical coefficient[15]

c_i neoclassical coefficient[15]

D particle diffusion coefficient

$\underline{E}$ electric field vector

e electron charge

I plasma current

m mass, toroidal mode number

n poloidal mode number

$\bar{n}$ average density

P (scalar) pressure

$\underline{\underline{P}}$ pressure tensor

q stability safety factor $\frac{2\pi a^2 B_T}{\mu_o R I}$

R plasma major radius, rate coefficient for recombination

$\underline{R}$ friction force vector

r_o, r_s minor radii for sawtooth oscillations

S ionization rate coefficient

T temperature

$\underline{u}$ plasma flow velocity

v_a Alfven speed $\frac{mn}{\sqrt{B^2/2\mu_o}}$

Z ionic charge

< > average over a poloidal flux surface

<u>subscripts</u>

e electron

i ion

p proton

z impurity

REFERENCES*

1. L. A. Artsimovich, "What Every Physicist Should Know About Plasmas," MIR Publishers, Moscow, 1979.
   B. Coppi and J. Rem, Scientific American, June 1972.
   H. Furth, Scientific American, August 1979.
2. M. Murakami and H. Eubank, Physics Today, 32 (25) 1979.
3. H. Bethe, Physics Today, 32 (44) 1979.
4. L. A. Artsimovich, Nuclear Fusion, 12 (265) 1972.
5. H. P. Furth, Nuclear Fusion, 15 (487) 1975.
6. S. I. Braginskii, Reviews of Plasma Theory, edited by M. A. Leontovich, (Consultants Bureau, N.Y.) V. 1, p. 205.
7. R. Hazeltine, F. Hinton, Rev. Mod. Phys. 48 (239) 1976.
8. S. V. Mirnov, IAEA 7, Innsbruck, 1978 (to be published by IAEA, Vienna).
9. IAEA, "World Survey of Major Facilities in Controlled Fusion Research," Vienna 1976.
10. PLT Group, Phys. Rev. Lett. 43 (270) 1979.
11. R. R. Parker, Bull. Am. Phys. Soc. 20 (1392) 1976.
12. M. Murakami et al., Phys. Rev. Lett. 39 (615) 1977.
13. ISX-A Group: IAEA 7, Innsbruck, 1978 (to be published by IAEA, Vienna).
14. B. Badger et al., University of Wisconsin Report UWFDM 330, March 1979, D. Steiner et al., ORNL Tech. Mem. 6720, 1979.
15. K. Burrel, K. Wong, General Atomic Co. Report GA-15131, 1979.
16. R. Hawryluk, S. Suckewer, S. Hirshman, Nuclear Fusion 19 (607) 1979.
17. K. Britov et al., IAEA 7, Innsbruck, 1978 (to be published by IAEA, Vienna).
18. B. Kadomtsev, Sov. Jour. Plasma Physics, 1 (389) 1975.
19. S. V. Mirnov (Private Communication).
20. DIVA Group, IAEA 7, and W. H. M. Clark et al., IAEA 7, Innsbruck, 1978 (to be published by IAEA, Vienna).
21. A. T. Mense, G. A. Emmert, Nuclear Fusion 19 (361) 1979.
22. K. H. Berkner et al., Phys. Rev. Lett. 41 (163) 1978.
23. R. C. Isler, E. C. Crume, Phys. Rev. Lett. 41 (1296) 1978.
24. P. Moriette, Fontenay-aux-Roses Report EURCEA 954, December 1977.
25. R. C. Isler, Phys. Rev. Lett. 38 (1359) 1977.
26. V. Abramov et al., JETP Letters 29 (550) 1979.
27. PLT Group, IAEA 7, Innsbruck, 1978 (to be published by IAEA, Vienna).

*IAEA 7 refers to the 7th International Conference on Controlled Nuclear Fusion and Plasma Physics, Innsbruck, Austria, 1978.

ATOMIC PROCESSES RELEVANT TO ENERGY BALANCE AND IMPURITY CONTROL IN FUSION REACTORS

Dr. Erol Oktay

U. S. Department of Energy

Washington, D. C. 20545

Introduction

General principles of magnetic fusion confinement and a general survey of Atomic and Molecular problems relevant to CTR have been discussed in lectures by Hogan and Harrison. In these three lectures I will discuss primarily atomic physics relevant to energy balance and impurity control in fusion reactors.

Consideration of energy balance involves identification of power input, output and loss processes in a reactor. A critical power loss process in plasmas is radiation especially by impurities. The presence of impurities and helium from D-T reaction in plasma introduces a second effect on the energy balance which is dilution of fuel ions (deuterium and tritium). Other detrimental effects of impurities that have already been observed in present experiments include current profile modification, density limitation, and enhanced charge exchange losses. Therefore impurity control methods such as divertors and gas blanket, have to be developed to reduce impurity levels in plasmas, and to remove heat load and helium ash from the fusion plasma core.

It should be noted in perspective that impurity problem manifests itself in present experiments, and its control is probably the most difficult issue for the design of next generation devices such as Engineering Test Facility (ETF) and International Tokamak Reactor (INTOR). Progress in fusion has been possible partially because methods have been developed along the way to reduce or manipulate impurity levels in plasmas. For example, change-over from glass to metallic vacuum chambers, development of various techniques to condition chamber walls, and utilization of carbon instead of tungsten limiters in PLT experiments last summer have all contributed to achievement of improved plasma parameters. These are passive methods of impurity control whereas in a reactor active control with divertors or divertor-like systems will be necessary.

The outline of these lectures is as follows: In the first lecture I will briefly review general principles of power balance, Q, ignition and derivation of Lawson criteria. The effect of impurities on ignition temperature, Q and n values will then be examined in the second lecture. The third lecture will be devoted to brief discussion of impurity sources and impurity control techniques.

There are a variety of concepts that can be used as the fusion plasma core of a reactor. I will limit my discussions to tokamak type concepts which are closed toroidal systems. The characteristics of the configurations (open and closed field lines) operating plasma regimes (density, temperature, confinement time), pulsed versus steady state nature of the reactors, and the types of fuels (D-T, D-D, or neutronless advanced fuels) impose different constraints on the power balance and on role of impurities in these systems.

Power Balance

A simplified diagram of power-flow in a fusion reactor system is shown in Fig. 1. P_c is the circulating power required to operate the plasma preparation, confinement and heating systems. These systems provide an input power, P_i, to the fusion power core for heating the plasma to thermonuclear temperatures. The total power output of the fusion core is P_o which is converted into electrical power, P_g, in the power conversion system with overall efficiency of η_T. Part of this electrical power, P_c, is circulated back into the reactor system to maintain its operation, and the rest, P_n, is fed into the power grid as net electrical output from the reactor. The ratio of P_i/P_c, defined as η_i, is the efficiency with which the circulating energy is converted to input power for the fusion core. Q is the factor with which this input

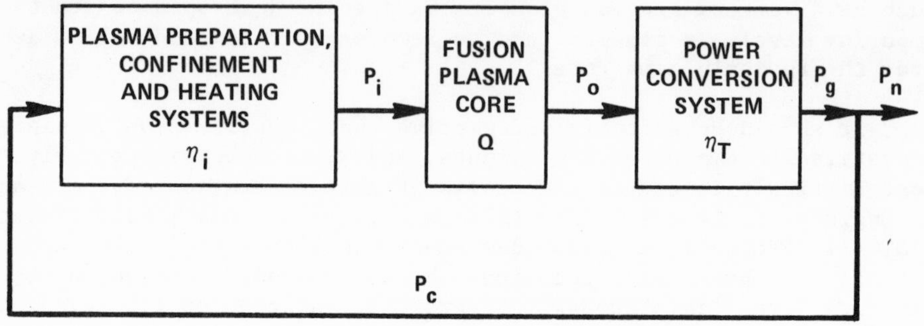

Fig. 1. Power Flow in a Fusion Reactor System

power is multiplied in the fusion core in order to derive net power from the reactor. It is defined as the ratio of nuclear power output to the power input.

The critical factors effecting the economies of fusion reactors are included primarily in the magnitude of Q, and in the value of circulating power fraction, P_c/P_g which are determined by the characteristics of the plasma heating and confinement systems. The role of the atomic physics in reactor energy balance manifests itself in the value of Q.

A second level detail of the power input to and power out of the fusion power source is shown in Fig. 2. The methods for getting external power into the fusion core to heat the plasma include ohmic heating, adiabatic compression, particle injection, photon injection, and rf heating. Non-convertible energy loss processes from plasma include radiation (bremsstrahlung, radiative recombination, line, and synchotron radiation), thermal heat conduction, particle diffusion, charge exchange losses, and other anomalous losses. The primary power output of a D-T fusion reactor is by neutrons and alpha particles. The neutron energy is recoverd in the reactor blanket whereas the alpha particle energy is deposited within the plasma.

In an ideal fusion reactor, external heating is required only for the start-up period during which the plasma parameters are raised to those required for igniton. The plasma burn conditions are then maintained by alpha particle heating only which compensates for nonconvertible power losses. The Q of an ignited reactor is ∞. There can be severe limitations, however, in reaching ignition conditions. Impurity radiation loss is one such limitation. In this case, the reactor can be operated in a driven mode in which external power

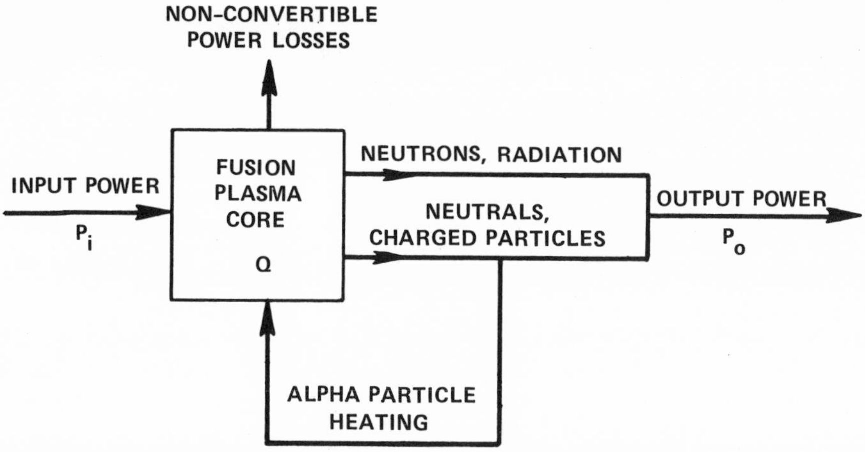

Fig. 2. Power Flow in a Fusion Plasma Core

source, such as neutral beam heating, is provided continuously to supplement the losses. This mode of operation of a reactor is analogous to a power amplifier. Another variation to this approach is the hot-ion mode operation of the reactor in which power is selectively provided to ions. In this case ion temperature is kept higher than that of electrons. Since the high ion temperature implies high reaction rates and low electron temperature reduces power loss, the energy balance in the reactor becomes more favorable and ignition may be achieved with less restrictions.

Lawson Criteria

For the present purposes, let us first consider power balance for an ignited fusion reactor. In 1957 John Lawson of U. K. developed a concept of power balance for fusion reactors that provides guidelines to reach breakeven and ignition in a fusion plasma. (Ref. 1) (Breakeven refers to the condition in which there is no net power output from the reactor.) These guidelines relate $n\tau$ and T values in plasma and has become somewhat of a roadmap for progress in fusion. (n, τ, and T are density, energy confinement time, and temperature, respectively.) The derivation of this power balance takes into consideration the energy input to plasma, energy output and losses, and the overall efficiency of converting thermal power into electrical power. The energy input to the plasma, E_i, goes into thermal energy of particles, nkT, and radiative power losses from the plasma, P_R. The primary radiative losses are synchotron, bremsstrahlung, and line radiation. In this treatment, with negligible impurities, hydrogen bremsstrahlung is the dominant radiative energy loss. Synchotron radiation becomes important only at high magnetic fields and temperatures and it is neglected here.

The energy output of the plasma is

$$E_o = nkT + \tau P_R + \tau P_N \tag{1}$$

where P_N is the thermonuclear power output. The efficiency of conversion of the output energy into electrical energy is denoted by η. For a net power producing system, the requirement is:

$$\eta E_o > E_i \; ; \quad \eta(nkT + \tau P_R + \tau P_N) > nkT + \tau P_R \tag{2}$$

The parametric dependence of P_R and P_N are

$$P_R = A n^2 T^{1/2}, \qquad P_N = B n^2 f(T)$$

where A and B are some constants, and f(T) is fusion cross-section which is a strong function of temperature. By substituting these functions into (2), we obtain

$$n\tau[AT^{1/2}(\eta-1)+\eta Bf(T)] > kT(1-\eta) \tag{3}$$

$$n\tau > \frac{kT(\frac{1}{\eta}-1)}{Bf(T)+(1-\frac{1}{\eta})AT^{1/2}} \tag{4}$$

Since $\eta < 1$, the following condition has to be satisfied for the above relationship to be valid:

$$Bf(T) > (1-\frac{1}{\eta})AT^{1/2} \tag{5}$$

This expression is only a function of temperature, and for a Maxwellian plasma and $\eta \approx 0.33$, it is satisfied for T ~ 4 keV, for D-T and ~ 20 keV for D-D fuels. These temperatures are referred to as the breakeven temperatures for these particular fuel cycles.

A minimum value for $n\tau$ can then be obtained for these temperatures from the expression (4). For the D-T fuel cycle, $n\tau$ value for breakeven conditions is about 3×10^{14} cm^{-3} sec. As either one of these two parameters, $n\tau$ and T, increase beyond the breakeven conditions, the gain Q of the system increases. These conditions are summarized in a universal curve for $n\tau$ versus T for different gain parameter Q. (Ref. 2).

General Impurity Effects

It should be noted here that radiative power loss (hydrogen bremmstralung) limits ignition to temperatures above 4 keV for D-T plasmas. As it will be shown in more detail in the next lecture, the primary effect of impurities in energy balance is in raising the ignition temperature to even higher values because of increase in radiative losses.

A second important effect of the impurities in power balance is its dilution of fuel ions. Quasi-charge neutrality is a basic property of plasmas. That is, $\sum_i n_i Z_i = n_e$ where n_i is the density of ions with charge Z_i. Full ionization of heavy ions such as iron, molybendum, tungsten, etc. would introduce many electrons per ion and thereby reduce the fraction of deuterium and tritium ions that can be sustained at such electron densities. For example, in the above discussion of power balance it

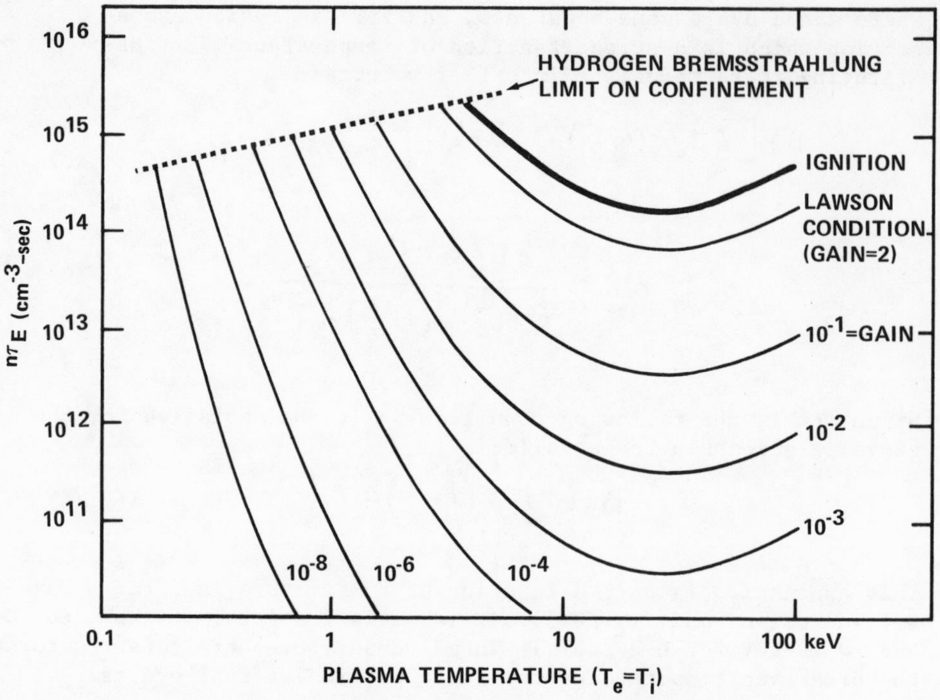

Fig. 3. $n\tau$ Versus T for Various Values of Gain

was assumed that $n_e = n_D + n_T$, and the n in the parameter $n\tau$ is the electron density. A reduction in n_D and n_T as a result of impurities will of course decrease fusion reaction rate because of its dependence on the value of $n_D n_T$.

There are other effects of impurities in plasmas that deserve a brief mention. First of these is the impact of fuel-ion depletion on the energetics of the plasma. Stability considerations impose certain limitations on the plasma beta (β) in any magnetic configuration. β is the ratio of plasma pressure to magnetic pressure, $\beta = nkT/(B^2/2\mu_0)$, where all the particles contribute to the nkT term. Since there is significant numbers of alpha particles (helium ions) produced in the D-T reacton, part of the plasma pressure will be made up of helium particle pressure if these particles are not removed effectively from the plasma as they give their energy to electrons and ions. Consequently, helium build-up either limits the stable operation of the plasma, or quenches the plasma burn by limiting fuel ion temperature or density.

Another impurity effect that has been observed in present-day experiments is current profile modifications that can lead to instabilities in certain cases. Low Z impurities such as oxygen or carbon near the walls increase the plasma resistivity in outer regimes of the plasma column. The current profile then becomes peaked and may lead to disruptive instabilities. On the other hand, heavy impurities such as iron, tungsten, molybendum radiate significant power from the plasma core which then results in hollow temperature profiles.

Neutral beam injection which has now become a powerful plasma heating tool is also affected by impurities. The neutral beam heating method is governed by charge exchange of fast injected neutrals with low energy fuel ions in the plasma. The charge exchange cross section is a direct function of the charge of the target ions. Consequently part of the injected neutral beam energy can be lost by their charge exchange with impurity ions. Similarly, there can be significant energy loss from the plasma by charge exchange of hot fuel ions with impurities. These particles leave the plasma core as energetic neltrals. Some of these topics will be discussed by other lecturers.

Effect of Impurities on $n\tau$, Q and Ignition Temperature

In the previous lecture, power balance was discussed in general terms. In this lecture, the effect of impurities will be included in the formulation of power balance in order to account for impurity radiation losses and fuel ion dilution. This presentration is based on the work done by Jensen, Post, Jassby (Ref. 3) and Jensen, et al (Ref. 4).

The calculation in Refs. 3 and 4 are aimed at determining the effects of most common impurities on the Q and $n\tau$ values of D-T reactors. The primary power inputs to the reactor are external neutral beam injection and internal alpha particle heating. Ohmic heating is negligible. It is assumed that all of the alpha particle energy is deposited in the plasma. These calculations apply to all closed toroidal magnetic confinement systems.

The power balance per unit volume is given by:

$$P_b + P_\alpha + P_{OH} = \frac{\frac{3}{2} n_e T_e + \sum_{ions} n_{ions} T_i}{\tau_E} + P_R \qquad (6)$$

where τ_E is energy confinement time,
P_R radiation power density
P_b neutral beam injection power density
P_α alpha particle heating power density
P_{OH} ohmic heating power density.

Using the definition $Q = P_f/P_b$ where P_f, the fusion power output density, is the sum of P_{TN}, the thermonuclear power density from Maxwellian plasma, and P_{fb}, the fusion power density of beam-target reactions, one can write $P_\alpha = 0.2\ QP_b$. (3.5 MeV α-particle energy represents 20% of 17.6 MeV total fusion energy released per reaction.)

The following definitions are given for particle densities:

$$n_T = \text{tritium ion density} = F\, n_i$$

$$n_D = \text{deuterium ion density} = (I\text{-}F)\, n_i \tag{7}$$

$$n_i = n_T + n_D = \text{total hydrogenic ion density}$$

$$n_Z = f_Z n_i = \text{density of impurity ions with charge } Z$$

$$n_e = n_i + \sum n_Z Z = n_i(1 + f_Z \bar{Z})$$

Defining $P_R = n_e n_Z L_Z + n_e n_i L_H$ where L_Z and L_H are impurity and hydrogenic radiation power density per ion and per electron, one arrives at the following expression for $n_e\tau_E$:

$$n\tau_E = \frac{\frac{3}{2}\left[T_e(1+f_z\bar{Z}) + T_i(1+f_z)\right)}{\dfrac{A\langle\sigma v\rangle F(1-F)E_f}{1+f_z\bar{Z}} - (f_z L_z + L_H)} \tag{8}$$

$$A = \frac{1+0.2Q}{Q - \dfrac{P_{bf}}{P_b}\left(\dfrac{F}{1+f_z\bar{Z}}\right)} \tag{9}$$

The fuel depletion effect of impurities is accounted for in eq. (7) and it results in an increase in the required $n\tau_E$ values by reducing the fusion power by a factor $(1 + f_Z Z)$ in the denominator of eq. (8). $\bar{Z}$, the average charge of impurity ions, have been caclulated as a function of temperature by Post and Jensen for impurity elements relevant to fusion plasmas. (Ref. 5). $\langle\sigma v\rangle$ is fusion reaction rate which is a product of fusion cross-section, σ, and relative deuterium-tritium velocity averaged over a Maxwellian velocity distribution. A plot of $\langle\sigma v\rangle$ as a function of T for D-T reaction is shown in Fig. 4. These can also be obtained from the expression (Ref 5)

$$\log_{10}\langle\sigma v\rangle = \sum_{k=1}^{5} A(k)\left[\log_{10} T(\text{keV})\right]^k \tag{10}$$

where A(o) = -20.20436 A(3) = 0.8458346
A(1) = 6.0977007 A(4) = -0.5944575
A(2) = -2,242023 A(5) = 0.1450558

for $1 \lesssim T_i(\text{keV}) \lesssim 100$.

L_H is the radiation rate per electron per hydrogenic ion obtained from the expression for bremsstrahlung radiation given by

$$P_b = n_e n_i L_H = 3\times10^{-15} n_e n_i Z_{eff}\sqrt{T_e(\text{keV})}\ \text{kev/cm}^3\text{-sec} \quad (11)$$

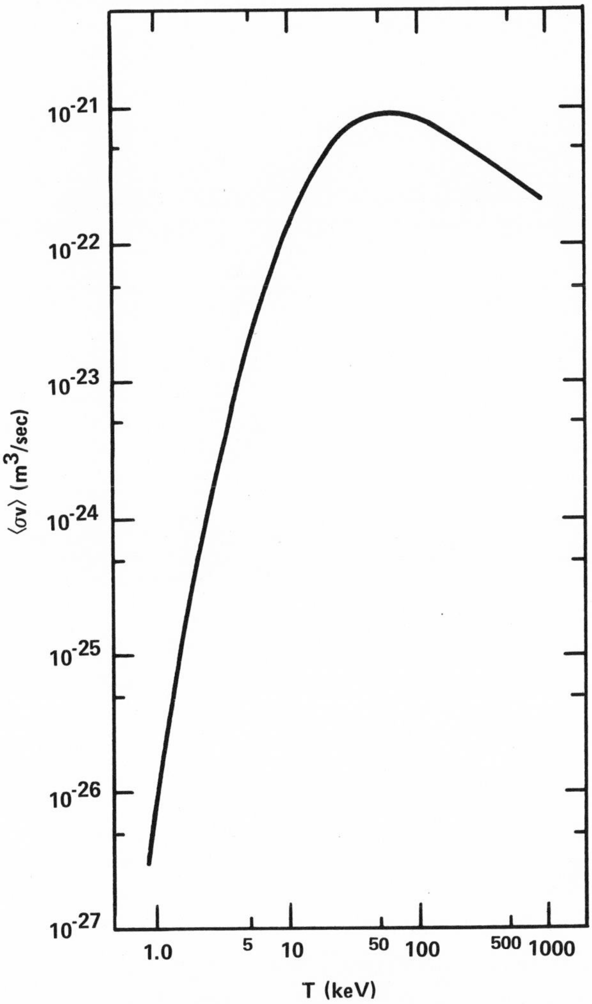

Fig. 4. $\langle\sigma v\rangle$ Versus T for D-T Reactions

L_Z is the impurity radiation per electron per impurity ion which includes bremsstrahlung, radiative and dielectronic recombination, and line radiation. The calculation of impurity radiation is very complicated because of the presence of a variety of impurities in many diffeent ionization stages in different regions of the plasma and at various stages of the discharge. Jensen, et al, (Ref. 4) have used the average ion model as an approximation to calculate the total radiated power as a function of electron temperature for 47 elements in the range of $2 \lesssim Z \lesssim 92$. Results of such calculations for some of the most common impurities are shown in Fig. 5.

The ratio of fusion power derived from beam-target reaction, P_{bf}, to the input beam power, P_b, is Q_b which is a function of beam energy and the target plasma temperature. The values of Q_b have been calculated by Jassby (Ref. 6) and the results are shown in Fig. 6.

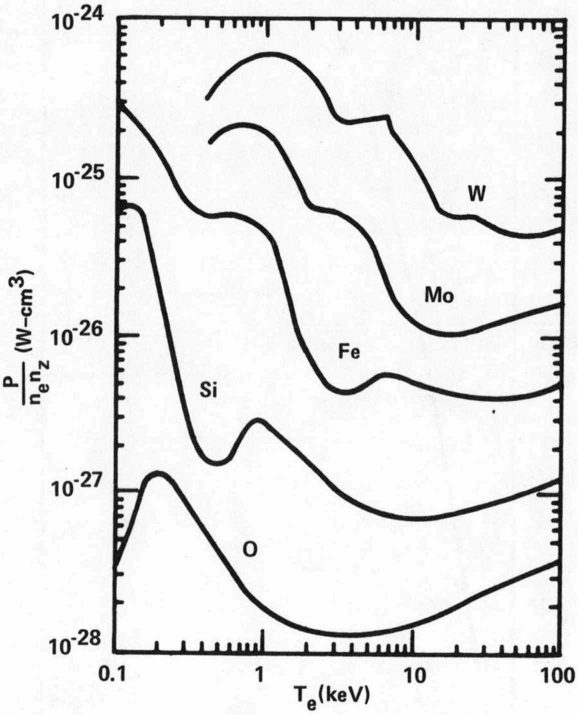

Fig. 5. Total Radiated Power as a Function of Electron Temperature for Some of the Common Impurities in Fusion Plasmas (From Ref. 4)

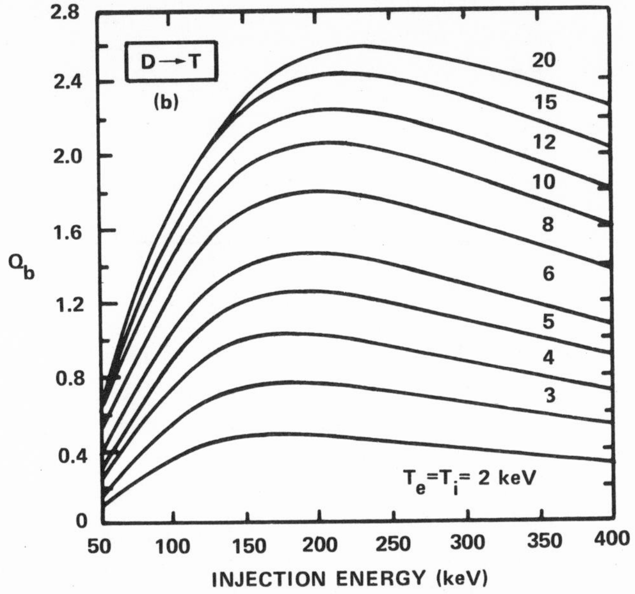

Fig. 6. Fusion Energy Multiplication Factor Q_b as a Function of Deuteron Injection Energy for Various T_e Values

(From Ref. 6)

Using these sets of information, Jensen, et al, have calculated the required values of $n\tau_E$ as a function of impurity fraction to achieve certain values of Q in a fusion reactor. These are shown in Fig. 7 for 200 keV deuterium beams injected into a D-T plasma with $T_e = T_i = 10$ keV. The value of F (tritium fraction) is taken as 0.5 for $Q \geq 3$, 0.65 for Q = 2, and 0.80 for Q = 1. These results indicated that the required $n\tau_E$ value increases dramatically if the impurity level is raised beyond a critical limit. Jensen, et al, have defined this critical impurity concentration, f_{Zc}, as that approximately twice the impurity fraction which would raise the $n\tau_E$ requirements by a factor of 2 from the requirements in an impurity-free reactor. The values of f_{Zc} are plotted as functions of Q for different impurities in Fig. 8 and as a function of impurity atomic numbers for igniton at different temperatures in Fig. 9.

The primary conclusions from these calculations are as follows:

- The permissible fraction of impurities f_{cz} decreases as Q and impurity atomic Z number increases.

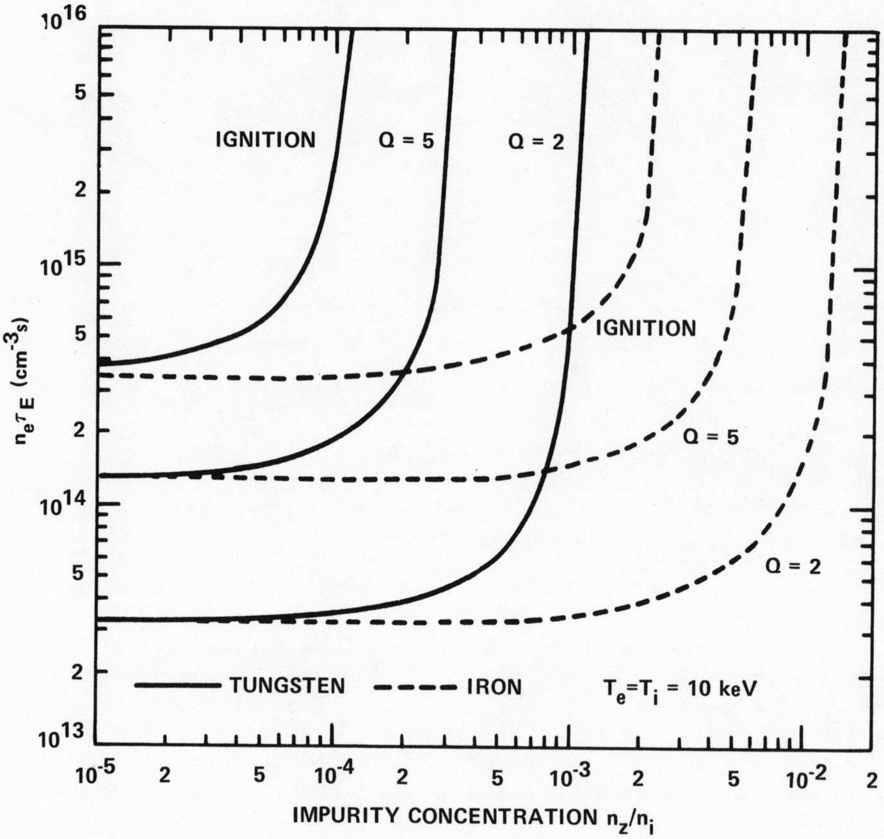

Fig. 7. $n_e\tau_E$ Requirements to Achieve Various Q-Values in D-T as a Function of Concentration of Tungsten or Iron Impurities (From Ref. 3)

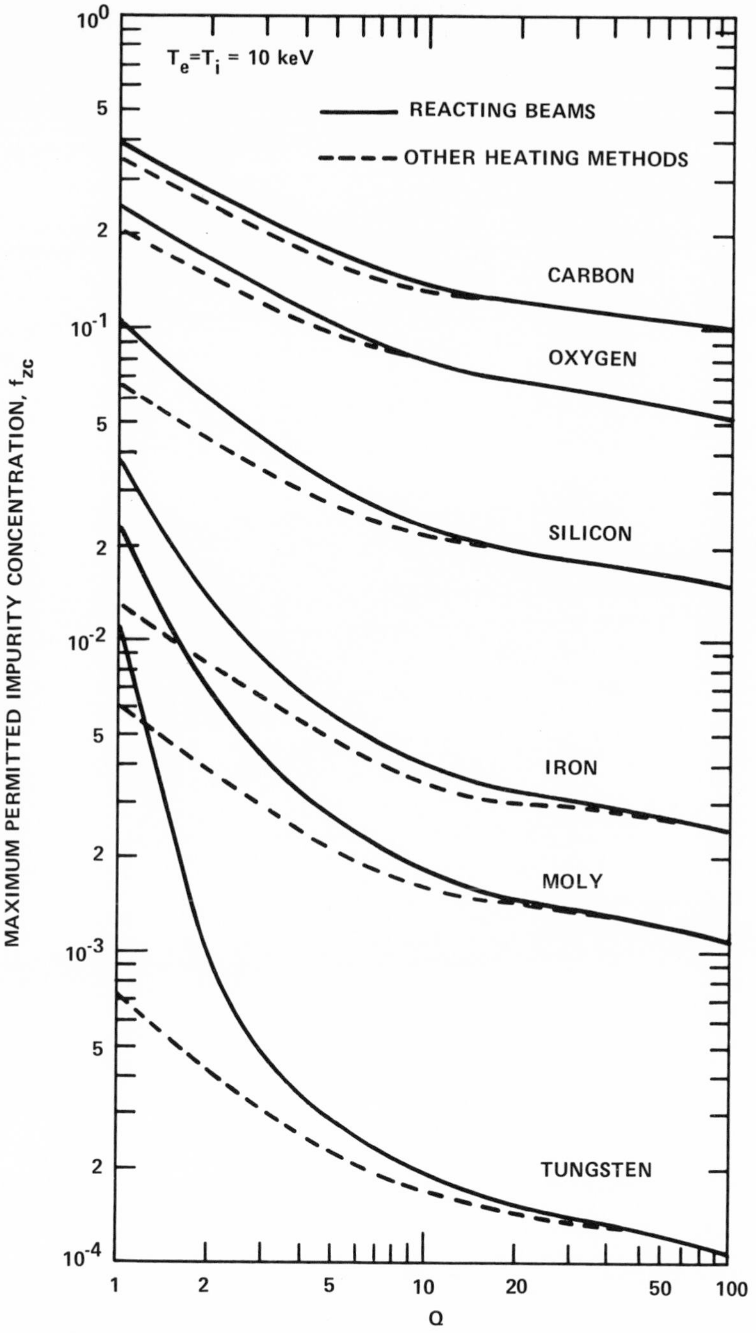

Fig. 8. Maximum Allowed Concentrations of Various Impurity Species for Attaining Given Q-Values in a 10-keV D-T Plasma. The Solid Lines are for Plasma Heating by 200-keV Deuterium Beams. (From Ref. 3)

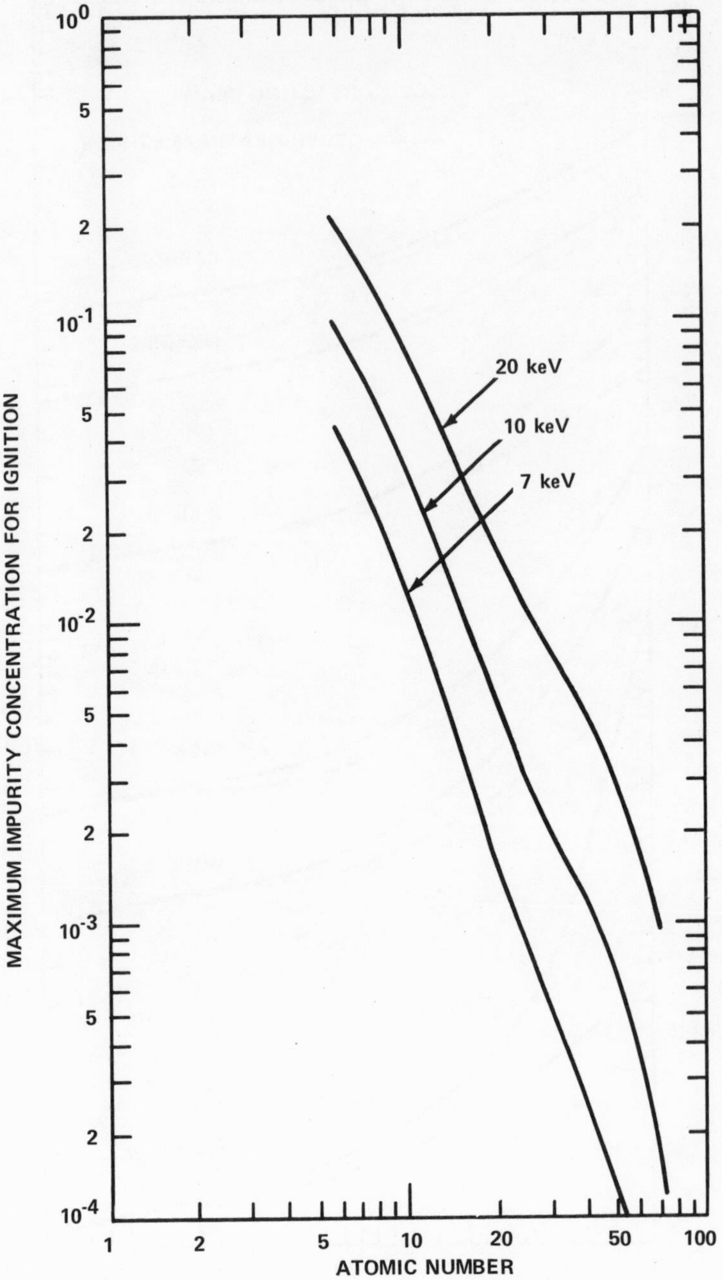

Fig. 9. Maximum Allowed Impurity Concentration Versus Z for Ignition at Various Plasma Temperatures ($T_e = T_i$), Assuming Zero Nonradiative Losses (From Ref. 3)

Large Q means large power amplification in the plasma core. To achieve higher Q, therefore, requires that non-convertible power loss from the plasma be reduced by lowering impurity fraction.

- The primary effect of low Z impurities is in fuel ion depletion while that of high Z impurites is in radiative losses. At lower Q values, however, high Z impurities also contribute significantly to fuel ion depletion.

- The permissible fraction of impurities increases with bulk plasma temperature because of increased fusion reactions.

In these calculations, fusion output from beam-plasma interaction has significant contribution to the energy balance in the core. This effect can be seen in Fig. 9 in which f_{cz} has been plotted for a system in which heating is other than neutral beams, such as rf heating. In this case, $P_{fb} = 0$ and the permissible impurity concentration is even lower.

In Table 1, estimated impurity levels are given for some present-day tokamaks (Ref. 3). These observed impurity fractions are too high for a ignited reactor. Therefore, methods have to be developed for impurity control in reactors in order to improve the economic viability of fusion reactors.

TABLE 1

ESTIMATED IMPURITY LEVELS IN SOME PRESENT-DAY TOKAMAK PLASMAS

	Oxygen n_z/n_i	Min. T_i (keV) for Ignition+	Other Species	n_z/n_i	Min. T_i (keV) for Ignition
ALCATOR*(a)	$2x10^{-3}$	4.6	Mo	$1x10^{-5}$	4.6
ORMAK(b)	$2x10^{-2}$	6.7	W	$1x10^{-3}$	20
PLT(c)	$1x10^{-2}$	5.6	W	$1x10^{-3}$	20
TFR(d)	$3x10^{-2}$	7.6	Mo	$1x10^{-3}$	9.7

*High-density regime.
+In reactor-sized plasma with indicated single-species impurity content.

(a) R. R. Parker, Trans. Am. Nucl. Soc. 25 (1977) in press.
(b) L. A. Berry, et al., in Plasma Physics and Controlled Nuclear Fusion Research (Proc. 6th Int. Conf., Berchtesgaden, 1976) 1, IAEA, Vienna (1977) 49.

(c) D. Grove, et al., ibid., 21; also E. Hinnov, private communication (1977).
(d) TFR Group, ibid., 35.

Impurity Sources

Let us first review the possible sources of impurities in fusion reactors. All non-hydrogenic elements are defined as impurities. A lage part of these originate from external sources such as walls, limiters, etc. and helium from the fusion reaction of deuterium and helium. Removal of helium from the plasma represent one of the difficult aspects of the general problem area defined as impurity control.

Impurities generated from external sources are usually discussed in two categories which are low Z and high Z impurities. Typical low Z impurities are carbon, oxygen, silicon, chlorine, aluminum, nitrogen; and high Z impurities are iron, titanium, copper, niobium, tantalum, tungsten, gold. Most of the low Z impurities are introduced into the plasma by gas feeds, pumps, surface oxidation, water vapor, alloy components, and chemical formations. The high Z impurities are charateristics of the material used in limiters and walls. Both categories of impurities are released into the plasma by plasma material interactions.

The limiter and wall surfaces are subject to radiation, heat conduction, electron, ion, neutral and neutron fluxes from the plasma. The type of impurities and their relative fraction released from the surfaces depend on the severity of the above mentioned fluxes from the plasma. The low Z impurities are loosely bound on the surfaces and they are easily released by minimal radiation, or low energy particle fluxes. At the end of a discharge these impurities settle back on the surface.

The principal mechanisms that introduce high Z impurities include sputtering by neutrals and ions, evaporation, arcing, melting, chemical erosion, and low energy run-away electron erosion. For example high energy ions from the plasma core charge-exchange at the plasma boundary where the neutral density is high, and continue to the wall as high velocity neutral particles. Sputtering by these neutrals are as important as sputtering by ions. Another source of high velocity neutrals is the neutral beam injection. A certain fraction of neutrals from injections will emerge from the plasma without being trapped and continue to the opposite wall at high velocities.

An important area of concern is the observed correlation between the increases in high Z impurity levels in the plasmas and the occurrence of disruptive instabilities. The plasma wall-interaction increases in such disruptions and high Z impurities are released from walls and limiter. There is also evidence that high levels of impurities increase disruptive activities. The evolution of disruptive instabilities and the processes that increase impurity levels to lead to such disruptions is complicated and requires both theoretical and experimental studies.

Impurity Transport

The next consideration is on the behavior of impurity migration after it has been released from surfaces as discussed above. This is referred to as impurity transport and it is a poorly understood topic. This is because impurities of different mass and at different ionization levels result in a large spectrum of collisionality regimes, gyro orbits, diffusion, and radiation, which all impact transport. These particles undergo evolution in their charge states through excitation, ionization, recombination and charge exchange as they move across the plasma with different n, T regimes. Rate coefficients for most of these species are not well known. Production rates of impurities by the multitudes of processes and spatial measures of population of these particles in different states are difficult. Even with these limitations, however, progress is being made both experimentally and theoretically, and much more work needs to be done.

Impurity Control

Impurity control is usually discussed in three contexts. These are (a) surface and plasma preparation methods to reduce impurities at their sources; (b) active impurity control by diverting impurities from plasma or from their source of origin to an external chamber; and (c) role of impurity control on the heat and ash removal from plasmas.

As it was mentioned in the introduction, the evolution of the surface preparation methods in fusion devices over the past twenty years have contributed significantly to the advances made in fusion research and development. In 1950's plasma chambers were usually made of dielectric material such as quartz or alumina, which are rich sources of low Z impurities. Temperature in these devices were limited to below about 100 eV because of line radiation by these impurities. The utilization of metal vacuum chambers has made it possible to reach much higher temperatures by elimination of low Z impurity radiation barrier. While high Z impurities are present in plasmas with metal vacuum chamber, the temperature reaches to keV range before these high Z impurities pose a serious radiation barrier.

There has been an evolution in the surface cleaning and preparation methods for metal vacuum chambers. Coating of surfaces with Gold (as it was done in ORMAK) baking, standard discharge cleaning, Taylor discharge cleaning, gettering are all examples of surface preparation techniques. The main purpose of these methods is to reduce the level of low Z impurities such as oxygen or carbons that are formed as a layer on the metal surfaces. Baking and discharge cleaning releases such surface layers of impurities by heat disorption and are pumped out of the system. In Taylor discharge cleaning, pumping out of oxygen is improved by formation of water vapor in the hydrogen plasma. Titanium has a good coefficient of sticking for hydrogen and oxygen. Therefore, inside of vacuum chamber is coated with titanium in the gettering method to reduce hyrogen neutral pressure and oxygen density to levels lower than that achieved by discharge cleaning techniques. The gettering is repeated every so often to maintain a fresh coat of titanium on the surfaces.

A second level of impurity control involes manipulation of gas density at the plasma edge. A high density neutral gas barrier between the plasma core and the wall acts as a shield to reduce the impact of plasma powers flux on the walls, and reduces impurity flux from the wall to the plasma core. Charge exchange and low Z impurity radiation are the two primary mechanisms by which the gas barrier distributes power flux from plasma uniformly over the wall surfaces. It prevents localized power flow from plasma that might result from field ripple, runaway electrons or MHD instabilities. Charge exchange is also the mechanism by which impurity transport from wall to plasma core is reduced. An incoming impurity neutral from the wall can charge exchange at the gas blanket and be tied to the field lines near the plasma edge.

While a gas blanket might appear attractive conceptually, there are various difficulties with this passive method of impurity control. At the present there is not much of an external control to manipulate the properties (density, thickness, etc.) of such a blanket in order to suit the impurity control requrements. It might be possible to consider high density operation of Alcator and a low temperature ($\sim$10 eV) toroidal arc discharge experiment at Jutphaas as examples of tokamaks with gas blanket impurity control. In these cases, however, the operation at high fill densities provides a natural gas blanket but it is difficult to extend these results to lower density standard tokamaks that operate in the $10^{14} cm^{-3}$ density regime.

Furthermore, it is unlikely that this method can be applied to longer pulse tokamaks. Maintenance of a gas blanket and propagation of edge cooling towards the plasma core in time are possible areas of difficulties. A gas blanket can become a barrier for neutral beam penetration also. The netural beam energy therefore might have to be raised in a reactor to accommodate beam penetration requirements. Finally, the vacuum chamber in a reactor has to have additional space to accommodate the blanket regime. Such a possibility is not desirable from an economic point of view.

Divertors

The consensus in the fusion community is that it will be necessary to have active impurity control in fusion reactors. It is already shown that presence of even a trace amount of high Z impurities can be detrimental to the operation of fusion reactors. A second factor is that economics of fusion reactors improve with increase in pulse length, and it is desirable to run these reactors in steady state. Impurity generation and build-up and helium ash accumulation pose a serious limitation to such long-pulse operation of the reactors unless an active impurity control method is developed.

In a closed field line configuration such as tokamaks, plasma-material contact occurs primarily at the limiters which are located a few centimeters off the vacuum chamber walls. The heat flux to the limiter is extremely high and its close location to the plasma makes cooling and pumping of neutralized particles impractical. In the present day tokamaks such as PLT, Alcator and others, damage to limiters resulting from melting, sputtering and runaway electrons have already been observed. In addition to the concern about the damage, the limiters also act as a large source of impurities in the plasma. Utilization of such limiters in reactors would, therefore, be even more of a concern. A magnetic divertor system is a feasible method to provide an active impurities control and heat removal system even though its presence in a reactor introduces significant engineering difficulties.

The concept of divertors is to move the plasma-material contact over from near-proximity of plasma column to a remote area where processes such as pumping and cooling can be used, and where the plasma-material contact area can be increased to reduce the power density on the material. The only feasible way to divert particles into a remote chamber is by designing proper magnetic field line configuration. There are toroidal concepts such as EBT and Stellarators which include open field lines at the periphery as an inherent design of such configuration. These concepts thus provide natural divertor fields for impurity control.

Tokamaks concept does not have such open field lines Therefore, auxiliary coils have to be used to open up the field lines. There are three principle designs for this purpose. These are toroidal divertor, poloidal divertor, and bundle divertor. The toroidal divertor has been tested successfully in the earlier stellarators, but it is not being considered seriously for tokamaks.

In both the poloidal and bundle divertor concepts a mull point is generated in either the poloidal or the toroidal fields respectively, with auxiliary field coils. This generates a separatix that separates the closed field lines from the open field lines. Particles on the closed field lines remain undiverted whereas particles on the open field lines are diverted into the divertor chamber. The region of the open field lines is referred to as the scrape-off region. There will be particle transfer from the interior of the plasma core to the scrape-off region thru cross-field diffusion across the separatix. The neutral particles on the periphery of the plasma will get ionized in the scrape-off region and get carried into the divertor. Thus, these open field lines serve as a barrier for impurity ions and other neutral particles from getting into the plasma.

The two divertor system differ considerable because of their action on different fields. In the poloidal divertor, the poloidal field generated by the plasma current is diverted into the burial chamber. The divertor coils have to be inside toroidal field coils and vacuum chamber. Some of the advantages of this design are that it provides a symmetric divertor action around the torus, and reduces divertor coil power requirements because poloidal field is weak in tokamaks. There are some disadvantages, however. Some of these are: operation and remote maintenance of internal coils and programmed divertor fields to manipulate null point location as the plasma current is ramped up. Internal coils, possibly made of super-conducting material, will require large space for neutron shielding. These coils will either have special joints or will require new design so that they can be remotely maintained in a reactor environment.

In the bundle divertor, some of the toroidal field lines on the periphery are diverted into a burial chamber at a discrete location at the toroidal circumference. The toroidal field is high, ~ 5 T, therefore large power is required for the bundle divertor coils. As a result, engineering aspects of this divertor system becomes difficult resulting from increased stress on the coils.

The bundle divertor is located at one or two points around the circumference. Therefore it results in assymetry in the magnetic configuration system. In addition, ripple field is generated in this design which might result in excessive loss of high energy particles.

The advantage of a bundle divertor system is that it is located outside the vacuum chamber, in between toroidal coils. There is easy access to it for pumping, maintenance, etc. It simplifies engineering aspects of divertors significantly.

The data base and understanding of the divertors is at a primitive stage at the present and a judgment cannot yet be made on the most suitable divertor for a reactor. Bundle divertors and axisymmetric divertors have been operated on DITE (UK) and DIVA (Japan) experiments respectively. In both of these systems reduction of impurities by divertor action have been observed. More significant results are required, however, to investigate a large number of physics and engineering issues with respect to divertors. Some of these are:

(1) How much of the plasma fuel ions and alpha particles are carried along into the divertor chamber with the impurities?

(2) What kind of instabilities are introduced by modification of closed confinement configuration and how do these affect plasma confinement properties?

(3) What is the efficiency of neutral particle and impurity ionization in the scrape-off region?

(4) What is the effect of the ionization process in the scrape-off region on the refueling of plasma in quasi steady-state plasmas?

(5) How much of the impurities that might be produced in the divertor chamber will get back into the plasma?

Large scale divertor experiments will be in operation within the next few years on PDX (PPPL), ASDEX (Garching), and ISX-B (ORNL). In addition design studies on ETF and INTOR are providing a focal point for the examination of the divertor issue. It is expected that with such increased activity on impurity control there will be significant advances made in this important area in the next few years.

REFERENCES

1. J. D. Lawson, Proc. Phys. Soc. London B70, 6 (1957).

2 D. W. Ignat, and R. A Blanken, Internal DOE Report, (1976).

3. R. W. Jensen, D. E. Post, and D. L. Jassby, Nucl Sci. Engr., 65, 282, (1978).

4. R. V. Jensen, D. E. Post, W. H. Grasberger, C. B. Tarter, and W. A. Lokke, Nucl. Fusion, 17, 1187 (1977).

5. R. V. Jensen, D. E. Post, C. B. Tarter, W. H. Grasberger, and W. A. Lokke, Atomic Data and Nuclear Tables, 20, 397 (1977).

6. D. L. Jassby, Nucl. Fusion, 17, 309 (1977).

WALL PROCESSES AND IMPURITY PRODUCTION DUE TO HYDROGEN SPECIES

K. J. DIETZ*,
I. ALI-KHAN†, F. WAELBROECK†, P. WIENHOLD†

*The JET Project, Abingdon, Oxfordshire OX14 3EA, England

†EUR/KFA, Kernforschungsanlage, 517 Jülich, Federal Republic of Germany

I INTRODUCTION

As the easily accessible energy resources of our planet decline and their drain is already within sight, the exploitation of new methods to generate energy at high temperatures is urgently needed. Fusion offers a great potential as such an energy source with nearly inexhaustible fuel resources.

The basic process for fusion is in the easiest scheme

$$d + t \rightarrow \alpha + 3.5\text{MeV} + n + 14.1\text{MeV} \qquad (1)$$

For net power output from fusion, the energy released in the plasma must be equal (break even) or larger than the power input needed to maintain the process. This is expressed by the Lawson criterion, which for 10 keV particles is given by[1,2]

$$(n_d + n_t)\ \tau = n\ \tau \geq 3{,}5 \times 10^{14}\ \text{cm}^{-3}\text{s} \qquad (2)$$

Here n_d and n_t is the particle density of deuterons and tritons respectively, and τ the energy confinement time, the ratio of the energy of the plasma to the power losses from the plasma.

The values of particle energy, density and confinement time necessary to produce energy from fusion involve the generation of hot plasmas. At the present time two different attempts are under study. These are the inertial[3] and the magnetic confinement[4]. The latter aims at low particle densities and confinement times in the order of seconds are required. The burn time of the plasma must be even longer. Particles and radiation emitted from the plasma release material from the wall surrounding the plasma. This material penetrates into the plasma and enhances then the power losses thus reducing the confinement time.

The dependence of nτ on the concentration of different impurities in the plasma is shown in figure 1 for mean energies of the plasma of 10 and 20 keV respectively.

During past years attention in plasma-wall interaction was mainly focussed on phenomena as sputtering[5,6,7,8], blistering[9,10,11] and backscattering[12,13], which depend primarily on the energy of the incident particles. The extensive data resulting from investigations of these processes cannot explain the temporal behaviour of the impurity influx into the plasma of a tokamak. Additional mechanisms of impurity generation must be looked after and be examined in detail experimentally.

The liberation of wall material due to thermal and current induced processes and due to energetic particles has been reviewed recently[14].

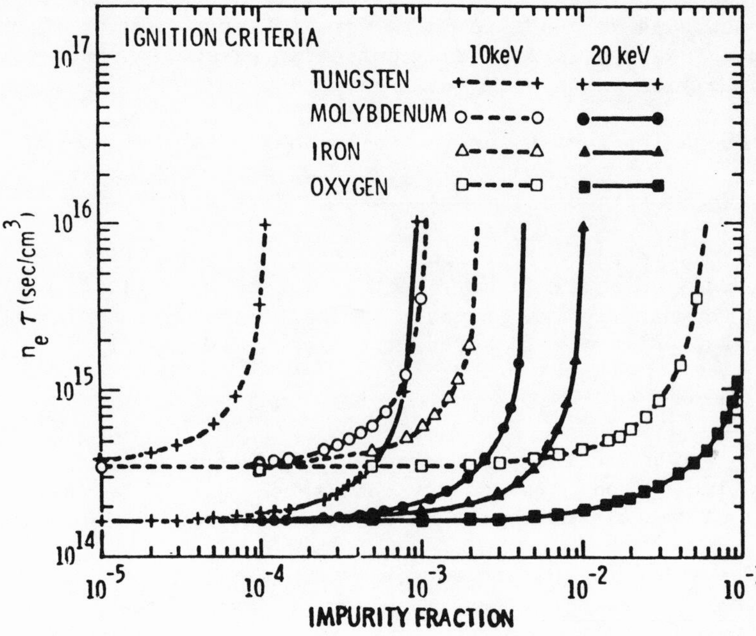

Figure 1: Plot[2] of nτ versus the impurity fraction for plasma energies of 10 and 20 keV

This paper deals with the impurity generation due to chemical and metallurgical effects caused by the large concentration of hydrogen in the walls of fusion devices.

The following processes in wall materials such as stainless steels and nickel base alloys depend on the hydrogen concentration in the bulk:

- the release of water, hydrocarbons and carbon oxides,
- the liberation of wall material due to surface embrittlement,
- recycling, isotope exchange processes, and tritium inventory.

II THE RECYCLING MODEL

The question in which form and at which energy hydrogen is released from the wall is important for the calculation of the energy distribution of the particles impinging on the wall, the concentration of the hydrogen retained in the wall, and the impurity release. During the current plateau in quasiequilibrium as many neutrals and ions as are lost from the plasma are substituted by particles released from the wall[15]. These can be hydrogen atoms that are reflected[12] or sputtered off[16] the wall, hydrogen molecules resulting from recombination due to the Eley Rideal[17] and the Langmuir-Hinshelwood[17] mechanism or, for wall temperatures relevant for fusion (i.e. $T > 500K$), hydrogen is released in the form of molecules due to one of the processes discussed by Ash and Barrer[18]. This has frequently been assumed for the computer calculations of the energy distribution[19,20,21] whereas for calculations of recycling processes also the ion induced desorption of atoms and particle reflection have been assumed to be the main release mechanism[22].

As it has been found in metallurgy the interaction of thermal hydrogen atoms with surfaces of stainless steel already increases the concentration of absorbed hydrogen in the bulk by orders of magnitude. The hydrogen atoms have been produced electrolytically in aqueous solutions[23], by dissociation of H_2 on hot tungsten filaments[24] at a pressure of 10^{-1} Pa, or in glow discharges[25,26].

To calculate the effects related to the hydrogen in the bulk the following assumptions have been made to establish the recycling model:

- Surface Cleanliness

Only "clean" surfaces are considered, as for confinement devices the surface contamination must be below 10^{-3} of a monolayer.

- Wall Temperature

The wall temperatures are so high, that for hydrogen the surface coverage θ, i.e. the ratio of sites occupied by hydrogen to the number of sites available, can be assumed to be small compared to one. For the larger confinement experiments, which are under construction now, the surface temperature during the discharge will indeed be high. In JET[27], temperatures in the range 500-800 K are expected during the discharges of 20s duration. For hydrogen θ is then small compared to one.

- Hydrogen Exchange between Surface and Bulk

The exchange of hydrogen atoms between the surface and the bulk is assumed to be a fast process and not rate determining for the hydrogen release. The concentration s of hydrogen at the surface is then proportional to the concentration c in the bulk as long as θ is small compared to one. This relationship is known as Henry's law.

$$s = \beta c \tag{3}$$

As observed experimentally[28] the surface concentration and the release rate of molecules are connected by

$$ds/dt = \nu n_o^2 \theta^2 \exp(-E/kT) \tag{4}$$

where $\nu_o = 8 \times 10^{-2} cm^2 atom^{-1} s^{-1}$ is the frequency factor for chemisorption, $n_o = 1.5 \times 10^{15} cm^{-2}$ the maximal number of sites available for chemisorption, $E = 1.25 \times 10^{-19}$ Ws the activation energy for the desorption of hydrogen, and T the wall temperature.

The surface coverage is finite which would not have been observed if the exchange surface bulk had been rate limiting. In this case θ would have been zero as the time required for the release of hydrogen would have been small compared to the time needed for the exchange of an atom between the surface and the bulk.

Investigations using the permeation method to study the hydrogen release mechanism show clearly that at least for the systems hydrogen-steel[29] and hydrogen-α-iron[30] either the diffusion is rate determining or the recombinative release of hydrogen molecules but never the exchange surface bulk.

- Solubilisation of Hydrogen Molecules

The rate of dissociative solubilisation of hydrogen molecules from the gas phase is assumed to be proportional to the ambient pressure p. The rate of solubilisation is given[30] by

$$(v_i)_2 = 2r\ k_s\ p \tag{5}$$

where r is the surface roughness, i.e. the ratio of the real surface (micro-structure) to the geometric wall area, k_s is the rate constant, and p the hydrogen pressure. It must be noted that equ. 5 is a rate equation and no equilibrium relation.

- Solubilisation of Hydrogen Atoms

The hydrogen atoms and atomic ions penetrate with a probability $\alpha(E)$, which depends on their energy, into the bulk. The rate $(v_i)_1$ at which the particles disappear in the lattice is given by

$$(v_i)_1 = \frac{1}{E}\int_0^E \alpha(E')\ dE' = \langle \alpha\phi \rangle \tag{6}$$

$\phi(E)$ is the flux density as a function of energy. As the exchange surface bulk is not rate limiting, $\alpha(E)$ for energies above 100 eV is given by the calculations of Oen and Robinson[13], above 500 eV by the measurements of Eckstein and Verbeck[12], and for energies below 1 eV by the measurements of Clausing and Emerson[31] and those of Waelbroeck et al[30].

- Implantation Depth of Hydrogen

For wall temperatures above 400 K the source of the incoming hydrogen is assumed to be localized at the surface. This is possible when the projected range of the impinging particles is small compared to the e-folding length of the proton distribution in the bulk resulting from particle diffusion.

The projected range[32] for H-atoms in iron, nickel or chromium is about 6×10^{-6} cm for energies of 10 keV. The diffusion constant D in austenitic stainless steels is[33]

$$D = 4.7 \times 10^{-3} \exp\ (-\ 8.9 \times 10^{-20}/kT)\ [cm^2 s^{-1}] \tag{7}$$

The time required for the hydrogen distribution to be broader than the projected range d is

$$t = d^2/D \tag{8}$$

For T = 400 K a value of $t = 7.7 \times 10^{-2}$ s results. This is short when comparing this time with the envisaged discharge duration of 20 s for JET. For T = 300 K a time of 16.6 s results, the above assumption is no longer strictly valid for discharges following extensive outgassing of the wall.

- Hydrogen Release

The release of hydrogen is a non linear process in θ. According to equ. 4

$$ds/dt = \nu\, n_o^2\, \theta^2 \exp(-E/kT) = k_r{}^* s^2 = v_r \tag{9}$$

where $k_r{}^*$ is the rate constant for surface recombination. Using equ. 3 the rate of recombinative release can be expressed in terms of the bulk concentration immediately underneath the surface.

$$v_r = 2r\, k_r\, c^2 \tag{10}$$

where k_r is the corresponding rate constant. The factor two is introduced for convenience.

The reason to relate the recombination to the concentration of hydrogen in the bulk is due to the easy access to this value.

- Ion Induced Desorption

The ion induced desorption is neglected with respect to the recombinative release for wall temperatures relevant to magnetic confinement devices (T > 400 K).

- Hydrogen Diffusion

The diffusion in the bulk is interstitial, the diffusion constant D depends neither on the concentration of hydrogen in the bulk, nor on the distance from the surface. Then, according to Fick's law, the rate v_D of the diffusion is

$$v_D = -D\, dc/dx \tag{11}$$

where x is the spatial coordinate.

Using the above assumptions the behaviour of the hydrogen distribution in the bulk during exposure of the wall to hydrogen atoms, the diffusion through the material, the hydrogen release, the retention (tritium) in the wall, and the isotope exchange can be calculated. The following equation holds

$$(v_i)_1 + (v_i)_2 - v_r - v_D = 0 \tag{12}$$

II.1 Equilibrium Behaviour

For a slab of material surrounded by hydrogen molecules and atoms in equilibrium dc/dx = 0 results and hence

$$<\alpha\phi> + 2\ r\ k_s\ p - 2\ r\ k_r\ c^2 = 0 \tag{13}$$

The equilibrium concentration c^*_{eq} of hydrogen dissolved in the material in the absence of atoms ($<\alpha\phi> \equiv 0$) can be calculated from equ. 13

$$c^*_{eq} = (k_s/k_r)^{\frac{1}{2}}\ p^{\frac{1}{2}} = K_s\ p^{\frac{1}{2}} \tag{14}$$

This relationship is also known as Sievert's law[34] with K_s being the solubility of hydrogen in the bulk.

In the presence of hydrogen atoms, the equilibrium concentration c_{eq} can be derived from equ. 13

$$c_{eq} = [(<\alpha\phi> + 2\ r\ k_r\ K_s^2\ p)/2\ r\ k_r]^{\frac{1}{2}} \tag{15}$$

Under tokamak conditions c_{eq} is always much larger than c^*_{eq}. This is illustrated by an example.

A stainless steel sample at T = 400 K is assumed to be introduced into molecular hydrogen at a pressure of 0.1 Pa and to be exposed to hydrogen ions with a rate of solubilisation of $<\alpha\phi> = 10^{15} cm^{-2} s^{-1}$ (for stainless steel α = 0.8 for 4 keV hydrogen ions[35]). The solubility K_s for austenitic steels is[33] (k in Ws/K)

$$K_s = 8.22 \times 10^{16} \exp(-1.39 \times 10^{-20}/kT) \text{ atoms cm}^{-3}\text{Pa}^{-\frac{1}{2}} \tag{16}$$

and the recombination rate constant[29]

$$k_r = 8.4 \times 10^{-18} \exp(-9.45 \times 10^{-20}/kT) \text{ molecules cm}^4\text{atom}^{-2}\text{s}^{-1} \tag{17}$$

For these values the ratio of the equilibrium concentrations is for r = 5

$$c_{eq}/c^*_{eq} = (<\alpha\phi>/2r\ k_r + K^2_s\ p)^{\frac{1}{2}}\ (K_s^2\ p)^{-\frac{1}{2}} = 2.7 \times 10^4 \tag{18}$$

The equilibrium concentration in the presence of the atoms amounts to $c_{eq} = 1.8 \times 10^{19}$ atoms cm^{-3}, this compares to 6.6×10^{14} atoms cm^{-3} if only the molecules are considered. To achieve this high concentration without atoms present, a pressure of 7.38×10^6 Pa is required. Such high hydrogen concentrations in the bulk lead to corrosion[35] and embrittlement processes[24,36,37,38].

II.2 Stationary Behaviour

When the material is exposed only at one side to mixtures of hydrogen molecules and atoms or atomic ions, the diffusion must be considered.

Previous evaluations[39] on the hydrogen distribution and the inventory in stainless steels assume that the concentration at the surface is zero, i.e. that the diffusion is rate determining. Richardson's formula is valid. That can be derived from equ. 11 for the permeation of hydrogen through a wall of thickness x_o, with the equilibrium concentration c^*_{eq} (equ. 14) on the high and $c = 0$ on the low pressure side, hence

$$v_D = D\ K_s\ x_o^{-1}\ p^{\frac{1}{2}} \qquad (19)$$

Deviations from this formula have been observed[40,41,42] especially for low hydrogen concentrations or for thin walls. In these cases v_D has been found to be proportional to p which has been attributed to surface contamination[40] but has also been observed for clean surfaces[30,43]. The transition from diffusion limited permeation (validity of Richardson's formula) to recombination limited permeation ($v_D \sim p$) has been shown theoretically and experimentally. A characteristic number W', which is the ratio of the recombination flux density (equ. 10) to the diffusion flux density (equ. 19) for $c = c^*_{eq}$, describes the transition regime.

$$W' = \frac{2r\ k_r\ (c^*_{eq})^2\ x_o}{DK_s\ \ c^*_{eq}} = K\ \frac{K_s}{D}\ x_o\ \ p^{\frac{1}{2}} \qquad (20)$$

with $K = 2r\ k_r$. For $W' >> 1$ Richardson's formula is valid, for $W' >> 1$ the permeation flux density is expressed by[43]

$$V_D = r\ k_r\ K_s^{\ 2}\ p \qquad (21)$$

Now in contrast to equ. 19 the permeation flux density is proportional to the pressure. Moreover, it depends on surface properties; for the diffusion limited process the permeation is described by bulk characteristics. Equs. 19 and 21 are derived for a molecular environment. They can be extended to a surrounding of atoms, atomic ions, and molecules by replacing p through an effective pressure p^*. That can be calculated by applying Sievert's law to the equilibrium concentration c_{eq} of equation 15:

$$p^* = p + \frac{\langle \alpha\phi \rangle}{K\ K_s^{\ 2}} \qquad (22)$$

The physical meaning of p^* is that such a pressure would yield the same concentration c in the bulk as the mixture of atoms and molecules. With equ. 22, the equs. 19 or 21, depending on W', can now be used to calculate the hydrogen permeation through the wall exposed to a hydrogen plasma.

II.3 Time Dependent Behaviour

As the tokamak discharges, especially at low wall temperatures, never last long enough to attain the stationary state; equ. 12 must be solved time dependent with adequate boundary conditions. A second order desorption process is involved thus isotope exchange processes in the wall and the formation of mixed molecules (e.g. DT molecules) can be calculated simultaneously.

Within the wall of thickness x_0 uncoupled diffusion of two hydrogen isotopes i = a, b is assumed. It is described by the one dimensional Fick's law.

$$\delta c_i (x,t)/\delta t = D_i \delta^2 c_i (x,t)/\delta x^2 \qquad (23)$$

for the concentrations c_i as a function of distance x from the surface and time t. D_i are the diffusion constants.

The irradiation by one or two hydrogen isotopes with flux densities ϕ_i (t) starts at t = 0 at x = 0. The fraction

$$\langle \alpha \phi_i (t) \rangle \equiv A_i (t) \qquad (24)$$

which penetrates through the surface is assumed to split into the diffusion flux density into the material, into the recombination flux density of equal molecules, and into the flux density of mixed molecules. For the isotope i:

$$K_i = 2r k_r \qquad (25)$$

At the other surface ($x = x_0$) the incoming diffusion flux is balanced by the outgoing flux of molecules.

Results are shown in figures 2 and 3.

Figure 2 illustrates the development of the hydrogen concentration during an irradiaition at constant flux density. The irradiation starts at t = 0. c_{eq} is obtained from equ. 15. The calculation is made for W' = 10. Reduced values for the coordinates are used, the parameter t/τ is a reduced time related to the rate A of solubilisation; $\tau = D/AK$. The concentration in the wall builds up slowly until stationary values are reached.

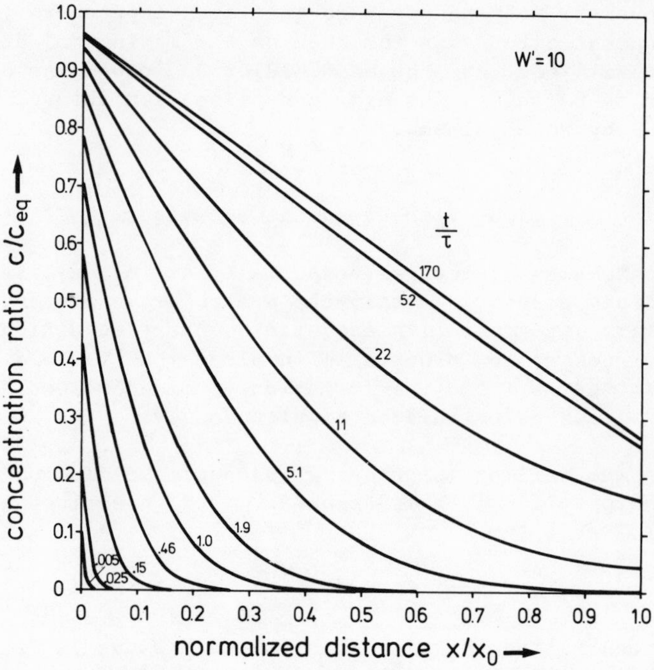

Figure 2: Evolution of the hydrogen distribution during irradiation: c_{eq}, τ, x_0, W' see text.

Due to the particle loss by diffusion the equilibrium value for the concentration cannot be attained.

Figure 3 shows the decrease of the hydrogen concentration in the wall after the stationary state has been reached and the irradiation has been stopped. The values for c(0) and $c(x_0)$ decrease with time and approach zero asymptotically. This is due to the time-dependent boundary condition.

From the known hydrogen distribution the quantities which characterize recycling and inventory can be calculated as a function of time: the flux density of the molecules $\phi_{rec}(t)$ and the average concentration of atoms in the wall

$$\overline{c}(t) = \frac{1}{x_0} \int_0^{x_0} c(x,t)\, dx \qquad (26)$$

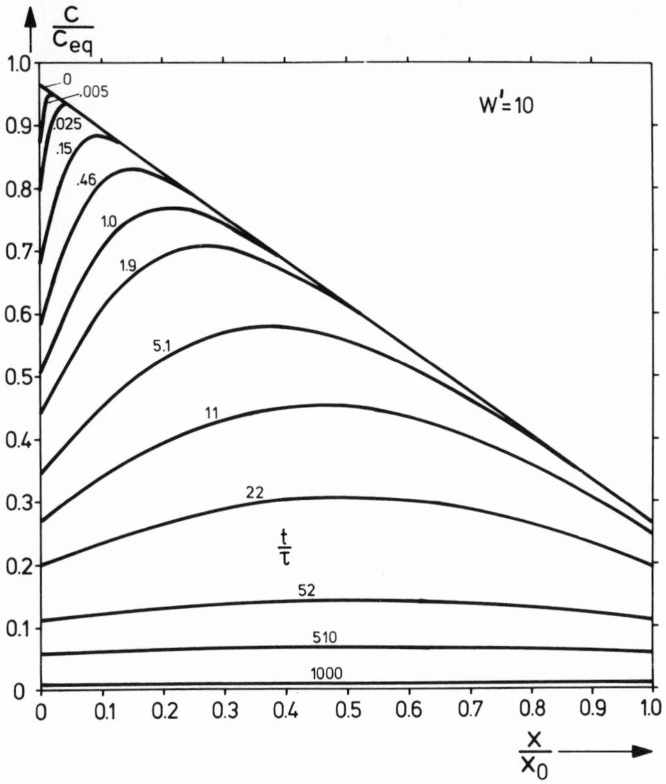

Figure 3: Evolution of hydrogen distribution after irradiation, c_{eq}, τ and W' see text.

III IMPURITY PRODUCTION PROCESSES RELATED TO THE HYDROGEN IN THE BULK

The knowledge of the hydrogen concentration in the bulk allows the evaluation of:

- the production rates of hydrogen compounds from surface contaminants[44,45],
- the mechanism of surface embrittlement and cracking[37,44],
- a possible triggering process for unipolar arcs[46],
- the recycling of hydrogen and its isotopes[47],
- the isotope exchange when discharges in one hydrogen isotope are followed by discharges in another[47,48],

- the hydrogen outgassing rates for confinement devices with hot walls between discharges[47,49,50],

- the tritium retention in the wall for devices operated with DT[47],

- the tritium permeation through the walls exposed to the plasma[30]

The first three processes listed here immediately contribute to the plasma contamination and will be discussed in detail.

III.1 The Oxygen Contamination of the Plasma

The walls of today's confinement devices are mostly fabricated of stainless steels or nickel base alloys. The surface of these materials is often strongly oxidized[51,52,53] and partially covered with carbon. The thickness of the oxide layer depends on the pretreatment but is usually of the order of 50-100 Å. This is illustrated in figure 4. The main constituents of stainless steel AISI type 304 are shown as a function of the dose during sputter etching. A dose of 5×10^{18} argon ions corresponds to a depth of about 0.5 μm.

The concentrations at this depth are compatible to those obtained from a chemical analysis of the bulk (at %): 67% Fe, 18% Cr, 9% Ni, 0.1% C, and oxygen below the lower limit of detection. At the surface oxygen (4.5%) and carbon (60%) are enriched. Other pretreatments, e.g. etching in an acidic bath can yield surface oxidation up to 50%.

Hydrogen atoms and atomic ions escaping from the confined plasma interact with the wall and form volatile hydrides with oxygen and carbon.

A model proposed by Dietz and Waelbroeck[46], which has been confirmed by experiments, yields for the rate v_{18} of water production

$$v_{18} = k_{18}\, c_{mo}\, c^2 \tag{27}$$

where k_{18} is the reaction rate constant, c_{mo} the concentration of metal oxide and c the concentration of hydrogen immediately underneath the surface.

As the hydrogen concentration decreases slowly between the discharges, as shown in figure 3, the water production continues during this time. For wall temperatures below 400 K most of the

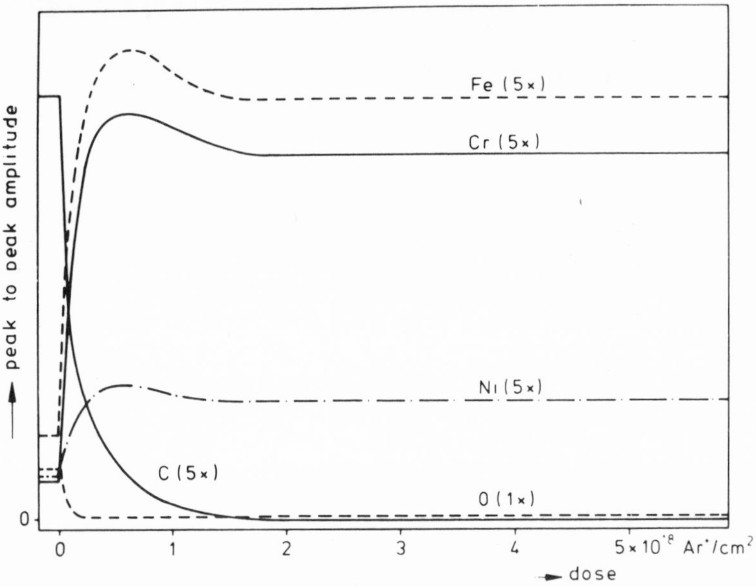

Figure 4: Depth profile of SS 304, lapped sample

water is not released from the wall but remains adsorbed ready for desorption during the start-up of the subsequent discharge. The desorption process can be thermal[54,55], radiation[56] - or particle[57] - induced.

The water release rate depends strongly on the temperature of the wall. This is illustrated in figure 5. The logarithm of the ratio of water molecules released to incident hydrogen atoms is plotted against the inverse wall temperature. As no exponential dependence results either the release is not activated or a second mechanism must be taken into account. That is the additional release of hydrogen in the form of water which must be taken into account in the particle balance (equ. 12). Then an Arrhenius plot results[45]. An activation energy of 1.4×10^{-19} Ws can be derived which agrees well with that for the adsorption energy for water on stainless steel of 1.56×10^{-19} Ws as found in the literature. That is strong evidence that the thermal desorption of water might be related to the reaction of hydrogen dissolved in the material with surface oxides.

From the experimental results gained from tokamaks[50,58,59,60,61,62], from simulation experiments with particles of high[62] or low[45] energies, from the interaction of glow discharges in hydrogen with walls of low-Z materials[63], and from permeation experiments, the following conclusions must be drawn:

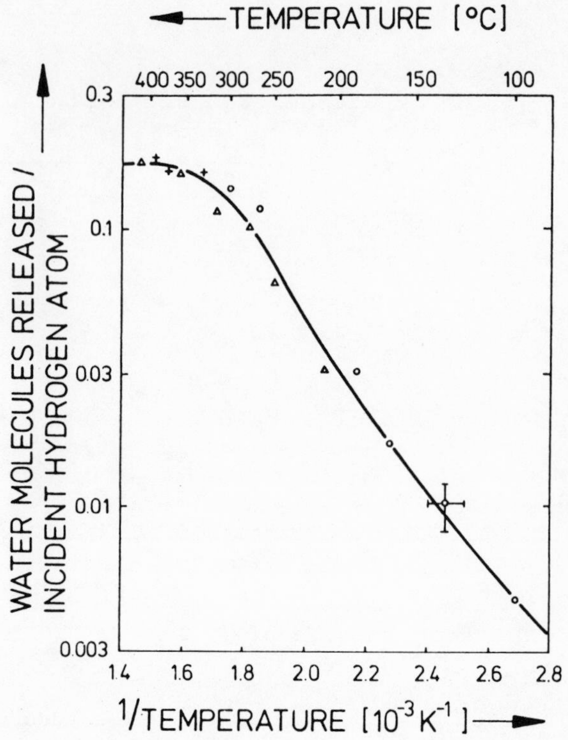

Figure 5: Released water molecules per incident hydrogen atom

- The walls of confinement devices are chemically attacked by hydrogen atoms.

- The rate of the formation of hydrogen compounds cannot be calculated by thermodynamics as no equilibrium processes are involved.

- The release of volatile hydrogen compounds from metallic surfaces is for wall temperatures above 400 K proportional to the hydrogen concentration underneath and independent of the energy of the incident protons.

- The chemical reactions do not occur during the slowing down time of energetic particles as the interaction time is too short.

- Materials of which constituents can form volatile hydrogen compounds cannot be tolerated as wall materials in tokamaks.

III.2 Surface Embrittlement and Cracking

In addition to surface processes due to the interaction of hydrogen ions and charge exchanged neutrals volume processes have recently been observed[36]. Deuterium blisters have been formed after exposing the nickel-base alloy PE-16 to 600eV ions produced in a shock tube. Fig. 6 shows the result of such interactions. The cracked and ruptured grain boundaries can be observed and the formation of hundreds of small blister-like marks can be seen.

The blisters are attributed to hydrogen concentrations of 5-10 ppm in the bulk and to the known embrittlement processes at such concentrations of hydrogen in stainless steel or nickel-base alloys[65]. Measurements in tokamaks in order to get data on the metallurgical behaviour of the wall have never been performed. The observations of Panayotou et.al. meet some difficulties in the interpretation and may not be directly transferable to magnetic confinement devices.

An attempt to investigate possible embrittlement processes in tokamaks has been made by using 0.2eV hydrogen atoms for the irradiation in order to investigate only those processes which are related to the atomic nature of the impinging hydrogen atoms and not to their momentum[29,37,38,46].

If it were possible to extrapolate equ. 15 directly down to ambient temperatures the hydrogen concentration in the walls of today's tokamaks could be calculated and according to equ. 22 the resulting internal pressure. At a wall temperature of 400 K and a rate of penetration of $10^{16} cm^{-2} s^{-1}$ into stainless steel an internal pressure of 7.4×10^7 Pa builds up in lattice defects. When the resulting forces overcome the strength of the material bubbles near the surface may burst open, releasing the contained gas and liberating wall material.

The explosive release of gas has been observed after treating a stainless steel wall at temperatures below 400 K with flux densities of 10^{14} atoms $s^{-1} cm^{-2}$ of 0.2eV hydrogen atoms. This phenomenon has been attributed to surface embrittlement and

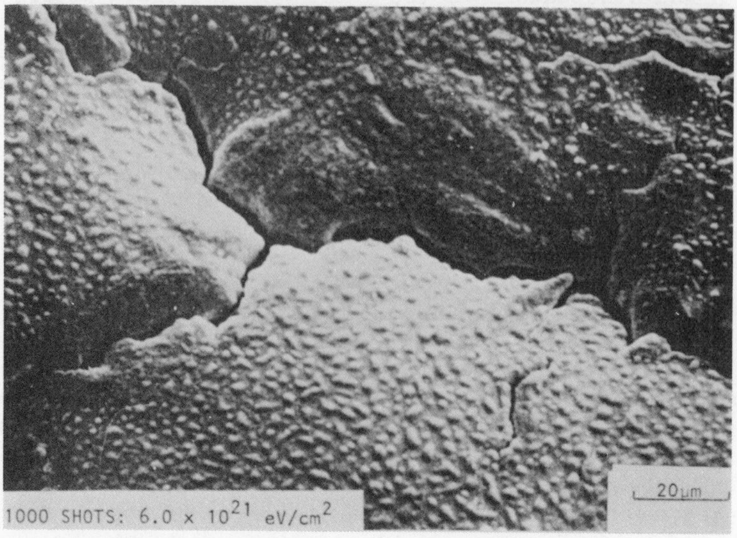

Figure 6: Scanning electron micrograph of PE-16 after exposure to 1000 discharges[36]

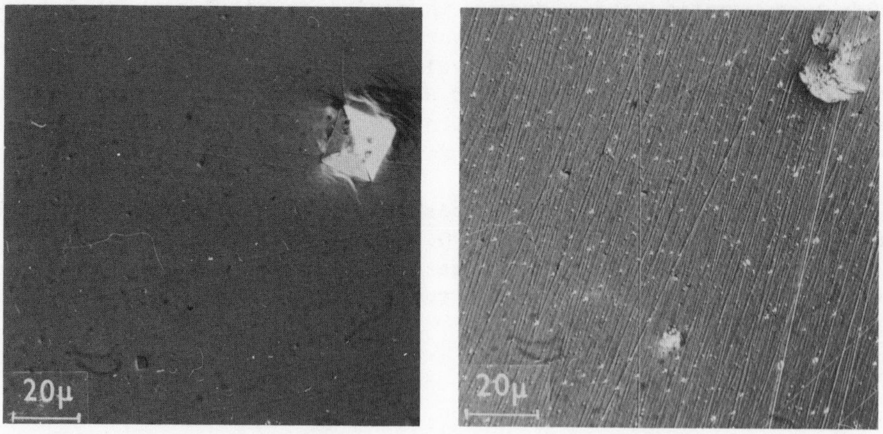

before after

Figure 7: Interference phase contrast micrographs of stainless steel (AISI type 321) before and after irradiation with 6.5 x 10^{20} hydrogen atoms cm^{-2}, T = 320 K. The marks result from measurements of the microhardness.

investigated more closely[29,37,38]. A typical result is presented in figure 7. The surface of a mechanically polished stainless steel sample (AISI type 321) is shown before and after the irradiation with a dose of 6.5 x 10^{20} hydrogen atoms cm^{-2}. Pits or bubbles are formed during the exposure. Their number amounts to $\simeq 10^7 cm^{-2}$. Some spots might be spinels or other oxidation products which, due to the small reaction velocity at 320 K, the temperature at which the irradiation has been performed, could not be reduced during the treatment. For electropolished samples no surface change could be observed.

The marks from measuring the surface hardness can be seen in fig. 7. The microhardness increased by 10-15% which is expected in the case of surface embrittlement. According to equ. 22, the effective pressure should decrease in materials with a high recombination rate constant. Palladium for example has a high solubility for hydrogen and accelerates, in comparison to stainless steels, the hydrogen release[65]. Palladium coatings should thus reduce the hydrogen concentration in the underlying steel and surface embrittlement should not occur. This has been demonstrated experimentally[37].

To avoid surface embrittlement in confinement devices with walls made of stainless steels or nickel-base alloys the wall temperature should be kept well above 400 K. Then burst release of water or methane might occur which must be suppressed by discharge cleaning procedures with hot walls.

Materials with a high solubility and recombination rate for hydrogen should be envisaged for the first wall. Metals or isolators in which these are even smaller than for stainless steels cannot be tolerated in magnetic confinement devices neither for the first wall nor for the limiter. Should it be necessary to use such materials for structural components suitable coatings must be applied to avoid surface embrittlement.

III.3 Unipolar Arcs

Localised melting of wall material can be caused by electrical interactions between the plasma and the wall. The thermal velocity of the electrons is larger than that of the ions provided that both have identical temperatures. This is correct, within a factor of two, for the averaged values in tokamaks. The electron fluxes to the wall are higher than those of the ions until a potential, the sheath potential, has built up which decelerates the electrons and maintains quasineutrality of the plasma. The plasma is charged positively with respect to the wall. The value U of the sheath potential is for a hydrogen plasma approximately given by[66]

$$U = 3\ kT_e/e \tag{28}$$

T_e is the electron temperature and e the elementary charge. In equ. 31 secondary ion emission, which reduces the sheath potential[67], has not been taken into account. For a thermonuclear plasma the sheath potential might reach tens of volts depending on the edge temperature of the electrons and the secondary electron emission coefficient.

The particle density in the region near the wall is very small[68,69]; gas breakdown cannot be expected. Unipolar arcs, however, have often been observed, thus a trigger mechanism exists. This could be either a hot spot at the wall which emits electrons, field emitted electrons from surface imperfections or also localized high density hydrogen pulses emitted from exploding gas pockets.

It has been observed in the JT-2 tokamak[70] that arcing causing visible arc tracks could always be related to MHD unstable discharges, moreover the arcing diappears with the continued operation of the machine. Similar observations have been made in ISX[71] as well as in simulation experiments for arcing[72]. In the tokamak experiments the production of high-Z impurities has been observed to occur at the beginning and the end of the discharge and has been attributed to arcing processes. Similar results have been obtained in DITE[73], the time behaviour of the impurity generation is not as pronounced as in ISX. Arcing in the simulation experiment[72] has been related to surface contamination and has disappeared for clean walls.

Voltages of the order of 10 volts are required to initiate arcing in the presence of a hot electron emitting spot at the wall. According to equ. 31 electron energies of some tens of eV are already sufficient. The hot spot[74] can result either from joule heating of field emitting points or from a high local wall load due to runaway electrons, from a poorly centred plasma, from trapped ions diffusing in magnetic mirrors or from loss of confinement at the end of a discharge. This is consistent with the observation that arcing occurs mainly at the start and end of a discharge, but cannot explain the disappearance with tokamak operation time or with the enhanced cleanliness of the walls.

The model of the release of gas pockets and the subsequent gas breakdown can explain most of the experimental observations. The disappearance of arcing with operation time can be attributed to the increasing wall roughness and the related decrease of the hydrogen pressure in the wall. The start-up and the end of discharges increase the surface temperature and are connected to mechanical shocks, thus marginally stable bubbles may explode at these times. The correlation of arcing to the surface cleanliness might be attributed to the observation that slightly oxidized or carbonized walls pump hydrogen more effectively[72] thus increasing the hydrogen concentration and the internal pressure. In the model, however, it is difficult to explain why in stable, well centred discharges arcing is not observed.

The mechanism for the triggering of unipolar arcs is not yet understood. From the experimental evidence gained in tokamaks it is obvious that to minimize the arcing the following measures should be taken:

- the vessel must be a bakeable UHV system,
- surface contaminations have to be removed by discharge cleaning,
- the wall material should have a high solubility for hydrogen,
- the plasma must be well centred with respect to the wall (limiter),
- the plasma edge must be kept cool, for example by radiation cooling due to Lyman-α,
- the sheath potential might be short circuited by using electron emitting devices at the wall[75].

IV CONCLUSIONS

In this paper aspects of the plasma wall interaction, which have only recently attracted adequate attention, have been discussed. The processes examined in detail have already been known for a long time; they deal with the atomic nature and the high flux densities of the hydrogen impinging on the wall.

It is acknowledged now that the impurity production in today's tokamaks mainly originates from the following sources:

- evaporation of limiter and wall material;
- chemical reactions which provide adsorbed hydrogen compounds, ready to be desorbed by particle- or radiation-induced processes;

- arcing phenomena which involve hot spots at the wall from where high-Z material is evaporated;

- sputtering processes which lead to ion induced desorption of low-Z hydrides and to the emission of atoms of the wall material;

- surface embrittlement releasing high-Z materials at low wall temperatures.

The contribution of the single processes to the plasma contamination is not yet established. It is most likely that all these phenomena contribute to the impurity production.

The tokamaks, operating today, produce plasmas which can be very clean. Z_{eff}-values near one have often been observed. This has been achieved either by discharge cleaning procedures with low plasma energies and hot walls, or, in machines which are not bakeable, by burying the impurities underneath evaporated layers of titanium.

The impurity confinement times in tokamaks are not known, it must be assumed that the burn times of the plasmas are not long enough compared to the impurity confinement time to allow for impurity recycling and for the build-up of a stationary impurity distribution. This will occur in the next generation of tokamaks and will certainly pose new difficulties in achieving clean plasmas. The methods which will be applied in the future are:

- Use of bakeable UHV-vessels with negligible leak rates,

- Discharge cleaning with low energy (T_e < 5eV) plasmas at elevated wall temperatures ($T \geq 500$ K),

- Control of the wall temperature during discharges,

- Generation of well centred discharges with negligible limiter contact,

- Modelling of the plasma edge by cooling (H_2-injection) or heating (neutral injection) the wall near layer; control of the sheath potential,

- Coating of walls and limiter with materials having the surface properties desired,

- Use of low-Z materials for limiters and walls.

It can be expected that these measures will decrease the

impurity level so far, that ignition of the plasma in the large tokamaks coming into operation during the early eighties will be achieved.

From this time on the impurity problems are much more complex, as the effects of neutrons and α-particles contribute to the material problems, moreover in the next generation of tokamaks the plasma burn time will be extended to minutes and impurity accumulation processes will become dominant.

REFERENCES

1. D.J. Rose and M. Clark, Jr., Plasmas and Controlled Fusion, (Wiley and Sons, New York 1961)

2. R.V. Jensen, D.E. Post, W.H. Grasberger, C.B. Tarter, and W.A. Lokke, PPPL-1334 (1977)

3. J.L. Emmett, 7th Internat. Conf. Plasma Physics and Controlled Fusion Research, IAEA, Innsbruck 1978, Paper B1

4. L. Artsimovich, Configurations des plasmas fermées, (Presses Universitaires de France 1968)

5. B. Emmoth, Ionic Impacts on Solids and Sputtering Phenomena, (Stockholm 1979, ISBN 91-7146-047-0)

6. R. Behrisch, W. Heiland, W. Poschenrieder, P. Staib, and H. Verbeek (eds.) Ion Surface Interaction, Sputtering and Related Phenomena (Gordon and Breach, London 1973)

7. B. Navinšek, Progr. Surface Science 7,49,1975

8. H.H. Andersen, Proc. 7th Yugoslav Symp. on Physics of Ionized Gases, 1974, Rovinj, Yugoslavia.

9. J. Biersack, J.Nucl. Mat. 63,253,1976

10. M.I. Guseva, V. Gusev, U.L. Krasulin, Yu.V.Martynenko, S.K. Das, and M. Kaminsky, J.Nucl. Mat. 63,245,1976

11. R. Behrisch, M. Risch, J. Roth, and B.M.U. Scherzer, Proc. 9th Symp. Fusion Techn. EUR 5602,531,1976

12. W. Eckstein and H. Verbeck, J.Nucl. Mat. 76 & 77,365,1978

13. O.S. Oen and M.T. Robinson, J.Nucl. Mat. 76 & 77,370,1978

14. G.M. McCracken and P.E. Stott, Culham report CLM-P-573

15. E.S. Marmar, J.Nucl.Mat. 76 & 77,59,1978

16. H.F. Winters and P. Sigmund, J.Appl.Phys. 45,4760,1974

17. H. Wise and B.J. Wood, React.Atom.Molec.Phys. 3,291,1967

18. R. Ash and R.M. Barrer, Phil.Mag. 4,1198,1959

19. J. Hackmann, Experimentelle und numerische Untersuchungen zur Wechselwirkung von Plasmen mit festen Wänden (Universität Düsseldorf 1977)

20. C.R. Parsons and S.S. Medley, Plasma Phys. 16,267,1974

21. J.T. Hogan and J.F. Clarke, J.Nucl.Mat. 53,1,1974

22. S.J. Fielding, G.M. McCracken, and P.E. Stott, J.Nucl.Mat. 76 & 77,273,1978

23. H. Coriou, L. Grall, A. Besnard and G. Pinard-Legry, Proc. Congrès Int.L'Hydrogène dans les Métaux, Paris 1972, p.241

24. H.G. Nelson, D.P. Williams, and A.S. Tetelman, Metall.Trans. 2,953,1971

25. W.J. Kass, Effect of Hydrogen on Behaviour of Metals, A.W. Thompson and I.M. Bernstein (eds.), Metallurgical Society of AIME 1976, p.327

26. B. Brandt, Proc.Int.Symp. Plasma Wall Interaction, EUR 5782e, 375,1977

27. The JET Project, Scientific and Technical Development, EUR 5791e,1977

28. K. Christman, O. Schober, G. Ertl, and M. Neumann, J.Chem.Phys. 60,4528,1974

29. I. Ali-Khan, K.J. Dietz, F. Waelbroeck, and P. Wienhold, J.Nucl.Mat. 74,132,1978 and 74,138,1978

30. F. Waelbroeck, I. Ali-Khan, K.J. Dietz and P. Wienhold, First Top. Meeting on Fusion Reactor Materials, Miami Beach, USA, 1979, paper D-14

31. R.E. Clausing, L.C. Emerson, and L. Heatherly, J.Nucl.Mat. 76 & 77,267,1978

32. H.H. Andersen and J.F. Ziegler, Hydrogen Stopping Power and Ranges in All Elements (Pergamon 1977)

33. M.R. Louthan and R.G. Derrick, Corrosion Science 15,556,1975

34. A. Sieverts, Z. Elektrochemie, 16,707,1910

35. A. Rahmel and W. Schwenk, Korrosion und Korrosionsschutz von Stählen (Verlag Chemie 1977)

36. N.F. Panayotou, J.K. Tien, and R.A. Gross, J.Nucl.Mat. 63, 137,1976

37. I. Ali-Khan, K.J. Dietz, F. Waelbroeck, and P. Wienhold, First Top. Meeting Fusion Reactor Materials, Miami Beach, USA, 1979 paper 0-4

38. I. Ali-Khan, K.J. Dietz, F.G. Waelbroeck, and P. Wienhold, J.Nucl.Mat. 76 & 77,263,1978

39. R. Broudeur, J.P. Fidelle, P. Tison, G. Roux and M. Rapin, Influence de l'hydrogène sur le comportement des métaux IV, Rapport CEA-R-4701 (Juin 1976)

40. H.D. Röhrig, R. Hecker, J. Blumensaat, and J. Schäfer, Nucl. Eng. and Des. 34,157,1975

41. C. Smithells, Proc.Roy.Soc. London A150,172,1935

42. H.L. Eschbach, F. Gross, and S. Schulien, Vacuum 13,543,1963

43. I. Ali-Khan, K.J. Dietz, F. Waelbroeck, and P. Wienhold, J.Nucl.Mat. 76 & 77,337,1978

44. K.J. Dietz, F. Waelbroeck, and P. Wienhold, Jül-1448,1977

45. K.J. Dietz, I. Ali-Khan, F. Waelbroeck, and P. Wienhold, IUPAC Communications, 4ème Symp.Int.Chimie des Plasmas, Zürich 1979, (in publication)

46. K.J. Dietz and F. Waelbroeck, Proc.Int.Symp. Plasma Wall Interaction, EUR 5782e,445,1977

47. P. Wienhold, I. Ali-Khan, K.J. Dietz, M. Profant, and F. Waelbroeck, First Top. Meeting Fusion Reactor Materials, Miami Beach, USA, 1979, paper 0-5

48. A. Pospieszczyk, J. Burt, S.J. Fielding, G.M. McCracken, and P.E. Stott, Proc.Int.Symp. Plasma Wall Interaction, EUR 5782e, 471,1977

49. J. Burt, G.M. McCracken, and P.E. Stott, ibid, EUR 5782e,457, 1977

50. TFR Group, ibid, EUR 5782e,465,1977

51. G. Betz, G.K. Wehner, L. Toth, and A. Jochi, J.Appl.Phys. 45, 5312,1974

52. J. Kirschner, K.J. Dietz, and F. Waelbroeck, Proc. 9th SOFT, EUR 5601,65,1976

53. K.J. Dietz, E. Geissler, F. Waelbroeck, J. Kirschner, E.A. Niekisch, K.G. Tschersich, G. Stöcklin, E. Vietzke, and K. Vogelbruch, J.Nucl.Mat. 63,167,1976

54. P.A. Readhead, Vacuum 12,203,1962

55. A.G. Mathewson, Proc.Int.Symp. Plasma Wall Interaction, EUR 5782e,517,1977

56. S. Brumbach and H. Kaminsky, J.Nucl.Mat. 63,188,1976

57. W. Bauer, J.Nucl.Mat. 76 & 77,3,1978

58. L. Oren and R.J. Taylor, PPG-294, March 1977

59. PLT Group, 7th Internat.Conf. Plasma Physics and Controlled Fusion Research, IAEA, Innsbruck 1978, paper A1

60. ISX Group, ibid, paper N4

61. Diva Group, ibid, paper T3-1

62. TFR Group, J.Nucl.Mat. 76 & 77,587,1978

63. S. Veprek and C.M. Braganza, First Top. Meeting Fusion Reactor Materials, Miami Beach, USA, 1979, paper G-5

64. A.W. Thompson and I.M. Bernstein, (eds.), Effect of hydrogen on Behaviour of Metals, Metallurgical Society of AIME 1976

65. H.J. König and K.W. Lange, Arch.Eisenhüttenwesen 46,237,1975

66. A.E. Robson and P.C. Thoneman, Proc.Phys.Soc. 73,508,1959

67. G.D. Hobbs and J.A. Wessen, Plasma Physics 9,85,1967

68. P.E. Stott, J. Burt, S.K. Erents, S.J. Fielding, D.H.J. Goodall, M. Hobby, J. Hugill, G.M. McCracken, J.W.M. Paul, A. Pospieszczyk, R. Prentice and D.D.R. Summers, Proc.Int. Symp. Plasma Wall Interaction, EUR 5782e,39,1977

69. A. Gibson, J.Nucl.Mat. 76 & 77,92,1978

70. Y. Gomay, N. Fujisawa, and M. Maeno, First Top. Meeting Fusion Reactor Materials, Miami Beach, USA, 1979, paper 0-2

71. A. Zuhr, B.R. Appleton, R.E. Clausing, L.C. Emmerson and L. Heatherly, ibid, paper 0-8

72. P. Mioduszczewski, R.E. Clausing, L. Heatherly, ibid, paper G-11

73. G.M. McCracken, G. Dearnaley, R.D. Gill, J. Hugill, J.W.M. Paul, B.A. Powell, P.E. Stott, J.F. Turner, and J.E. Vince, J.Nucl.Mat. 76 & 77,431,1978

74. T.H. Lee and A. Greenwood, J.Appl.Phys. 32,916,1961

75. R.J. Taylor and L. Oren, Phys,Rev. Letters 42,446,1979

THEORETICAL METHODS FOR ATOMIC COLLISIONS

A GENERAL SURVEY

C. J. Joachain

Physique Théorique, Faculté des Sciences

Université Libre de Bruxelles, Belgium

1. INTRODUCTION

The theory of atomic collisions is a branch of quantum collision theory which is concerned with collisions between two atomic systems, or between an atomic system and an "elementary particle" such as an electron, a positron or a proton. A comprehensive account of the subject, prior to the seventies, may be found in the books by Mott and Massey[1] and Bransden[2]. In recent years, however, considerable progress has been made in developing new methods. This is due to a number of factors. Firstly, there is an increasing demand for collision cross sections in other fields such as astrophysics, laser physics and plasma physics. Secondly, important advances have occurred on the experimental side. These include absolute measurements of cross sections[3] as well as experiments using tunable dye lasers[4], polarized beams or targets[5] and coincidence techniques[6-9] which provide stringent tests of the theory. Finally, the availability of increasingly more powerful computers has made it possible to carry out calculations which would have been untractable a decade ago.

In these lectures I shall discuss various theoretical methods which have proved to be particularly useful in analysing atomic collisions. I shall begin by considering electron scattering by atoms (or ions) and then proceed to analyse atom (ion)-atom collisions. In both cases it will be convenient to divide the energy range into "slow" and "fast" collisions. All quantities will be expressed in atomic units.

2. ELECTRON-ATOM COLLISIONS

2.1. Low-energy electron-atom scattering

In this section we shall assume that the incident electron energy is low enough so that only a few target states can be excited, or in other words only a few channels are open. In this case it is possible to represent explicitly all open channels in the total electron-atom scattering wave function, and we shall discuss shortly several methods[10] which exhibit this feature. Before we do so, however, it is convenient to introduce a representation in which the total electron-atom wave function is expanded in a complete set of target atom eigenfunctions. For the sake of illustration we shall consider first the simple case of scattering by a one-electron atom or ion. We choose the nucleus of the atom as the origin of our coordinate system, denote by $\mathbf{r}_1$ and $\mathbf{r}_2$ the position vectors of the two electrons and expand the spatial part of the full wave function as

$$\Psi^{\pm}(\mathbf{r}_1,\mathbf{r}_2) = \sum_n [F_n^{\pm}(\mathbf{r}_1)\,\psi_n(\mathbf{r}_2) \pm F_n^{\pm}(\mathbf{r}_2)\psi_n(\mathbf{r}_1)] \tag{2.1}$$

where the superscripts + and - refer to the singlet (S = 0) and triplet (S = 1) cases, respectively, and $\psi_n(\mathbf{r}) \equiv \langle \mathbf{r}|n\rangle$ is an eigenfunction of the target corresponding to an eigenenergy w_n, namely

$$(-\frac{1}{2}\nabla^2 - \frac{Z}{r} + w_n)\,\psi_n(\mathbf{r}) = 0. \tag{2.2}$$

The Schrödinger equation satisfied by $\Psi^{\pm}(\mathbf{r}_1,\mathbf{r}_2)$ is

$$(-\frac{1}{2}\nabla_1^2 - \frac{1}{2}\nabla_2^2 - \frac{Z}{r_1} - \frac{Z}{r_2} + \frac{1}{r_{12}} - E)\,\Psi^{\pm}(\mathbf{r}_1,\mathbf{r}_2) = 0. \tag{2.3}$$

By projecting this equation with the basis functions ψ_n and using (2.2) we find that the functions $F_n^{\pm}$ must satisfy the infinite set of coupled integro-differential equations

$$(\nabla^2 + k_n^2)\,F_n^{\pm}(\mathbf{r}) = \sum_m U_{nm}^{\pm}(\mathbf{r})\,F_m^{\pm}(\mathbf{r}) \tag{2.4}$$

with $k_n^2 = 2(E-w_n)$. The potential operators $U_{nm}^{\pm}$ are such that

$$U_{nm}^{\pm}(\mathbf{r})F_m^{\pm}(\mathbf{r}) = 2[V_{nm}(\mathbf{r})F_m^{\pm}(\mathbf{r}) \pm \int W_{nm}(\mathbf{r},\mathbf{r}')F_m^{\pm}(\mathbf{r}')d\mathbf{r}']. \tag{2.5}$$

Here V_{nm} is a local direct potential given by

$$V_{nm}(\mathbf{r}) = -\frac{Z}{r}\,\delta_{nm} + \int \psi_n^*(\mathbf{r}')\,\frac{1}{|\mathbf{r}-\mathbf{r}'|}\,\psi_m(\mathbf{r}')d\mathbf{r}' \tag{2.6}$$

and $W_{nm}(\underset{\sim}{r},\underset{\sim}{r}')$ is a non-local exchange kernel such that

$$W_{nm}(\underset{\sim}{r},\underset{\sim}{r}') = \psi_n^*(\underset{\sim}{r}')\psi_m(\underset{\sim}{r})[\frac{1}{|\underset{\sim}{r}-\underset{\sim}{r}'|} + w_n + w_m - E]. \qquad (2.7)$$

For scattering by a neutral atom which is initially in the state $|0\rangle$ the coupled equations (2.4) must be solved subject to the boundary conditions

$$F_n^{\pm}(\underset{\sim}{r}) \underset{r \to \infty}{\to} e^{i\underset{\sim}{k}_n \cdot \underset{\sim}{r}} \delta_{no} + f_{no}^{\pm}(\theta,\phi) \frac{e^{ik_m r}}{r} \qquad (2.8)$$

for open channels. In the closed channels $F_n^{\pm}(r)$ vanishes asymptotically. If the target is an ion it is necessary to modify the exponents in (2.8) by including the logarithmic phase factors due to the distortion by the Coulomb potential. Denoting by $\underset{\sim}{k_i}$ the momentum of the incident electron and by $\underset{\sim}{k_f}$ that of the scattered one, we find (after averaging over initial spins and summing over final ones) that the differential cross section for a transition $|\underset{\sim}{k_i},o\rangle \to |\underset{\sim}{k_f},m\rangle$ is given by

$$\frac{d\sigma}{d\Omega} = \frac{k_f}{k_i} (\frac{1}{4} |f_{mo}^+|^2 + \frac{3}{4} |f_{mo}^-|^2). \qquad (2.9)$$

The close-coupling approximation

In practice, only a few target eigenfunctions ψ_n corresponding to discrete levels can be retained in the expansion (2.1). Since we are dealing here with a low-energy situation for which only a few channels are open, it is reasonable to retain in (2.1) all the open channels and possibly also some closely coupled channels. If the eigenfunctions ψ_n kept in (2.1) are labelled from $n = 1$ to $n = M$, we obtain the system of M coupled integro-differential equations

$$(\nabla^2 + k_n^2)F_n^{\pm}(\underset{\sim}{r}) = \sum_{m=1}^{M} U_{nm}^{\pm}(\underset{\sim}{r})F_m^{\pm}(r), \qquad n = 1,2,...M. \qquad (2.10)$$

These equations are known as the close-coupling equations. After a partial wave decomposition[11] they may be solved either by direct numerical integration[12], or by using variational or R-matrix methods which will be discussed below.

For electron scattering by an N-electron atom or ion the close-coupling equations (in partial wave form) may be obtained by starting from a total electron-atom wave function of the form

$$= A \sum_{n=1}^{M} F_n(r_{N+1})\phi_n(q_1,q_2,\ldots q_n,\sigma_{N+1}\ \hat{r}_{N+1}) \tag{2.11}$$

and projecting the Schrödinger equation $(H-E)\Psi = 0$ onto the eigenchannel functions ϕ_n. In the above equation A is the antisymmetrization operator, q_i denotes the combined space and spin coordinates of the ith electron, and the eigenchannel functions ϕ_n are determined by coupling the target eigenstates with the spin-angle functions of the scattered electron.

The close coupling method is a good approximation for strong transitions between low-lying states whose energies are well separated from all other states. In other cases the convergence is often slow and the method must be improved by adding terms to the expansion (2.11). In order to gain insight into this problem it is convenient to use the Hulthén-Kohn variational method[1,2], to which we now turn our attention.

The variational method

Let us begin by considering the problem of electron scattering by atomic hydrogen. We want to obtain the scattering amplitudes $f^{\pm}_{mo}(\mathbf{k}_i,\mathbf{k}_f)$ corresponding to a transition $|\mathbf{k}_i,o\rangle \rightarrow |\mathbf{k}_f,m)$. Omitting the superscripts $\pm$, we define two solutions of the Schrödinger equation Ψ_1 and Ψ_2 such that

$$\Psi_1(r_1,r_2) \underset{r_1\to\infty}{\sim} \exp(i\mathbf{k}_i\cdot\mathbf{r}_1)\psi_o(\mathbf{r}_2) + \sum_s f_{os}(\hat{\mathbf{k}}_i\cdot\mathbf{r}_1)\mathbf{r}_1^{-1}\exp(ik_s r_1)\psi_s(\mathbf{r}_2) \tag{2.12a}$$

and

$$\Psi_2(\mathbf{r}_1,\mathbf{r}_2) \underset{r_1\to\infty}{\sim} \exp(i\mathbf{k}_f\cdot\mathbf{r}_1)\psi_m(\mathbf{r}_2) + \sum_u f_{mu}(\hat{\mathbf{k}}_f\cdot\mathbf{r}_1)r_1^{-1}\exp(-ik_u r_1)\psi_u(\mathbf{r}_2) \tag{2.12b}$$

Defining

$$I[\Psi_1,\Psi_2] = 2\int \Psi_2^{*}(E-H)\Psi_1\, d\mathbf{r}_1\, d\mathbf{r}_2 \tag{2.13}$$

we find that

$$\Delta I = I[\Psi_1^t,\Psi_2^t] = -4\pi\Delta f_{mo}(\mathbf{k}_i,\mathbf{k}_f) + I[\Delta\Psi_1,\Delta\Psi_2] \tag{2.14}$$

where Ψ_1^t and Ψ_2^t are trial functions which satisfy boundary conditions of the same type as Ψ_1 and Ψ_2, but with scattering amplitudes $(f_{os}+\Delta f_{os})$ and $(f_{mu}+\Delta f_{mu})$ rather than f_{os} and f_{mu}. Thus the error in f_{mo} will be of second order if suitable trial functions are chosen such that $\Delta I = 0$.

As a first application of the Hulthén-Kohn method, let us take as trial functions truncated eigenfunction expansions such as

$$\Psi^{\pm}(\mathbf{r}_1,\mathbf{r}_2) = \sum_{n=1}^{M} [F_n^{\pm}(\mathbf{r}_1)\psi_n(\mathbf{r}_2) \pm F_n^{\pm}(\mathbf{r}_2)\psi_n(\mathbf{r}_1)]. \tag{2.15}$$

On varying each of the unknown functions $F_n^{\pm}$ (subject to the boundary conditions discussed above) so that $\Delta I = 0$, one finds a set of M coupled integro-differential equations

$$\int d\mathbf{r}_1\, \psi_n^{*}(r_1)[H - E]\Psi^{\pm}(\mathbf{r}_1,\mathbf{r}_2) = 0, \qquad n = 1,2,\ldots M \tag{2.16}$$

which projects the Schrödinger equation onto the eigenfunction ψ_n and hence are equivalent to the close-coupling equations (2.10).

The Hulthén-Kohn variational principle may of course be used with other trial functions which are particularly suitable to describe certain important features of the collision process. For example, we may include in the trial function some terms which explicitly represent polarization effects, in the spirit of the polarized orbital method first introduced by Temkin[13]. These polarization effects, which are characteristic of the interaction between a charged particle and a neutral polarizable system, give rise in the elastic channel to an attractive long range potential that varies like r^{-4} for large r. Together with exchange effects, they play a central role in low-energy electron-atom scattering and remain also important at higher energies.

As in the case of bound state problems, we may take advantage of the variational principle to perform systematic Rayleigh-Ritz type calculations, in which square integrable (L^2) functions are included in the trial function. This approach is capable of giving excellent results if enough L^2 functions are retained. For example, in the case of e^--H scattering, where Hylleraas-type L^2 functions can be used, very accurate elastic s-wave phase shifts have been obtained by Schwartz[14].

We have already shown that the close coupling equations are obtained if the truncated eigenfunction expansion (2.15) is used as a trial function in the Hulthen-Kohn variational principle. A more general trial function is obtained by adding L^2 functions to the expansion (2.15). The resulting set of mixed integro-differential and algebraic equations[15,16] may be solved either by direct numerical integration[16] or by using variational techniques such as the matrix variational method[17]. This approach, which is known as the correlation approximation, is particularly useful when the remaining strongly coupled channels - not represented in (2.15) - are closed, since in that case they can be adequately represented by L^2 functions. It is worth noting that polarization effects are not always taken into account accurately in the correlation approximation, in particular for atoms where a sizeable fraction of the dipole polarizability arises from continuum intermediate states. In this case it is necessary to represent these polarization effects by adding extra terms in the trial function.

Modified close-coupling expansions. Pseudostates

The foregoing discussion indicates that a general electron-atom collision wave function, suitable for the low-energy range, may be written as[10]

$$\Psi = A \sum_{n=1}^{M} F_n(r_{N+1})\phi_n(q_1,q_2,\ldots q_N,\sigma_{N+1}\hat{r}_{N+1})$$

$$+ A \sum_{n=M+1}^{M+P} F_n(r_{N+1})\overline{\phi}_n(q_1,q_2,\ldots q_N,\sigma_{N+1}\hat{r}_{N+1})$$

$$+ \sum_{n=1} c_n\, \chi_n(q_1,q_2,\ldots q_{N+1}). \qquad (2.17)$$

The first expansion on the right of this equation is simply the close coupling expansion (2.11). The second expansion is written in terms of pseudochannel functions $\overline{\phi}_n$, which are obtained by coupling the spin-angle functions of the scattered electron with pseudostates $\overline{\psi}_n$. These pseudostates are constructed so that they represent polarization effects; they can be determined by using methods[18,19] which generalise the polarized orbital approximation of Temkin. Finally, the third expansion in (2.17) is made in terms of L^2 functions. The approach which retains terms in all three expansions of (2.17) is known as the polarized pseudostate approximation.

Integro-differential equations for the functions $F_n(r)$, coupled to linear equations for the coefficients c_n can be derived either by using Ψ as a trial function in the Hulthén-Kohn variational principle, or by projecting the Schrödinger equation $(H - E)\,\Psi = 0$ onto the functions ϕ_n, $\overline{\phi}_n$ and χ_n. These equations may then be solved by using direct numerical integration, variational techniques or the R-matrix method, which we shall now briefly describe.

The R-matrix method

Originally introduced by Wigner and Eisenbud[20] in the theory of nuclear reactions, the R-matrix method was first applied to atomic collisions by Burke et al[21,22]. It has been developed extensively in recent years to describe a broad range of atomic and molecular processes[23].

The basic idea is to realise that the dynamics of the projectile-target system differs depending on the relative distance r between the two colliding particles, so that configuration space for the system may be divided into an internal region $(r < a)$ and an external region $(r > a)$. For electron-atom (ion) collisions r is the radial distance between the electron and the nucleus of the atom (ion) and a is chosen in such a way that the charge distribution of the target states of interest is contained within the sphere $r = a$. It is clear that in the internal region the electron-target interaction is strong and difficult to handle, since both electron exchange and electron correlation effects are important. On the other hand, in the external region exchange between the scattered electron and the target may be neglected, and the solution to the collision problem can be readily obtained.

The R-matrix is now defined as follows. We begin by imposing logarithmic boundary conditions on the surface of the internal region, namely

$$\frac{a}{F_n(a)} \left. \frac{d\,F_n(r)}{dr} \right|_{r=a} = b_n \tag{2.18}$$

where $F_n(r)$ are the radial functions describing the motion of the scattered electron in channel n, and b_n are arbitrary constants which may depend on the channel quantum numbers. With these boundary conditions the spectrum of the Hamiltonian describing the electron-target system in the internal region consists of discrete energy levels E_k, and the corresponding eigenstates Ψ_k form in this region a complete set which may be used to expand the wave function Ψ_E for any energy (for $r < a$) as

$$\Psi_E = \sum_n A_{Ek}\,\Psi_k\,. \tag{2.19}$$

By substituting this expansion into the Schrödinger equation $(H - E)\,\Psi_E = 0$ and using the boundary conditions (2.18) satisfied by the radial parts of the functions Ψ_k, it is found that

$$F_n(a) = \sum_m R_{nm}(E)\left(a\frac{dF_m}{dr} - b_m F_m\right)_{r=a} \tag{2.20}$$

where

$$R_{nm}(E) = \frac{1}{2a}\sum_k \frac{w_{nk}(a)\,w_{mk}(a)}{E_k - E} \tag{2.21}$$

is the R-matrix. Here the surface amplitudes $w_{nk}(a)$ are the values of the radial parts of the functions Ψ_k in channel n on the boundary. The central problem in the R-matrix method is therefore to calculate the surface amplitudes $w_{nk}(a)$ and the eigen-energies E_k. The K-matrix (or S-matrix) and cross sections are then related to the R-matrix (2.21) through the solution in the external region.

A very interesting feature of the R-matrix theory is that the functions Ψ_k are energy-independent, and can be obtained by standard bound state procedures. In this respect the R-matrix method is similar to recent L^2 approaches to scattering problems[24-28]. Another important property of the R-matrix is that it is a real meromorphic function of the energy with simple poles lying on the real axis, so that $R_{nm}(E)$ does not contain any branch points. This fact provides the basis of the quantum defect theory[29], in which analytically known properties of electrons moving in a pure Coulomb field are used to describe electron-ion collisions in terms of a few parameters.

The optical potential

The basic idea of the optical potential method is to analyse the elastic scattering of a particle from a complex target by replacing the complicated interactions between the projectile and the target particles by an optical potential (or pseudopotential) in which the incident particle moves[30]. Once the optical potential V_{opt} is determined, the original many-body elastic scattering problem reduces to a one-body situation. However, this reduction is in general a difficult task, and approximation methods are necessary. We shall describe below a multiple scattering approach which is very convenient for the case of fast collisions. At low energies, a particularly useful method is the Feshbach[31] projection operator formalism which has been applied[32,33] to study the resonances arising in electron scattering by atoms and ions. Of course the other elaborate low-energy methods we have described above are also capable of giving a detailed account of resonances in low-energy electron-atom (ion) scattering.

Many-body methods

Before leaving the subject of low-energy electron-atom scattering, I would like to mention that the many-body Green's function methods developed by Taylor et al[34] have also been applied successfully to a variety of electron-atom scattering processes. For elastic scattering an approximation scheme called the "generalised random phase approximation" has been proposed to calculate in a self-consistent way the response function for the target and obtain an optical potential. For inelastic scattering the approach developed by Taylor et al from the Green's function formalism is similar to the distorted wave approximation discussed below.

2.2. Intermediate and high-energy electron-atom collisions

We shall now consider various methods which have been proposed for the case where the energy of the projectile electron is higher than the first ionization energy of the target atom or ion. Detailed discussions of several of these methods may be found in the review articles of Joachain and Quigg[35], Bransden and McDowell[36] and Byron and Joachain[37]. It is convenient to still distinguish two regions of energy: a "high-energy" domain which extends from a few times the first ionization threshold upwards[38], and an "intermediate energy" region ranging from the low-energy region considered above to a few times the ionization energy of the target.

The Born series

Let us begin by considering the high-energy region, for which it is reasonable to try an approach based on perturbation theory. We shall first discuss direct scattering. We denote respectively by $\mathbf{k}_i$ and $\mathbf{k}_f$ the initial and final momentum of the projectile, with $|\mathbf{k}_i| = k$. The quantity $\mathbf{\Delta} = \mathbf{k}_i - \mathbf{k}_f$ is the momentum transfer. The free motion of the colliding particles before the collision is described by the direct arrangement channel Hamiltonian $H_d = K + h$, where K is the kinetic energy operator of the projectile and h the internal target Hamiltonian, such that $h|n\rangle = w_n|n\rangle$. The full Hamiltonian of the system is $H = H_d + V_d$, where V_d is the interaction between the electron and the target in the initial (direct) arrangement channel. For the sake of illustration we shall assume here that the target is a neutral atom of atomic number Z, in which case we have

$$V_d = -\frac{Z}{r} + \sum_{j=1}^{Z} \frac{1}{|\mathbf{r} - \mathbf{r}_j|} \; . \qquad (2.22)$$

Here $\mathbf{r}$ denotes the coordinates of the projectile and $\mathbf{r}_j$ those of the target electrons.

We now write the Born series for the direct scattering amplitude as[39]

$$f = \sum_{n=1}^{\infty} \bar{f}_{Bn} \qquad (2.23)$$

where the nth Born term $\bar{f}_{Bn}$ contains n times the interaction

V_d and $(n - 1)$ times the direct Green's operator $G_d^{(+)} = (E - H_d + i\varepsilon)^{-1}$, $\varepsilon \to o^+$. It is also convenient to define the quantity f_{Bj} as the sum of the first j terms of (2.23), so that f_{Bj} is the jth Born approximation to the direct scattering amplitude.

The first term $\bar{f}_{B1}$ is the familiar first Born amplitude, which has been evaluated for a large number of electron-atom elastic, excitation and ionization scattering processes[40]. One of the basic goals of the theory is to obtain systematic improvements over the first Born approximation. To this end, let us first consider the second Born term $\bar{f}_{B2}$. For the direct transition $|\underset{\sim}{k}_i,o\rangle \to |\underset{\sim}{k}_f,m\rangle$, this term reads[39]

$$\bar{f}_{B2} = 8\pi^2 \int d\underset{\sim}{q} \sum_n \frac{\langle \underset{\sim}{k}_f,m|V_d|\underset{\sim}{q},n\rangle\langle \underset{\sim}{q},n|V_d|\underset{\sim}{k}_i,o\rangle}{q^2 - k^2 + 2(w_n - w_o) - i\varepsilon}, \quad \varepsilon \to o^+ \qquad (2.24)$$

A useful approximation for $\bar{f}_{B2}$ at sufficiently high energies may be obtained[41] by replacing the energy differences $(w_n - w_o)$ by an average excitation energy $\bar{w}$, so that the sum on the intermediate target states $|n\rangle$ can be done by closure. An improvement over this approximation consists in evaluating exactly the first few terms in the sum, while treating the remaining states by closure[42]. This method has been widely used in recent years, along with further improvements[43]. It is also worth noting that the sum on intermediate states in (2.24) has been evaluated "exactly" (without using closure) for elastic $(e^- -H)$ scattering in the forward direction[44]. This calculation, which can be generalised to other transitions, yields useful information about the determination of the average excitation energy $\bar{w}$.

Since we are dealing in this section with fast electron-atom collisions, it is important to examine the behaviour of the terms $\bar{f}_{Bn}$ for large values of the projectile wave number k (expressed in a.u.). Referring to Table 1, we see that for direct scattering at small momentum transfers the quantity Re $\bar{f}_{B2}$ (which is governed by polarization effects) gives the dominant contribution (of order k^{-1}) to the first Born differential cross section. The situation at large Δ is easily understood since in this limit the terms $\bar{f}_{Bn}$ $(n \geq 2)$ are dominated by processes in which the atom remains in its initial state $|o\rangle$ in all intermediate states. This is equivalent to scattering by the static potential $V_{st} = \langle o|V_d|o\rangle$ of the atom. It is worth noting that for direct elastic scattering the dominant contribution at large k is given by the first Born term $\bar{f}_{B1}$ at all momentum transfers.

The situation is different for direct inelastic collisions where for large Δ the first Born term falls off rapidly and the Born series is dominated by the second Born term $\bar{f}_{B2}$. This is

TABLE 1

Dependence of various terms of the Born and Glauber multiple scattering series for the direct elastic scattering amplitude, as a function of (large) k and Δ. The dominant contributions are framed. The terms located above the dashed line contribute through order k^{-2} to the differential cross section.[45]

Order of Pert. Theory	Term	Small Δ ($\Delta < k^{-1}$)	Interm. Δ ($k^{-1} < \Delta < 1$)	Large Δ ($\Delta > k$)
First	$\bar{f}_{B1}=\bar{f}_{G1}$	[1]	[1]	[Δ^{-2}]
Second	Re $\bar{f}_{B2}$	k^{-1}	k^{-2}	$k^{-2}\ \Delta^{-2}$
	Re $\bar{f}_{G2}$	0	0	0
	Im $\bar{f}_{B2}$	$k^{-1}\ \ell nk$	k^{-1}	$k^{-1}\ \Delta^{-2}\ \ell n\Delta$
	Im $\bar{f}_{G2}$	$k^{-1}\ \ell n\Delta$	k^{-1}	$k^{-1}\ \Delta^{-2}\ \ell n\Delta$
Third	Re $\bar{f}_{B3}$	k^{-2}	k^{-2}	$k^{-2}\ \Delta^{-2}\ \ell n^2\Delta$
	Re $\bar{f}_{G3}$	k^{-2}	k^{-2}	$k^{-2}\ \Delta^{-2}\ \ell n^2\Delta$
	- - - - - - -	- - - - -	- - - - -	- - - - - - -
	Im $\bar{f}_{B3}$	k^{-3}	k^{-3}	$k^{-3}\ \Delta^{-2}\ \ell n\Delta$
	Im $\bar{f}_{G3}$	0	0	0
n^{th}	$\bar{f}_{Bn}$	$(ik)^{1-n}$	$(ik)^{1-n}$	$(ik)^{1-n}\ \Delta^{-2}\ \ell n^{n-1}\Delta$
	$\bar{f}_{Gn}$	$(ik)^{1-n}$	$(ik)^{1-n}$	$(ik)^{1-n}\ \Delta^{-2}\ \ell n^{n-1}\Delta$

illustrated in Table 2 for the case of inelastic s - s transitions. The fact that $\bar{f}_{B2}$ falls off more slowly than $\bar{f}_{B1}$ at large momentum transfers is due to the possibility of off-shell elastic scattering in intermediate states, where the projectile can experience the Coulomb potential of the nucleus. We remark that since the values $\Delta < 1$ correspond to angles $\theta < k^{-1}$, the angular domain in which the first Born approximation is valid shrinks as the energy increases. However, because the dominant contribution to the integrated cross section comes precisely from the region $\Delta < 1$, the first Born values for integrated inelastic cross sections should be reliable at high energies.

The terms $\bar{g}_{Bn}$ of the Born series for exchange scattering are much more difficult to analyse than the direct Born terms $\bar{f}_{Bn}$ we have considered above. We simply mention here that for large k the term $\bar{g}_{B2}$ falls off more slowly than $\bar{g}_{B1}$, except for elastic exchange scattering at small Δ, where the Ochkur amplitude g_{Och} (which is the leading piece of $\bar{g}_{B1}$) is of order k^{-2}.

Our discussion of the Born series has been restricted so far to

TABLE 2

Dependence of various terms of the Born and Glauber multiple scattering series for the direct scattering amplitude corresponding to inelastic (s-s) transitions, as a function of (large) k and Δ. The dominant contributions are framed. The terms located above the dashed line contribute through order k^{-2} to the differential cross section.[45]

Order of Pert. Theory	Term	Small Δ ($\Delta < k^{-1}$)	Interm. Δ ($k^{-1} < \Delta < 1$)	Large Δ ($\Delta > k$)
First	$\bar{f}_{B1}=\bar{f}_{G1}$	$\boxed{1}$	$\boxed{1}$	Δ^{-6}
Second	Re $\bar{f}_{B2}$	k^{-1}	k^{-2}	$k^{-2}\ \Delta^{-2}$
	Re $\bar{f}_{G2}$	0	0	0
	Im $\bar{f}_{B2}$	$k^{-1}\ \ell nk$	k^{-1}	$\boxed{k^{-1}\ \Delta^{-2}}$
	Im $\bar{f}_{G2}$	$k^{-1}\ \ell n\Delta$	k^{-1}	$\boxed{k^{-1}\ \Delta^{-2}}$
Third	Re $\bar{f}_{B3}$	k^{-2}	k^{-2}	$k^{-2}\ \Delta^{-2}\ \ell n\Delta$
	Re $\bar{f}_{G3}$	k^{-2}	k^{-2}	$k^{-2}\ \Delta^{-2}\ \ell n\Delta$
	- - - - -	- - -	- - -	- - -
	Im $\bar{f}_{B3}$	k^{-3}	k^{-3}	$k^{-3}\ \Delta^{-2}\ \ell n\Delta$
	Im $\bar{f}_{G3}$	0	0	0
n^{th}	$\bar{f}_{Bn}$	$(ik)^{1-n}$	$(ik)^{1-n}$	$(ik)^{1-n}\ \Delta^{-2}\ \ell n^{n-2}\Delta$
	$\bar{f}_{Gn}$	$(ik)^{1-n}$	$(ik)^{1-n}$	$(ik)^{1-n}\ \Delta^{-2}\ \ell n^{n-2}\Delta$

elastic and inelastic processes of the type $|\mathbf{k}_i, o\rangle \rightarrow |\mathbf{k}_f, m\rangle$, with two fragments in the initial and final channels. For ionization collisions, in which three or more particles emerge in the final channel, it is very difficult to perform calculations of the Born terms beyond first order. Nevertheless, by generalising the arguments given above for inelastic (two-fragment) collisions, it is possible to show[46] that in certain kinematical situations the second Born term $\bar{f}_{B2}$ again falls off more slowly than $\bar{f}_{B1}$ for large k.

The Glauber approximation

The Glauber method is a many-body generalisation of the eikonal approximation[47]. It was first proposed[48] to analyse high-energy hadron-nucleus collisions, but has also been applied in recent years to study atomic collision processes. For a direct collision leading from an initial target state $|o\rangle$ to a final state $|m\rangle$ the Glauber scattering amplitude is given by[39,48]

$$f_G = \frac{k}{2\pi i} \int d^2\mathbf{b}\ e^{i\mathbf{\Delta}\cdot\mathbf{b}} \langle m| \{ e^{i\chi_G(k,\mathbf{b},X)} - 1\} |o\rangle \qquad (2.25)$$

where the symbol X denotes the ensemble of the target coordinates, and we use a cylindrical coordinate system, with $\mathbf{r} = \mathbf{b} + z\hat{\mathbf{z}}$. The Glauber phase shift function χ_G is given in terms of the direct interaction (2.22) between the projectile and the target by

$$\chi_G(k,\mathbf{b},X) = -\frac{1}{K}\int_{-\infty}^{+\infty} V_d(\mathbf{b},z,X)dz, \qquad (2.26)$$

the integration being performed along a z-axis perpendicular to $\mathbf{\Delta}$.

Detailed discussions of the Glauber approximation and related methods may be found in the review articles of Joachain and Quigg[35] and Byron and Joachain[37]. We shall only mention here a few important points concerning the Glauber approach. Firstly, it may be viewed as an eikonal approximation to a "frozen target" model proposed by Chase[49], in which closure is used with an average excitation energy $\bar{w} = 0$. Secondly, considerable insight into the properties of the Glauber method may be gained by expanding the Glauber amplitude (2.25) in powers of V_d, namely

$$f_G = \sum_{n=1}^{\infty} \bar{f}_{Gn} \qquad (2.27)$$

where

$$\bar{f}_{Gn} = \frac{k}{2\pi i}\frac{i^n}{n!}\int d^2\mathbf{b}\; e^{i\mathbf{\Delta}\cdot\mathbf{b}} \langle m|[\chi_G(k,\mathbf{b},X)]^n|o \qquad (2.28)$$

and comparing the terms $\bar{f}_{Gn}$ with those of the Born series $\bar{f}_{Bn}$. We note at once that $\bar{f}_{B1} = \bar{f}_{G1}$ because of our choice of z-axis. We also remark that the terms $\bar{f}_{Gn}$ are alternately real or purely imaginary, while the corresponding Born terms $\bar{f}_{Bn}$ are complex for $n \geq 2$. This special feature of the Glauber amplitude leads to several defects such as (i) the absence of the important term $\mathrm{Re}\,\bar{f}_{B2}$ for elastic scattering and (ii) identical cross sections for electron- and positron-atom scattering. Other deficiencies of the Glauber amplitude (2.25) include a logarithmic divergence for elastic scattering in the forward direction (which is due to the choice $\bar{w} = 0$ made in obtaining (2.25) and may be traced to the behaviour of $\bar{f}_{G2}$ at $\Delta = 0$, as shown in Table 1) and a poor description of inelastic collisions invokving non-spherically symmetric states. Despite these limitations, the Glauber approximation has been applied to a variety of atomic scattering processes. However, its major role in atomic collision theory has been to stimulate interest in eikonal methods[35,37] such as the "eikonal-Born series" (EBS) theory[50] which we shall now discuss.

The eikonal-Born series method

The basic idea of the EBS approach consists in analysing the terms of the Born series (2.23) and the Glauber series (2.27) with the aim of obtaining a consistent expansion of the scattering amplitude in powers of k^{-1}. The main results are summarised in Tables 1 and 2 for elastic and inelastic (s - s) transitions, respectively. We note that for these processes the Glauber term $\bar{f}_{Gn}$ gives in each order of perturbation theory the leading piece of the corresponding Born term (for large k) for all momentum transfers, except in second order where the long range of the Coulomb potential is responsible for the anomalous behaviour of $\bar{f}_{G2}$ at small Δ. We also remark from Tables 1 and 2 that neither the second Born amplitude $f_{B2} = \bar{f}_{B1} + \bar{f}_{B2}$ nor the Glauber amplitude f_G are correct through order k^{-2}. In fact, a consistent calculation of the direct scattering amplitude through that order requires the terms $\bar{f}_{B1}$, $\bar{f}_{B2}$ and Re $\bar{f}_{B3}$ (or $\bar{f}_{G3}$). Since Re $\bar{f}_{B3}$ is very difficult to evaluate, and because it is a good approximation to Re $\bar{f}_{B3}$ for large enough k, it is reasonable to use $\bar{f}_{G3}$ in place of Re $\bar{f}_{B3}$. Thus we obtain in this way the "eikonal-Born series" direct scattering amplitude

$$f_{EBS} = \bar{f}_{B1} + \bar{f}_{B2} + \bar{f}_{G3} \quad (2.29)$$

In addition, exchange effects are taken into account by using the Ochkur amplitude g_{Och}.

A detailed account of the EBS method and its application to various electron-atom collision processes at intermediate and high energies may be found in the review article of Byron and Joachain[37]. It is apparent from the foregoing discussion that the EBS theory represents an improvement over the second Born or Glauber approximations. We recall, however, that the EBS method is a perturbative approach and it is clear from Tables 1 and 2 that the convergence of the Born series for the direct amplitude is slower at large than in the small Δ region. Thus an "all-order" treatment would be clearly desirable at large Δ, and we shall now turn out attention to optical model and target eigenfunction expansion methods which can provide (approximately) such "all-order" treatments.

Optical potentials for fast electron-atom scattering

We have already discussed the basic idea of the optical model method in dealing with low-energy electron-atom scattering. At intermediate and high energies it is particularly convenient to use the multiple scattering[30,35] approach developed by Mittleman and Watson[51,52]. We begin by considering direct elastic scattering, for which the direct optical potential V^d_{opt} is given to second order in the projectile-target interaction V_d by

$$V^d_{opt} = V^{(1)} + V^{(2)} + \ldots \quad (2.30)$$

Here $V^{(1)} = V_{st} = \langle o|V_d|o\rangle$ is the static potential while the second order part reads

$$V^{(2)} = \sum_{n\neq o} \frac{\langle o|V_d|n\rangle\langle n|V_d|o\rangle}{k^2/2 - K - (w_n - w_o) + i\varepsilon} \quad , \qquad \varepsilon \to o^+ \tag{2.31}$$

The static potential V_{st} is readily evaluated for simple target atoms, or when an independent particle model is used to describe the target state $|o\rangle$, which we assume here to be spherically symmetric. We note that V_{st}, which is real and of short range, does not account for polarization and absorption effects which play an important role in the energy range considered here. However, for small values of the projectile coordinate r we note that V_{st} correctly reduces to the Coulomb interaction $-Z/r$ acting between the incident electron and the target nucleus, and hence should give a good account of large angle direct elastic scattering, as we already remarked in our discussion of the Born series.

Although the second order part $V^{(2)}$ of the direct optical potential is in general a complicated non-local, complex operator, at sufficiently high energies a useful local approximation of $V^{(2)}$ may be found by introducing an average excitation energy $\bar{w}$ and using eikonal methods[53-55]. The resulting $V^{(2)}$ may be written as

$$V^{(2)} = V_{pol} + i\, V_{abs} \tag{2.32}$$

where V_{pol} and V_{abs} are real and central but energy-dependent. The term V_{pol} (which falls off like r^{-4} at large r) accounts for polarization effects and $i\, V_{abs}$ for absorption effects due to the open channels.

With $V^{(2)}$ determined from (2.32), the corresponding direct optical potential is given by $V^d_{opt} = V_{st} + V_{pol} + i\, V_{abs}$. Finally, exchange effects may be taken into account by using a local "exchange pseudopotential"[52,54-56] V^{ex}_{opt}, so that the full optical potential is given by $V_{opt} = V^d_{opt} + V^{ex}_{opt}$. An "exact" (partial-wave) treatment of this potential is then carried out. It is worth noting that in performing such an exact, full-wave treatment of the optical potential V_{opt}, one generates approximations to all terms of perturbation theory, a feature which is an important advantage for large angle scattering.

Target eigenfunction expansions

It is also possible to formulate the problem of fast electron-atom collisions within the framework of the target eigenfunction expansions discussed at the beginning of Section 2.1. It is clear, however, that at incident electron energies above the ionization threshold the low-energy methods described in that section - and in particular the close-coupling approximation - must be modified. This may be done by using pseudostates[57] which describe the loss of flux from the M coupled channels retained in the close-coupling expansion.

Another modification of the close coupling method, called the second order potential (SOP) method, has been applied by Bransden et al[58] to a variety of electron-atom scattering processes. The modified close-coupling equations read

$$(\nabla^2 + k_n^2)\, F_n(\underset{\sim}{r}) = \sum_{m=1}^{M} [U_{nm}^{\pm}(\underset{\sim}{r}) + K_{nm}^{\pm}(\underset{\sim}{r})]F_m^{\pm}(\underset{\sim}{r}) \qquad (2.33)$$

where $U_{nm}^{\pm}$ is defined by (2.5) and $K_{nm}^{\pm}(\underset{\sim}{r})$ is a potential matrix which accounts approximately for the coupling with the states $n' \geq M+1$. Neglecting exchange, it is given in lowest order by $K_{nm} \simeq 2V_{nm}^{(2)}$, where

$$V_{nm}^{(2)} = \sum_{n' \geq M+1} \frac{\langle n|V_d|n'\rangle\langle n'|V_d|m\rangle}{k^2/2 - K - (w_{n'} - w_o) + i\varepsilon}, \quad \varepsilon \to o^+ \qquad (2.34)$$

is the expected generalisation of the second order part $V^{(2)}$ of the optical potential, given by (2.31) for elastic scattering. A detailed discussion of the second order potential method and of the pseudostate approach mentioned above may be found in the review article of Bransden and McDowell[36].

Distorted waves

The basic idea of distorted wave treatments is to break the interaction in two parts, one which is treated exactly and the other which is handled by perturbation theory. Distorted wave methods are therefore conveniently discussed within the framework of the two-potential formalism[59]. We assume that the interaction potentials in the initial and final arrangement channels may be split as

$$V_i = U_i + W_i, \quad V_f = U_f + W_f. \qquad (2.35)$$

We also suppose that we know the distorted waves

$$\chi_a^{(+)} = \Phi_a + (E - H_i - U_i + i\varepsilon)^{-1}\, U_i\, \Phi_a, \quad \varepsilon \to o^+ \qquad (2.36a)$$

and

$$\chi_b^{(-)} = \Phi_b + (E - H_f - U_f - i\varepsilon)^{-1} U_f \Phi_b, \quad \varepsilon \to o^+ \tag{2.36b}$$

where Φ_a and Φ_b are "free" waves, such that $H_i\Phi_a = E\Phi_a$ and $H_f\Phi_b = E\Phi_b$, with $H_i = H - V_i$ and $H_f = H - V_f$. The T-matrix elements corresponding to the transition a b are then given by

$$T_{ba} = \langle\chi_b^{(-)} | V_i - W_f | \Phi_a\rangle + \langle\chi_b^{(-)} | W_f | \Psi_a^{(+)}\rangle \tag{2.37a}$$

and

$$T_{ba} = \langle\Phi_b | V_f - W_i | \chi_a^{(+)}\rangle + \langle\Psi_b^{(-)} | W_i | \chi_a^{(+)}\rangle \tag{2.37b}$$

with

$$\Psi_a^{(+)} = \chi_a^{(+)} + (E - H_i - U_i + i\varepsilon)^{-1} W_i \Psi_a^{(+)}, \quad \varepsilon \to o^+ \tag{2.38a}$$

and

$$\Psi_b^{(-)} = \chi_b^{(-)} + (E - H_f - U_f - i\varepsilon)^{-1} W_f \Psi_b^{(-)}, \quad \varepsilon \to o^+ . \tag{2.38b}$$

The two-potential formulae (2.37) simplify when the distorting potentials U_i and U_f cannot induce the transition $a \to b$, for example if U_i and U_f only generate elastic scattering and the transition $a \to b$ is an inelastic process or a rearrangement collision. In this case the first term on the right of the equation (2.37) vanishes, and one has

$$T_{ba} = \langle\chi_b^{(-)} | W_f | \Psi_a^{(+)}\rangle = \langle\Psi_b^{(-)} | W_i | \chi_a^{(+)}\rangle . \tag{2.39}$$

If we want to treat exactly the interactions U_i and U_f but to use perturbation theory for the interactions W_i and W_f we may solve the equations (2.38) by iteration and substitute the corresponding expansions in (2.39). In this way we generate the distorted-wave Born series, the first term of which being the distorted-wave Born approximation (DWBA)

$$T_{ba}^{DWBA} = \langle\chi_b^{(-)} | W_i | \chi_a^{(+)}\rangle = \langle\chi_b^{(-)} | W_f | \chi_a^{(+)}\rangle . \tag{2.40}$$

A variety of distorted-wave treatments have been applied to inelastic electron-atom (ion) collisions. These include the DWBA calculations of Shelton et al[60], where static distorting potentials are used for both the incident and scattered electrons, the eikonal DWBA calculations[61,62] performed by using eikonal approximations for the distorted waves $\chi_a^{(+)}$ and $\chi_b^{(-)}$, and the distorted-wave polarized orbital (DWPO) method of McDowell et al[63]. Finally, we mention that distorted-wave approximations have also been applied recently to ionization processes[64-67]. In particular, McCarthy et al[65] have developed a distorted-wave impulse approximation which is accurate under kinematical conditions such that the residual ion can be treated as a spectator. This method has been very useful in performing $(e^-, 2e^-)$ spectroscopy studies[65].

3. ATOM-ATOM COLLISIONS

We shall now discuss some important theoretical methods which have been proposed to study atom- (or ion-) atom collisions[68]. Although the general theory is formally similar to that of electron-atom collisions, there are important differences. Firstly, because the incident particle is now in general a "composite" object, atom-atom collisions are more difficult to analyse from first principles. Secondly, since the projectile is heavy, at practically all velocities for which experiments are performed many channels are open. In particular, excitation or ionization of the target and (or) the projectile can occur, as well as charge transfer.

However, a great simplification can often be made in the theory of atom (ion)-atom collisions, since the wave number k associated with the relative motion of the colliding particles is usually very large (in atomic units). For example, in the case of proton-atomic hydrogen collisions an incident proton energy of 1 keV in the laboratory system corresponds to a relative wave number $k \simeq 184$ a.u. Thus the condition $k \gg 1$ is satisfied at all but the lowest colliding energies, so that semi-classical or even classical methods may be used to describe the relative motion, except for very slow collisions ($E \lesssim 1$ eV) which we shall not discuss here. On the other hand the motion of the atomic electrons must be described by quantum mechanics.

It is worth noting that the relation $k \gg 1$ does not necessarily imply that the colliding velocity v (expressed in atomic units) is large. Thus, returning to our above example of $p + H$ collisions, we remark that an incident proton having a laboratory energy of 1 keV (corresponding to a relative wave number $k \simeq 184$ a.u.) has only a (laboratory) velocity $v \simeq 0.2$ a.u., so that the collision may be considered as slow. On the other hand an incident proton energy of 400 keV corresponds to a velocity $v \simeq 4$ a.u. and therefore to a fast collision. More generally, if v_0 is a typical velocity of the active electron(s) in the collision, we shall say that the collision is slow when the colliding velocity v is such that $v < v_0$, and fast when $v > v_0$.

3.1. Fast atom-atom collisions

Let us begin by analysing fast atom (ion)-atom collisions. For the sake of illustration we shall consider a model system in which a nucleus A, of mass M_A and charge Z_A, is incident on an atom

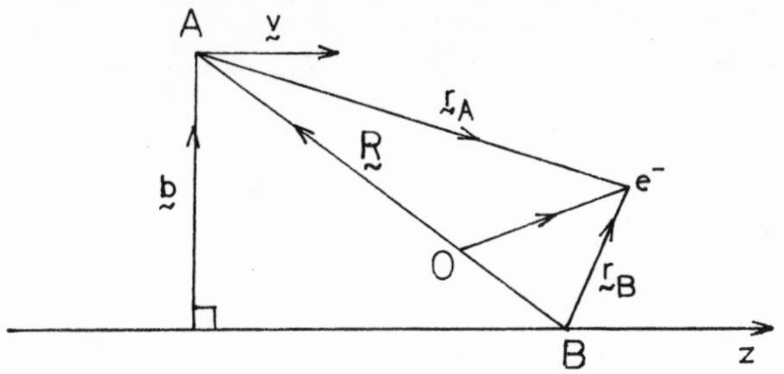

Figure 1. The coordinate system for the collision of the nucleus A with the atom $(B + e^-)$.

$(B + e^-)$ containing a nucleus B, of mass M_B and charge Z_B, and an electron. The coordinate system is shown in Fig. 1, where $\mathbf{R}$ is the position vector of A relative to B, $\mathbf{r}_A$ and $\mathbf{r}_B$ are the position vectors of the electron relative to A and B, respectively, and $\mathbf{r}$ is a vector joining an arbitrary origin O located along the internuclear axis to the position of the electron. Thus we have

$$\mathbf{r}_A = \mathbf{r} - p\,\mathbf{R}, \qquad \mathbf{r}_B = \mathbf{r} + q\,\mathbf{R} \tag{3.1}$$

with $p + q = 1$. The classical path followed by the nucleus A is specified by the dependence of $\mathbf{R}$ on the time t for each value of the impact parameter vector $\mathbf{b}$, which lies in the (xy) plane. Since we are dealing here with fast collisions, we shall assume that the classical path is a straight line trajectory, so that

$$\mathbf{R}(t) = \mathbf{b} + \mathbf{v}\,t, \qquad \mathbf{b} \cdot \mathbf{v} = 0 \tag{3.2}$$

and the velocity $\mathbf{v}$ of A is constant. In this impact parameter formalism the electronic transitions are determined by the time-dependent Schrödinger equation

$$\left(H_{e\ell} - i\,\frac{\partial}{\partial t}\right)\,\Psi(\mathbf{r}_B, \mathbf{b}, t) = 0 \tag{3.3}$$

where $H_{e\ell}$ is the Hamiltonian of the electron in the field of the two nuclei, and we have set the origin O at B. Thus

$$H_{e\ell} = -\frac{1}{2}\nabla^2_{r_B} + V_{Ae}(r_A) + V_{Be}(r_B) + V_{AB}(R) \tag{3.4}$$

with

$$V_{Ae} = -\frac{Z_A}{r_A}, \quad V_{Be} = -\frac{Z_B}{r_B}, \quad V_{AB} = \frac{Z_A Z_B}{R}. \tag{3.5}$$

It is clear that $H_{e\ell}$ is time-dependent since $\underset{\sim}{R}$ depends on t through (3.1) and $\underset{\sim}{r}_A = \underset{\sim}{r}_B - \underset{\sim}{R}(t)$.

Expansions in atomic eigenfunctions

During a fast collision, the electron, for most of the time, will be bound to one or other of the nuclei in an atomic orbital. Therefore an "atomic picture", based on expansions of the scattering wave function in atomic eigenfunctions, is physically reasonable in this case. We shall start by considering direct collisions, thus neglecting for the moment charge-exchange (electron capture) reactions which, as we shall see below, have much smaller cross sections when the velocity v is large. We denote by $\psi_m^B(\underset{\sim}{r}_B)$ the eigenfunctions of the atom $(B + e^-)$, which correspond to the energies ε_m^B and form a complete orthonormal set. Using the well-known method of the variation of the constants we write the wave function $\Psi(\underset{\sim}{r}_B, \underset{\sim}{b}, t)$ as

$$\Psi(\underset{\sim}{r}_B, \underset{\sim}{b}, t) = \sum_m a_m(\underset{\sim}{b}, t)\, \psi_m^B(\underset{\sim}{r}_B)\, e^{-i\varepsilon_m^B t} \tag{3.6}$$

and we find after substitution in (3.3) that the amplitudes a_m must satisfy the set of coupled first-order differential equations

$$i\,\dot{a}_k = \sum_m V_{km}(t)\, e^{i(\varepsilon_k^B - \varepsilon_m^B)t}\, a_m \tag{3.7}$$

where

$$V_{km}(t) = \int \psi_k^{B*}(\underset{\sim}{r}_B)[V_{AB} + V_{Ae}]\, \psi_m^B(\underset{\sim}{r}_B)\, d\underset{\sim}{r}_B \tag{3.8}$$

is the matrix element of the direct interaction $V_{AB} + V_{Ae}$ between the projectile A and the target $(B + e^-)$.

If the atom $(B + e^-)$ is initially in the state $\psi_o^B(\underset{\sim}{r}_B)$, with energy ε_o^B, we must have

$$\Psi(\underset{\sim}{r}_B, \underset{\sim}{b}, t) \underset{t \to -\infty}{\to} \Phi_o(\underset{\sim}{r}_B, t) = \psi_o^B(\underset{\sim}{r}_B)\, e^{-i\varepsilon_o^B t} \tag{3.9}$$

so that $a_m(b, t = -\infty) = \delta_{mo}$. For a collision leaving the target in the state $\psi_m^B(\underset{\sim}{r}_B)$, we have

$$\Psi(\underset{\sim}{r}_B, \underset{\sim}{b}, t) \underset{t \to +\infty}{\to} \Phi_m(\underset{\sim}{r}_B, t) = \psi_m^B(\underset{\sim}{r}_B)\, e^{-i\varepsilon_m^B t} \tag{3.10}$$

and the probability amplitude for the direct transition $0 \to m$ is given by

$$a_m(\underset{\sim}{b}, t=+\infty) = \lim_{t \to +\infty} \int \Phi_m^*(\underset{\sim}{r}_B, t)\ \Psi(\underset{\sim}{r}_B, \underset{\sim}{b}, t)\ d\underset{\sim}{r}_B \ . \qquad (3.11)$$

From the knowledge of the amplitudes $a_m(\underset{\sim}{b}, t=+\infty)$ it is easy to calculate total (integrated) direct cross sections[2,69]. In particular, the total direct (D) cross section for an inelastic transition $0 \to m$ $(m \neq 0)$ is given by

$$\sigma_{mo}^{D} = \int d^2\underset{\sim}{b}\ |a_m(\underset{\sim}{b}, t=+\infty)|^2 \ . \qquad (3.12)$$

It is also possible to correct[48] the zero order approximation in which the incident particle is undeflected and obtain angular distributions[69,70].

If we retain only M discrete terms in the expansion (3.6), the system (3.7) reduces to a set of M coupled differential equations which must be solved in order to obtain the amplitudes a_m and the cross sections. At high velocities $v \gg v_o$ a first order perturbation solution may be found by noting that all the amplitudes a_m apart from a_o remain small as time evolves, and a_o remains close to unity. The coupled equations (3.7) are then approximated by

$$i\ \dot{a}_k = V_{ko}\ e^{i(\varepsilon_k^B - \varepsilon_o^B)t} \qquad (3.13)$$

and the probability amplitude that the target atom $(B + e^-)$ will be left in the state $k \neq 0$ after the collision is given in first order by

$$a_k(\underset{\sim}{b}, t=+\infty) = -i \int_{-\infty}^{+\infty} V_{ko}(t)\ e^{i(\varepsilon_k^B - \varepsilon_o^B)t}\ dt. \qquad (3.14)$$

The result (3.14) is just the impact parameter version of the first Born approximation for the direct transition $o \to k$. The corresponding total direct first Born cross sections exhibit at high energies the characteristic fall-off corresponding to direct collisions of fast charged particles by atoms. Thus, as in the case of direct electron-atom collisions, these cross sections decrease like v^{-2} for $s-s$ transitions, while those corresponding to $s-p$ transitions fall off like $v^{-2}\ \ell n\ v$ for large v. We note that the foregoing treatment may be extended to direct ionization[1], the expression (3.14) giving a first-order approximation to the transition amplitude for ionization provided that the wave functions $\psi_k^B(\underset{\sim}{r}_B)$ are taken to be the proper positive energy eigenfunctions of the system $(B + e^-)$.

So far we have neglected charge transfer collisions and we shall now indicate how the above analysis must be modified in order to include these processes. Let us denote by $\phi_n^A(\underset{\sim}{r}_A)$ the eigenfunctions of the atom $(A + e^-)$, with eigenenergies ε_n^A. For a charge transfer reaction $A + (B + e^-) \to (A + e^-) + B$ such that the atom $(A + e^-)$ is left in the state ϕ_n^A after the collision, we have

$$\Psi(\underset{\sim}{r}_B, \underset{\sim}{b}, t) \underset{t\to+\infty}{\to} X_n(\underset{\sim}{r}_b, \underset{\sim}{b}, t) = \phi_n^A(\underset{\sim}{r}_A)\, e^{-i\varepsilon_n^A t}\, e^{i(\underset{\sim}{v}\cdot\underset{\sim}{r}_B - \frac{1}{2}v^2 t)} . \tag{3.15}$$

The electron translation factor $\exp[i(\underset{\sim}{v}\cdot\underset{\sim}{r}_B - v^2t/2)]$ which appears in (3.15) is required because the nucleus A is moving with respect to the origin of the coordinate system, which we have located here at the nucleus B. The necessity of including such translation factors was first recognised by Bates and McCarroll[71]. Taking into account (3.15), we must modify our atomic expansion (3.6), which now reads

$$\Psi(\underset{\sim}{r}_B, \underset{\sim}{b}, t) = \sum_{m=1}^{M} a_m(\underset{\sim}{b},t)\, \psi_m^B(\underset{\sim}{r}_B)\, e^{-i\varepsilon_m^B t}$$

$$+ \sum_{n=1}^{N} c_n(\underset{\sim}{b},t)\, \phi_n^A(\underset{\sim}{r}_A)\, e^{-i\varepsilon_n^A t}\, e^{i(\underset{\sim}{v}\cdot\underset{\sim}{r}_B - \frac{1}{2}v^2 t)} \tag{3.16}$$

where the sum on m and n runs over discrete states.

Assuming that the atom $(B + e^-)$ is initially in the state ψ_o^B, the probability amplitude for a charge exchange collision leaving the atom $(A + e^-)$ in the state ϕ_n^A is given by

$$c_n(\underset{\sim}{b}, t=+\infty) = \lim_{t\to+\infty} \int X_n^*(\underset{\sim}{r}_B, \underset{\sim}{b}, t)\, \Psi(r_B, b, t)\, d\underset{\sim}{r}_B \tag{3.17}$$

and the corresponding total charge exchange cross section is

$$\sigma_{no}^{cE} = \int d^2\underset{\sim}{b}\, |c_n(\underset{\sim}{b}, t=+\infty)|^2 . \tag{3.18}$$

In order to obtain the required generalisation of the coupled equations (3.7), a variational method (similar to the Hulthén-Kohn approach discussed in Section 2.1) may be used. The result is[69]

$$\begin{aligned} i(\dot{\underset{\sim}{a}} + \underline{\underline{N}}\,\dot{\underset{\sim}{c}}) &= \underline{\underline{H}}\,\underset{\sim}{a} + \underline{\underline{K}}\,\underset{\sim}{c} \\ i(\underline{\underline{N}}^\dagger\,\dot{\underset{\sim}{a}} + \dot{\underset{\sim}{c}}) &= \overline{\underline{\underline{K}}}\,\underset{\sim}{a} + \overline{\underline{\underline{H}}}\,\underset{\sim}{c} \end{aligned} \tag{3.19}$$

where $\underset{\sim}{a}$ and $\underset{\sim}{c}$ are respectively M and N-dimensional column vectors with elements a_m and c_n, and $\underline{N}$, $\underline{H}$, $\underline{\overline{H}}$, $\underline{K}$ and $\underline{\overline{K}}$ are matrices with elements[69]

$$N_{mn} = \int \psi_m^{B*}(\underset{\sim}{r}_B)\, e^{i\underset{\sim}{v}\cdot\underset{\sim}{r}}\, \phi_n^A(\underset{\sim}{r}_A)\, d\underset{\sim}{r}_B\, e^{i(\varepsilon_m^B - \varepsilon_n^A)t}$$

$$H_{mn} = \int \psi_m^{B*}(\underset{\sim}{r}_b)\, \left[\frac{Z_A Z_B}{R} - \frac{Z_A}{r_A}\right] \psi_n^B(\underset{\sim}{r}_B)\, d\underset{\sim}{r}_B\, e^{i(\varepsilon_m^B - \varepsilon_n^B)t}$$

$$\overline{H}_{mn} = \int \phi_m^{A*}(\underset{\sim}{r}_A)\, \left[\frac{Z_A Z_B}{R} - \frac{Z_B}{r_B}\right] \phi_n^A(\underset{\sim}{r}_A)\, d\underset{\sim}{r}_B\, e^{i(\varepsilon_m^A - \varepsilon_n^A)t} \qquad (3.20)$$

$$K_{mn} = \int \psi_m^{B*}(\underset{\sim}{r}_B)\, e^{i\underset{\sim}{v}\cdot\underset{\sim}{r}} \left[\frac{Z_A Z_B}{R} - \frac{Z_A}{r_A}\right] \phi_n^A(\underset{\sim}{r}_A)\, d\underset{\sim}{r}_B\, e^{i(\varepsilon_m^B - \varepsilon_n^A)t}$$

$$\overline{K}_{mn} = \int \phi_m^{A*}(\underset{\sim}{r}_A)\, e^{-i\underset{\sim}{v}\cdot\underset{\sim}{r}} \left[\frac{Z_A Z_B}{R} - \frac{Z_B}{r_B}\right] \psi_n^B(\underset{\sim}{r}_B)\, d\underset{\sim}{r}_B\, e^{i(\varepsilon_m^A - \varepsilon_n^B)t}\,.$$

The factors $\exp(\pm i\underset{\sim}{v}\cdot\underset{\sim}{r})$ which occur in N_{mn}, K_{mn} and $\overline{K}_{mn}$ represent the gain in momentum of the captured electron. As the velocity increases, these oscillatory factors lead to a rapid decrease of the matrix elements K_{mn} and $\overline{K}_{mn}$, so that charge exchange cross sections fall off much more rapidly at high energies than excitation cross sections.

The problem of the high-energy limit of charge-exchange cross sections has in fact been the subject of numerous investigations. Following the work of Drisko[72], it has been shown[73-75] that this high-energy limit is not given by the first Born approximation, but by the second Born term[76]. For ground state capture of an electron by a nucleus A of charge Z_A from a nucleus B of charge Z_B the leading term of the second Born approximation is given by

$$\sigma_{B2} = \sigma_{BK}\left[0.295 + \frac{5\pi v}{2^{11}(Z_A + Z_B)}\right] \qquad (3.21)$$

where

$$\sigma_{BK} = \pi 2^{18} \frac{(Z_A Z_B)^5}{5v^{12}} \quad \text{(a.u.)} \qquad (3.22)$$

is the asymptotic Brinkman-Kramers[78] first order cross section. Thus we see that σ_{B2} decreases like v^{-11} at large v, in contrast to σ_{B1} which falls off like v^{-12}. Drisko[72] also gave arguments to show that the third Born term modifies the coefficient 0.295 of the first term in (3.21) to the value 0.315, but does not alter the v^{-11} behaviour at large v. It should be noted, however, that the approach to the high-energy limit is slow, and that the second term in (3.21) becomes dominant only at energies $E \simeq 40(Z_A + Z_B)^2$ MeV/nucleon, whereas for $E > 9$ MeV/nucleon the radiative capture process $A + (B + e^-) \rightarrow (A + e^-) + B + h\nu$ is the dominant charge transfer mechanism[79].

Rotating coordinates

So far we have assumed that the axis of quantization of the atomic wave functions is fixed in space. However, it is also possible to make another choice, and in particular to take this axis of quantization to be the internuclear line AB (see Fig. 2). Let $\mathbf{r}'_A$ and $\mathbf{r}'_B$ refer to the position vectors of the electron from A and B, respectively, in a new (primed) coordinate system fixed with respect to AB. The trial function (3.16) will now contain functions of the type $\psi_m^B(\mathbf{r}'_B)$ and $\phi_n^A(\mathbf{r}'_A)$ instead of $\psi_m^B(\mathbf{r}_B)$ and $\phi_n^A(\mathbf{r}_A)$. Since the axis AB rotates during the collision through an angle of π, the corresponding rotation of the primed coordinate system must be taken into account in establishing the coupled equations which govern the dynamics of the collision. In particular, it should be remembered that the operation $\partial/\partial t$ must be performed with the vector $\mathbf{r}_B$ fixed in the laboratory system. Defining θ as the angle between the rotating internuclear line AB and the z-axis (which is the direction of incidence), so that $\dot{\theta} = bv/R^2$ is the rate of rotation of the internuclear axis, one finds[69] that the matrix elements H_{mn} and K_{mn} are now given respectively by

$$H_{mn} = \int \psi_m^{B*}(\mathbf{r}'_B) \left[\frac{Z_A Z_B}{R} - \frac{Z_A}{r'_A} + \dot{\theta} L_{yB'}\right] \psi_n^B(\mathbf{r}'_B) d\mathbf{r}'_B \, e^{i(\varepsilon_m^B - \varepsilon_n^B)t} , \quad (3.23)$$

$$K_{mn} = \int \psi_m^{B*}(\mathbf{r}'_B) \, e^{i\mathbf{v}\cdot\mathbf{r}} \left[\frac{Z_A Z_B}{R} - \frac{Z_A}{r'_A} + \dot{\theta} L_{yB'}\right] \phi_n^A(\mathbf{r}'_A) d\mathbf{r}'_B \, e^{i(\varepsilon_m^B - \varepsilon_n^A)t} \quad (3.23)$$

where $L_{yB'} = -i(z'_B \, \partial/\partial x'_B - x'_B \, \partial/\partial z'_B)$ is the y' component of the electronic angular momentum in the rotating frame. Similar expressions may be derived for the matrix elements $\overline{H}_{mn}$ and $\overline{K}_{mn}$. We note that in this rotating coordinate system the transitions are induced not only by the Coulomb forces but also by rotational coupling.

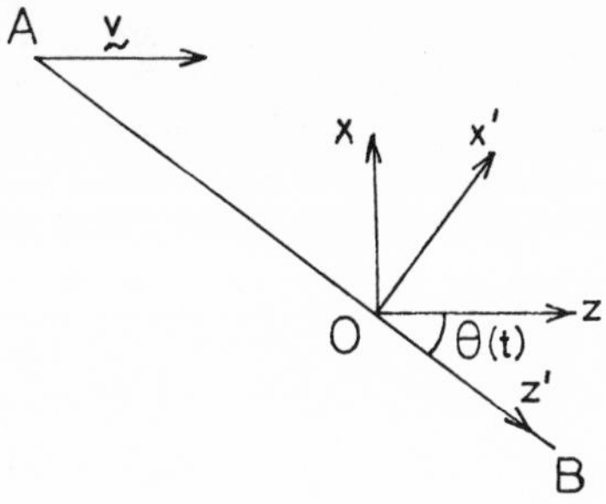

Figure 2. The fixed and rotating coordinate systems

Improved treatments of fast collisions

The atomic expansion (3.16) which we have considered above only contained truncated summations over a certain number of discrete atomic states $\psi_m^B(\mathbf{r}_B)$ and $\phi_n^A(\mathbf{r}_A)$. Unfortunately, this truncation of the expansion is often an important limitation, partly because it does not take into account the continuum states, and partly because atomic expansions are slowly convergent at short internuclear distances, where the wave function resembles that of the united atom.

Several methods have been proposed to overcome these difficulties. One possibility, introduced by Gallaher and Wilets[80], and developed recently by Shakeshaft[81] for p + H collisions, is to make the expansion in terms of Sturmian functions. These functions have the merit of forming an infinite discrete set which, in its entirety, is complete. In particular, since the Sturmian functions overlap the continuum eigenfunctions of the hydrogen atom, the coefficients of the basis vectors also contain information about the probability for ionization. However, the boundary conditions are not easily satisfied with Sturmian expansions.

Another approach which improves the flexibility of the trial function consists in adding pseudostates to the atomic expansion (3.16). Extensive calculations of this kind have been done for the p + H system by Cheshire et al[82]. An interesting alternative[83] is to use atomic wave functions with variable charges $Z(t)$ which are determined in a variational way. It is also possible to expand the scattering wave function about three, rather than two, centres, for example about the two nuclei A and B and their centre of charge[84]. A rather different method, in which a modified system of elliptical coordinates is introduced and the wave function is expanded in terms of orthogonal polynomials, has also been used recently[85].

Other modifications of the trial function include the use of "switching factors"[86] which are chosen in order to remove the difficulties associated with the electron translation factors at short internuclear distances. To see how this comes about, let us return to the atomic expansion (3.16). It is convenient to change the origin of the coordinates from B to an arbitrary point along the internuclear axis. Using (3.1), we then have

$$\Psi(\mathbf{r}_B, \mathbf{b}, t) = \sum_m a_m(\mathbf{b}, t)\, \psi_m^B(\mathbf{r}_B)\, e^{-i\varepsilon_m^B t}\, e^{-i(q\mathbf{v}\cdot\mathbf{r} + \frac{1}{2} q^2 v^2 t)}$$

$$+ \sum_n c_n(\mathbf{b}, t)\, \phi_n^A(\mathbf{r}_A)\, e^{-i\varepsilon_n^A t}\, e^{i(p\mathbf{v}\cdot\mathbf{r} - \frac{1}{2} p^2 v^2 t)} \quad . \quad (3.24)$$

Now the factors $\exp(ip\mathbf{v}.\mathbf{r})$ and $\exp(-\,iq\mathbf{v}.\mathbf{r})$, although required at large R, have the undesirable property of associating the electron with one or the other nucleus at small R. To avoid this difficulty, Schneiderman and Russek[86] suggested to use more complicated electron translation factors that reduce to the plane wave form as $R \to \infty$. This can be done by replacing in (3.24) the factors $\exp(ip\mathbf{v}.\mathbf{r})$ and $\exp(-\,iq\mathbf{v}.\mathbf{r})$ by expressions of the form $\exp[if(\mathbf{r},\mathbf{R})\mathbf{v}.\mathbf{r}]$, with the boundary condition that $f \to p$ when $R \to \infty$ with r_A finite and $f \to -\,q$ when $R \to \infty$ with r_B finite. Elsewhere the switching factors f are chosen on physical grounds in order to give the best dynamical picture of the electron motion. We shall return to this point below in discussing the "molecular picture" of slow atom-atom collisions.

The calculations described above are usually very lengthy, but under certain conditions important simplifications can be made. For example, if the charge Z_A of the projectile is much less than the charge Z_B of the target nucleus, the interaction between the incident particle A and the electron can be treated as a perturbation, and the scattering wave function can be expanded in terms of atomic states and pseudostates centred on the target atom. This method has been used by Reading et al[87-90] to calculate ionization and charge-transfer cross sections when a light, fully-stripped projectile (proton, α particle, etc...) is incident on a moderately heavy neutral target atom.

To conclude this section on fast atom (ion)-atom collisions, we remark that at relatively high energies the "continuum distorted wave" (CDW) approximation introduced by Cheshire[91] has been successfully applied to various charge-exchange reactions[92-96]. Detailed discussions of fast charge-exchange processes may be found in the review articles of Bransden[69,97,98], Basu et al[99], Shakeshaft and Spruch[75] and Belkic et al[100].

3.2. Slow atom-atom collisions

When the velocity v of the incident particle is less than the orbital velocity v_0 of the active electron(s) in the collision, it is reasonable to use a "molecular-picture" of the collision, in which the scattering wave function is expanded in molecular eigenstates of the "compound system" made of the projectile and the target. This way of treating slow atom (ion)-atom collisions, introduced by Massey and Smith[101], is known as the "perturbed stationary states" (PSS) method and will now be briefly described. Detailed accounts of the thoery of slow atom (ion)-atom collisions may be found in the books by Mott and Massey[1] and Bransden[2], and in the reviews of Bransden[69,97,98] and Briggs[103].

The perturbed stationary states method

Let us consider again our simple model of one electron interacting with two nuclei A and B. We write the total Hamiltonian of the system as

$$H = K_R + H_{e\ell} . \tag{3.25}$$

Here $K_R = -\nabla_R^2/2M$ is the kinetic energy operator of the nuclei ($M = M_A M_B/M_A + M_B$) being their reduced mass) and

$$H_{e\ell} = -\frac{1}{2}\nabla_r^2 + V(\mathbf{r},\mathbf{R}) \tag{3.26}$$

is the electronic Hamiltonian, where $V = V_{Ae} + V_{Be} + V_{AB}$ is the full potential energy[104].

According to the Born-Oppenheimer approximation, stationary molecular states are described by wave functions having the form

$$\Psi_n(\mathbf{r},\mathbf{R}) = {}_n(\mathbf{r},\mathbf{R})\ \xi_n(\mathbf{R}) \tag{3.27}$$

where $\chi_n(\mathbf{r},\mathbf{R})$ is the electronic wave function and $\xi_n(\mathbf{R})$ the nuclear wave function. At a fixed nuclear separation R the electronic molecular orbitals (MO) χ_n are solutions of the electronic Schrödinger equation

$$H_{e\ell}\ \chi_n(\mathbf{r},\mathbf{R}) = E_n(R)\ \chi_n(\mathbf{r},\mathbf{R}). \tag{3.28}$$

For each fixed value of R the solutions of (3.28) form a complete set, the eigenvalues E_n and the eigenfunctions χ_n depending parametrically on $\mathbf{R}$. For an arbitrary state of given total energy E_{tot}, the total wave function Ψ_{tot} solution of the Schrödinger equation

$$(H - E_{tot})\,\Psi_{tot} = 0 \tag{3.29}$$

may be expanded as

$$\Psi_{tot} = \sum_n \Psi_n \tag{3.30}$$

Substituting (3.30) into (3.29) and using (3.25)-(3.28), we find that

$$\sum_n \{\chi_n[K_R + E_n(R) - E_{tot}]\xi_n + \xi_n K_R \chi_n - \frac{1}{M}\,\nabla_{\mathbf{R}}\chi_n \cdot \nabla_{\mathbf{R}}\xi_n\} = 0. \tag{3.31}$$

Upon projection onto any other state χ_k of the orthonormal set of electronic eigenfunctions, one finds that the nuclear motion is governed by the equation

$$[K_R + E_k(R) - E_{tot}]\xi_k(\mathbf{R}) + \sum_n \Gamma_{kn}\,\xi_n(\mathbf{R}) = 0 \tag{3.32}$$

where

$$\Gamma_{kn} = \langle\chi_k|[K_R,\chi_n]|\rangle = -\frac{1}{2M}\,[\langle\chi_k|\nabla_R^2|\chi_n\rangle + 2\langle\chi_k|\nabla_{\mathbf{R}}|\chi_n\rangle\cdot\nabla_{\mathbf{R}}] \tag{3.33}$$

are electron-nuclear coupling terms. If these terms are neglected the equation (3.32) reduces to the nuclear equation of the Born-Oppenheimer approximation, namely

$$[K_R + E_k(R)]\,\xi_k(\mathbf{R}) = E'_{tot}\,\xi_k(\mathbf{R})\,. \tag{3.34}$$

The electron-nuclear coupling terms, arising from the presence of the nuclear kinetic energy K_R, are non-diagonal in the electronic basis of states χ_n. From (3.31) we see that these terms lead to transitions between electronic Born-Oppenheimer states, the corresponding perturbations being known as electronic-vibrational couplings in the quantum theory of molecules. In dealing with the atom-atom scattering problem, Massey and Smith[101] recognised that the Born-Oppenheimer (adiabatic) electronic states may still be used to describe the electronic states in slow atom-atom collisions. If the electron-nuclear coupling terms (3.33) are neglected, no modification of the electronic Born-Oppenheimer states can happen during the collision, so that only elastic scattering is allowed. On the other hand, if the coupling terms (3.33) are retained, transitions to other electronic states can occur.

We have already pointed out that even for slow collisions, where the colliding velocity v is smaller than a typical electronic orbital velocity v_0, the wave number k associated with the nuclear motion is often large (in a.u.), so that a classical description of this nuclear motion can be made. The total scattering wave function may then be written as[103,105]

$$\Psi_{tot}(\mathbf{r},\mathbf{R},t) = \xi(\mathbf{R},t)\ \Psi(\mathbf{r},t) \tag{3.35}$$

where the function $\xi(\mathbf{R},t)$ represents a narrow wave packet describing the nuclear motion due to some effective internuclear potential[106,107] $U(\mathbf{r})$ and the electronic wave function $\Psi(\mathbf{r},t)$ satisfies the time-dependent Schrödinger equation

$$(\tilde{H}_{e\ell} - i\frac{\partial}{\partial t})\ \Psi(\mathbf{r},t) = 0 \tag{3.36}$$

with

$$\tilde{H}_{e\ell} = -\frac{1}{2}\nabla_r^2 - \frac{Z_A}{r_A} - \frac{Z_B}{r_B} + W(\mathbf{R}) \tag{3.37}$$

and

$$W(\mathbf{R}) = \frac{Z_A Z_B}{R} - U(\mathbf{R}) \tag{3.38}$$

The potential $W(\mathbf{R})$ does not influence transition probabilities, but a consistent choice of W must be made in order to obtain angular distributions. At energies larger than 1 keV/nucleon the potential $\mathbf{U}(R)$ can be neglected, so that the classical path corresponding to the nuclear motion is a straight line trajectory. In what follows we shall assume for the sake of illustration that this simplification is allowed.

According to the (generalised) perturbed stationary state method, an approximate solution of (3.36) is attempted by choosing a trial function of the form[98,103]

$$\Psi(\mathbf{r},t) = \sum_m a_m(\mathbf{b},t)\ F_m^B(\mathbf{r},t) + \sum_n c_n(\mathbf{b},t)\ G_n^A(\mathbf{r},t) \tag{3.39}$$

where the linearly independent functions F_m^B and G_n^A are "travelling" molecular orbitals, i.e. adiabatic MO functions (or combinations of MO's) times the appropriate electron translation factors required to satisfy the boundary conditions[71]. For example, in the case of an asymmetric collision, we may choose

$$F_m^B = \chi_m^B(\mathbf{r},\mathbf{R})\ e^{-i\varepsilon_m^B t}\ e^{-i(q\mathbf{v}\cdot\mathbf{r} + \frac{1}{2}q^2v^2t)} \tag{3.40a}$$

and

$$G_n^A = \chi_n^A(\mathbf{r},\mathbf{R})\ e^{-i\varepsilon_n^A t}\ e^{i(p\mathbf{v}\cdot\mathbf{r} - \frac{1}{2}p^2v^2t)} \tag{3.40b}$$

where $\chi_m^B \to \psi_m^B(\mathbf{r}_B)$ and $\chi_n^A \to \phi_n^A(\mathbf{r}_A)$ as $R \to \infty$. The coupled equations resulting from the choice (3.39) of trial function may again be written in the form (3.19), where typical direct and exchange coupling matrix elements are given respectively by

$$H_{mn} = \langle F_m^B | H_{e\ell} - i \frac{\partial}{\partial t} | F_n^B \rangle,$$
$$K_{mn} = \langle F_m^B | H_{e\ell} - i \frac{\partial}{\partial t} | G_n^A \rangle. \qquad (3.41)$$

As in the case of the atomic expansion (3.24) the plane wave factors $\exp(i p \mathbf{v} . \mathbf{r})$ and $\exp(- i q \mathbf{v} . \mathbf{r})$ are only appropriate in the MO expansion (3.39) when $R \to \infty$. For finite R they present the undesirable feature of constraining the electron to move with one or the other nucleus. Following the suggestion of Schneiderman and Russek[86], already discussed above in connection with the expansion (3.24), this difficulty can be avoided by introducing more complicated electron translation factors of the form $\exp[if(\mathbf{r},\mathbf{R})\mathbf{v}.\mathbf{r}]$, where f is a switching factor such that $f \to p$ when $R \to \infty$ and $r_A \ll r_B$, while $f \to -q$ if $R \to \infty$ and $r_B \ll r_A$. In addition, one may also require that $f \to 0$ when $R \to 0$, so that at short distances the electron occupies an MO with zero velocity with respect to the centre of mass. The conditions to be imposed on the switching factors, together with specific choices of them, have been analysed by various authors[86,108,111].

Turning now to the evaluation of coupling matrix elements, we note that when acting on molecular wave functions χ_k it should be remembered that the operation $\partial/\partial t$ is to be performed with the electron coordinate $\mathbf{r}$ fixed in the laboratory frame. Thus, if we denote by $\chi_k(\mathbf{r}',\mathbf{R})$ the MO's quantized along the molecular axis AB we have[103]

$$\frac{\partial}{\partial t} \chi_k(\mathbf{r}',R) = (v_R \frac{\partial}{\partial R} - i \dot{\theta} L_{y'}) \chi_k(\mathbf{r}',R) \qquad (3.42)$$

where $v_R = dR/dt$ is the radial velocity, $\dot{\theta} = bv/R^2$ is the angular velocity of the internuclear axis (v being the initial collision velocity) and $L_{y'}$ the y' component of the electronic angular momentum in the rotating coordinate system. The two parts $\partial/\partial R$ and $L_{y'}$ of the coupling are called the radial and rotational coupling, respectively. We note that the radial coupling connects MO's having the same angular symmetry (i.e. $\sigma \leftrightarrow \sigma$, $\pi \leftrightarrow \pi$, etc...) while the rotational coupling connects states of differing angular symmetry, e.g. $\sigma \leftrightarrow \pi$. The radial coupling often varies rapidly near an avoided crossing of two molecular potential curves in the adiabatic basis. It is then useful to work in a new, "diabatic" basis chosen in such a way that the radial coupling remains small. A detailed discussion of this problem may be found in the reviews by Briggs[102] and Sidis[112].

Molecular correlation diagrams, the promotion model and inner-shell excitation

In order to decide which MO's are important in the coupled-state calculations, it is useful to draw a correlation diagram, which relates the energy levels in the separated-atom (SA) and united-atom (UA) limits, and therefore summarises the qualitative features of the energy levels as a function of the internuclear distance R. A simple example of a correlation diagram is shown in Fig. 3 for the case of the H_2^+ molecular ion.

A key feature of correlation diagrams for the analysis of slow atom (ion)-atom collisions is the occurrence of the promotion phenomenon, i.e. the increase in principal quantum number of certain of the MO's during the transition from $R = \infty$ to $R = 0$. This is illustrated in Fig. 3, where we see that the 1s separated-atom level is correlated with the 2p $(m = 0)$ level of the united atom by the $2p\sigma_u$ MO. As the two protons approach each other, the $2p\sigma_u$ MO is therefore "promoted" from 1s in the SA to 2p in the UA. The importance of this promotion is that the $2p\sigma_u$ and $2p\pi_u$ MO's come very close in energy as $R \to 0$, their energy

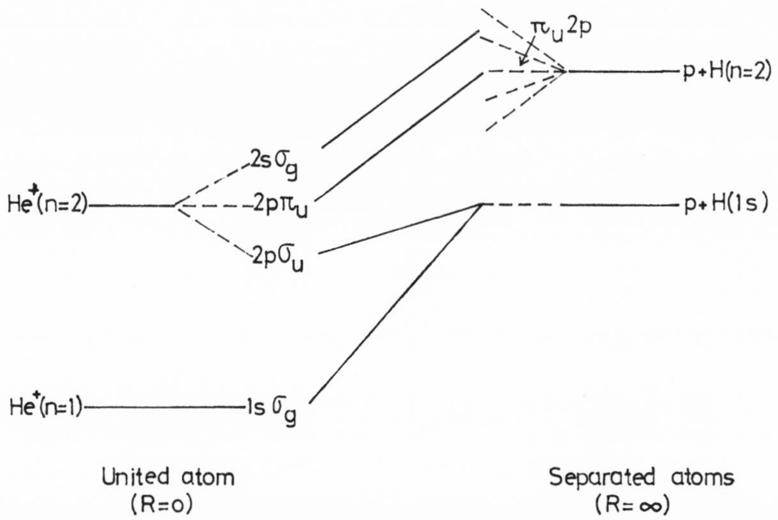

Figure 3. The correlation diagram for the low-lying energy levels of H_2^+. The internuclear repulsion energy is not included.

separation decreasing dramatically from the hydrogen 1s - 2p energy difference in the SA limit to the degeneracy of the 2p (m = 0) and 2p (|m| = 1) levels of He^+ in the UA limit. This decrease in energy separation enhances the rotational coupling between the two MO's. Since the $2p\pi_u$ MO in turn dissociates to p + H(2p) in the SA limit (see Fig. 3), the promotion mechanism leads in this case to sizeable H(2p) excitation at very low energies[113-114].

The promotion phenomenon is a major ingredient of the Fano-Lichten model which has been applied successfully to estimate the excitation probability for inner-shell electrons in slow homonuclear[115-116] and heteronuclear[117] heavy-ion atom collisions. Fano and Lichten recognised that the inner-shell electrons, being close to the nuclei, move essentially independently during the collision under the influence of the field of the colliding nuclei. This means that these inner-shell electrons occupy one-electron MO's corresponding to the molecular ion formed during the collision. Thus the results of the one-electron, two nuclei collision problem discussed above can be used directly to analyse inner-shell excitation processes in slow ion-atom collisions. In particular, according to the "promotion model", excitation of inner-shell electrons occurs with highest probability by transitions from an inner-shell MO into an empty MO with which it may become degenerate during the collision. Another interesting consequence of the similarity between the one-electron, two nuclei collision problem and inner-shell collision processes is the existence of scaling laws, as shown by Briggs and Macek[118] for symmetrical ($Z_A = Z_B$) inner-shell excitation and extended by Taulbjerg et al[119] to asymmetric (heteronuclear) collisions.

Coupled-state molecular calculations

The Fano-Lichten analysis may be considered as a first step within the framework of the perturbed stationary states method, in which the important MO's for inner-shell excitation are identified and the dominant transition probabilities are assessed qualitatively. However, coupled state molecular calculations must in general be performed in order to obtain quantitative results for inner-shell excitation processes, as well as for other slow atom (ion)-atom collisions.

The simplest non-trivial case is the two-state problem, for which various approximate analytical solutions such as the Landau-Zener-Stückelberg[120] and the Demkov[121] model have been proposed. Unfortunately, these approximate treatments are often inaccurate, so that a numerical solution of the PSS coupled equations is necessary. Moreover, in many cases of interest the two-state approximation itself is not adequate, and multi-state PSS coupled equations must be solved. Detailed discussions of such coupled-state calculations may be found in the review articles of Bransden[69,97,98] and Briggs[103].

REFERENCES

1. N. F. Mott and H. S. W. Massey, The Theory of Atomic Collisions (3rd ed, Oxford, 1965).
2. B. H. Bransden, Atomic Collision Theory (Benjamin, New York, 1970).
3. See for example R. H. J. Jansen, F. J. de Heer, H. J. Luyken, B. van Wingerden and H. J. Blaauw, J. Phys. B. 9, 185 (1976).
4. I. V. Hertel and W. Stoll, Adv. Atom. Molec. Phys. 13, 113 (1977).
5. J. Kessler, Polarized Electrons (Springer-Verlag, Berlin, 1976).
6. M. Eminyan, K. B. MacAdam, J. Slevin and H. Kleinpoppen, Phys. Rev. Letters 31, 576 (1973); J. Phys. B. 7, 1519 (1974).
7. G. Vassilev, G. Rahmat, J. Slevin and J. Baudon, Phys. Rev. Letters 34, 444 (1975).
8. D. Paul, K. Jung, E. Schubert and H. Ehrhardt, in The Physics of Electronic and Atomic Collisions, ed. by J. S. Risley and R. Geballe (Univ. of Washington Press, Seattle, 1975), p.194.
9. E. Weigold, S. T. Hood, I. Fuss and A. J. Dixon, J. Phys. B. 10, L 623 (1977).
10. A general survey of theoretical methods for electron-atom scattering may be found in P. G. Burke and J. F. Williams, Phys. Reports 34 C, 325 (1977).
11. L. Castillejo, I. Percival and M. J. Seaton, Proc. Roy. Soc. A 254, 259 (1960).
12. P. G. Burke and M. J. Seaton, Methods in Comput. Phys. 10, 1 (1971).
13. A. Temkin, Phys. Rev. 107, 1004 (1957); see also R. J. Drachman and A. Temkin, in Case Studies in Atomic Collision Physics II, ed. by M. R. C. McDowell and E. W. McDaniel (North-Holland Publ. Cy, Amsterdam, 1972), p.399.
14. C. Schwartz, Phys. Rev. 124, 1468 (1961).
15. M. Gailitis, Soviet Physics (JETP) 20, 107 (1965).
16. P. G. Burke and A. J. Taylor, Proc. Phys. Soc. 88, 549 (1966).
17. R. K. Nesbet, Adv. Atom. Molec. Phys. 13, 315 (1977).
18. R. Damburg and E. Karule, Proc. Phys. Soc. 90, 637 (1967).
19. Vo Ky Lan, M. Le Dourneuf and P. G. Burke, J. Phys. B. 9, 1065 (1976).
20. E. P. Wigner, Phys. Rev. 70, 15, 606 (1946); E. P. Wigner and L. Eisenbud, Phys. Rev. 72, 29 (1947).
21. P. G. Burke, A. Hibbert and W. D. Robb, J. Phys. B. 4, 153 (1971).
22. P. G. Burke and W. D. Robb, J. Phys. B. 5, 44 (1972).
23. P. G. Burke and W. D. Robb, Adv. Atom. Molec. Phys. 11, 143 (1975).

24. A. V. Hazi and H. S. Taylor, Phys. Rev. A 1, 1109 (1970).
25. H. S. Taylor, Adv. Chem. Phys. 18, 91 (1970).
26. E. J. Heller and H. A. Yamani, Phys. Rev. A 9, 1201 (1974); A 9, 1209 (1974).
27. J. T. Broad and W. P. Reinhardt, J. Phys. B. 9, 1491 (1976), Phys. Rev. A 14, 2159 (1976).
28. T. N. Rescigno and W. P. Reinhardt, Phys. Rev. A 10, 1584 (1974).
29. M. J. Seaton, Adv. Atom. Molec. Phys. 11, 83 (1975).
30. A detailed discussion of the optical potential theory may be found in C. J. Joachain, Quantum Collision Theory (North-Holland Publ. Cy, Amsterdam, 1975), Chapter 20.
31. H. Feshbach, Ann. Phys. (N.Y.) 5, 357 (1958); 19, 287 (1962).
32. T. F. O'Malley and S. Geltman, Phys. Rev. 137, A 1344 (1965)
33. A. K. Bhatia, A. Temkin and J. F. Perkins, Phys. Rev. 153, 177 (1967); A. K. Bhatia and A. Temkin, Phys. Rev. 182, 15 (1969); A. K. Bhatia, P. G. Burke and A. Temkin, Phys. Rev. A 8, 21 (1973); A 10, 459 (1974); A. K. Bhatia and A. Temkin, Phys. Rev. A 8, 2184 (1973); A 10, 458 (1974), A 11, 2018 (1975).
34. Gy Csanak, H. S. Taylor and R. Yaris, Adv. Atom. Molec. Phys. 7, 287 (1971); L. D. Thomas, Gy Csanak, H. S. Taylor and B. S. Yarlagadda, J. Phys. B. 7, 1719 (1974).
35. C. J. Joachain and C. Quigg, Rev. Mod. Phys. 46, 279 (1974).
36. B. H. Bransden and M. R. C. McDowell, Phys. Reports 30 C, 207 (1977); 46 C, 249 (1978).
37. F. W. Byron, Jr. and C. J. Joachain, Phys. Reports 34 C, 233 (1977).
38. We shall not consider here the case of relativistic collisions.
39. C. J. Joachain, Quantum Collision Theory (North-Holland Publ. Cy, Amsterdam, 1975), Chapter 19.
40. K. L. Bell and A. E. Kingston, Adv. Atom. Molec. Phys. 10, 53 (1974).
41. H. S. W. Massey and C. B. O. Mohr, Proc. Roy. Soc. A 146, 880 (1934).
42. A. R. Holt and B. L. Moiseiwitsch, J. Phys. B. 1, 36 (1968); A. R. Holt, J. Hunt and B. L. Moiseiwitsch, J. Phys. B. 4, 1318 (1971).
43. See for example F. W. Byron, Jr. and C. J. Joachain, J. Phys. B. 10, 207 (1977); Phys. Reports 34 C, 233 (1977); D. P. Dewangan and H. R. J. Walters, J. Phys. B. 10, 637 (1977).
44. A. R. Holt, J. Phys. B. 5, L 6 (1972).
45. C. J. Joachain, in Electronic and Atomic Collisions, ed. by G. Watel (North Holland Publ. Cy, Amsterdam, 1978), p.71.
46. F. W. Byron, Jr., C. J. Joachain and B. Piraux, to be published.
47. Detailed discussions of the eikonal approximation and related methods may be found in F. W. Byron, Jr., C. J. Joachain and E. H. Mund, Phys. Rev. D 8, 2622 (1973); D 11, 1662 (1975); F. W. Byron, Jr. and C. J. Joachain, Phys. Reports 34 C, 233 (1977).

48. R. J. Glauber, in Lectures in Theoretical Physics, Vol. 1, ed. W. E. Brittin (Interscience, New York, 1959), p.315.
49. D. M. Chase, Phys. Rev. 104, 838 (1956).
50. F. W. Byron, Jr. and C. J. Joachain, Phys. Rev. A 8, 1267 (1973); A 8, 3266 (1973); J. Phys. B. 7, L 212 (1974); J. Phys. B. 8, L 284 (1975); J. Phys. B. 10, 207 (1977).
51. M. H. Mittleman and K. M. Watson, Phys. Rev. 113, 198 (1959).
52. M. H. Mittleman and K. M. Watson, Ann. Phys. (N.Y.) 10, 268 (1960).
53. C. J. Joachain and M. H. Mittleman, Phys. Rev. A 4, 1492 (1971); F. W. Byron, Jr. and C. J. Joachain, Phys. Rev. A 9 2559 (1974).
54. F. W. Byron, Jr. and C. J. Joachain, Phys. Letters A 49, 306 (1974); Phys. Rev. A 15, 128 (1977); C. J. Joachain, R. Vanderpoorten, K. H. Winters and F. W. Byron, Jr., J. Phys. B. 10, 227 (1977).
55. R. Vanderpoorten, J. Phys. B. 8, 926 (1975).
56. J. B. Furness and I. E. McCarthy, J. Phys. B. 6, 2280 (1973).
57. P. G. Burke and T. G. Webb, J. Phys. B. 3, L 131 (1970); P. G. Burke and J. F. B. Mitchell, J. Phys. B. 6, 620 (1973); J. Callaway, M. R. C. McDowell and L. A. Morgan, J. Phys. B. 8, 2181 (1976); J. Phys. B. 9, 2043 (176).
58. B. H. Bransden and J. P. Coleman, J. Phys. B. 5, 537 (1972); K. H. Winters, C. D. Clark, B. H. Bransden and J. P. Coleman, J. Phys. B. 6, L 247 (1973); J. Phys. B. 7, 788 (1974); see also the reviews by B. H. Bransden and M. R. C. McDowell, Phys. Reports 30 C, 207 (1977); 46 C, 249 (1978).
59. See for example C. J. Joachain, loc. cit. (ref. 39), Chapter 17.
60. W. N. Shelton, E. S. Leherissey and D. H. Madison, Phys. Rev. A 3, 242 (1971); D. H. Madison and W. N. Shelton, Phys. Rev. A 7, 499 (1973).
61. J. C. Y. Chen, C. J. Joachain and K. M. Watson, Phys. Rev. A 5, 2460 (1972).
62. C. J. Joachain and R. Vanderpoorten, J. Phys. B. 6, 622 (1973); J. Phys.B. 7, 817 (1974).
63. M. R. C. McDowell, L. A. Morgan and V. P. Myerscough, J. Phys. B. 6, 1435 (1974); J. Phys. B. 7, L 195 (1974); T. Scott and M. R. C. McDowell, J. Phys. B. 8, 1851 (1975), J. Phys. B. 8, 2369 (1975), J. Phys. B. 9, 2235 (1976).
64. K. L. Baluja and H. S. Taylor, J. Phys. B. 9, 829 (1976).
65. I. E. McCarthy and E. Wiegold, Phys. Reports 27 C, 275 (1976); E. Wiegold and I. E. McCarthy, Adv. Atom. Molec. Phys. 14, 127 (1978).
66. D. H. Madison, T. C. Calhoun and W. N. Shelton, Phys. Rev. A 16, 552 (1977).
67. B. H. Bransden, J. J. Smith and K. H. Winters, J. Phys. B. 11, 3095 (1978); J. J. Smith, K. H. Winters and B. H. Bransden, J. Phys. B. 12, 1723 (1979).
68. A detailed discussion of several of the methods considered in this section may be found in N. F. Mott and H. S. W. Massey, loc. cit. (ref. 1), B. H. Bransden, loc. cit. (ref. 2) and in

J. P. Coleman and M. R. C. McDowell, Introduction to the Theory of Ion-Atom Collisions (North-Holland Publ. Cy, Amsterdam, (1970).
69. B. H. Bransden, Rep. Progr. Phys. 35, 949 (1972).
70. R. McCarroll and A. Salin, J. Phys. B. 1, 163 (1968).
71. D. R. Bates and R. McCarroll, Proc. Roy. Soc. A 245, 175 (1958).
72. R. M. Drisko, Ph.D. Thesis, Carnegie Institute of Technology (1955).
73. A. M. Brodskii, V. S. Potapov and V. V. Tolmadev, Soviet Physics (J.E.T.P.) 31, 144 (1970).
74. K. Dettmann, in Springer Tracts in Modern Physics, ed. G. Hohler (Springer-Verlag, Berlin, 1971), Vol. 58, p.119.
75. R. Shakeshaft and L. Spruch, Rev. Mod. Phys. 51, 369 (1979).
76. This result is not surprising if we recall our discussion of high-energy rearrangement (exchange) electron-atom scattering of Section 2.2.
77. J. S. Briggs, J. Phys. B. 10, 3075 (1977).
78. H. C. Brinkman and H. A. Kramers, Proc. Acad. Sci. Amst. 33, 973 (1930).
79. J. S. Briggs and K. Dettmann, Phys. Rev. Letters 33, 1123 (1974), J. Phys. B. 10, 1113 (1977).
80. D. F. Gallaher and L. Wilets, Phys. Rev. 169, 139 (1968).
81. R. Shakeshaft, Phys. Rev. A 14, 1626 (1976).
82. I. M. Cheshire, D. F. Gallaher and A. J. Taylor, J. Phys. B. 3, 813 (1970).
83. I. M. Cheshire, J. Phys. B. 1, 428 (1968).
84. D. G. M. Anderson, M. J. Antol and M. B. McElroy, J. Phys. B. 7, L 118 (1974).
85. H. G. Morrison and U. Opik, J. Phys. B. 11, 473 (1978).
86. S. B. Schneiderman and A. Russek, Phys. Rev. 181, 311 (1969).
87. J. F. Reading, A. L. Ford and E. Fitchard, Phys. Rev. Letters 36, 573 (1976).
88. A. L. Ford, E. Fitchard and J. F. Reading, Phys. Rev. A 16, 133 (1977).
89. J. F. Reading and A. L. Ford, J. Phys. B. 12, 1367 (1979).
90. A. L. Ford, J. F. Reading and R. L. Becker, J. Phys. B. 12, 2905 (1979).
91. I. M. Cheshire, Proc. Phys. Soc. (London) 84, 89 (1964).
92. A. Salin, J. Phys. B. 3, 937 (1970).
93. R. Gayet, J. Phys. B. 5, 483 (1972).
94. Dz Belkic and R. K. Janev, J. Phys. B. 6, 1020 (1973).
95. Dz Belkic and R. Gayet, J. Phys. B. 10, 1911 (1977), J. Phys. B. 10, 1923 (1977); Dz Belkic and R. McCarroll, J. Phys. B. 10, 1933 (1977).
96. K. E. Banyard and B. J. Szuster, Phys. Rev. A 15, 435 (1977).
97. B. H. Bransden, Physicalia Mag. 1 (4), 11 (1979).
98. B. H. Bransden, "Theoretical Models for Charge Exchange", this volume.
99. D. Basu, S. C. Mukherjee and D. P. Sural, Phys. Reports 42 C, 145 (1978).

100. Dz Belkic, R. Gayet and A. Salin, Phys. Reports (to be published).
101. H. S. W. Massey and R. A. Smith, Proc. Roy. Soc. A 142, 142 (1933).
102. For a detailed account of the perturbed stationary states method, see for example N. F. Mott and H. S. W. Massey, loc. cit. (ref. 1), and B. H. Bransden, loc. cit. (ref. 2, 69).
103. J. S. Briggs, Rep. Progr. Phys. 39, 217 (1976).
104. The internuclear potential V_{AB}, which depends only on R, may be neglected in the electronic problem.
105. F. Drepper, unpublished results (1975), quoted in J. S. Briggs, loc. cit. (ref. 103).
106. C. Gaussorgues, C. Le Sech, F. Masnou-Seeuws, R. McCarroll and A. Riera, J. Phys. B. 8, 239 (1975); B 8, 253 (1975).
107. J. K. Cayford and W. R. Fimple, J. Phys. B. 9, 3055 (1976).
108. W. B. Thorson and H. Levy, Phys. Rev. 181, 240 (1969).
109. M. H. Mittleman and H. Tai, Phys. Rev. A 8, 1880 (1973).
110. K. Taulbjerg, J. Vaaben and B. Fastrup, Phys. Rev. A 12, 2325 (1975).
111. W. R. Thorson and J. B. Delos, Phys. Rev. A 18, 117 (1978); A 18, 156 (1978).
112. V. Sidis, in The Physics of Electronic and Atomic Collisions, ed. by J. S. Risley and R. Geballe (Univ. of Washington Press, Seattle, 1975), p.295.
113. D. R. Bates and D. A. Williams, Proc. Phys. Soc. 83, 425 (1964).
114. H. Rosenthal, Phys. Rev. Letters 27, 635 (1971).
115. U. Fano and W. Lichten, Phys. Rev. Letters 14, 627 (1965).
116. W. Lichten, Phys. Rev. 164, 131 (1967).
117. M. Barat and W. Lichten, Phys. Rev. A 6, 211 (1972).
118. J. S. Briggs and J. H. Macek, J. Phys. B. 6, 982 (1973).
119. K. Taulbjerg, J. S. Briggs and J. Vaaben, J. Phys. B. 9, 1351 (1976).
120. L. D. Landau, Z. Phys. Sowjet 2, 46 (1932); C. Zener, Proc. Roy. Soc. A 137, 696 (1932); E. C. G. Stückelberg, Helv. Phys. Acta 5, 320 (1932).
121. Yu. N. Demkov, Sov. Phys. (J.E.T.P.) 18, 138 (1964); see also R. E. Olson, Phys. Rev. A 6, 1822 (1972).

THEORETICAL MODELS FOR CHARGE EXCHANGE

B.H. Bransden

The University of Durham

Durham City, England

INTRODUCTION

In developing the theory of charge exchange[1], it is convenient to concentrate on a one electron system containing two nuclei, A and B. If the electron is initially attached to the nucleus B, the reactions occuring are of the form

$$A + (B + e^-) \rightarrow A + (B+e^-)^*, \text{ excitation,} \quad (1)$$

$$\rightarrow (A + e^-)^* + B, \text{ charge exchange,} \quad (2)$$

$$\rightarrow A + B + e^-, \text{ ionisation.} \quad (3)$$

In ionisation, the cross-section is enhanced by the final state interaction between the ejected electron and the incident nucleus A and when the electron moves with a small relative velocity to A, the process is often called 'charge exchange into the continuum'[2], but we will not deal with this process here.

The energy range of interest, in terms of the laboratory energy of the nucleus A, incident on a stationary target is from zero to 10 MeV per nucleon. At greater energies, charge exchange occurs predominantly by the radiative process[3]

$$A + (B + e^-) \rightarrow (A + e^-)^* + B + h\nu. \quad (4)$$

At sufficiently high velocities, the binding of the electron to the target nucleus can be neglected, and the cross-section approaches that for capture of a free electron.

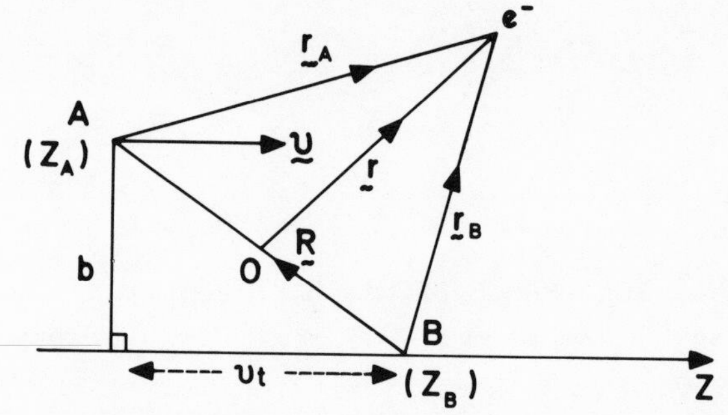

Fig.1: The coordinate system for a system of one electron and two nuclei.

The one electron theory, which we shall describe, can be generalised to many electron system, in an independent electron approximation; but for some processes, such as two-electron capture, correlation effects may be significant. It is interesting to note that in heavy ion collisions, the inner shell electrons are relatively uninfluenced by the outer electrons, and K shell processes can be described quantitatively by the one electron theory[4].

Provided the kinetic energy of the ion A, is greater than $\sim$ 1 eV per nucleon, the wavelength associated with its motion, $(h/M_A v)$, is much smaller than the interaction region, which is determined by the dimensions of the atoms $(A + e^-)$ and $(B + e^-)$. Under these circumstances nuclear motion can be described by a narrow wavepacket, the centre of which follows a classical path defined by some effective internuclear potential $U(\mathbf{R})$. For each value of the impact parameter (see Fig.1), the relative position vector $\mathbf{R}$ of the two nuclei, will be some function of t, the time, which can be calculated once $U(\mathbf{R})$ is known,

$$\mathbf{R} = \mathbf{R}\,(b,t). \tag{5}$$

The electronic wavefunction $\Psi(\mathbf{r},t)$ satisfies the time dependent Schrödinger equation (atomic units are used unless otherwise stated):

$$(H - i\frac{\partial}{\partial t})\,\Psi(\mathbf{r},t) = 0, \tag{6}$$

where

$$H = -\tfrac{1}{2}\nabla_r^2 - \frac{Z_B}{r_B} - \frac{Z_A}{r_A} + W(\mathbf{R}),$$

and

$$\mathbf{r}_B(t) = \mathbf{r} + \mathbf{R}(t)/2; \; \mathbf{r}_A(t) = \mathbf{r} - \mathbf{R}(t)/2. \tag{7}$$

The origin of the coordinate system is the mid-point of the internuclear line AB, and the operation $\partial/\partial t$ is to be taken with the position vector $\mathbf{r}$, fixed in the laboratory frame. Since the potential $U(\mathbf{R})$ has already been taken into account when defining the trajectory $\mathbf{R}(t)$, the effective potential $W(\mathbf{R})$ is to be identified with $\left[Z_A Z_B/R - U(\mathbf{R})\right]$. The potential $W(\mathbf{R})$ can be removed by a phase transformation, and for this reason does not influence transition probabilities, which depend on the modulus of Ψ; but a consistent choice of W is necessary if angular distributions are to be calculated accurately.

The effective internuclear potential U(R), is itself determined by the electronic wave-function. It is of the form[5]:

$$U(R) = \frac{1}{\langle\Psi|\Psi\rangle} \langle\Phi|H - i\frac{\partial}{\partial t}|\Psi\rangle - \frac{1}{2\mu}\langle\Psi|\nabla_R^2|\Psi\rangle - \frac{1}{\mu}\langle\Psi|\nabla_{\mathbf{R}}|\Psi\rangle . \nabla_{\mathbf{R}}, \tag{8}$$

where the terms containing ∇_R^2 and ∇_R can usually be neglected. Equations (5), (6) and (8) form a coupled system, which is rather complicated (but not impossible) to solve. Fortunately, it is often sufficiently accurate to employ a simple screened Coulomb potential, independent of Ψ, for U(R), and at the higher energies (> 1 keV per nucleon), U(R) can be ignored, in which case the nuclear motion is rectilinear and is defined by:

$$\mathbf{R}(t) = \mathbf{b} + \mathbf{v}\,t; \; \mathbf{b}.\mathbf{v} = o, \tag{9}$$

where $\mathbf{v}$ is the (constant) velocity of A, and $\mathbf{b}$ is a two dimensional impact parameter vector.

The unperturbed solutions of (6) can be expressed in terms of wave-functions $\phi_n(\mathbf{r}_B)$ representing the atom ($B + e^-$) and $X_n(\mathbf{r}_A)$ representing the atom ($A + e^-$) and satisfying respectively:

$$\begin{aligned} &(-\tfrac{1}{2}\nabla_{r_B}^2 - Z_B/r_B - \varepsilon_n)\,\phi_n(\mathbf{r}_B) = o, \\ &(-\tfrac{1}{2}\nabla_{r_A}^2 - Z_A/r_A - \eta_m)\,\chi_m(\mathbf{r}_A) = o. \end{aligned} \tag{10}$$

The functions

$$\Phi_n(\mathbf{r},t) = \phi_n(\mathbf{r}_B)\,\exp{-i}\{\varepsilon_n t + \tfrac{1}{8}v^2 t - \tfrac{1}{2}\mathbf{v}.\mathbf{r}\}, \tag{11}$$

satisfy equation (6) in the limit $t \to \pm\infty$ and $r_A >> r_B$, and the functions

$$X_m(\underset{\sim}{r},t) = \chi_m(\underset{\sim}{r}_A) \exp{-i} \{\eta_m t + \tfrac{1}{8} v^2 t + \tfrac{1}{2} \underset{\sim}{v}.\underset{\sim}{r}\} \tag{12}$$

satisfy equation (6) in the limit $t \to \pm\infty$ and $r_B >> r_A$.

The factors $\exp i\, (\pm \tfrac{1}{2} \underset{\sim}{v}.\underset{\sim}{r} - \tfrac{1}{4} v^2 t)$ are required to account for the momentum of the electron, bound to A or B, when A and B move relative to the origin O, and without these factors the boundary conditions cannot be satisfied and the Gallilean invariance of the theory is destroyed.

If the system is initially in the ith level of $(B + e^-)$ the boundary condition is

$$\Psi\,(\underset{\sim}{r},t) \to \Phi_i\,(\underset{\sim}{r},t) \text{ as } t \to -\infty \tag{13}$$

The probability amplitude for finding the system in the jth level of $(A + e^-)$ after the collision is

$$C_{ji}(b) = \underset{t\to+\infty}{\ell t} \int d\underset{\sim}{r}\; X_j^*(\underset{\sim}{r},t)\, \Psi\,(\underset{\sim}{r},t) \tag{14}$$

and the cross-section for capture into the level j is

$$Q_{ji} = 2\pi \int_0^\infty |C_{ji}(b)|^2\, b db \tag{15}$$

This completes the general formulation of the problem, using the semi-classical impact parameter method. Alternative formulations, in which the nuclear motion is described wave-mechanically can be given, but except at very low energies, these are equivalent to one presented here.

COUPLED STATE APPROXIMATIONS

An approximate solution of equation (6) may be attempted by expressing $\Psi(\underset{\sim}{r},t)$ as a combination of linearly independent functions F_n and G_m, which are required to have the asymptotic forms Φ_n and X_m respectively:

$$F_n \to \Phi_n;\; G_m \to X_m;\; t \to \pm\infty. \tag{16}$$

If we write, a trial function Ψ_τ as

$$\Psi_\tau(\underset{\sim}{r},t) = \sum_{n=1}^{N} a_{ni}(t) F_n(\underset{\sim}{r},t) + \sum_{m=1}^{M} c_{mi}(t)\, G_m(\underset{\sim}{r},t), \tag{17}$$

coupled first order differential equations for the coefficients $a_{ni}(t)$ and $c_{mi}(t)$ can be found by requiring

$$\int d\underset{\sim}{r}\; F_n^*(\underset{\sim}{r},t)\;(H-i\frac{\partial}{\partial t})\;\Psi_\tau(\underset{\sim}{r},t) = 0\,,\; n = 1,2\;..N\,, \tag{18}$$

$$\int d\underset{\sim}{r}\; G_m^*(\underset{\sim}{r},t)\;(H-i\frac{\partial}{\partial t})\;\Psi_\tau(\underset{\sim}{r},t) = 0,\; m = 1,2\;..M.$$

These equations, which are consisted with the variational principle, constitute a set of (N+M) equations of the form

$$i\;(\underset{\sim}{S}_{11}\dot{\underset{\sim}{a}} + \underset{\sim}{S}_{12}\,\dot{\underset{\sim}{c}}) = \underset{\sim}{M}_{11}\,\underset{\sim}{a} + \underset{\sim}{M}_{12}\,\underset{\sim}{c}\,, \tag{19}$$

$$i\;(\underset{\sim}{S}_{21}\dot{\underset{\sim}{a}} + S_{22}\,\dot{\underset{\sim}{c}}) + \underset{\sim}{M}_{21}\,\underset{\sim}{a} + \underset{\sim}{M}_{22}\,\underset{\sim}{c}\,,$$

where a and c and M dimensional column vectors with elements $a_{ni}(t)$ and $c_{ni}(t)$, and the boundary conditions are:

$$a_{ni}(t) \to \delta_{ni};\; c_{ni}(t) \to o, \qquad t \to -\infty,$$

$$c_{ni}(t) \to c_{ni}(t), \qquad t \to +\infty. \tag{20}$$

The matrix $\underset{\sim}{S}$ is known as the overlap matrix and the elements of the sub-matrices $\underset{\sim}{S}_{ij}$ are

$$(\underset{\sim}{S}_{11})_{nm} = \langle F_n|F_n\rangle;\; (\underset{\sim}{S}_{12})_{nm} = \langle F_n|G_m\rangle;\; (\underset{\sim}{S}_{21})_{nm} = \langle G_m|F_n\rangle;$$

$$(\underset{\sim}{S}_{22})_{nm} = \langle G_n|G_m\rangle\;. \tag{21}$$

The interaction matrix $\underset{\sim}{M}$ has sub-matrices $\underset{\sim}{M}_{ij}$ with elements

$$(\underset{\sim}{M}_{11})_{nm} = \langle F_n|H-i\frac{\partial}{\partial t}|F_m\rangle;\; (\underset{\sim}{M}_{12})_{nm} = \langle F_n|H-i\frac{\partial}{\partial t}|G_m\rangle$$

$$(\underset{\sim}{M}_{22})_{nm} = \langle G_m|H-i\frac{\partial}{\partial t}|F_n\rangle;\; (\underset{\sim}{M}_{21})_{nm} = \langle G_n|H-i\frac{\partial}{\partial t}|F_n\rangle. \tag{22}$$

Provided no further approximations are made, the coupled equations preserve conservation of probability, and if Ψ_t is normalised at $t = t_o$, it remains normalised for all t:-

$$\frac{d}{dt}\;\langle\Psi_t|\Psi_t\rangle = o \tag{23a}$$

From this condition it can be shown that

$$-\,i\,\dot{\underset{\sim}{S}} = \underset{\sim}{M} - \underset{\sim}{M}^\dagger \tag{23b}$$

and this is a useful check on numerical work.

LOW ENERGIES

When the velocity of the incident ion v, is less than the Bohr velocity v_B, of the electron in the target atom, it is usual to express the functions F_n and G_m as combinations of molecular orbitals (MO), which are defined as solutions of the electronic Schrödinger equation at a fixed nuclear separation

$$\left[-\tfrac{1}{2}\nabla_r^2 - \frac{Z_A}{r_A} - \frac{Z_B}{r_B} + W(R) - \varepsilon_\lambda(R)\right]\psi_\lambda(R,\underset{\sim}{r}) = 0. \tag{24}$$

The MO's and eigenenergies depend parametrically upon the internuclear separation R. Exact solutions of this equation can be obtained, for the single electron system, but this is not the case for many electron systems and even for one electron system, it is convenient to express the ψ_λ in terms of Slater functions or of Gaussian functions to enable the various matrix elements to be evaluated easily.

Each F_n or G_m functions can usually be expressed in terms of a combination of one or two MO's, multiplied by the momentum transfer factors, for example

$$F_n = \left[\psi_\lambda + \psi_\mu\right] \exp -i(\varepsilon_n t + \tfrac{1}{8} v^2 t + \tfrac{1}{2} \underset{\sim}{v}\cdot\underset{\sim}{r}), \tag{25}$$

where $(\psi_\lambda + \psi_\mu)$ is asymptotic to the atomic function, $\phi_n(r_B)$, centred on the nucleus B, for large R.

The most serious practical difficulty lies in the calculation of exchange matrix elements, of the type $\langle F_n|\ldots|G_m\rangle$, because such matrix elements contain the momentum transfer factors exp $(\pm\, i\, \underset{\sim}{v}\cdot\underset{\sim}{r})$. Even for very small velocities, it is not strictly correct to replace these factors by unity[7], because the boundary conditions are not then satisfied. However if this simplication is made, the coupled equations can be reduced to the form

$$i\,\dot{b}_\lambda(t) = \sum_{\mu\neq\lambda} \langle\psi_\lambda|-i\frac{\partial}{\partial t}|\psi_\mu\rangle \exp\left[+i\int_{-\infty}^{t}\{\varepsilon_\lambda(R)-\varepsilon_\mu(R)dt'\right] b_\mu(t) \tag{26}$$

where $b_\lambda(t)$ is the amplitude of the MO ψ_λ. Because of the oscillating factor depending on $\left[\varepsilon_\lambda(R) - \varepsilon_\lambda(R)\right]$, we see that the transitions between MO's only occur with a high probability, when the energy difference between two orbitals is small for a certain range of values of R.

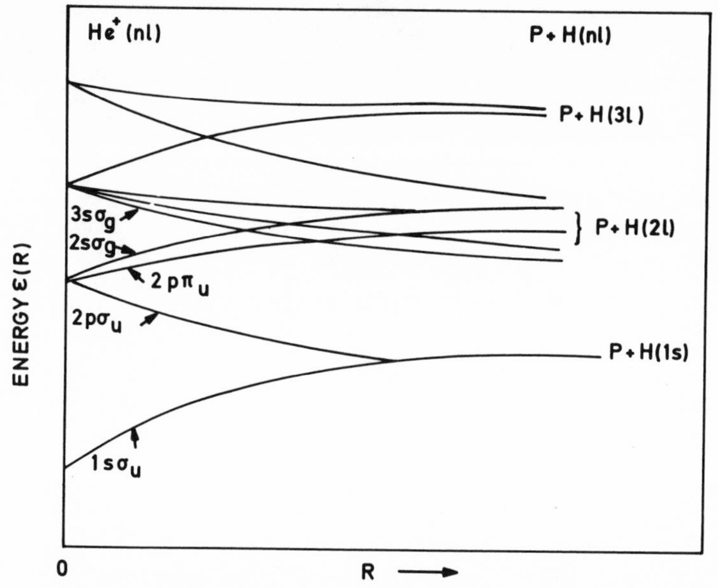

Fig.2: A correlation diagram for the proton-atomic hydrogen system.

To decide which MO's are of importance it is useful to draw a correlation diagram, relating the functions in the separated and united atom limits. For the homonuclear system p+H such a diagram is shown in figure 2. At large distances the wave-function of the p+H(1s) system is an equal mixture of the $1s\sigma_g$ and $2p\sigma_u$ molecular orbitals. In the lowest approximation in which just these two orbitals are retained, both charge exchange and elastic scattering results from the phase difference that develops during the collision between these two orbitals. However, the 2-state approximation is not adequate because at small distances the energy difference between the $2p\sigma_u$ and $2p\pi_u$ levels becomes vanishingly small. As the probability of a transition between two orbitals is large when the energy difference is small, the $2p\pi_u$ level becomes populated rather easily during the collision, and this MO must also be included in a serious study.

An interesting technical point arises from the examination of the coupling matrix(omitting the momentum transfer factor):

$$<\psi_\lambda(R,\underset{\sim}{r})\ |\frac{\partial}{\partial t}|\psi_\mu(R,\underset{\sim}{r})>. \qquad (27)$$

The operation $\frac{\partial}{\partial t}$ is to be taken with $\underset{\sim}{r}$ fixed in the laboratory coordinate system. If the adiabatic functions ψ_u are quantised along the internuclear axis, (the usual case), then

$$\frac{\partial}{\partial t} = v_R \frac{\partial}{\partial R} - i\dot{\theta}\, L_y , \tag{28}$$

where $v_R = dR/dt$, $\dot{\theta} = bv/R^2$ and L_y is the component of the electronic angular momentum perpendicular to the plane of scattering and the operations are taken with $\underset{\sim}{r}$ fixed in the molecular (rotating) frame. The matrix elements containing $\frac{\partial}{\partial R}$ and L_y are known as the radial and rotational coupling terms respectively. It often happens that the radial coupling changes rapidly near an avoided crossing of two molecular potential curves in the adiabatic representation. In this case, it may be convenient to make an R-dependent unitary transformation to a new 'diabatic' basis, chosen so that the radial coupling is small or zero.

Apart from the technical difficulties of evaluation of the matrix elements, the factors exp $(\pm\frac{1}{2}i\underset{\sim}{v}.\underset{\sim}{r})$, which occur in F_n and G_m, although necessary at large R, have the undesirable effect of associating the electron with one or other of the nuclei A and B at small separations, which is unphysical. These difficulties can be removed by introducing orbitals of the form

$$\phi_\lambda = \psi_\lambda(\underset{\sim}{R},\underset{\sim}{r}) \exp\left[\tfrac{1}{2} i f(\underset{\sim}{r},\underset{\sim}{R})\ \underset{\sim}{v}.\underset{\sim}{r}\right], \tag{29}$$

where $f(r,R)$ is a 'switching factor' with the properties

$$\begin{aligned} f(\underset{\sim}{r},\underset{\sim}{R}) &\to +1 \text{ if } R \to \infty \text{ and } r_A >> r_B , \\ &\to -1 \text{ if } R \to \infty \text{ and } r_A << r_B , \\ &\to 0 \quad \text{as} \quad R \to 0 . \end{aligned} \tag{30}$$

A particular function with these properties is

$$f(r,R) = \cos\Theta \left[1 + a^2/R^2\right]^{-1} , \tag{31}$$

where Θ is the angle between $\underset{\sim}{r}$ and $\underset{\sim}{R}$ and a is a parameter. If the same function f is used for all the MO's, no difficult exponential factors occur in the matrix elements; but at the expense of introducing more coupling terms[9].

A long standing problem has been to explain the experimental angular distribution for the proton-hydrogen system at energies less than 1 keV. For energies of a few hundred electron volts, the electronic wave-function is well described by the three states $1s\sigma$, $2p\sigma_u$ and $2p\pi_u$, and the principal problem is to determine what internuclear trajectory will be adequate. The first thing to try is a simple straight line trajectory, and this was done in the work of McCarroll and Piacentini[10]. Subsequently, Gaussorques et al[11] employed a trajectory calculated from a predetermined mean potential.

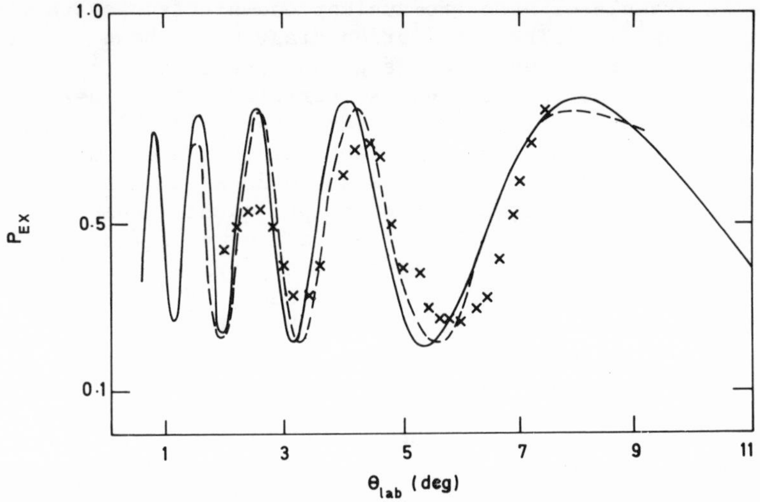

Fig.3: The electron capture probability for 250 eV protons incident on atomic hydrogen in the ground state.
- - - Three state calculation[10] (rectilinear trajectory)
——— Three state calculation Cayford and Fimple[12]
x x x Experimental data[13]

Recently, Cayford and Fimple[12] have introduced a self consistent method based on the coupled system of equations (5), (6) and (8). Having calculated the amplitudes a_{ij} $(b,t=\infty)$, c_{if} $(b,t=\infty)$, there is still some difficulty in defining the appropriate quantal scattering amplitude $f_{if}(\theta)$. Starting from the semi-classical approximation for potential scattering

$$f(\theta) = \left|\frac{b}{\sin\theta}\ \frac{db}{d\theta}\right|^{\frac{1}{2}} \exp\,(iS)\ , \tag{32}$$

where θ is the classical angle of scattering corresponding to the impact parameter b, and S is the classical action, Cayford and Fimple propose the generalisation

$$f(\theta) = \left|\frac{b}{\sin\theta}\ \frac{db}{d\theta}\right|^{\frac{1}{2}} C_{ji}\ (b)\ , \tag{33}$$

where c(b) is the probability amplitude for charge exchange.

In figure 3 we show the probability of electron capture from atomic hydrogen by protons at 250 eV as a function of angle. Although for larger angles, the approximations differ appreciably, it is seen that in the measured region ($\theta<7^{o}$), there is nothing to choose between the earlier results of McCarroll and Piacentini and those based on non-rectilinear trajectories.

As an example of a hetero-nuclear one-electron system, we take the case of He^{++}+H. The correlation diagram is shown in fig. 4. The 2pσ state correlates with the ground state of H in the separated atom limit, and the 2pπ 3dσ and 2sσ correlate to the n=2 states of He^{+}. As all four states have the same energy in this limit, they are expected to be the most important ones. Winter and Lane[14] made detailed calculations for energies up to 20 keV, using up to 20 basis states, including all states of principal quantum number n<4 in the united atom limit. It was found that at low energies <1 keV, it was sufficient to employ a three-state approximation retaining 2pσ, 2pπ and 3dσ orbitals. By 10 keV, it is necessary to add the 3pσ, 3pπ, 3dπ, 3dσ and 1sσ orbitals, and by 20 keV the 4fσ, 4dσ and 4fπ orbitals are also important. At the lower energies, some allowance was made for the non-linear nature of the nuclear trajectory. The total cross-section for charge exchange is shown in figure 5, in the 20 state approxomation compared with the 3 state (2pσ, 2pπ,3dσ) results of Piacentini and Salin[15]. The agreement is quite good above 10 keV, although the energy dependence of the calculated cross-sections appears to differ from the data at lower energies. This is even more marked for a calculation by Rapp[16] based on an atomic rather than a molecular expansion.

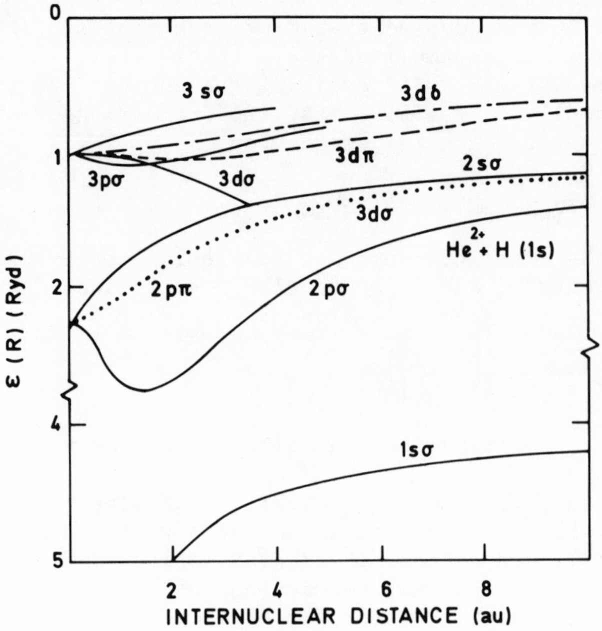

Fig.4: A correlation diagram for He^{++}+H (after ref. 15).

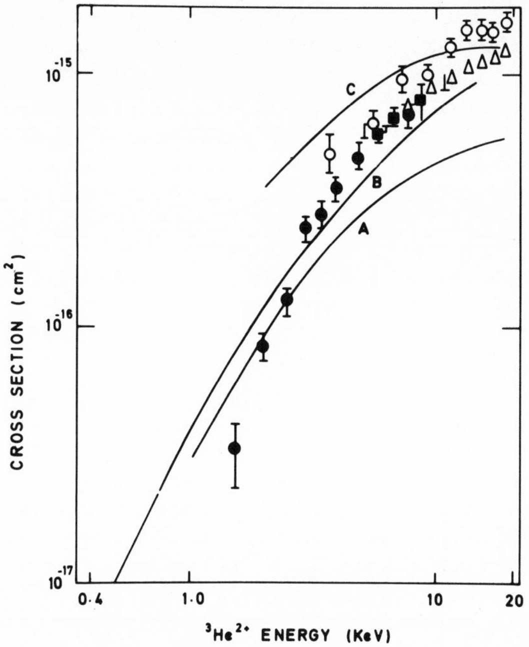

Fig.5: Total cross section for electron capture by $^3He^{++}$ from atomic hydrogen.
Approximation: A 3-state[15]; B 20-state[14]; C 8-state atomic expansion[16]
Experimental data: Δ from ref. [17] O from ref. [18].

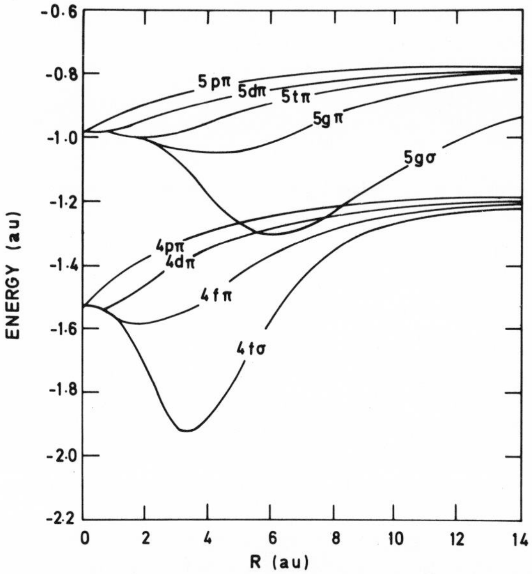

Fig.6: Correlation diagram for the C^{+6}+H system (after ref. [20])

A subject of current interest is charge exchange between highly charged ions and atomic hydrogen or deuterium. In the reaction,

$$A^{q+} + H(1s) \rightarrow A^{(q+1)+} + p, \quad (34)$$

capture at low velocities is generally into excited states, which have a high probability of decaying by photon emission. This process leads to significant cooling of fusion plasmas, the ions concerned including for example C^{q+}, O^{q+} and Fe^{q+}. Let us consider the case of the reaction

$$C^{6+} + H(1s) \rightarrow C^{5+*} + p. \quad (35)$$

The correlation diagram is shown in fig. 6. The (1s) ground state of atomic hydrogen correlates with the (6hσ) level in the united atom limit. However, there is an avoided crossing at about R=16 a.u. between the (6hσ) and (5gσ) levels, and effectively the system completely transfers to the (5gσ) level at this point. The other levels correlate with levels of C^{5+*} in the separated atom limit. The major source of charge transfer is the avoided crossing between the (5gσ) and (4fσ) levels near R=8 a.u. The results of an eleven state calculation[19] are shown in fig. (7), in which the (5gσ), (4fσ), (4sσ), (4pσ,π), (4dσ,π,δ) and (4fπ,δ,ϕ) levels were retained in the region near R=8 a.u. The couplings to the (3dσ) level are small and can be ignored. The internuclear potential was taken to be the unscreened Coulomb potential at small R, while at large R a straight line trajectory was sufficiently accurate. The results shown are in harmony with a six state calculation of Salop and Olson[20], which included the 5gσ, 4fσ, 4fπ, 5gπ, 4dπ and 4pπ levels. Also shown in fig. 7 are the results of a 2-state (5gσ)-(4fσ) calculation. It is seen that while the 2-state calculation is perfectly satisfactory, if a rough order of magnitudes, result is required, it does not provide an accurate cross-section for this system and the oscillations observed in the calculated cross-section are dumped out by rotational couplings to π levels in the more accurate approximation.

There have been many attempts to obtain approximate analytical solutions to the 2-state coupled equations; but, in general these cannot be relied upon, even to give a result which is of the correct order of magnitude. The most famous of the approximations, the Landan-Zenner approximation is accurate only if the transition occurs in a very narrow region about a real, or avoided, crossing of two potential curves. Details of the method are discussed in the text-books[22,23]. In the present case the result is fairly satisfactory[21] and has been included in fig. 7. A better approximation, showing the oscillations in the 2-state cross-section has been given by Greenland[24]; but, it cannot be emphasised too strongly that the conditions under which analytic approximations are accurate are often violated.

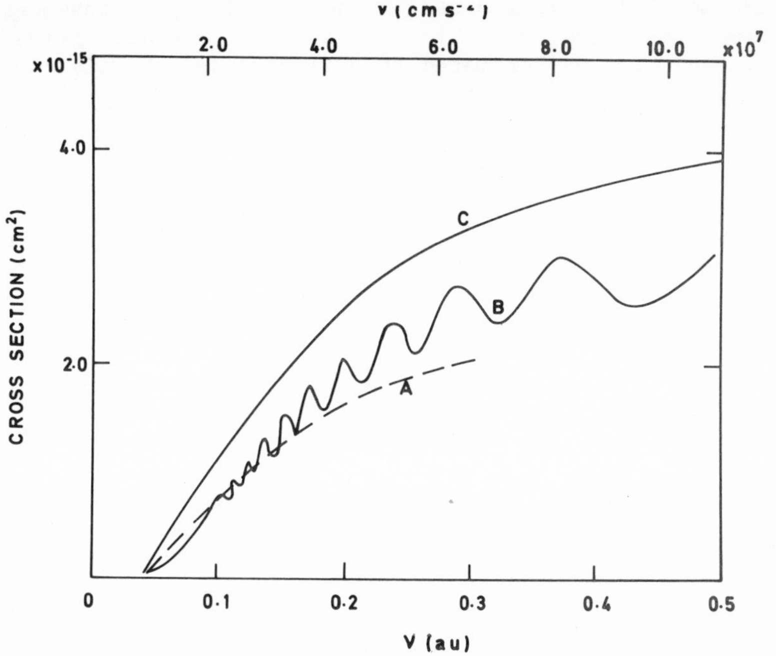

Fig. 7: The total cross-section for electron capture by C^{6+} from H(1s)
A The Landau-Zener approximation to the 2-state model[21]
B The 2-state model[19]
C The 11-state model[19]

Charge exchange can take place more easily, when the final state of the atom (A^+e^-) has an energy which is not too dissimilar from the energy of the initial state of the atom ($B + e^-$). As a consequence, as the charge on the incident ion increases, capture into states with larger values of the principal quantum number n becomes increasingly important. At low energies capture is often very selective[25], for example in the reaction:

$$O^{8+} + H(1s) \rightarrow O^{7+} + p, \tag{36}$$

the final state of O^{7+} is mainly in the n=5 level, while in the carbon reaction (35), C^{5+} is obtained mainly in the n=4 level. This may be of practical importance, as this way of selectively populating excited states might form the basis of an X-ray laser[26].

The generally satisfactory agreement between the predictions of the MO model and experiment for the one-electron system, suggests that the model is well-founded at low energies. In the work we have been discussing, the velocity dependent momentum transfer factors have been approximated (or ignored) and this is one of the main limitations, preventing the theory from being applied in this form for velocities much greater than ∿0.5 a.u.

The MO expansion has been used to calculate charge exchange cross sections for systems containing several, or many, electrons. The trial function ψ_t must now be antisymmetrical in the electron coordinates, and the Coulomb interactions between the electrons must be included in the Hamiltonian. The two electron system was discussed some time ago by Colegrave and Stephens[27], who calculated the cross-section for capture into the ground state by protons incident on helium. The three electron symmetrical system He^{+}+He has been studied by Rai Dastider and Bhattacharyya[28] and by McCarroll and Piacentini[29] in the impact parameter formation, and also by Stern et al[30], who used a quantal close coupling formalism. The heteronuclear system B^{3+}+H and C^{4+}+H have been studied by Olson et al[31] (see also Harel and Salin[25]). The four-electron systems B^{3+}+He and C^{4+}+He have also received attention[32] A simplified theory has been proposed by Bottcher[33], for charge transfer between systems whose ionisation potentials are nearly equal, such as Na^{+}+Li, Li^{+}+Ne, while the existence of elaborate MO computer codes has allowed the study of even larger systems such as He^{+}+Ar[34] and the symmetrical systems of rare gas atoms[35] (He+He, Ar+Ar, Ne+Ne).

INTERMEDIATE ENERGIES

At energies such that the velocity of the incident ion is comparable or greater than the orbital velocity of the electron in the target atom, the momentum transfer of the captured electron cannot be ignored. The introduction of the momentum transfer factors exp $(i\underset{\sim}{v}.\underset{\sim}{r})$ causes a large reduction in the size of the matrix elements of M_{12} and M_{21} compared with those of M_{11} and M_{22} (see eq. 21) which rapidly reduces the importance of cross-sections for charge exchange compared with those for excitation at high velocities. However, the evaluation of M_{12} and M_{21} becomes very time consuming and may limit seriously the number of terms used in the expansions.

The most straightforward approach is to identify the expansion functions, with either atomic functions ϕ_n, χ_m centred on the nuclei A amd B, or with atomic pseudo states $\bar{\phi}_n$, $\bar{\chi}_m$, again centred on A and B. Pseudo-state functions are a set of linearly independent functions, which are normalisable and which overlap the hydrogenic continuum functions. The functions F_n and G_m are expressed as:

$$\begin{aligned} F_n(\underset{\sim}{r},t) &= \phi_n(\underset{\sim}{r}_B)\ \exp\left(-i\varepsilon_n t - i\tfrac{1}{8}v^2 t - \tfrac{1}{2}\underset{\sim}{v}.\underset{\sim}{r}\right), n<n' , \\ &= \bar{\phi}_n(\underset{\sim}{r}_B)\ \exp\left(-i\varepsilon_n t - i\tfrac{1}{8}v^2 t - \tfrac{1}{2}\underset{\sim}{v}.\underset{\sim}{r}\right), n>n' , \\ G_m(\underset{\sim}{r},t) &= \chi_m(\underset{\sim}{r}_A)\ \exp\left(-i\eta_m t - i\tfrac{1}{8}v^2 t + \tfrac{1}{2}\underset{\sim}{v}.\underset{\sim}{r}\right), m<m' , \\ &= \bar{\chi}_m(\underset{\sim}{r}_A)\ \exp\left(-i\eta_m t - i\tfrac{1}{8}v^2 t + \tfrac{1}{2}\underset{\sim}{v}.\underset{\sim}{r}\right), m>m' . \end{aligned} \tag{37}$$

The addition of pseudo-states, is partly to represent ionized states, and partly because the atomic expansion is slowly convergent when the nuclei are close and the wave-function resembles that of the united atom. An alternative and promising approach is to expand the wave-function about three, rather than two, centres:- for example, about the two nuclei and the centre of charge[36]. In the pioneering work of Cheshire et al[37], on the p-H system, the pseudo-state wave functions were taken to be Slater orbitals, chosen to be orthogonal to the hydrogenic 1s, 2s and 2p wave-functions which were represented explicitly in the expansion. The parameters in the pseudo-state orbitals, which were designated $\bar{3}$s and $\bar{3}$p, were chosen to maximise the overlap with the wave-functions of the He^+ united atom. Another choice, introduced by Gallaher and Wilets[38], and recently thoroughly investigated by Shakeshaft[39], is to make the expansion in terms of Sturmian functions, which form a complete, discrete set of which the radial functions satisfy the equation

$$\left[-\frac{1}{2}\frac{d^2}{dr^2} + \frac{\ell(\ell+1)}{2r^2} - \frac{\alpha_{\lambda\ell}}{r}\right] S_{n\ell}(r) = -\frac{1}{2(\ell+1)} S_{n\ell}(r) , \tag{38}$$

where the parameter $\alpha_{\lambda\ell}$ is treated as an eigenvalue. The most extensive results of Shakeshaft include six s and six p states centred about each proton, giving rise to a large computing problem.

A rather different approach has recently been introduced by Morrison and Öpik[40], which involves the introduction of a modified system of elliptical co-ordinates. The wave-function is then expanded in terms of sets of orthogonal polynomials, which are chosen so that the asymptotic wave-functions, as $t\to\pm\infty$, can be expressed exactly using a few terms. The method is rather economical and large basis sets of up to 30 functions can be used.

The results of Morrison and Öpik, of the pseudo-state expansion, and of the Sturmian expansion agree rather well over the energy region in which calculations have been made (25 to 100 keV). Experimental data is scanty, and a meaningful comparison for capture into excited state is difficult to make;but the total capture cross-section, which is dominated by ground state capture agrees well the MO approach at the lower energies (down to 1keV) and with experiment.

Atomic expansion methods[40,41] have also been applied extensively to the case of the $(He^{2+}+H)$ system, and the results agree reasonably with experiment. The two electron system (p+He) was studied in detail a few years ago by Winter and Lin[42] using an atomic expansion with up to eleven states. Cross-sections for capture into the ground state and the n=2 and n=3 levels of hydrogen were computed, for energies up to 1 MeV. In general, good agreement was obtained with the experimental data.

If many states are required in the expansion method, the calculations become extremely lengthy, however as Lin has recently emphasised, there are circumstances in which it is possible to obtain useful accuracy with a two state coupled system, retaining just the initial atomic state centred on B and the final state of interest centred on A. Except for the case of resonance, the cross-section for capture into the ground state exhibits a maximum, near incident velocities v, which are close to the orbital velocity of the target electron, v_B. At these velocities, capture occurs at large impact parameters, for which the wave-functions can be approximated by the two-state expansion. For ground state capture by protons from helium the result of the two state calculations and eleven state calculations agree closely near the cross-section maximum. This does not imply that the two-state model is adequate at high energies, because second order terms become important for $v \gg v_B$ and these require continuum intermediate states to be represented in the expansion. For example, in the case of helium the experimental and calculated cross sections agree reasonably up to 100 keV, but at 500 keV the coupled state values are about twice those given by the data. Making use of these ideas, Lin[43,44] has calculated a series of cross-sections for capture into the K shell, by fully stripped ions from the K shell of the inert gases. Some of his results are shown in Table 1, compared with the data.

Table 1 Electron capture from the K shell into the K shell, per target electron in the two-state approximation (from ref. 43). Units of $10^{-20}cm^2$

Process	Energy(MeV)	v/v_B	σ(Theory)	σ(Expt)
N^{7+} + Ne	14	0.79	368	355
	19	0.92	342	350
O^{8+} + Ne	24	0.96	368	435
	30	1.08	269	330
	35	1.17	195	300
F^{9+} + Ne	20	0.81	516	440
	25	0.90	440	430
	30	1.00	360	410
C^{6+} + Ar	12.6	0.42	1.30	1.79
	19.0	0.52	2.52	2.34
	22.6	0.57	5.40	5.83

The agreement is remarkable, and it should be noted that first order perturbation cross-sections are from ten to several hundred times the experimental values.

Another interesting situation of some simplicity is when the projectile has a charge z_A, which is much less than the nuclear charge of the target z_B. Under these circumstances[45], the projectile-electron interaction can be treated as a perturbation, and the complete wave-function approximated by an expansion in atomic states and pseudo-state centred on the target. Thus, the probability of charge exchange is

$$c_{fi}(b) = -i \int_{-\infty}^{\infty} dt \int d\underset{\sim}{r}\; X_f^*(\underset{\sim}{r},t)\, \left(\frac{z_A}{r_A}\right)\, \psi_i^+(\underset{\sim}{r},t), \tag{39}$$

where $\psi(r,t)$ is expanded about the nucleus B

$$\psi_i^+(\underset{\sim}{r},t) = \sum_n a_n(t)\, \phi_n(\underset{\sim}{r}_B)\, \exp\left(-i\varepsilon_n t - \tfrac{1}{8}iv^2 t - i\underset{\sim}{v}.\underset{\sim}{r}/2\right),$$

and where X_f is one of the unperturbed functions (12). Reading et al[45] have noted that charge transfer takes place most rapidly when there is an energy balance between the direct and rearranged systems. The final state has energy $(\frac{1}{2}v^2 + n_f) \simeq \frac{1}{2}v^2$ and an important mechanism is that in which the electron is first excited to a continuum state of the target with energy $\sim\frac{1}{2}v^2$ and then is captured in a resonant charge transfer process. The energy interval in the continuum spectrum of the target in which this occurs is of width $\Delta E \sim 1/t = (vz_A)$. In a pseudo-state treatment, there is a danger that if the energy of a pseudo-state is $(\sim\frac{1}{2}v^2)$ the process will be overestimated, and if it is far from $(\frac{1}{2}v^2)$ the process will be seriously underestimated. The difficulty can be overcome by an optimum choice of pseudo-states. For example, it is necessary to represent the Green's function $\theta(t-t')$ exp i $H_T(t-t')$ where H_T is the target Hamiltonian as well as possible. The expansion of exp $(iH_T\tau)$ in terms of hydrogenic functions is

$$e^{iH_T\tau} = \sum_n \phi_n(\underset{\sim}{r})\phi_n^*(\underset{\sim}{r}')e^{i\varepsilon_n\tau} + \int_0^\infty dE\,\phi_E(\underset{\sim}{r})\phi_E^*(\underset{\sim}{r}')e^{iE\tau}. \tag{40}$$

The pseudo-states Φ_n with energies E_n which diagonalise H_T in a finite sub-space can be chosen, together with weights ω_n, so that,

$$\int_0^\infty dE\; \phi_E(\underset{\sim}{r})\; \phi_E^*(\underset{\sim}{r}')e^{iE\tau} \simeq \sum_n \phi(\underset{\sim}{r})\phi_n^*(\underset{\sim}{r}')e^{iE_n\tau}\,\omega_n, \tag{41}$$

is satisfied as closely as possible. Reference [45] shows how this can be achieved and how the accuracy can be tested. Applications of this method to a number of inner shell capture processes has been made with encouraging results.

An entirely different approach which appears to provide reasonably accurate (within 20%) total charge exchange cross-sections at intermediate energies is to use the model of Abrines and Percival[46]. In this the problem is treated as a purely classical three-body problem, a statistical distribution of orbits being assumed for the electron in the initial state. The model works well for the proton-hydrogen system (see fig. 8) in a limited energy region; the predicted cross-section falling below the data when $v<v_B$, and when $v>2v_B$. Assuming the model is valid in a limited velocity range above $v=v_B$, Olson and Salop[47] have calculated charge exchange and ionisation cross-sections for fully stripped ions with $1<z<36$ colliding with hydrogen atoms, in the range $v = 2\text{-}7\text{x}10^8$cm/c. Experimental data is rather limited but no large discrepancies appear to occur. A further application by Berkner et al[48] has been to electron capture and ionisation by iron ions Fe^{q+}, q=10, 15, 20, 25 incident on atomic hydrogen, at energies in the range 50 to 1,200 keV per nucleon. If the experimental data, obtained for molecular hydrogen, is divided by two, good agreement is found with the calculated cross-sections.

HIGH ENERGIES

For rearrangement collisions, various forms of first order approximations can be written down. A Born approximation, further amplitude in the compact parameter formalism, is given by the expression

$$C(b) = -i\int_{-\infty}^{\infty} dt \int d\underset{\sim}{r}\ X_f^*(\underset{\sim}{r},t) \left[+ \frac{Z_A}{r_A} \dot{-} W(R) \right] \Phi_i(\underset{\sim}{r},t), \qquad (42)$$

where X_f and Φ_i are the unperturbed functions of (11) and (12). Brinkman and Kramers suggested that because the large contribution from W(R) must be cancelled by higher order terms, it should be omitted. The resulting approximation gives rise to the Brinkman-Kramers cross-section σ_{BK}. Unfortunately neither the Born nor the Brinkman-Kramers cross-sections are reliable[1], because higher order terms are large at all energies. Indeed following the work of Drisco, it has been shown that in the extreme high energy limit the cross-section is given by the seconed Born approximation[1,49] and in that limit (for ground state capture)

$$\sigma_{B2} = \sigma_{BK} \{0.295 + 5\pi v/2^{11}(Z_A + Z_B)\} \text{ a.u.}, \qquad (43)$$

where

$$\sigma_{BK} = \pi 2^{18}(Z_A Z_B)^5 / 5\, v^{12} .$$

This high energy form is however not valid in the region of experimental interest, where third and higher order terms are generally of importance.

The most satisfactory high energy model was introduced by Cheshire[50] and later put on a semi theoretical basis by Gayet[51]. It is known as the contimuum distorted wave model, and it attempts to allow for the distortion of the electronic wave-function by the

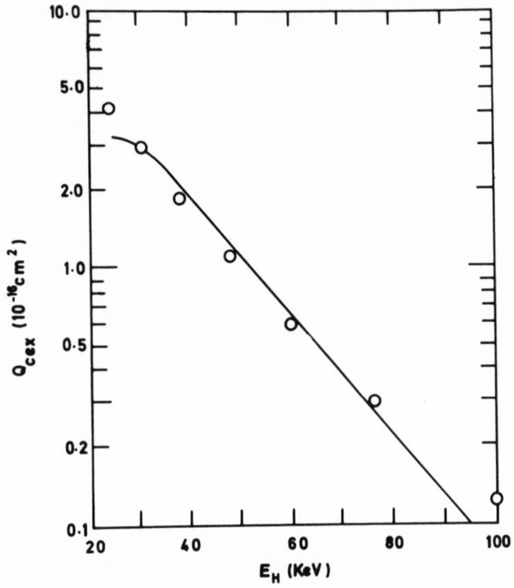

Fig.8: Total electron capture cross-sections for p+H in the classical model

——— Theoretical cross-section[47]

o o o Experimental data

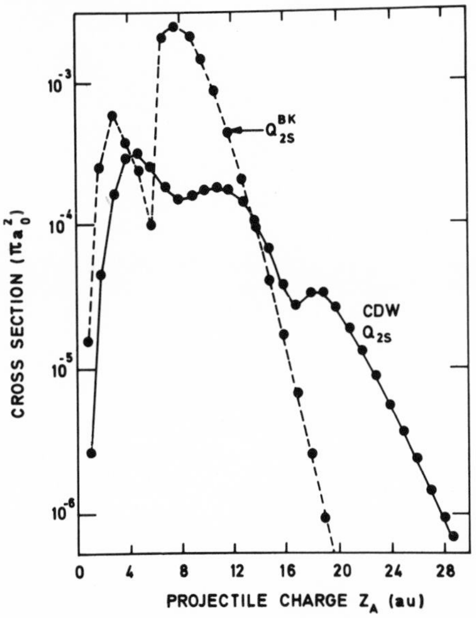

Fig.9: Total cross-sections for electron capture into the (2s) level by fully stripped ions of charge z_A from atomic hydrogen in the ground state, at v=5 a.u.

——— CDW model[53]

- - - Brinkman-Kramers first order approximation[53]

Coulomb interaction with both ions, in a symmetrical fashion. It takes the form

$$C(b) = -i\int_{\infty}^{\infty} dt\, N(v_A)\, N(v_B)\, \langle X_f, {}_1F_1(iv_B,1;\underset{\sim}{v}.\underset{\sim}{r}_B+ivr_B)$$
$$|A|\Phi_i, {}_1F_1(iv_A,1;\, \underset{\sim}{v}.\underset{\sim}{r}_A+ivr_A)\rangle \;, \qquad (44)$$

with $v_A = z_A/v$, $v_B = z_B/v$ and where A is a certain differential operator.

Recently, Belkić and Gayet have applied the CDW model to electron capture from hydrogen by protons and alpha particles for energies above 25 keV and up to 10 MeV per nucleon[52,49]. Good agreement with the experimental data was found for the total capture cross-sections and for capture into the individual s states of hydrogen. The data for capture into excited p and d states of hydrogen is rather uncertain and definite conclusions could not be drawn in that case. Similar calculations for electron capture by protons and alpha particles from helium also agree well with the high energy data. This work has been extended by Belkić and McCarroll[53], who have calculated cross-sections for capture by highly charged ions ($1<z_A<30$)from atomic hydrogen. For those cases for which experiments have been performed, the theoretical results are in good agreement with the data. The capture cross-section into the (2s) state as a function of projectile charge (z_A) is shown in fig. 9, where it is compared with the Brinkman-Kramers first order cross-sections. The failure of the first order approximation is seen clearly.

CONCLUSION

In several areas, very satisfactory theoretical modes of charge exchange now exist. The independent particle MO model provide a good description of the low velocity data and at high energies the Continuum Distorted Wave and related models are, perhaps surprisingly, successful. In general, the intermediate energy region is the most difficult one in which to assess the reliability of proposed approximation, although it appears in favourable circumstances quite simple two state approximations can be adequate. In the very low energy region, there are still problems in defining the best classical trajectory to be used in an impact parameter treatment, although the problem can be avoided by carrying out a fully quantal approximation.

REFERENCES

Articles marked with an asterisk are review articles containing extensive bibliographies.

1. * D.R. Bates, and R. McCarroll, Adv.Phys. 11, 39 (1962).
* B.H. Bransden, Adv.Atom.Molec.Phys. 1, 85 (1965).
* B.H. Bransden, Lectures in Theoretical Physics XIc, Ed. S. Geltman et al., (Gorden and Breach, New York, 1969), p. 139-192.
* B.H. Bransden, Rep.Prog.Phys. 35, 949 (1972).
* R. McCarroll, Atomic and Molecular Physics and the Interstellar Matter, Ed. R. Balian et al., (North Holland, Amsterdam, 1975) p. 155-175.
* D. Basu, S.C. Mukherjee, and D.P. Sural, Phys.Reps. 42C, 145 (1978).
2. * M.E. Rudd and J.H. Macek, Case Studies in Atomic Physics 3, 47 (1972).
Y.B. Band, J.Phys.B. 7, 2557 (1974).
3. J.S. Briggs, and K. Dettmann, J.Phys.B. 10, 1113 (1977).
C.M. Lee, Phys.Rev.A. 17, 566 (1978).
4. * J.S. Briggs, Rep.Prog.Phys. 39, 217 (1976).
5. M.H. Mittleman, Phys.Rev. 122, 499 (1961).
6. T.A. Green, Proc.Phys.Soc. 86, 1017 (1965).
7. A. Riera, and A. Salin, J.Phys.B. 9, 2877 (1976).
8. S.B. Schneiderman, and A. Russek, Phys.Rev. 181, 311 (1969).
9. K. Taulbjerg, J. Vauben and B. Fastrup, Phys.Rev. A. 12, 2325 (1975).
10. R. McCarroll, and R.D. Piacentini, J.Phys.B. 3, 1336 (1970). Corrigenaum, J.Phys.B. 4, 886 (1971).
11. C. Gaussorgues, C. LeSech, F. Masnou-Seeuers, R. McCarroll, and A. Riera, J.Phys.B. 8, 239, 253 (1975).
12. J.K. Cayford, and W.R. Fimple, J.Phys.B. 9, 3055 (1976).
13. J.C. Houver, J. Fayeton, and M. Barat, J.Phys.B. 7, 1358 (1974).
14. T.G. Winter, and N.F. Lane, Phys.Rev. A17, 66 (1978).
15. R.D. Piacentini, and A. Salin, J.Phys.B. 7, 1666 (1974); 9, 563 (1976); 10, 1515 (1977).
16. D. Rapp, J.Chem.Phys. 61, 3777 (1974).
17. M.B. Shah, and H.B. Gilbody, J.Phys.B. 7, 630 (1974).
18. W.L. Nutt, R.W. McCullough, K. Brady, M.B. Shah, and H.B. Gilbody, J.Phys.B. 11, 1457 (1978).
19. J. Vaaben, and J.S. Briggs, J.Phys.B. 10, L521 (1977).
20. A. Salop, and R.E. Olson, Phys.Rev. 16, 1811 (1977).
21. A. Salop, and R.E. Olson, Phys.Rev. 13, 1312 (1976).
22. B.H. Bransden, Atomic Collision Theory (Benjamin, New York, 1970).
23. M.R.C. McDowell and J.P. Coleman, Introduction to the Theory of Ion-Atom Collisions (North Holland, Amsterdam, 1970).
24. P.T. Greenland, J.Phys.B. 11, L191 (1978).
25. C. Harel, and A. Salin, J.Phys.B. 10, 3511 (1977).
26. W.H. Louisell, M.O. Scully, and W.B. McKnight, Phys.Rev.A11, 989 (1975).
27. R.K. Colgrave, and D.B.L. Stephens, J.Phys.B. 1, 856 (1968).

28. T.K. Dastider, and S.S.Bhattacharyya, Ind.J.Phys. 50, 731 (1976).
29. R. McCarroll and R. Piacentini, J.Phys.B. 4, 1026 (1971).
30. B. Stern, J.P. Gauyacq, and V. Sidis, J.Phys.B. 11, 653 (1978).
31. R.E. Olson, E.J. Shipsey, and J.C. Browne, J.Phys.B. 11, 699 (1978).
32. E.J. Shipsey, J.C. Browne, and R.E. Olson, Phys.Rev. 15, 2166 (1977).
33. C.J. Bottcher, J.Phys.B. 11, 507 (1978).
34. R. Albat, and B. Wirsam, J.Phys.B. 10, 81 (1977)
35. J.P. Gauyacq, J.Phys.B 9, 2289, 3067 (1976); 11, 85 (1978).
36. D.G.M. Anderson, M.J. Antol, and M.B. McElroy, J.Phys.B. 7, L118 (1974).
37. I.M. Cheshire, D.F. Gallaher, and A.J. Taylor, J.Phys.B. 3, 813 (1970).
38. D.F. Gallaher, and L. Wilets, Phys.Rev. 169, 139 (1968).
39. R. Shakeshaft, Phys.Rev. A14, 1626 (1976).
40. H.G. Morrison, and U. Öpik, J.Phys.B. 11, 473 (1978).
41. D. Rapp, and D. Dinwiddie, J.Chem.Phys. 57, 4919 (1972)
42. T.G. Winter, and C.C. Lin, Phys.Rev. A10, 2141 (1974).
43. C.D. Lin, J.Phys.B. 11, L185 (1978).
44. C.D. Lin, S.C. Soong, and L.N. Tunnell, Phys.Rev. A 17, 1646 (1978).
45. J.F. Reading, A.L. Ford, G.L. Swafford and A. Fitchard, Preprint Texas A&M University (1978).
46. R. Abrines, and I.C. Percival, Proc.Phys.Soc. 88, 861, 873, (1966).
47. R.E. Olson, and S. Salop, Phys.Rev. A 16, 531 (1977).
48. K.H. Berkner, W.G. Graham, R.V. Pyle, A.S. Scholachter, J.W. Stearns, and R.E. Olsen, J.Phys.B. 11, 875 (1978).
49. *D. Belkić, R. Gayet and A. Salin, Phys.Reps. (in the press).
50. I.M. Cheshire, Proc.Phys.Soc. 84, 89 (1964).
51. R. Gayet, J.Phys.B 5, 483 (1972).
52. D. Belkić, and R. Gayet, J.Phys.B. 10, 1911 (1977).
53. D. Belkić, and R. McCarroll, J.Phys.B 10, 1933 (1977).

THEORETICAL METHODS FOR IONIZATION

I.E. McCarthy and A.T. Stelbovics

Institute for Atomic Studies, The Flinders University of South Australia, Bedford Park, S.A. 5042, Australia

1. FORMAL SCATTERING THEORY

In an atomic system we assume that the effective potentials are the pairwise Coulomb potentials between the particles concerned, nuclei with charges Z_j and N electrons. The total energy is E.

The total Hamiltonian H for the system is partitioned into two.

$$H = \hat{K} + \hat{V} , \tag{1}$$

where $\hat{K}$ includes at least the Hamiltonian H_o of all subsystems that remain bound during the reaction and the kinetic energy K of the centres of mass of the interacting subsystems. The remainder of $\hat{K}$ is an arbitrary potential V_D, which is chosen to suit a particular calculation. V_D is of course subtracted from the total potential V between the interacting subsystems.

$$\hat{K} = K + H_o + V_D , \quad \hat{V} = V - V_D . \tag{2}$$

We are concerned with the eigenstates $\Psi^{(\pm)}$ of the whole system

$$[E^{(\pm)} - H]\Psi^{(\pm)} = 0 , \tag{3}$$

and with the eigenstates $\Phi^{(\pm)}$ of $\hat{K}$

$$[E^{(\pm)} - \hat{K}]\Phi^{(\pm)} = 0 , \tag{4}$$

where the superscripts ± represent outgoing or ingoing spherical wave boundary conditions respectively.

The distorting potential V_D is always chosen so that the Hamiltonian $\hat{K}$ separates in the coordinates of the centres of mass of all interacting subsystems. $\Phi^{(\pm)}$ then separates into a product of a state vector ϕ_μ for the internal motion of bound subsystems in the channel μ, defined by the eigenstates of the bound subsystems, and distorted waves $\chi^{(\pm)}(\underline{k}_j)$ for the motion of the subsystems j relative to the total centre of mass.

$$\Phi^{(\pm)} = \phi_\mu \, \chi^{(\pm)}(\underline{k}_1) \, \chi^{(\pm)}(\underline{k}_2) \ldots , \tag{5}$$

The Pauli exclusion principle for the electrons is strictly taken into account by including in $\Phi^{(\pm)}$ appropriate linear combinations of terms (5) with all possible permutations of electron indices.

The amplitude[1] for a reaction in which the system in state $\Psi_I^{(+)}$ is detected in a configuration denoted by F is

$$M_{FI} = \langle \Phi_F^{(-)} | \hat{V} | \Psi_I^{(+)} \rangle = \langle \Psi_F^{(-)} | \hat{V} | \Phi_I^{(+)} \rangle \quad . \tag{6}$$

For short-range potentials, such as that between an electron and a neutral atom, the wave functions for motion of the bound subsystems are plane waves, provided we choose $V_D = 0$. If there are Coulomb potentials between subsystems, the motion never becomes a plane wave, but we may still use a plane-wave representation for formal simplicity, and include the Coulomb effect in the operator T describing the transition.

$$M_{FI} = \langle \underline{k}_1', \underline{k}_2', \ldots ; \ \phi_{\mu'} | T | \phi_\mu ; \ \underline{k}_0 \rangle \tag{7}$$

$$\equiv \langle \Phi_F^{(-)} | T | \Phi_I^{(+)} \rangle \ . \tag{8}$$

Here primes indicate final-state quantities.

The wave function $\Psi^{(\pm)}$ is a solution of the Lippmann-Schwinger equation

$$\Psi^{(\pm)} = \Phi^{(\pm)} + G^{(\pm)} \hat{V} \Phi^{(\pm)} = \Phi^{(\pm)} + G_o^{(\pm)} \hat{V} \Psi^{(\pm)} \ , \tag{9}$$

where the Green's functions are defined by

$$G^{(\pm)} = [E^{(\pm)} - H]^{-1}; \quad G_o^{(\pm)} = [E^{(\pm)} - \hat{K}]^{-1}. \tag{10}$$

The definition (8) of the T-matrix may be combined with (6), (9) and (10) to give the Lippmann-Schwinger equations for T or G

$$T = \hat{V} + \hat{V} \, G_o \, T \ , \tag{11}$$

$$G = G_o + G_o \, \hat{V} \, G \ . \tag{12}$$

Where the superscript + or - is omitted, the physical value + is assumed.

2. THE ELECTRON-IMPACT IONIZATION PROBLEM

In the ionization problem to be discussed there is only one bound subsystem, the target atom or ion in its ground state $\mu = 0$ initially, and the residual ion in a state μ' finally. We make the independent-particle model for the electrons in these subsystems. This can be generalized by taking the appropriate linear combinations of configurations in an independent-particle representation if necessary, but we consider only the lowest-energy (Hartree-Fock) configuration.

In the independent-particle model we consider the incident electron interacting with one target electron whose orbital wave function is ψ_i. The cross section for a reaction is the sum of the cross sections for electrons in occupied orbitals i. Each is given by solving a three-body problem.

The incident electron is labeled by the subscript 1, the target electron by 2 and the residual ion by 3. The potential v_3 is the Coulomb potential between the two electrons, v_1 and v_2 are the potentials between electrons 1 and 2 respectively and the ion 3. The indices used for the potentials label each two-body subsystem. [Note the change from the cyclic notation used more usually[2] for the three-body problem. Since we are not interested in rearrangement collisions it is clearer to label a subsystem containing one electron by the label of that electron].

In order to calculate the quantity $\hat{V}|\Psi_I^{(+)}>$ in the ionization amplitude (6) we have to solve the three-body Schrödinger equation with boundary conditions appropriate to the problem. We partition the Hamiltonian thus

$$\hat{K} = K_1 + (K_2 + v_2) \ , \quad \hat{V} = v_1 + v_3 \ , \tag{13}$$

where K_1 and K_2 are the kinetic energy operators of the electrons 1 and 2. The eigenstate $|\Phi_I^{(+)}>$ of $\hat{K}$ is

$$|\Phi_I^{(+)}> = |\psi_i, \underline{k}_1> \ , \tag{14}$$

where ψ_i is the orbital of the target electron and $\underline{k}_1$ is the incident momentum. We have made the approximation that the mass of the ion is infinite, so that K_3 is zero. This approximation may be removed by redefining $\underline{k}_1$ and $\underline{k}_2$ as the appropriate Jacobi momentum coordinates after removing the centre-of-mass motion from the problem (see equation (55)).

The T-matrix for the 3-body problem is obtained by solving the Schrödinger equation in an appropriate form[2]. The Lippmann-Schwinger equation (11), (12) becomes

$$T = v_1 + v_3 + (v_1 + v_3)G(v_1 + v_3) , \tag{15}$$

$$G = [E - (K_1 + v_1) - (K_2 + v_2) - v_3]^{-1} . \tag{16}$$

The simplest solution formally is the Born series obtained by iterating equation (12) for G and substituting in (15). The first-order term in the interaction potentials gives the Born approximation.

$$M_{FI} = \langle \underline{k}_1', \chi^{(-)}(\underline{k}_2') | v_1 + v_3 | \psi_i, \underline{k}_1 \rangle . \tag{17}$$

We keep in mind the fact that matrix elements must be antisymmetric in the electron coordinates, but do not show it explicitly until it becomes necessary to discuss it explicitly. The orbital ψ_i and the continuum orbital (distorted wave) $\chi^{(-)}(\underline{k}_2')$ are both eigenstates of the Hamiltonian $K_2 + v_2$ for the two-body subsystem 2.

The Born series is often useful for a pictorial consideration of the reaction, but in general it is divergent. This is not to say that the problem is insoluble. It is conceivable that a re-arranged series is convergent. A possible rearrangement for example is the distorted-wave Born series obtained by defining the distorting potential V_D of (2)

$$V_D = v_1 . \tag{18}$$

Now we have

$$T = v_3 + v_3 \, G \, v_3 , \tag{19}$$

and the first iteration gives the distorted-wave Born approximation

$$M_{FI} = \langle \chi^{(-)}(\underline{k}_1'), \chi^{(-)}(\underline{k}_2') | v_3 | \psi_i, \chi^{(+)}(\underline{k}_1) \rangle . \tag{20}$$

Here the wave functions $\chi^{(-)}(\underline{k}_i)$ are finite quantities obtained by solving the Lippmann-Schwinger equation for the two-body subsystem j.

$$|\chi^{(-)}(\underline{k}_j)\rangle = |\underline{k}_j\rangle + G_o v_j |\chi^{(-)}(\underline{k}_j)\rangle = [1 + G_o t_j]|\underline{k}_j\rangle , \tag{21}$$

which has a solution, even though it may not be found by iteration if the Born series for the two-body t-matrix t_j is divergent.

In (20) the potential v_3 is the Born term in the T-matrix t_3 for the two-electron system. A further rearrangement is to collect the terms in the Born series involving the remainder of t_3, which is given by the two-body Lippmann-Schwinger equation

$$t_3 = v_3 + v_3 G_o t_3 . \tag{22}$$

The two-body Born series for t_3 may be divergent, but we know that scattering amplitudes (and hence t_3) are in general finite. The resulting approximation is the distorted-wave impulse approximation

$$M_{FI} = \langle \chi^{(-)}(\underline{k}_1'), \chi^{(-)}(\underline{k}_2') | t_3 | \psi_i, \chi^{(+)}(\underline{k}_1) \rangle \tag{23}$$

$$= \langle \underline{k}_1', \underline{k}_2' | [1+t_1 G_o][1+t_2 G_o] t_3 [1+G_o t_1] | \psi_i, \underline{k}_1 \rangle . \tag{24}$$

This incorporates several low-order terms of the Watson[3] multiple-scattering series for T.

In fact we do not know *a priori* if any of these series are convergent or if the reaction amplitude is well-represented by any of the approximations (17), (20) or (23) or by lower-order terms in any of the corresponding series.

The Lippmann-Schwinger equation (11) is not even the appropriate form of the Schrödinger equation to yield the correct finite three-body amplitudes. This is because it does not incorporate boundary conditions describing all the possible partitions of the three-body system into asymptotic two-body subsystems. The problem of formulating the correct integral equations was first solved independently by Mitra[4] and Faddeev[5]. The Watson multiple-scattering series is the series obtained by iterating these equations, which are most conveniently expressed for short-range potentials in the form due to Alt, Grassberger and Sandhas[6]. For our ionization problem the AGS equations are

$$U_{ji} = (1-\delta_{ji})(E-K) + \Sigma_{\ell\neq j} t_\ell G_o U_{\ell i} , \tag{25}$$

where K is the kinetic energy operator, G_o is the free-particle propagator and U_{ji} is the amplitude for going from the partition i to the partition j. The ionization amplitude is

$$M_{FI} = \langle \underline{k}_1', \underline{k}_2' | \Sigma_\ell t_\ell G_o U_{\ell i} | \psi_i, \underline{k}_1 \rangle . \tag{26}$$

The Watson multiple-scattering series for ionization is obtained by substituting for $U_{\ell i}$ in (26) the result of iterating (25).

The AGS equations can be solved for short-range potentials but for Coulomb potentials, although there have been recent formal developments, a practical solution of the positive-energy three-body problem is at present impossible. We must try low-order approximations such as (17), (20) or (23) in as many experimental situations as possible to see if they have any validity.

3. DIFFERENTIAL CROSS SECTION FOR IONIZATION

There are two main types of experiment in which the kinematically-complete (e,2e) differential cross section is measured. In the first[7], one electron emerges with its momentum very little different from the incident momentum. The angular distribution of the lower-energy electron is measured for a fixed momentum transfer $\underline{K}$ to the higher-energy electron. For these experiments, on closed-shell atoms, the Born approximation (17) gives a qualitatively-correct description.

The second type of experiment[8] is one in which the momentum of the residual ion is small. The emerging electrons have comparable energies. The main difference from the point of view of reaction mechanism is that the two electrons have a distant interaction in the first type and a close collision in the second type.

The second type of experiment has been performed on hydrogen[9], which we will take as our example. The theory of ionization for hydrogen contains all the principles of the independent-particle model for other systems, but there are advantages for testing a three-body theory, such as simplicity and certain knowledge of the three potentials involved in the final state. For the initial state it is necessary to approximate the potential that governs the interaction of the electron with the hydrogen atom. The static potential due to the charge distribution of a free hydrogen atom must be supplemented by a complex polarization term due to possible real and virtual excitations of the atom. This optical potential will be discussed in detail in Section 7 and following sections.

Of the three low-order approximations discussed, the distorted-wave impulse approximation (23) is the most complete, since every factor in it is a complete description of the interaction of its corresponding two-body subsystem. This approximation is a 12-dimensional integral. It has not yet been completely evaluated. At incident energies of a few keV the distorted waves are quite closely approximated by plane waves, giving the plane-wave impulse approximation. The close two-electron collision amplitude is of course given quite accurately by t_3. The amplitude (23) factorizes into a plane-wave matrix element of t_3 (which is easily antisymmetrized) and the momentum-space wave function for the ion recoil momentum $\underline{q}$.

$$\phi_i(\underline{q}) = (2\pi)^{-3/2}\int d^3r \, \exp(i\underline{q}.\underline{r}) \, \psi_i(\underline{r}) \, , \tag{27}$$

$$\underline{q} = \underline{k}_1 - \underline{k}_1' - \underline{k}_2' \, . \tag{28}$$

This factorization is assumed to be valid at lower energies where $\phi_i(\underline{q})$ is replaced by the distorted-wave transform corresponding to (27)

$$\tilde{\phi}_i(\underline{q}) = (2\pi)^{-3/2}\int d^3r\chi^{(-)*}(\underline{k}_1',\underline{r})\chi^{(-)*}(\underline{k}_2',\underline{r})\chi^{(+)}(\underline{k}_1,\underline{r})\psi_i(r) \quad (29)$$

The squared two-electron t-matrix element with antisymmetrization operator A is

$$A|\langle\underline{k}'|t_3|\underline{k}\rangle|^2 = \frac{1}{4\pi^4}\left\{\frac{1}{|\underline{k}'-\underline{k}|^4} + \frac{1}{|\underline{k}'+\underline{k}|^4} - \frac{1}{|\underline{k}'-\underline{k}|^2}\frac{1}{|\underline{k}'+\underline{k}|^2}\right.$$

$$\left. \times \cos\left[\nu\ell n\frac{|\underline{k}'+\underline{k}|^2}{|\underline{k}'-\underline{k}|^2}\right]\right\} , \quad (30)$$

$$\underline{k}' = \tfrac{1}{2}(\underline{k}_1' - \underline{k}_2') , \quad \underline{k} = \tfrac{1}{2}(\underline{k}_1 + \underline{q}) , \quad \nu = 1/2k'. \quad (31)$$

The factorized impulse approximation and the corresponding plane-wave impulse approximation are compared in fig. 1 with the data of Weigold, Noble, Hood and Fuss[9].

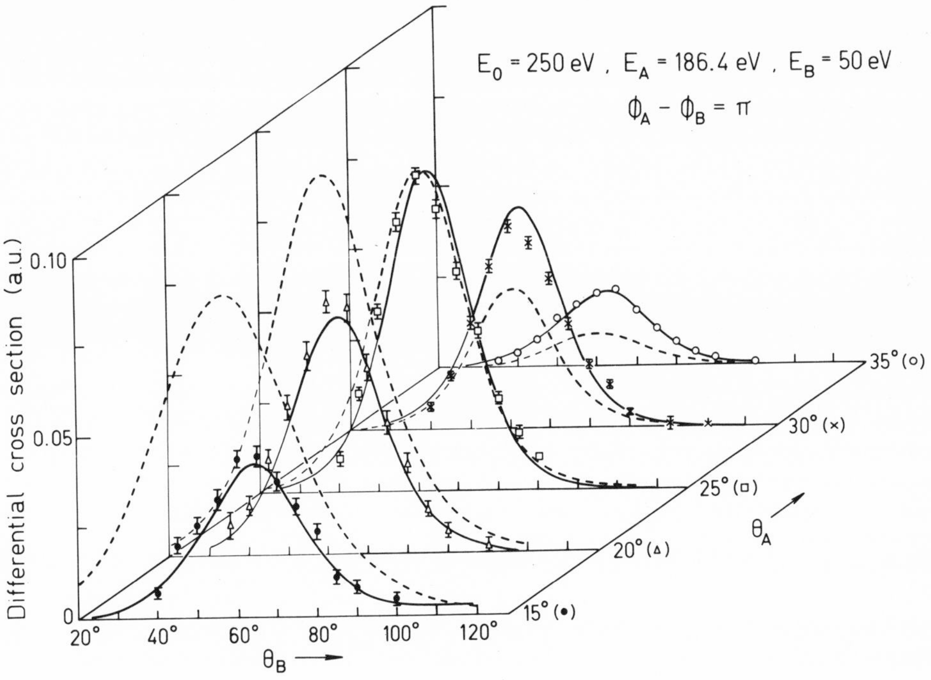

Fig. 1. Differential cross section for (e,2e) on hydrogen[9]. Theoretical curves are the factorized distorted-wave impulse approximation (full line) and plane-wave impulse approximation (dashed line).

Note that the direct part of the squared t-matrix element (30) is proportional to K^{-4}, where K is the momentum transfer to the faster emerging electron. Although the impulse approximation is quite good for the more-symmetric (e,2e) experiments, and much better than the Born approximation[10], the absolute cross section is much lower than for the asymmetric (e,2e) experiment, our first type.

The first type of (e,2e) experiment, therefore, dominates the total ionization cross section and relevant approximations will be discussed in the context of the total cross section.

4. THE COULOMB TWO-BODY PROBLEM

The Coulomb two-body problem has been reviewed at length by Chen and Chen[11]. It has some peculiar properties that make it very difficult to handle by methods used for potentials of finite range. For example the partial-wave expansion of the t-matrix for positive energy E is divergent. The problem however has simplifications of its own. The t-matrix and wave function are known in closed form. The calculations to be described make use of the momentum representation. The closed forms to be used are due to Stelbovics[12].

The t-matrix for the interaction of two electrons is given by

$$\langle \underline{k}'|t(E)|\underline{k}\rangle = \frac{1}{2\pi^2 P^2}\left[1 + \frac{1}{(1+\varepsilon)^{\frac{1}{2}}}\left\{e^{-i\nu \ln|t|}\frac{2\pi\nu C}{e^{\pi\nu}-e^{-\pi\nu}} + 1 - 2\mathrm{Re}\;{}_2F_1(1,-i\nu,1-i\nu,t)\right\}\right], \qquad (32)$$

where (using atomic units throughout)

$$\begin{aligned}
P &= |\underline{k}' - \underline{k}|, & \varepsilon &= (p_o^2 - k'^2)(p_o^2 - k^2)/p_o^2P^2, \\
\nu &= -1/2p_o, & C &= 1 \quad : (k'>p_o, k<p_o) \text{ or} \\
p_o^2 &= E, & & \qquad\;\; (k'<p_o, k>p_o), \\
t &= \frac{(1+\varepsilon)^{\frac{1}{2}} - 1}{(1+\varepsilon)^{\frac{1}{2}} + 1}, & &= e^{-\pi\nu} \;: k'>p_o, k>p_o, \\
& & &= e^{\pi\nu} \;\;\,: k'<p_o, k<p_o. \qquad (33)
\end{aligned}$$

The wave function for positive energy is defined in coordinate space[11]. For an electron interacting with a heavy nucleus of charge Z,

$$\langle \underline{r}|\psi^{(+)}(\underline{k})\rangle = \Gamma(1-i\nu)e^{\pi\nu/2}(2\pi)^{-3/2}e^{i\underline{k}.\underline{r}}\;{}_1F_1(i\nu,1,ikr-i\underline{k}.\underline{r}), \qquad (34)$$

where

$$k^2 = 2E,$$

$$\nu = Z/k. \tag{35}$$

The momentum-space wave function is formally the Fourier transform of (34). However (34) is very like a plane wave and its Fourier transform is defined only in the same sense as the delta function. For Z = 0, (34) is a plane wave. Relevant integrals are convergent and some useful ones are known in closed form.

$$\langle \underline{p}|\psi^{(+)}(\underline{k})\rangle \equiv \lim_{\eta\to o}\langle \underline{p}|\psi^{(+)}(\underline{k}),\eta\rangle$$

$$= \lim_{\eta\to o}\int d^3r e^{-\eta r}\langle \underline{p}|\underline{r}\rangle\langle \underline{r}|\psi^{(+)}(\underline{k})\rangle, \tag{36}$$

$$\langle \underline{p}|\psi^{(+)}(\underline{k}),\eta\rangle = \pi^{-2}\Gamma(1-i\nu)e^{\pi\nu/2}\left[\frac{-\nu(k+i\eta)}{A} + \frac{\eta(-1+i\nu)}{B}\right]\frac{1}{B}\left[\frac{A}{B}\right]^{-i\nu}, \tag{37}$$

$$A = p^2 - (k+i\eta)^2,$$

$$B = |\underline{p}-\underline{k}|^2 + \eta^2. \tag{38}$$

Note that

$$\langle \underline{p}|\psi^{(+)}(\underline{k})\rangle = \delta(\underline{p}-\underline{k}) \text{ for } Z = 0. \tag{39}$$

The form (37) is essentially different from the delta function since, in addition to the denominator B which vanishes in the limit $\underline{p} = \underline{k}$, it has a phase which diverges logarithmically in the same limit. The t-matrix (32) suffers from the same phase problem in the limit $\underline{k} \to \underline{k}'$ and in both half-shell limits $k = p_0$ and $k' = p_0$. The logarithmic phase vanishes for Z = 0. These divergences pose possible numerical difficulties, but do not affect the convergence of relevant integrals.

An integral that is very useful in the theory of ionization of hydrogen-like ions is the overlap of the Coulomb function $\psi^{(+)}(\underline{k})$ with the ground state ϕ_o of the hydrogen atom. The relevant overlap function is

$$\int d^3p\langle\phi_o|\underline{p}\rangle\langle\underline{p}+\underline{q}|\psi^{(+)}(\underline{k})\rangle = 2^{3/2}\pi\langle\underline{q}|\psi^{(+)}(\underline{k}),1\rangle. \tag{40}$$

A numerical method of calculating momentum-space integrals with the Coulomb function (36) as a factor in the integrand is to calculate with (37) for a few values of η and extrapolate the result to $\eta = 0$. This method is very successful for the left-hand side of

(40), which may be checked by the closed-form expression on the right-hand side. The three-dimensional integral may be performed directly by Monte Carlo techniques.

5. COULOMB BOUNDARY CONDITIONS IN THE THREE-BODY PROBLEM

The final state in the ionization problem is a three-body system interacting, at least at long range, by Coulomb potentials. One of the main difficulties of the Coulomb three-body problem is to find the asymptotic boundary condition for the final state.

A variational method has been developed by Peterkop[13] and by Rudge and Seaton[14].

The Hamiltonian $\hat{K}$ of (2) is chosen to be separable in the 1 and 2 subsystems. Each interaction is taken to be a Coulomb potential screened by the presence of the other electron. In the coordinate-space representation

$$\hat{K}(z_1,z_2) = K - z_1/r_1 - z_2/r_2. \tag{41}$$

The corresponding amplitude $M_{FI}(z_1,z_2)$, given by equation (6), has a logarithmically-divergent phase factor. It is related to the actual ionization amplitude M_{FI} by[14]

$$M_{FI}(z_1,z_2) = (2\pi)^{5/2} M_{FI}(\sin\beta)^{-2iz_2/k_2'}(\cos\beta)^{-2iz_1/k_1'}$$

$$\times \lim_{\rho\to\infty} \exp\{i[1/k_1'+1/k_2'-1/|\underline{k}_1'-\underline{k}_2'|-z_1/k_1'-z_2/k_2']\ln(2X\rho)\}, \tag{42}$$

where the hyperspherical coordinates X, β are defined by

$$k_1' = X\cos\beta,\ k_2' = X\sin\beta,\ 0 \leqslant \beta \leqslant \pi/2. \tag{43}$$

If it were not for the Pauli exclusion principle the logarithmic phase would not matter, since it would disappear on taking the absolute square. However the exchange amplitude involves the function (42) with k_1' and k_2' interchanged, so that there is a divergent phase governing the superposition of direct and exchange amplitudes.

We may choose the variational charge parameters z_1, z_2 so that the phase in (42) vanishes

$$\frac{z_1}{k_1'} + \frac{z_2}{k_2'} = \frac{1}{k_1'} + \frac{1}{k_2'} - \frac{1}{|\underline{k}_1'-\underline{k}_2'|}\ . \tag{44}$$

The ionization amplitude becomes

$$M_{FI} = (2\pi)^{5/2} \exp[i\Delta(\underline{k}_1', \underline{k}_2')] \langle\Phi_F^{(-)}(z_1,z_2)|E-\hat{K}|\Psi_I^{(+)}\rangle \pm \text{exchange}, \tag{45}$$

$$\Delta(\underline{k}_1',\underline{k}_2') = 2[(z_1/k_1')\ell n(k_1'/X)+(z_2/k_2')\ell n(k_2'/X)]. \tag{46}$$

The condition (44) gives only one condition for choosing the two effective charges z_1, z_2. A second condition must be chosen on grounds of physical reasonableness. For example in the limit $k_2' \to 0$ the effective charge z_2 should be 1. Also for $\underline{k}_1' \to \underline{k}_2'$ the indistinguishability of electrons requires $z_1 = z_2$. A suitable condition is

$$\frac{z_2}{z_1} = \left(\frac{k_1'}{k_2'}\right)^n . \tag{47}$$

The power n may be treated as a parameter. Larger values of n imply that z_2 becomes closer to 1 and z_1 to 0 for $k_2' \to 0$.

6. TOTAL IONIZATION CROSS SECTION: DISTORTED-WAVE METHOD

The total ionization cross section is given in terms of the three-body amplitude M_{FI} of equation (6) for each electron in the independent-particle system.

$$\sigma = \frac{(2\pi)^4}{k_1} \int_{E_B \leqslant E_A} dE_B \int d\hat{k}_A \int d\hat{k}_B k_A k_B |M_{FI}|^2 . \tag{48}$$

The emerging electrons are physically distinguished by their energies. We have labeled the faster one by A. The amplitude M_{FI} includes an exchange term with the roles of A and B reversed. We again consider the case of hydrogen as an easy, but not essentially oversimplified, example. We try various low-order approximations for M_{FI} in (48).

We can see from (30) that the total cross section is dominated by ionizations with small momentum transfer $|\underline{k}_1-\underline{k}_A|$. In such cases the energy of electron B is very small and one would not expect a plane-wave approximation such as the plane-wave impulse approximation obtained by neglecting t_1 and t_2 in (24) to be valid. Indeed this approximation is too large by an order of magnitude.

One obtains the correct order of magnitude by including the exact Coulomb function $\psi^{(-)}(\underline{k}_B)$ as the distorted wave $\chi^{(-)}$ in the Born approximation (17). Here the final-state wave function is not antisymmetrized. This is justified on the approximate physical grounds that electrons of disparate energy are distinguishable.

The calculation has been reported for example by Rudge[15]. It has been repeated by Stelbovics and McCarthy as a check on the direct multidimensional integration method of LePage[16], which is the standard integration technique used in calculations reported here.

The factorized distorted-wave impulse approximation (29), (30), which is highly successful for the differential cross section in the case of comparable electron energies[9] (fig. 1), again underestimates the total cross section by an order of magnitude. The factorization is valid only if both outgoing distorted waves may be approximated by plane waves. This is clearly invalid for very low energies, particularly in view of the phase factor in (37) which oscillates very rapidly for large ν. The logarithmic phase is the essential difference between Coulomb waves and plane waves.

The comparative success of the Born approximation in the highly unsymmetric case is significant in the design of approximations for total ionization cross sections. It is evidently necessary to have an accurate representation of the Coulomb character of the distorted waves. It is less important to have a closed-form description of the electron-electron interaction. A reasonable speculation is that the unsymmetric ionization occurs through distant electron-electron collisions in which the effect of the electron-electron potential v_3 is small. In these conditions an expansion in powers of v_3 is a valid approximation. For symmetric ionization (which has a small total cross section) the close electron-electron collision requires the full interaction t_3. This point could be cleared up by a full unfactorized calculation of the distorted-wave impulse approximation (23), (24), but this has not yet been done. Clearly the phase oscillation of t_3 is important in reducing the value of the integral. The phase vanishes in the factorized approximation.

An approximation that represents the Coulomb character of the distorted waves is the distorted-wave Born approximation (20) with the initial state (non-Coulomb) distorted wave $\chi^{(+)}(\underline{k}_1)$ approximated by a plane wave, and the final states represented by Coulomb waves $\psi^{(-)}(\underline{k}')$. This may be called the Coulomb-Born approximation. For non-hydrogenic targets the final-state distorted waves have Coulomb boundary conditions, so that this approximation reproduces one important feature although it neglects, for example, absorption into non-ionization or multiple-ionization channels in the final state. For hydrogenic targets the final state is a three-body Coulomb state. The Coulomb-Born approximation is

$$M_{FI} = \langle\psi^{(-)}(\underline{k}_1'),\psi^{(-)}(\underline{k}_2')|v_3|\psi_i,\underline{k}_1\rangle \ . \tag{49}$$

Because of the screening of the Coulomb potentials it is desirable to make the variational approximation (41), which involves the condition (44) and one other condition, for example (47). Because of the term $1/|\underline{k}_1'-\underline{k}_2'|$ in (44) it has been necessary in previous numerical approximations to spherically average in some way. Several such calculations are reported by Rudge[15]. There is no significant improvement over the Born approximation. The cross section for hydrogen is overestimated below 100eV, but all approximations are reasonably good at higher incident energies.

The Coulomb-Born approximation for hydrogen shares with the full distorted-wave Born approximation (20) the advantage that the potential operator does not include v_1. The term in v_1 vanishes because of the orthogonality of the bound state ψ_i and one of the distorted waves. This does not happen in the Born approximation (17), where the exchange amplitude is

$$M_{FI}^{E}(\text{Born}) = \langle\chi^{(-)}(\underline{k}_1')|v_1|\chi^{(+)}(\underline{k}_1)\rangle\langle\underline{k}_2'|\psi_i\rangle + \langle\chi^{(-)}(\underline{k}_1'),\underline{k}_2|v_3|\psi_i,\chi^{(+)}(\underline{k}_1)\rangle \ , \qquad (50)$$

which involves a nonorthogonality term in v_1. This term has the same form as the heavy-particle knockout term, which occurs in the exchange amplitude if we take $m_3 \neq 0$. Nonorthogonality terms have been calculated by some authors[15] in variants of the Born approximation.

We prefer to take the point of view that there is no particular virtue in being consistent with the Born approximation, since we have no real idea of the convergence of any particular rearrangement of the Born series. Approximations are justified by their results. We may further approximate the distorted-wave Born approximation by choosing approximations for the distorted waves that are not eigenstates of the target. Examples are plane waves, Coulomb waves in the case of a multi-electron target, or distorted waves with effective charges (44), (47) chosen to simulate the effect of v_3. Nonorthogonality terms are not present in these variants of the distorted-wave Born approximation, since they are not present in the original.

One such variant, which is not calculated at this stage, but will be taken up again in Sections 12 and 13, is to approximate the distorted wave by a plane wave for the faster of the two electrons in the final state, say $\underline{k}_1'$. Direct and exchange amplitudes are

$$M_{FI}^{D} = \langle\underline{k}_1',\chi^{(-)}(\underline{k}_2')|v_3|\psi_i,\chi^{(+)}(\underline{k}_1)\rangle \ ,$$

$$M_{FI}^{E} = \langle\chi^{(-)}(\underline{k}_2'),\underline{k}_1'|v_3|\psi_i,\chi^{(+)}(\underline{k}_1)\rangle \ . \qquad (51)$$

The differential cross section is given for a state with particular total angular momentum by

$$|M_{FI}|^2 = |M_{FI}^D + \alpha M_{FI}^E|^2 \; , \tag{52}$$

where the exchange constant α is determined by the spin multiplicity of the target shell in the independent-particle model. For example $\alpha = +1$ for a singlet state, -1 for a triplet state in a one-electron target. For a closed shell $\alpha = \frac{1}{2}$.

The differential cross section (52) is a sum of products of direct and exchange amplitudes D and E as follows

$$|M_{FI}|^2 = DD + \alpha(DE + ED) + \alpha^2 EE \; . \tag{53}$$

For a hydrogen target we have singlet and triplet cross sections σ_s, σ_t.

$$\sigma_I = \tfrac{1}{4}\sigma_s + \tfrac{3}{4}\sigma_t \; . \tag{54}$$

For initial orientation we display here the results of two approximations, described in terms of the direct amplitude M_{FI}^D of (51). The first, curve A of fig. 2, simply uses a plane wave approximation for $\chi^{(+)}(\underline{k}_1)$. The second, curve B of fig. 2, uses this amplitude when $k_2' < k_1'$, but sets the amplitude equal to zero for $k_2' > k_1'$. In this way the distorted wave is always used for the slower particle, so that the approximation is symmetric in the final particles. The term EE of (53) is equal to DD in a symmetric approximation, but the exchange term DE = ED must be calculated independently. Method B is clearly an improvement over method A and is just as easy to compute. The total ionization cross sections calculated by the two methods coincide and agree quite well with experiment at energies above about 100eV. Method B is about 25% too large at the maximum and near threshold. Also the maximum is at too low an energy.

The 25% error of method B may be almost acceptable, but we would like to know whether it can be improved and how the method is related to a more complete calculation. We want to be sure that oversimplification is not causing compensating errors so that we may expect our method to work for a highly-charged ion where the plane wave assumption for $\chi^{(+)}(\underline{k}_1)$ and $\chi^{(-)}(\underline{k}_2)$ may be worse and where we have the additional complication that there is a non-Coulomb short-range potential.

In order to calculate the distorted wave $\chi^{(+)}(\underline{k}_1)$ for the entrance channel we must solve a scattering problem which includes the possibility of excitation and ionization. We turn to this next and defer our considerations of approximations for ionization.

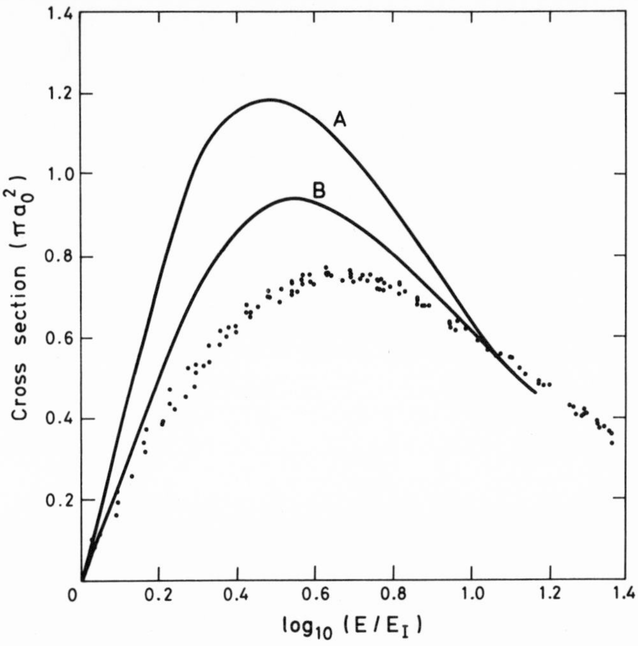

Fig. 2. Total ionization cross section. Theoretical curves A and B are described in the text. The experimental data are due to Boksenberg[23].

7. THE OPTICAL MODEL

Instead of considering ionization to a particular final state, we may consider all the possible reaction channels excited when an electron interacts with an atomic system. A discrete set of channels involves the residual system in a bound state. The remainder are ionization channels. It is possible to find the total cross section for ionization by considering the contribution from the ionization channels to the effective one-body potential governing the interaction, the optical potential. It will be shown in Section 12 that this method is equivalent to the previous differential method.

For formal purposes the ionization continuum is represented by a discrete notation (for example by imposing box normalization), but for calculations it is correctly treated as a continuum. We consider a three-body system consisting of two electrons and a core, which is assumed to be inert in the independent-particle model. We will not consider rearrangements of the system, so the incident electron will always be labeled by 1.

The Jacobi momentum coordinates are expressed in terms of the particle momenta $\underline{k}_1$, $\underline{k}_2$, $\underline{k}_3$ as

$$\underline{p} = [m_3(m_2+m_3)]^{-\frac{1}{2}}(m_3\underline{k}_2-m_2\underline{k}_3)$$
$$\underline{q} = [(m_2+m_3)(m_2+m_1+m_3)]^{-\frac{1}{2}}[(m_2+m_3)\underline{k}_1-m_1(\underline{k}_2+\underline{k}_3)] \ . \quad (55)$$

The relative momentum of the target particles 2 and 3 is $\underline{p}$. The momentum of the incident particle relative to the target centre of mass is $\underline{q}$. If m_3 is very large

$$\underline{p} \cong \underline{k}_2, \quad \underline{q} \cong \underline{k}_1 . \quad (56)$$

The corresponding kinetic energy operators are denoted by K_p and K_q. In practice we make the approximation that m_3 is infinite, thus avoiding heavy-particle knockout terms in exchange amplitudes.

The optical potential V_{ij} is the nonlocal one-body operator that describes the scattering of particle 1 by the target system in a state ϕ_j, leaving it in a state ϕ_i. In the absence of structure in the projectile (e.g. neglecting spin-orbit coupling but taking the Pauli principle into account) each target state i labels a physical channel.

The Pauli principle results in a multiplicity of similar but independent problems, depending on the multiplicity of spin-projection states in which the reaction can occur for each independent-particle shell. For example in the case of a spin ½ target such as hydrogen, we have independent singlet and triplet problems. For a closed-shell target there is only one problem. Each problem depends on a direct potential and an exchange potential with a multiplicative exchange factor α_{ij} which depends on the target shells. This factor is +1 for singlet channels, -1 for triplet channels and ½ for closed shells when i = j = 0.

The basic formalism of the optical model has been given by Feshbach[17] in terms of an operator P, which projects onto a finite set of discrete channels including the entrance channel j = 0, and its complementary operator Q, which projects onto the remaining channels including the continuum. The scattering calculation is truncated to P-space by including amplitudes for reactions involving Q-space in the optical potential.

We are interested in calculating the optical potential for ionization, in which P projects onto all the discrete states and Q projects onto the continuum. We must also discuss cases in which P projects onto a small number of low-energy channels in order to establish a simple approximate method of calculating the total ionization cross section from the ground-state scattering equation involving one optical potential.

The main results in the formal development of multichannel scattering theory will be given. The momentum representation will be used throughout.

The wave function $\Psi^{(+)}(\underline{q},\underline{p})$ for the three-body independent-particle-model system is expanded in eigenstates $\phi_i(\underline{p})$ of the target subsystem. For incident momentum $\underline{K}$ we have

$$\Psi^{(+)}(\underline{q},\underline{p}) = \Sigma_i f_i^{(+)}(\underline{K},\underline{q})\phi_i(\underline{p}) \ . \tag{57}$$

An index labelling the spin-projection state is implied on each function $\Psi^{(+)}$ and $f_i^{(+)}$, but since these states are independent it is not necessary to include them all in the treatment. The example of hydrogen is treated explicitly by McCarthy and Weigold[8].

The independent-particle orbital $\phi_i(\underline{p})$ is given by

$$[\varepsilon_i - K_p - v_2]\phi_i(\underline{p}) = 0. \tag{58}$$

The Schrödinger equation (3) becomes an infinite set of coupled equations. For this expansion we must use the discrete notation for the continuum that was referred to at the beginning of this section. The channels i are numbered in ascending energy order.

$$[E^{(+)}-\varepsilon_i-K_q]f_i^{(+)}(\underline{K},\underline{q}) = \Sigma_j\int d^3q' v_{ij}(\underline{q}',\underline{q})f_j^{(+)}(\underline{K},\underline{q}') \ , \tag{59}$$

where the potential matrix element includes the appropriate exchange term. In the approximation of an infinitely-massive nucleus we have

$$v_{ij}(\underline{q}',\underline{q}) = \langle\underline{q}'|v_1|\underline{q}\rangle\delta_{ij}+\int d^3p\int d^3p'\langle\phi_i|\underline{p}'\rangle[\langle\underline{q}',\underline{p}'|v_3|\underline{p},\underline{q}\rangle$$

$$+ \alpha_{ij}\langle\underline{q}',\underline{p}'|v_3|\underline{q},\underline{p}\rangle]\langle\underline{p}|\phi_j\rangle \ . \tag{60}$$

$$= \langle\underline{q}'|v_1|\underline{q}\rangle\delta_{ij}+A\int d^3p\int d^3p'\langle\phi_i|\underline{p}'\rangle\langle\underline{q}',\underline{p}'|v_3|\underline{p},\underline{q}\rangle\langle\underline{p}|\phi_j\rangle. \tag{61}$$

The antisymmetrization operator A is defined by (60), (61).

The coupled equations for Q-space ($i \geqslant i_0$) are formally solved and the solutions are substituted in the remaining (P-space) equations to give a finite set of coupled equations for P-space. The channels in P-space are physically discrete by definition.

$$[E^{(+)}-\varepsilon_i-K_q]f_i^{(+)}(\underline{K},\underline{q}) = \Sigma_j\int d^3q' V_{ij}^{(Q)}(\underline{q}',\underline{q})f_j^{(+)}(\underline{K},\underline{q}');i,j\varepsilon P. \tag{62}$$

The scattering amplitude for channel j with outgoing momentum $\underline{K}'$ is

$$F_{0j}(\underline{K}',\underline{K}) = \int d^3q \int d^3q'\delta(\underline{K}'-\underline{q}')V_{0j}^{(Q)}(\underline{q}',\underline{q})f_j^{(+)}(\underline{K},\underline{q}) \ , \tag{63}$$

where $V_{ij}^{(Q)}$ is the optical potential connecting channels i and j.

In order to write the optical potential in terms of amplitudes representing the interaction of the individual two-body subsystems we write equation (59) in a matrix notation

$$<i|E-K|i>f_i^{(+)} = \Sigma_j<i|v|j>f_j^{(+)} \ , \tag{64}$$

where

$$K = K_q + K_p, \qquad v = A(v_1 + v_2 + v_3) \ , \tag{65}$$

and $|i>$ is a state of the subsystem 2.

Separating the coupled equations (64) into a P-space set and a Q-space set is achieved by use of the projection operators P and Q.

$$P(E-K-v)P = PvQ \ ,$$

$$Q(E-K-v)Q = QvP \ . \tag{66}$$

Solving the Q-space equations and substituting in the P-space equations gives

$$(E-K-V^{(Q)})P = 0 \ , \tag{67}$$

where the optical potential matrix $V^{(Q)}$ is

$$V^{(Q)} = PvP + PvQ\frac{1}{Q(E-K-v)Q}QvP \tag{68}$$

Using the properties of the projection operators and the fact that

$$PvQ = Pv_3Q \tag{69}$$

because of the orthogonality of states in P and Q spaces, equation (68) becomes

$$V^{(Q)} = PT_QP \ , \tag{70}$$

where

$$T_Q = v_1 + v_3 + v_3\frac{1}{E^{(+)}-K_q-K_p-v_1-v_2-v_3}Qv_3 \ . \tag{71}$$

The projection operator Q is represented by introducing a set of eigenstates of the operator $(K_q+v_1) + (K_p+v_2)$ and restricting it to states in the required Q-space. For ionization the set is restricted to the continuum. We define time-reversed distorted waves as follows:

$$\lim_{\varepsilon\to o+} [\tfrac{1}{2}p''^2 - i\varepsilon - K_j - v_j]\chi^{(-)}(\underline{p}'') = 0; \quad j = 1,2 . \tag{72}$$

The operator T_Q for ionization ($Q \equiv I$) is

$$T_I = A[v_1+v_3+v_3\int d^3q''\int d^3p'' \frac{1}{E^{(+)}-\tfrac{1}{2}(p''^2+q''^2)-v_3}|\chi^{(-)}(\underline{q}'')>$$

$$\times\ |\chi^{(-)}(\underline{p}'')><\chi^{(-)}(\underline{p}'')|<\chi^{(-)}(\underline{q}'')|v_3] . \tag{73}$$

The potential v_3 may be included variationally by using effective charges for the distorted waves defined by (44) and (47). In this case it does not appear explicitly in the Green's function. The matrix elements of the optical potential $V^{(I)}$ are written using equations (60), (70) and (73).

$$V_{ij}(\underline{q}',\underline{q}) = \int d^3p'\int d^3p<\phi_i|\underline{p}'><\underline{q}',\underline{p}'|T_I|\underline{p},\underline{q}><\underline{p}|\phi_j> . \tag{74}$$

8. THE WEAK-COUPLING APPROXIMATION FOR THE TOTAL IONIZATION CROSS SECTION

The weak-coupling approximation neglects off-diagonal potentials in the coupled equations (62). The scattering functions $f_j^{(+)}$ for channels j in P-space become elastic scattering functions for independent channels and the expression (63) for the amplitudes F_{0j} becomes the distorted-wave Born approximation. The elastic scattering equation is decoupled from the remainder:

$$[E^{(+)} - K_q]f_o^{(+)}(\underline{K},\underline{q}) = \int d^3q'V_{oo}^{(Q)}(\underline{q}',\underline{q})f_o^{(+)}(\underline{K},\underline{q}') . \tag{75}$$

From this equation we can calculate complex phase shifts $\delta_\ell^{(Q)}$ and a reaction cross section for channels in Q-space:

$$\sigma_R^{(Q)} = (\pi/K^2)\Sigma_\ell(2\ell+1)[1-|\exp(2i\delta_\ell^{(Q)})|^2] . \tag{76}$$

The total non-elastic cross section for P and Q spaces is given by

$$\sigma_R = \sigma_R^{(Q)} + \Sigma_{j\varepsilon P}\sigma_j , \tag{77}$$

where σ_j is the total cross section for exciting the channel j, and each σ_j is independent of the partition of the space into P and Q.

This approximation has been tested by McCarthy and McDowell[18]

TABLE 1

The cross section $\sigma_R^{(Q)}$ for hydrogen, where the P-space contains the first 3 channels, calculated (a) by solving the 3-channel problem with a phenomenological potential $V_{ij}^{(Q)}$ and (b) by solving only the elastic (one-channel) problem using the same potential[8].

E(eV)	30	50	100	150	200
(a)	.483	.684	.760	.615	.506
(b)	.463	.640	.767	.631	.528

using a phenomenological imaginary part for the operator T_Q for electron scattering on hydrogen. The real part of the potential is the first-order term, with the exchange potential given by the equivalent-local approximation of Bransden, Crocker, McCarthy, McDowell and Morgan[19]. The imaginary part of the potential is proportional to the density matrix, with a strength chosen to fit the total reaction cross section σ_R. Different partitions were tried, with the P-space containing the lowest 1, 3, 4 and 6 channels. The weak-coupling approximation was confirmed within a few percent. The cross section for R-space calculated in the three-channel model using a potential $V_{ij}^{(3)}$ is compared by way of illustration with the cross section for R-space calculated in the weak coupling model (one channel with the same potential) in Table 1.

The important consequence of the weak-coupling approximation is that if Q is the ionization space I, then the total ionization cross section is given simply by (76):

$$\sigma_I = \sigma_R^{(I)} \quad . \tag{78}$$

This is the approximation used in the calculations by Stelbovics and McCarthy, which will be described.

9. THE FIRST-ORDER POTENTIAL

In order to calculate the ionization cross section in the weak-coupling approximation (75), we need only the ground-state optical potential $V_{oo}^{(I)}$. In the independent-particle model the potential has a contribution from each electron orbital ϕ in the Hartree-Fock ground state. We will consider one such contribution.

The first-order potential is given by keeping only first-order terms in v_1+v_3 in equation (73). Applying the antisymmetrization operator A gives us a direct term V_{oo}^D and an exchange term V_{oo}^E. The direct term corresponds to scattering from the Hartree-Fock charge density.

$$V^D_{oo}(\underline{q}',\underline{q}) = \int d^3p' \int d^3p <\phi|\underline{p}'><\underline{q}',\underline{p}'|v_1+v_3|\underline{p},\underline{q}><\underline{p}|\phi> , \qquad (79)$$

where

$$<\underline{q}',\underline{p}'|v_1+v_3|\underline{p},\underline{q}> = <\underline{q}'|v_1|\underline{q}>\delta(\underline{p}'-\underline{p}) + <\underline{k}'|v_3|\underline{k}>\delta(\underline{q}'+\underline{p}'-\underline{q}-\underline{p}) . \qquad (80)$$

The relative momentum coordinates of the two electrons are, in the approximation of an infinitely-massive nucleus,

$$\underline{k}' = \tfrac{1}{2}(\underline{q}'-\underline{p}'), \quad \underline{k} = \tfrac{1}{2}(\underline{q}-\underline{p}) . \qquad (81)$$

The momentum-space Coulomb potential is

$$<\underline{k}'|v|\underline{k}> = [2\pi^2|\underline{k}'-\underline{k}|^2]^{-1} . \qquad (82)$$

The direct potential is local. In momentum space this means that it is a function only of the momentum transfer P, where

$$\underline{P} = \underline{q}-\underline{q}' . \qquad (83)$$

This is clearly true for the v_1 term of (80).

The v_3 term of (80) is also local, as we can see using the kinematic condition of translational invariance given by $\delta(\underline{q}'+\underline{p}'-\underline{q}-\underline{p})$.

$$\underline{k}'-\underline{k} = \underline{q}'-\underline{q} . \qquad (84)$$

The exchange potential, however is nonlocal, since reversing the roles of $\underline{q}$ and $\underline{p}$ gives

$$\underline{k}' + \underline{k} = \underline{q}' - \underline{p} , \qquad (85)$$

so that $V^E_{oo}(\underline{q}',\underline{q})$ depends on q and $\underline{q}.\underline{P}$.

10. THE EQUIVALENT LOCAL POTENTIAL

The optical potential V_{oo} is nonlocal in the sense that it depends on q and $\underline{q}.\underline{P}$. A local potential is obtained by spherically averaging, which means integration over $\hat{\underline{q}}.\hat{\underline{P}}$. We will show in what sense the local approximation is equivalent to the nonlocal potential, using the first-order exchange potential as an example. There is no numerical difficulty in solving the elastic scattering problem (75) for a nonlocal potential. However if an equivalent local potential can be found, one needs only to calculate a one-dimensional table for the optical potential, saving much computing time, since the multi-dimensional integration involved in the calculation of V_{oo} is by far the most time-consuming part of the procedure.

An equivalent local potential was found in coordinate space by Furness and McCarthy[21], using a Taylor expansion of the scattering wave function. The coordinate-space equivalent of (75) contains the following potential term

$$\int d^3r'V(\underline{r}',\underline{r})f_o^{(+)}(\underline{K},\underline{r}') = \int d^3r'V(\underline{r}',\underline{r})e^{\underline{\nabla}.(\underline{r}'-\underline{r})}f_o^{(+)}(\underline{K},\underline{r}) \ . \tag{86}$$

The equivalent local potential is thus formally

$$V(r) = \int d^3r'V(\underline{r}',\underline{r})e^{\underline{\nabla}.(\underline{r}'-\underline{r})}. \tag{87}$$

This is approximated by replacing the gradient operator by the self-consistent local value of the momentum, since this is locally the eigenvalue of $-i\underline{\nabla}$ with the eigenfunction being the local plane wave that approximately represents $f_o^{(+)}(\underline{K},\underline{r})$ at $\underline{r}$.

$$e^{\underline{\nabla}.(\underline{r}'-\underline{r})} = e^{i\underline{\bar{K}}.(\underline{r}'-\underline{r})} \ , \tag{88}$$

$$\bar{K}^2 = 2[E - V(r)] \ . \tag{89}$$

This self-consistent local potential gives very accurate phase shifts. For illustration we use a first-order calculation for hydrogen by Vanderpoorten[22]. This is denoted in Table 2 by the column SCL, where it is compared with an exact solution EX of the nonlocal first-order Schrödinger equation.

The self-consistent local method is simple only in coordinate space, since it depends essentially on the coincidence that the Taylor expansion operator $\underline{\nabla}$ is also the coordinate-space representation of the momentum operator. The difficulty in momentum space is due to the difficulty in obtaining the Fourier transform of $\exp[i\underline{\bar{K}}.(\underline{r}'-\underline{r})]$. This difficulty disappears if we replace $\bar{K}$ by a constant as in the averaged eikonal approximation

$$\bar{K}^2 = 2(E + \bar{V}) \ . \tag{90}$$

The averaged eikonal approximation gives an excellent representation of the distorted waves in (e,2e), where the appropriate value of $\bar{K}$ is given quite closely by the single-particle separation energy ε of the bound state concerned[10].

$$\bar{V} = \varepsilon. \tag{91}$$

In the averaged eikonal approximation the equivalent local potential in momentum space is

TABLE 2

The first-order approximation for electron-hydrogen phase shifts. The exact (EX) and self-consistent local (SCL) values were obtained by Vanderpoorten. The remaining columns use the momentum-space approximation (92) of Stelbovics and McCarthy.

Case	ℓ	EX	SCL	$\bar{V}=0$	$\bar{V}=13.6\text{eV}$
100eV singlet	0	.532	.539	.51	.52
	1	.187	.193	.179	.188
	2	.0796	.0809	.077	.082
	3	.0364	.0365	.036	.039
100eV triplet	0	.667	.662	.69	.68
	1	.302	.297	.31	.30
	2	.145	.144	.149	.143
	3	.0712	.0711	.072	.070
50eV singlet	0	.541	.547	.44	.50
	1	.0953	.109	.077	.108
	2	.0275	.0277	.021	.035
	3	.00898	.00797	.0087	.0133
50eV triplet	0	.854	.838	.95	.90
	1	.334	.321	.37	.34
	2	.124	.123	.135	.120
	3	.04663	.0474	.048	.043

$$V(P) = \int d(\hat{\underline{K}}.\hat{\underline{P}})V(\underline{P}+\underline{\bar{K}},\underline{\bar{K}}) \ . \tag{92}$$

The phase shifts for hydrogen obtained using the first-order equivalent local potential of (92) are compared with the exact phase shifts in Table 2 for two cases $\bar{V} = \varepsilon$ and 0. As would be expected $\bar{V} = 0$ is better than $\bar{V} = \varepsilon$ for large ℓ (distant collisions) and worse for small ℓ. On balance we have decided to use $\bar{V} = 0$, since larger ℓ are more important for ionization and the approximation improves as ℓ increases.

The local potential

$$V(P) = \int d\hat{\underline{q}}.\hat{\underline{P}} \ V(\underline{q}',\underline{q}), \quad P = \underline{q}'-\underline{q} \tag{93}$$

is thus equivalent to the nonlocal potential $V(\underline{q}',\underline{q})$ in the sense that it produces very similar first-order phase shifts. We will make the approximation that it is equivalent to the nonlocal potential in general. This approximation is easily tested by calculating the higher multipoles

$$V^{(\lambda)}(P) = \int d\hat{\underline{q}}.\hat{\underline{P}}(2\lambda+1)P_\lambda(\hat{\underline{q}}.\hat{\underline{P}})V(\underline{q}',\underline{q}) \ . \tag{94}$$

The dipole (λ=1) potential has been calculated for the direct term, which is the largest contribution to our ultimate approximation, equation (95), for $V_{oo}^{(1)}$. It is less than the monopole term by two orders of magnitude for small P.

11. THE SECOND-ORDER OPTICAL POTENTIAL AND IONIZATION CROSS SECTION

Since low orders in the distorted-wave Born approximation seem to be a good approximation to differential cross sections for ionization, it makes sense to use the corresponding approximation for the optical potential. We therefore keep only the lowest-order terms in v_3 for the power series expansion of T_I which will give us a description of ionization. We need the second-order term, since the first-order optical potential is real and gives no reaction cross section. The first-order potential has been treated in Section 9. We are interested in the second-order potential for the ground state. Note, however, that v_3 may be partially included in the definition of the intermediate distorted waves by defining effective charges according to the variational procedure (44), (47). In this case higher orders are approximately included.

$$V_{oo}^{(2)}(\underline{q}',\underline{q}) = A\langle\phi_o|\underline{p}'\rangle\langle\underline{q}',\underline{p}'|v_3|\underline{p}_1,\underline{q}_1\rangle\langle\underline{p}_1|\chi^{(-)}(\underline{p}'')\rangle\langle\underline{q}_1|\chi^{(-)}(\underline{q}'')\rangle$$
$$\times \frac{1}{E^{(+)}-\frac{1}{2}(p''^2+q''^2)}\langle\chi^{(-)}(\underline{q}'')|\underline{q}_2\rangle\langle\chi^{(-)}(\underline{p}'')|\underline{p}_2\rangle\langle\underline{q}_2,\underline{p}_2|v_3|\underline{p},\underline{q}\rangle\langle\underline{p}|\phi_o\rangle \tag{95}$$

The notation of equation (95) uses a convention implying integration over repeated momentum coordinates. The antisymmetrization operator A is defined by (60), (61). It amounts to the substitution

$$\langle\underline{q}',\underline{p}'|v_3|\underline{p},\underline{q}\rangle \to \langle\underline{q}',\underline{p}'|v_3|\underline{p},\underline{q}\rangle + \alpha_{oo}\langle\underline{q}',\underline{p}'|v_3|\underline{q},\underline{p}\rangle \tag{96}$$

for all matrix elements of v_3.

It is convenient to separate $V_{oo}^{(2)}$ into two parts

$$V_{oo}^{(2)} = -U - iW \ , \tag{97}$$

where U contains the real part of the Green's function and W contains the imaginary part of the Green's function. Both U and W are hermitian and W is positive definite. For example

$$W(\underline{q}',\underline{q}) = W^*(\underline{q},\underline{q}') \ . \tag{98}$$

The phase shift for a scattering problem involving a nonlocal potential is real if the potential is hermitian. The antihermitian part iW is responsible for the imaginary part of the phase shift, and hence for the reaction cross section. The dependence of the ionization cross section on W can be derived from the probability current, which is a coordinate-space concept.

In the weak-coupling approximation the Schrödinger equation for the elastic channel is the coordinate-space analogue of (75).

$$[\nabla^2+k^2]f_o^{(+)}(\underline{K},\underline{r}) = 2\int d^3r'[U(\underline{r}',\underline{r})+iW(\underline{r}',\underline{r})]f_o^{(+)}(\underline{K},\underline{r}') \qquad (99)$$

The ionization cross section is the integral over all space of the divergence of the probability current $\underline{j}(\underline{r})$, divided by the incident current K.

$$\sigma_I = \frac{1}{K}\int d^3r\underline{\nabla}.\frac{1}{2i}\left\{f_o^{(+)*}(\underline{K},\underline{r})\underline{\nabla}f_o^{(+)}(\underline{K},\underline{r})-f_o^{(+)}(\underline{K},\underline{r})\underline{\nabla}f_o^{(+)*}(\underline{K},\underline{r})\right\} \qquad (100)$$

Substituting for ∇^2 from the Schrödinger equation (99) and using the hermitian property of $W(\underline{r}',\underline{r})$, we find, using momentum-space normalization,

$$\sigma_I = (2/K)(2\pi)^3\langle f_o^{(+)}(\underline{K})|W|f_o^{(+)}(\underline{K})\rangle \ . \qquad (101)$$

We may find the total ionization cross section by solving the scattering problem (75). We either calculate the phase shifts and use (76) or calculate the distorted wave and use (101). The former method is computationally simpler, but equation (101) is very useful for understanding formal relationships.

The equivalent local optical potential in momentum space is given by spherically averaging $V_{oo}^{(2)}$. The equivalent local potential W must be real. However since the continuum wave functions in (95) are complex, the definition (97), (98) of W makes it complex. Our procedure for finding the equivalent local potential involves the additional prescription that we take the real part only. That this is the correct procedure will be formally verified in the next section.

12. EQUIVALENCE OF SECOND-ORDER OPTICAL AND DISTORTED-WAVE METHODS

The total ionization cross section is given in the weak-coupling approximation by substituting the antihermitian part of $V_{oo}^{(2)}$, equation (95), into the integral of the divergence of the probability current, equation (101). Using the convention that repeated momentum coordinates are integrated over, we have

$$\sigma_I = (2/K)(2\pi)^3 \langle f_o^{(+)}(\underline{K})|\underline{q}'\rangle\langle\phi_o|\underline{p}'\rangle\langle\underline{q}',\underline{p}'|v_3|\underline{p}_1,\underline{q}_1\rangle\langle\underline{p}_1|\chi^{(-)}(\underline{p}'')\rangle$$
$$\times \langle\underline{q}_1|\chi^{(-)}(\underline{q}'')\rangle\pi\delta(E_F-E_I)\langle\chi^{(-)}(\underline{q}'')|\underline{q}_2\rangle\langle\chi^{(-)}(\underline{p}'')|\underline{p}_2\rangle\langle\underline{q}_2,\underline{p}_2|v_3|\underline{p},\underline{q}\rangle$$
$$\times \langle\underline{p}|\phi_0\rangle\langle\underline{q}|f_0^{(+)}(\underline{K})\rangle \ . \tag{102}$$

Redefining the coordinates

$$\underline{k}_1 = \underline{K}, \quad \underline{k}_A = \underline{q}'', \quad \underline{k}_B = \underline{p}'' , \tag{103}$$

and performing one of the energy integrations using the energy-conserving delta function we have the expression for the integrated differential cross section:

$$\sigma_I = \frac{(2\pi)^4}{k_1}\int_{E_B \leq E_A} dE_B \int d\hat{\underline{k}}_A \int d\hat{\underline{k}}_B k_A k_B |M_{FI}|^2 , \tag{104}$$

where we have defined E_B to be the lesser of E_A and E_B and

$$M_{FI} = \langle\chi^{(-)}(\underline{k}_A),\chi^{(-)}(\underline{k}_B)|v_3|\phi_0,f_0^{(+)}(\underline{K})\rangle \ . \tag{105}$$

The matrix element M_{FI} is the distorted-wave Born approximation for the ionization amplitude, if we consider $f_0^{(+)}(\underline{K})$ as the elastic-scattering wave function. This is not strictly correct, since the second-order part of the optical potential represents scattering only to ionized states. However in the weak-coupling approximation limit the second-order potential has little effect on distorted waves. Compare (104), (105) with (48), (20).

We see that ionization is given by only the imaginary part of the Green's function in the second-order potential. This justifies the prescription of taking the real part of the spherically-averaged hermitian potential W in the equivalent-local approximation.

Since the optical and distorted-wave methods are essentially equivalent, it is interesting to compare them from the point of view of computational difficulty. The method computed in Section 6 used the plane-wave approximation for the entrance-channel wave function

$$f_0^{(+)}(\underline{k}_1) \equiv \chi^{(+)}(\underline{k}_1) , \tag{106}$$

and distorted waves for the final-state electrons. The consideration of screening in the asymptotic form of the final-state distorted waves amounts to going further than the second-order approximation in the optical method, where $\chi^{(-)}(\underline{q}'')\chi^{(-)}(\underline{p}'')$ would be a variational approximation to an eigenstate of the three-body Green's function. This approximation is equally feasible in both methods.

The optical method, however, is computationally easier since the computation of W requires two fewer three-dimensional integrations than the distorted-wave method. Also it does not require the computation of an entrance-channel distorted wave. This computation is essentially equivalent to the solution of the elastic scattering problem with the second-order potential $-iW$, which must be performed to obtain the reaction cross section from the complex phase shifts. The real second-order potential $-U$ is of little consequence.

Note that the initial-plane-wave approximation, which is the Coulomb-Born approximation (48), (49) for hydrogenic targets, is simply the momentum-space imaginary potential for $P = 0$ (see (101)).

A full computation of the distorted-wave method is essentially not feasible, while the equivalent optical calculation is feasible. Results are reported in the following section.

13. APPROXIMATIONS TO THE SECOND-ORDER OPTICAL METHOD

The approximations that can be tried within the framework of the second order optical model can all be understood with reference to the operator T_I (71) and the amplitudes M_{FI} (105). They amount to replacing different distorted waves by plane waves, including second-order exchange or not, or adopting different effective charges according to (44) and (47).

In all the present calculations by Stelbovics and McCarthy only the first-order direct part of the real potential is used. First-order exchange and second-order polarization (the real part of $V_{00}^{(2)}$) play secondary, but not quite negligible roles. These roles are well understood. The second-order imaginary potential is responsible for ionization and is the overwhelming factor in determining its cross section.

Approximating T_I only to second order in $v_1 + v_3$ means omitting this quantity from the denominator of the Green's function in (71) and hence from the projection operator. Only $\chi^{(-)}(\underline{k}_B)$ is then a distorted wave. $\chi^{(-)}(\underline{k}_A)$ is a plane wave. We call the projection operator the Born projector. The ionization cross section in the Born approximation (17) is given according to (101) by

$$\sigma_I = (2/K)(2\pi)^3 \, W(0) \; . \qquad (107)$$

For the Born approximation we use only the DD part of W(0) in the notation (53).

By calculating the full equivalent-local optical potential $W_{DD}(P)$ for this case we can solve the elastic scattering problem and calculate the ionization cross section from the complex phase shifts according to (76). This corresponds to using a distorted wave for $\chi^{(+)}(\underline{k}_1)$, approximation (51). We are thus able to answer the question whether the Born approximation (method A of Section 6) is considerably improved by distorting the entrance-channel wave.

The same can be done for the method B of Section 6, in which we considered only $k_A > k_B$. We call the projection operator the truncated Born projector. Since method B gave a reasonably close approximation to experiment, it could be hoped that distorting the entrance-channel wave would largely eliminate the discrepancy.

Curves A and B of fig. 2 are repeated in fig. 3. In addition we show the corresponding optical-model ionization cross sections, labelled AO and BO. Unfortunately distorting the entrance-channel wave considerably worsens both approximations, although there is evidence that the optical model is an improvement above about 300eV. Approximation BO also peaks nearer the experimental peak energy than the others.

We now consider what must be done to improve on the versions A and B of the optical model using the Born projector. The use of a plane wave considerably simplifies the computation, so the Born projector is not lightly discarded.

Perhaps the equivalent local potential is not a good enough approximation to the nonlocal optical potential. This may be immediately tested by calculating further terms in the multipole expansion (94) and solving the Lippmann-Schwinger equation for the resulting nonlocal potential in momentum space. Discrepancies are expected to be greatest at low energies. We have therefore calculated the $\lambda = 1$ contribution to the ionization cross section for the Born projector (method A) at 20eV. We obtain

$$\sigma_I^{(0)} = 2.40 \text{ a.u.},$$

$$\sigma_I^{(1)} = .20 \text{ a.u.}$$

This 10% discrepancy is not any more important than other errors in the calculation (see Section 14). We therefore regard the equivalent local model as justified.

We could consider second-order exchange for the Born projector at this stage, although there is no point in considering it for the truncated Born projector, where the truncation itself is a crude way of including exchange. We prefer, however, to go straight to the replacement of the plane wave in the projector by a distorted wave and consider exchange for the fuller theory.

Approximating T_I to second order in v_3 means including v_1 in the Green's function of (71) and hence having two distorted waves in the projection operator. We call it the distorted-wave

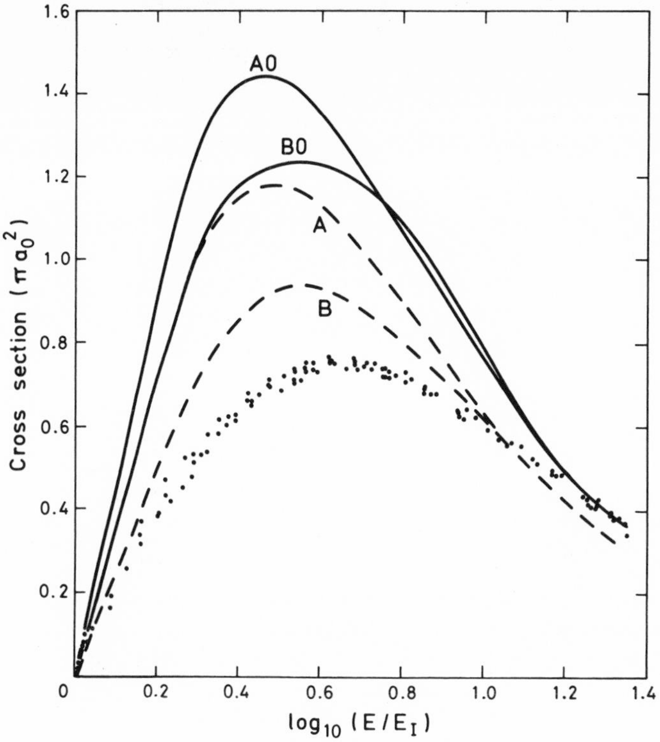

Fig. 3. Total ionization cross section. Curves A and B of Section 6 are compared with the curves A0 and B0 obtained by distorting the entrance-channel waves.

projector or method C.

Because of the ordering of the integration variables in (102), methods A and B required the computation of Coulomb distorted waves only in the form of overlap integrals with bound states (40). Such functions may be expressed analytically. However the integral (102) for the distorted-wave projector (or for exchange terms in the Born projector model) requires the computation of a Coulomb distorted wave (36) in the integrand for a finite value of the smoothing parameter η. Because of the rapidly-oscillating phase for $\underline{p}\rightarrow\underline{k}$ this is not nearly as simple as a delta function, although the integral is still convergent. We find that the multidimensional integration method converges too slowly for $\eta \lesssim .4$, so we perform calculations at $\eta = .4$ and consider methods of extrapolation to $\eta = 0$.

The second-order model C is illustrated for $\eta = .4$ in fig. 4. At most energies, computing-time considerations have forced us to

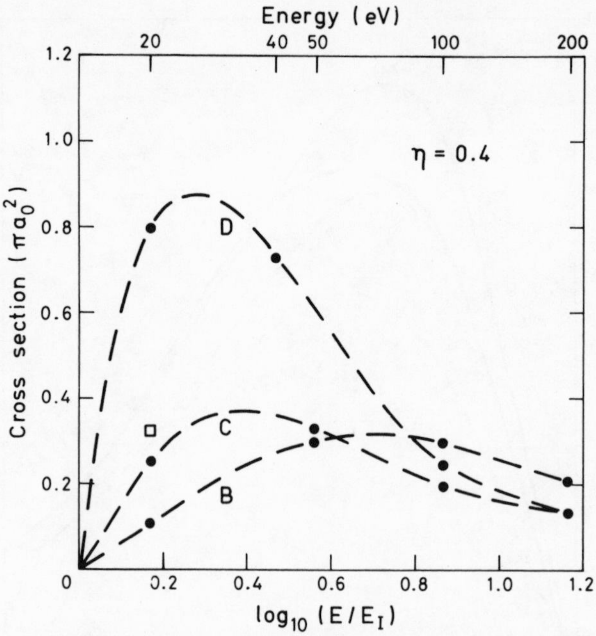

Fig. 4. Total ionization cross section computed with the smoothing parameter η = .4. The computations described in the text were performed only at the points indicated by filled circles. The dashed curves are merely to guide the eye.

calculate only the Coulomb-Born approximation (49), including second-order exchange. For this we evaluate the momentum-space optical potential only for P = 0 as in (107). We expect the full optical-model calculation to increase the cross section by about 20% as it did for the truncated Born projector B. The optical model C has been fully evaluated at 20eV with the expected result, as shown by the square in fig. 4.

In fact the imaginary part of the optical potential does not change much in shape as the incident energy is varied. This is illustrated in fig. 5 for the Born projector. The potential is fairly closely proportional to W(0) and the energy-variation of the ionization cross section calculated from W(0) gives a good idea of the energy-variation of the optical model ionization cross section.

The distorted-wave projector model C shares with the Born-

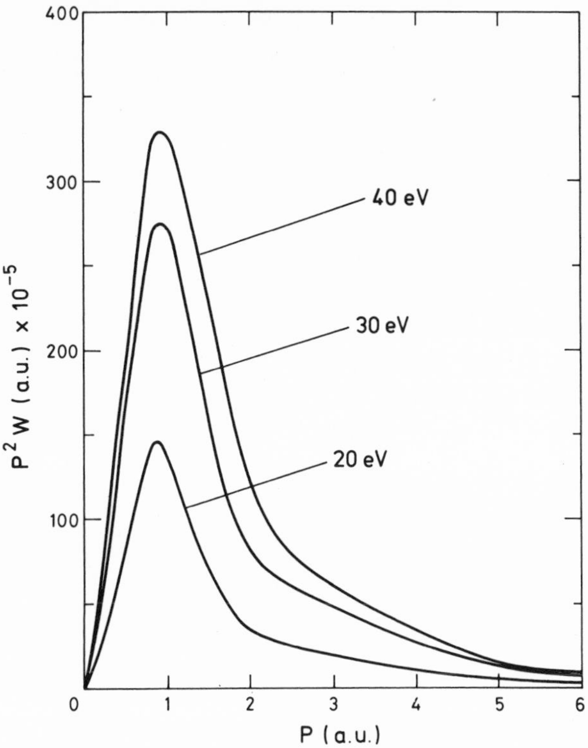

Fig. 5. The quantity $P^2W(P)$ plotted against P for incident energies 20eV, 30eV and 40eV.

projector model A the disadvantage of relatively overestimating the low-energy cross section. However in C the second-order exchange terms have been fully included, eliminating exchange as a possible cause of serious improvement at low energy. The relative importance of exchange at different energies is illustrated in Table 3, where the ratio of the triplet cross section σ_t to the singlet cross section σ_s is given.

$$\sigma_t = 2(DD-DE),$$
$$\sigma_s = 2(DD+DE). \tag{108}$$

The shapes of the curves in fig. 4 are proportional to the corresponding ionization cross section only if extrapolation from $\eta = .4$ to $\eta = 0$ involves an energy-independent scale factor. This is not the case for the truncated Born model B, which can be evaluated using the analytic expression (40) for finite η by

TABLE 3

The ratio of triplet to singlet ionization cross sections for hydrogen at various energies calculated by the optical model using the distorted-wave projector C and the $n = \infty$ distorted-wave projector D. The ionization cross section is evaluated in the W(0) approximation (107).

E(eV)	C	D
20	.18	.21
40	-	.20
50	.22	-
100	.30	.23
200	.46	.41

replacing the argument η in the definition (37) of the Coulomb distorted wave by $1 + \eta$. The extrapolation factor decreases from 3.9 to 2.2 as E increases from 20eV to 200eV. The $\eta = .4$ version of model B is illustrated in fig. 4.

Our next attempt to improve the model involves going beyond second order by making an attempt to include some effects of v_3 in the asymptotic screening approximation. The disadvantage of this is that the variational method yields only one relationship (44) for the effective charges. The auxiliary condition (47) has a screening parameter n which must be arbitrarily chosen.

The maximum effect of screening is to use a very large value of n (designated $n = \infty$), for which

$$z_1 = 1, \quad z_2 = 0; \quad k_1' < k_2' ,$$

$$z_1 = 0, \quad z_2 = 1; \quad k_2' < k_1' . \tag{109}$$

The resulting model is the $n = \infty$ distorted-wave model D, shown for the W(0) approximation (107) including second-order exchange in fig. 4. The effect of (109) in both direct and exchange amplitudes (105) is to replace the distorted wave by a plane wave for the faster particle. The elimination of the phase averaging in the Coulomb distorted wave causes an immediate exaggeration of the low-energy cross section, although this effect disappears at 200eV where models C and D are similar. The importance of exchange in this model is seen in Table 3, where the ratios of triplet to singlet cross sections are illustrated at various energies.

In order to see the effect of less-extreme screening we have calculated the W(0) approximation (107) for $n = 1$ at 20eV and 200eV, with $\eta = .4$. The results were essentially the same as for $n = \infty$, within the statistical errors of the Monte Carlo integration.

We now return to the second-order distorted-wave projector, method C. The smoothing parameter η must be extrapolated to zero. This has been done by calculating W(0) for η = .2, .4 and .6 and extrapolating by means of a cubic spline. Statistical errors in W(0) were between 5% and 8%. Results are shown in Table 4.

At 200eV the extrapolated cross section in the W(0) approximation (107) is essentially the same for the distorted-wave projector as in the direct Born projector methods A and B. It also agrees well with experiment. However the cross section is peaked at far too small an energy value and is too large at 20eV by a factor of 4.

The extrapolation method has been checked for the direct Born projector. In this case the potential can be calculated either by performing the overlap integrals (40) numerically for finite η according to (37) and extrapolating to $\eta = 0$ or independently by using the analytic expression (40). The method checks in this case to within 10%. It is essentially simpler than for the distorted-wave projector because the replacement of a distorted wave by a plane wave eliminates two of the three-dimensional integrations. Since the Born models give essentially the same results as the distorted-wave models above 200eV, we may assume that the extrapolation is valid at high energies. The same is not true near threshold and the extrapolation method has some unsatisfactory aspects.

14. SUMMARY OF APPROXIMATIONS AND ERROR ESTIMATES

In the calculations that have been reported for hydrogen the basic approximation is the weak-coupling approximation of Section 8. This approximation has been tested in the semi-phenomenological

TABLE 4

Extrapolation of W(0) for the second order model C to $\eta = 0$. The last two columns compare the ionization cross section with experiment (units πa_0^2).

E(eV)	η = .6	η = .4	η = .2	η = 0	σ(C)	σ(exp)
15	.00467	.01012	.02827	.0590	.46	.1
20	.01779	.03714	.09872	.2025	1.38	.34
50	.03447	.07632	.1485	.2510	1.08	.73
100	.03317	.06414	.1419	.2664	.81	.68
200	.03017	.06045	.1265	.2283	.49	.49

calculation of McCarthy and McDowell[18], whose results are illustrated for a three-channel example in Table 1. Errors are of the order of 5%.

The object has been to investigate the direct and exchange parts of the second-order imaginary potential. The first-order exchange and real second-order polarization potentials have been omitted for convenience, although there is no essential computational difficulty in including them. There is some evidence in an optical-model calculation of ionization by Coulter and Garrett[20] that first-order exchange has an effect of the order of a 10% reduction in the ionization cross section.

The omission of multipoles higher than monopole in the momentum-space optical potential (94) (the equivalent-local potential) has been justified by calculating the dipole contribution to the ionization cross section for the Born projector at a low energy (20eV), where the effect is expected to be greatest. This gives a 10% enhancement. The more-rapid oscillation of the integrand for higher multipoles of course ensures the rapid convergence of the multipole expansion.

The evaluation of the multidimensional integrals (102) for the optical potential by the Monte Carlo method[16] involves accuracy which can be arbitrarily increased by extending computing time. The Born projector models A and B have been evaluated to better than 3%. The distorted-wave projector models have been evaluated to better than 6% for small P and about 10% for larger P, where W(P) is several orders of magnitude smaller.

The largest possible error in the distorted-wave projector models is involved in the extrapolation for the smoothing parameter η to 0. This parameter can be made arbitrarily small, but the rapid phase oscillation so introduced produces a requirement for more integration points in the Monte Carlo method.

15. CONCLUSIONS, GENERALIZATIONS AND IMPROVEMENTS

All the models we have discussed for the total ionization cross section may be expressed in the language of the equivalent local optical model in momentum space. The models with entrance-channel plane waves give the ionization cross section in terms of the forward-scattering imaginary potential W(0), according to (107). The use of optical model distorted waves in the entrance channel is achieved by evaluating the potential for a range of the momentum coordinate P, solving the Schrödinger equation, and calculating the cross section from (76) or (101).

The treatment of the final-state distorted waves determines

the projection operator used in the optical potential. We have always used one distorted wave. If the other wave is a plane wave we have a Born projector, if it is a distorted wave in the full potential due to the residual ion we have the second-order distorted-wave projector. We have also investigated the effect of using effective charges determined by the three-body asymptotic conditions to simulate the partial inclusion of the electron-electron interaction v_3, thus simulating higher-order effects.

There are two outstanding features of the investigation. The first is that all the models converge on the experimental value at about 200eV. The second is that the models involving the direct Born propagators are apparently better at low energy than the more-complete distorted-wave models.

In the Born projector model A we have no reason to take the direct (DD) term seriously. Second-order exchange cannot just be ignored. We have evaluated W(O) with exchange at 20eV, using the value .3 for the smoothing parameter η. The DE term is greater than DD by a factor 2. The EE term is 15 times DD. Thus the apparent superiority of model A over the distorted-wave models is an illusion.

The truncated Born projector model B gives the best results at low energy, although distorting the entrance channel worsens the results. This model takes exchange into account in a semiclassical way by distinguishing the faster electron, whose wave is undistorted, from the slower electron, whose wave is calculated in the potential of the residual ion.

If there were any significance in the success of the Born projector models, one would expect refinements to improve them. In fact the most refined theories possible in the second-order context fail badly at low energy. We must conclude that the success of the Born-projector models for hydrogen is purely fortuitous and that there is no reason for expecting them to work as well for other targets.

Above 200 eV we can claim that the optical model for hydrogen is well-understood. The success of the Born-projector models is due to the reduced effect of distortion at higher energy and to the fact that second-order exchange is relatively insignificant. The success of the Born-projector models at high energies makes the optical potential at these energies extremely easy to calculate. It is the essential negligibility of second-order exchange that provides the greatest simplification. The distorted wave in the DD term always occurs only in an overlap integral with the bound-state wave function of the form (40). This can be calculated analytically for a Coulomb wave and does not require a smoothing parameter η.

We have noticed that it is possible to obtain the correct cross section at low energies for a finite value of the smoothing parameter η, about .3 at 20eV. This parameter was introduced as the inverse range of an exponential cutoff in the coordinate-space Coulomb function in the definition (26) of the smoothed momentum-space Coulomb function. This cutoff could be considered in some way as accounting for higher-order effects and it is possible to conjecture that choosing its value to fit a known piece of external data such as the static polarizability (calculated from $\mathrm{Re}V_{oo}^{(2)}$) may lead to acceptable ionization cross sections. The test of this is left to the future.

Another possible way of improving the second-order model is to assume that the potentials v_1 and v_2 in the three-body Green's function (71) are locally constant. The potential v_3 is then included by omitting it from the Green's function but replacing v_3 on the right-hand end of the expression (71) by t_3, given by (32). With fast computers it is possible to calculate this model. Preliminary tests have shown that it reduces the low-energy cross section considerably.

It seems clear that the use of effective charges to improve the three-body Coulomb boundary condition asymptotically is not an improvement. This was found also by Weigold et al.[9] for the differential ionization cross section of hydrogen. The cross section is sensitive to wave functions in the interaction region, not to asymptotic conditions.

In order to generalize the calculation to targets other than hydrogen it is necessary to use Hartree-Fock single-particle orbitals for the ground state of the target and distorted waves calculated either in the Hartree-Fock potential or, preferably, in a self-consistently-determined complex optical potential. At high energies it is likely that the Coulomb approximation is reasonably good for distorted waves, so that the analogue of the Coulomb overlap function (40) is the only required generalization. This is easily obtained analytically for Slater-type orbitals.

Non-Coulomb distorted waves with Coulomb boundary conditions are easily expressed in momentum space with no essential increase in the computational difficulty. Choosing the Coulomb part of the two-body potential as the distorting potential V_D of (2), leaving the residual short-range potential v, one multiplies the two-body Lippmann-Schwinger equation (21) on the left by $\underline{k}$ to obtain

$$\langle \underline{k}|\chi^{(-)}(\underline{p})\rangle = \langle \underline{k}|\psi^{(-)}(\underline{p})\rangle + \frac{1}{E^{(-)}-\frac{1}{2}k^2} \langle \underline{k}|v|\chi^{(-)}(\underline{p})\rangle \, , \qquad (110)$$

where $\psi^{(-)}(\underline{p})$ is the Coulomb wave and $\langle \underline{k}|v|\chi^{(-)}(\underline{p})\rangle$ is the half-shell t-matrix. This is calculated by solving the Schrödinger equation in coordinate-space with Coulomb boundary conditions, which is a computationally-fast procedure resulting in a table accurate enough for interpolation to give the Monte-Carlo optical potential integral. The first term in (110), the Coulomb distorted wave, gives the same trouble as it has in the case of hydrogen.

We can therefore assert that the high-energy ionization problem is understood for any target, since the main difficulty is the Coulomb distorted wave, which has been thoroughly investigated for hydrogen. It remains to be seen whether it is possible to find a parameter, such as the smoothing parameter η, whose value can be independently chosen to give a predictive model for the low-energy ionization cross section. Alternatively, it may turn out that inclusion of higher-order terms in the form of the Coulomb t-matrix gives sufficient improvement at low energies.

REFERENCES

1. M. Gell-Mann and M.L. Goldberger, Phys. Rev. 91, 398 (1953).
2. I.R. Afnan and A.W. Thomas, *Modern Three-hadron Physics*, A.W. Thomas (ed.), Springer, Berlin (1977), P.1.
3. K.M. Watson, Phys. Rev. 105, 1388 (1957).
4. A.N. Mitra, Nucl. Phys. 32, 529 (1962); A.N. Mitra and V.S. Bhasin, Phys. Rev. 131, 1265 (1963).
5. L.D. Faddeev, Zh. Eksp. Teor. Fiz 39, 1459 (1960) [English translation; Soviet Phys. JETP 12, 1014 (1961)].
6. E.O. Alt, P. Grassberger and W. Sandhas, Nucl. Phys. B2, 167 (1967).
7. H. Ehrhardt, K.H. Hesselbacher, K. Jung and K. Willmann, Case Studies in Atomic Physics 2, 159 (1971).
8. I.E. McCarthy and E. Weigold, Phys. Reports 27C, 275 (1976).
9. E. Weigold, C.J. Noble, S.T. Hood and I. Fuss, J. Phys. B 12, 291 (1979).
10. R. Camilloni, A. Giardini-Guidoni, I.E. McCarthy and G. Stefani, Phys. Rev. A 17, 1634 (1978).
11. J.C.Y. Chen and A.C. Chen, Adv. in Atom. and Molec. Phys. 8, 71 (1972).
12. A.T. Stelbovics, unpublished.
13. R.K. Peterkop, Zh. Eksp. Teor. Fiz 43, 616 (1962).
14. M.R.H. Rudge and M.J. Seaton, Proc. Phys. Soc. (London) 85, 607 (1965).
15. M.R.H. Rudge, Rev. Mod. Phys. 40, 564 (1968).
16. G.P. LePage, J. Comp. Phys. 27, 192 (1978).
17. H. Feshbach, Ann. Phys. (N.Y.) 19, 287 (1962).
18. I.E. McCarthy and M.R.C. McDowell, J. Phys. B (in press).
19. B.H. Bransden, M. Crocker, I.E. McCarthy, M.R.C. McDowell and L.A. Morgan, J. Phys. B 11, 3411 (1978).
20. P.W. Coulter and W.R. Garrett, Phys. Rev. A 18, 1902 (1978).

21. J.B. Furness and I.E. McCarthy, J. Phys. B 6, 2280 (1973).
22. R. Vanderpoorten, J. Phys. B 8, 926 (1975).
23. A. Boksenberg, Ph.D. Thesis, University of London, 1961.

THEORETICAL STUDIES OF ELECTRON IMPACT EXCITATION OF POSITIVE IONS

W. Derek Robb

Theoretical Division, Group T-4
Los Alamos Scientific Laboratory
Los Alamos, NM 87545 USA

1. INTRODUCTION

Electron impact excitation of impurity ions is a dominant mechanism for producing the radiation emitted by CTR-type plasmas.[1] In addition, the relative intensities of impurity lines excited by electron impact provide a sensitive diagnostic of temperature and density in the plasma. For the densities and temperatures found in magnetically confined plasmas we can assume, as a good approximation, that the ions behave as if they were isolated, and thus neglect any plasma effects on either their structure or their interactions with electrons.

In power loss studies on a model Fe-seeded Tokamak plasma Merts et al.[1] have shown that the radiative power loss from the Fe-ions is apportioned 76.5% to electron collisional excitation, 18% to dielectronic and radiative recombination and 5.5% to bremmstrahlung at an electron temperature of 2 keV. Power loss studies generally involve the summation and averaging over large numbers of individual lines of the ion and so the accuracy to which individual line rate coefficients is required is not critical, provided the summed rate is good to say 20%. Diagnostic calculations on the other hand, often depend critically on the relative accuracy of line rate coefficients and here a knowledge of each of them to better than 20% is desired.

The remainder of the paper is set out as follows, in Sec. 2 we define some basic aspects of collisional excitation processes, Sec. 3 describes the principal theoretical techniques for calculating excitation collision strengths, and in Sec. 4 we examine the sensitivity of the collision strength to target state wavefunctions. Section 5 deals with resonant contributions to excitation while Sec. 6 examines analytic expressions that may be used to fit numerical collision strength data.

2. BASIC THEORETICAL CONSIDERATIONS

The rate coefficient for excitation R_{ij} of an initial atomic state i to a final state j, under the assumption of a Maxwellian electron energy distribution is

$$R_{ij} = \frac{8.01 \cdot 10^{-8}\, \Delta E_{ij}}{\omega_i\, T^{3/2}} \int_1^{\infty} \Omega_{ij}(X) e^{-\frac{\Delta E_{ij}}{T} X}\, dX \;\; cm^3\, s^{-1} \tag{1}$$

where ΔE_{ij} is the threshold energy (in eV) for the transition, ω_i is the statistical weight of the initial state and T is the electron temperature (in eV).

Two fundamental entities within the integral in Eq. (1) are the collision strength for the transition Ω_{ij} which is related to the collisional excitation cross section Q_{ij} by

$$\Omega_{ij}(X) = \omega_i\, E\, Q_{ij}(E) \tag{2}$$

where Q_{ij} has units of πa_o^2, and the scaled electron energy X is,

$$X = \frac{E}{\Delta E_{ij}} \tag{3}$$

where E is the kinetic energy of the incident electron.

In the remainder of the paper we shall be concerned with ways of calculating the collision strength Ω_{ij} as a function of energy, and so it is instructive to examine overall trends in its behaviour relative to the conditions encountered in CTR plasmas.

First of all we shall assume that we are dealing with ions whose nucleii are not extremely heavy, say with Z << 50, and whose electronic structure makes them iso-electronic with first and second row elements. In this case we may assume that the electronic states are accurately described by LS term values within the Russell-Saunders[2] coupling scheme, with the splitting into individual SLJ levels being a perturbation. Within this scheme there are three major categories of transitions to be considered, they are:

i) Optically dipole-allowed transitions (which we write figuratively, though not strictly accurately as $\Delta L = 1$, $\Delta S = 0$).

ii) Optically allowed transitions, non-dipole ($\Delta L \neq 1$, $\Delta S = 0$).

iii) Spin forbidden transitions ($\Delta S \neq 0$).

A further demarcation of transitions arises through the change of principal quantum number n involved in the transition, namely those transitions for which $\Delta n = 0$, and those for which $\Delta n \neq 0$. It is easiest to explain the relevance of these various types of transition by example and so we show in Table I transitions in the ions Fe XXIV and Fe XXV which would be the predominant ion species in a 2 keV iron seeded plasma. The last column in Table I represents the exponential weighting of the collision strength in the integral of Eq. (1) and so we observe immediately that the $\Delta n = 0$ transitions require knowledge of the collision strength to quite high X values. However, for $\Delta n \neq 0$ transitions the rate is determined predominantly from the collision strength near threshold (X = 1). The shapes of the collision strength for the transition types (i)-(iii) are shown in Fig. I. The dipole allowed

TABLE I

Ion	Transition	ΔE_{ij}^{*}	$\frac{\Delta E_{ij}}{T}$
Fe XXIV	$1s^2 2s\ ^2S - 1s^2 2p\ ^2P$	50 eV	0.025
Fe XXIV	$1s^2 2s\ ^2S - 1s^2 3d\ ^2D$	1200 eV	0.6
Fe XXV	$1s^2\ ^1S - 1s2s\ ^3S$	6640 eV	3.3

*Rounded experimental values.[3]

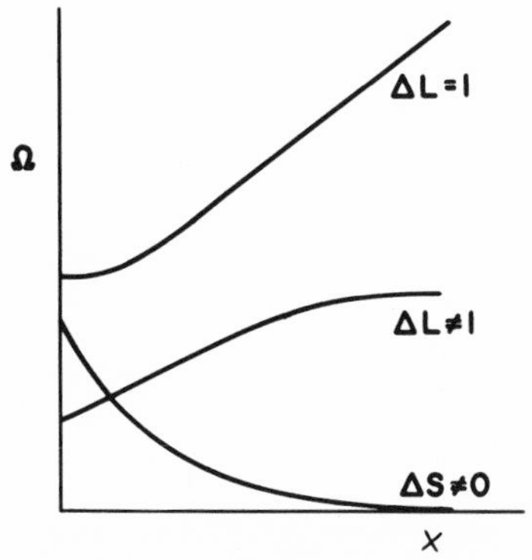

Fig. I. Basic trends of collision strengths.

transitions (Type (i)) show the characteristic linear dependence on the logarithm of the energy at high energy. The slope of the asymptotic collision strength is directly proportional to the optical oscillator strength for the transition, and so many of the semi-empirical methods of computing collision strengths for these transitions use purely radiative data. For the spin allowed non-dipole transitions (Type (ii)), the collision strength must go to a constant value at high energies i.e., the cross section falls off as E^{1}. This constant asymptotic value can readily be obtained from Plane-Wave Born approximation calculations. Finally the collision strength for the spin-forbidden transitions (Type (iii)) fall of inversely as the square of the energy (i.e., the cross section falls off as E^{3}), and this behaviour appears to set in rather rapidly (X = 2-5). These are the basic patterns of behaviour of the collision strength within a pure LS-coupling, non-resonant scheme. They are useful as basic guides to our calculations, however as we shall see, collision strengths for many ions of CTR interest are considerably more complicated in behaviour.

3. METHODS FOR CALCULATING COLLISION STRENGTHS

Let us consider the collision of an electron with an N-electron ion of nuclear charge Z whose target state wave functions Φ_i^N we assume are known. We will then start with the most exact quantum mechanical prescription possible for the solution of collision problem and proceed, by successive approximations until we end up with semi-empirical formalisms.

A. The Close-Coupling Method

<u>i) The Close-Coupling Equations.</u> The exact wave function for the (N+1)-electron system can be written as an expansion over the target state wave functions as:

$$\Psi^{N+1} = A \sum_{ij} \Phi_i^N F_{ij}^{N+1} + \sum_k c_k \chi_k^{N+1} \qquad (4)$$

The close-coupling (CC) approximation truncates the expansion in i to a finite number of atomic states, principally those of and closest to the transition of interest. In Eq. (4) A is an operator which antisymmeterizes the continuum electron orbital with the target state orbitals, and the functions χ_k^{N+1} are N+1-electron functions that can be constructed out of the bound state orbitals alone. The unknown continuum orbitals F_{ij} are determined by requiring that the total wavefunction Ψ^{N+1} satisfy the non-relativistic Schroedinger equation for the system,

$$H^{N+1} \Psi^{N+1} = E \Psi^{N+1} \tag{5}$$

where

$$H^{N+1} = \sum_{j=1}^{N+1} \left(-\tfrac{1}{2}\nabla_i^2 - \frac{Z}{r_i}\right) + \sum_{j=1}^{N+1} \sum_{k<j} \frac{1}{r_{jk}} \tag{6}$$

Since H^{N+1} commutes with the parity operator (π) and the L^2 and S^2 momentum operators, we construct the angular couplings in Ψ^{N+1} so that it is simultaneously an eigenfunction of given πLS symmetry. Since we expand the functions $F_{ij}(r)$ in an infinite summation of spherical harmonics (labelled ℓ) about the nucleus, this extra condition of angular symmetry limits this infinite summation to a small finite number of ℓ-values or channels. A full derivation of the close-coupling equations may be found in the review article of Burke and Seaton.[4] We shall only give a brief description of their nature here, and will concentrate more on methods for their solution and their inherent limitations.

The CC equations for the radial parts of the functions F_{ij} are obtained by substituting Eq. (4) into Eq. (5) and projecting onto the angular component of a given channel i.e.,

$$\int \Phi_i^N Y_{\ell_i}(\hat{\underline{r}}_{N+1}) (H^{N+1}-E) \Psi^{N+1} d\underline{x}_1 \ldots d\underline{x}_N d\hat{\underline{r}}_{N+1} = 0 \tag{7}$$

which give the resulting CC-equations:

$$\sum_j (\nabla_i^2 \delta_{ij} + V_{ij}(r) + k_i^2 \delta_{ij}) F_{jk}(r) = W_{ik}(\underline{\Phi},\underline{F};r) \tag{8}$$

In Eq. (8) ∇_i^2 and k_i^2 represent respectively the kinetic energy operator and kinetic energy of the incident electron in channel i. The potential matrix $V_{ij}(r)$ represents the static or no-exchange interaction of an electron with the target states while $W_{ik}(r)$ represents the exchange interaction of the electron with the target states.

The scattering or collision data is extracted from the solutions $F_{ij}(r)$ by examining their asymptotic form. We have that:

$$F_{ij}^{\pi LS}(r) \underset{r\to\infty}{\sim} f_i(r)\delta_{ij} + K_{ij}^{\pi LS} g_j(r) \tag{9}$$

where $K_{ij}^{\pi LS}$ is the reactance matrix which is related to the collision cross section through the equations:

$$\underline{T} = -2i\underline{K}\,(\underline{1} + i\,\underline{K})^{-1} \tag{10}$$

$$Q_{ij}^{\pi LS} = \frac{\pi}{k_i^2} \sum_{\ell_i \ell_j} \frac{(2L+1)(2S+1)}{(2L_i+1)(2S_i+1)} |T_{ij}|^2 \tag{11}$$

We have reintroduced the superscript πLS in Eq. (9) and (11) in order to emphasize that a single solution of Eq. (8) produces only one partial cross section and that the total cross section is the sum of these.

ii) Methods of Solution. The principal difficulty in solving Eq. (8) is the treatment of the exchange terms W_{ik} which involve integrals over the solution matrix $\underline{F}(r)$. The first CC solutions were obtained using iterative techniques,[5] which were subsequently improved to include variational corrections.[6] Non-iterative procedures were developed by Marriott[7] within the differential equations formulation and by Sams and Kouri[8] within the integral equations formulation. These methods expand the coupled integrodifferential Eq. (8) into larger sets of coupled differential equations. In practice they work well provided the number of equations remains small, but when more than a few target states are coupled and additional correlation terms are included in Eq. (4) their solution becomes intractable, especially for a large number of incident electron energies. Two approaches in current use, the R-matrix Method[9] and the Algebraic Linear Equations[10] Method, are designed to circumvent the problem of tractability, in particular with regard to the number of incident electron energies that can be treated. For complete details of these methods the reader is referred to review articles,[4,9] and to descriptions of the accompanying computer codes.[11,12] In principle all the basic interactions, such as exchange, correlation, induced polarization, and resonances, can be included in the CC method. The only limitations arise from the computing power available. However, for most applications which involve highly stripped ions with high energy electrons in CTR-plasmas, the Coulomb interaction dominates all the above-mentioned interactions and so the full power of the CC method is not required. So we must turn to other less sophisticated approximations to obtain the bulk of the data required for CTR purposes.

b. The Distorted Wave Method

The Distorted-Wave (DW) method has as many variants as it has proponents. The basic assumption however in all procedures is one of weak coupling[13] i.e., the diagonal terms of the potential matrices in Eq. (8) are much larger than the off-diagonal or coupling terms. In this case one can assume that the $\underline{F}$ matrix diagonal solutions are all important and thus their radial parts are obtained with reasonable accuracy as solutions of the single channel scattering equation

$$\left(\frac{d^2}{dr^2} - \frac{\ell_i(\ell_i+1)}{r^2} + V_{ii}(r) + k_i^2\right) f_i(r) = 0 \tag{12}$$

The potential $V_{ii}(r)$ is usually taken as the static potential of the state connected to the channel concerned.

The variational expression for the K-matrix due to Kohn, is then given by:

$$K_{ij} = K_{ij}^t - (\Psi_i^t | H^{N+1} - E | \Psi_j^t) \tag{13}$$

where we write

$$\Psi_i^t = A\, \Phi_i^N\, f_i^{N+1} \tag{14a}$$

$$\Psi_j^t = A\, \Phi_j^N\, f_j^{N+1} \tag{14b}$$

The calculation of the excitation cross sections follows from the use of Eqs. (10) and (11).

In practice the r.h.s. of Eq. (13) is evaluated in two ways. In the first and more traditional way, only the two states of the transition are involved and since only the off-diagonal elements of the K-matrix are required we have $K_{ij}^t = 0$, $i \neq j$ and so

$$K_{ij} = -\, (f_i | V_{ij} + W_{ij} | f_j) \tag{15}$$

where V_{ij} and W_{ij} are the direct and exchange potentials between the initial and final target states. A further assumption or condition involved in Eq. (15) is that the bound and continuum orbitals with the same ℓ are orthogonal to one another. This condition is not always strictly satisfied. The second approach, which has been widely used for astrophysical applications, has been developed by the University College Group.[14] This method uses a trial wave function of the form of Eq. (4) with several target states, and the continuum portions written as in Eq. (14). The functions f_i are obtained using a Thomas-Fermi potential in Eq. (12). This potential was used to obtain the bound target orbitals and though not terribly accurate it does guarantee orthogonality of bound and continuum orbitals. Equation (13) is used to obtain a complete K-matrix. The advantage of this procedure is that many correlation functions χ_k^{N+1} can be included to allow for short range effects.

c. The Coulomb-Born Approximation

In practice the Coulomb-Born (CB) approximation[15a] proceeds as for the traditional DW method. The additional approximation involved is to say that the electron does not see any distortion effects due to the charge cloud of the target ion. As a result we obtain the single channel continuum functions $f_i(r)$ from Eq. (12) using the asymptotic Coulomb potential

$$V_{ii}(r) = \frac{z}{r} \tag{15}$$

where z = Z-N is the residual charge of the target ion. Since the CB approximation assumes nothing other than a long range interaction for the continuum electron it may be expected that it breaks down severely for transitions which proceed purely via short

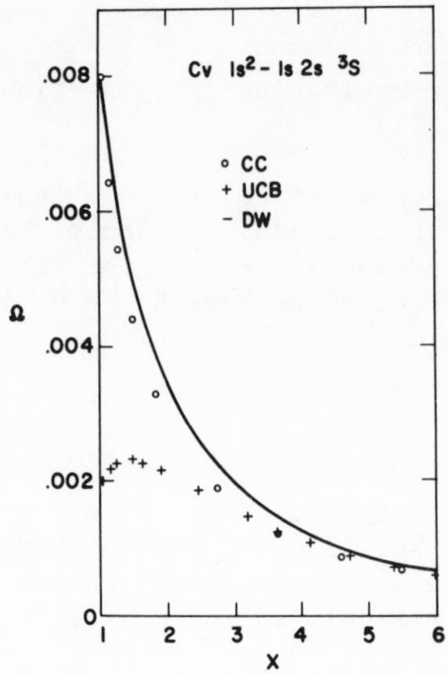

Fig. II. Pure exchange transition showing effects of distortion.

range interactions. A good example is the $1s^2\ {}^1S$ - $1s2s\ {}^3S$ transition in the C V ion, for which results are shown in Fig. II.

A significant amount of work using the $Z \to \infty$ limit of the CBO approximation has been performed by Sampson and his coworkers.[15b] This work includes exchange and target correlation effects and provides simple formulae for collision strengths for transitions in ions with $Z \geq 3N$.

d. Further Approximations Associated with the DW and CB Methods

It is important to discuss additional approximations commonly made within the DW and CB procedures.

Many calculations omit the exchange terms in Eq. (15), which corresponds to not anti-symmeterizing the wave functions in Eq. (14). This approximation of course produces zero cross sections for the type (iii) transitions described in Sec. 2. For type (i) and (ii) transitions the approximation is not too serious above 5-10 times the threshold energy, but close to threshold may produce error factors of 2 or more. As a rough general rule, no-exchange collision strengths are too high.

The CB approximation, on examining Eq. (9), really assumes a zero K^t matrix and a very small K matrix. This has prompted a further approximation to Eq. (10), namely that the inverse matrix is well approximated by a unit matrix, and so

$$\underline{T} \cong - 2i\ \underline{K} \tag{17}$$

Use of Eq. (17) for the T-matrix means that the scattering problem does not satisfy time-reversal and as a result the scattering matrix is non-unitary. The effect of non-unitarization can be quite large especially near threshold for ions which have small residual charge, however for highly stripped ions at energies well above threshold the effect is negligible.

A further approximation arises through the use of non-exact target state wave functions. It is called the post-prior discrepancy[16] and occurs in the choice of potential in the exchange term of Eq. (13). The difference between the post and prior forms may be quite large for low z ions near threshold, but it is in general small compared to the total exchange contribution.

The effects of these various approximations is typically illustrated in Fig. III for the excitation of the 2s-2p transition in Be II.[17] The symbols CB-Coulomb Born without exchange and CBO-Coulomb Born-Oppenheimer or CB with exchange, indicate results for which the approximation of Eq. (17) has been applied. The symbols UCB, UCBO and UDW imply that the full expression for $\underline{T}$ (Eq. (10)) has been used.

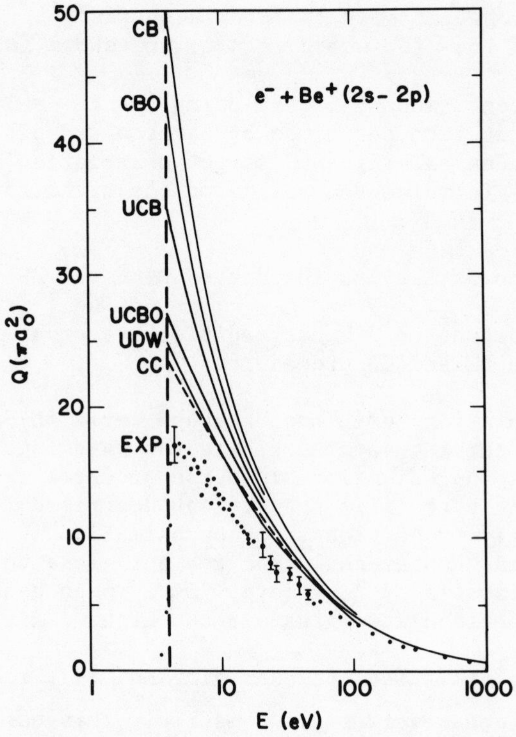

Fig. III. Comparison of various approximations with experiment.

e. Coulomb-Bethe Approximation

In the Coulomb-Bethe approximation[18] (CBe) the colliding electron-target electron interaction in the Coulomb-Born approximation is approximated by

$$\sum_{j=1}^{n} \frac{1}{|\underline{r}_j - \underline{r}_{N+1}|} \qquad \sum_{j=1}^{n} \sum_{\lambda=0}^{\infty} \frac{r_j^{\lambda}}{r_{N+1}^{\lambda+1}} P_\lambda(\hat{\underline{r}}_j \cdot \hat{\underline{r}}_{N+1}) \tag{18}$$

This approximation is valid for r_{N+1} > all r_j which is true for high incident ℓ-values. Classically this corresponds to scattering at large distances (impact parameters) and so it is only valid for ℓ values satisfying the classical turning point condition

$$\ell > (k^2\bar{r}^2 + 2z\bar{r})^{\frac{1}{2}} - \tfrac{1}{2} \tag{19}$$

where $\bar{r}$ is the average, or hard sphere, size of the target ion.

The importance of the CBe Approximation comes in completing the sum of partial collision strengths discussed after Eq. (11). If we write

$$\Omega_{ij} = \sum_{\ell=0}^{\ell_o} \Omega_{ij}^{\ell} + \sum_{\ell=\ell_o+1}^{\infty} \Omega_{ij}^{\ell} \tag{20}$$

where ℓ_o is a value satisfying Eq. (19), then we may use either the CC, DW or CB methods to obtain the partial collision strengths of the first summation and the CBe approximation for the second summation. Burgess[19] has given a closed expression for the second summation within the CBe approximation, but for dipole allowed transitions only.

f. The Plane Wave Born Approximation

The Plane Wave Born approximation[20] (PWBA) assumes that the electron energy k^2 is large enough to dominate even the asymptotic potential $2z/r$. Classically this corresponds to the incident electron having a straight line rather than a hyperbolic trajectory, and this is valid if $k^2 \bar{r} >> 2z$. The expression for PWBA collision strength is

$$\Omega_{ij} = \frac{2\omega_i}{\Delta E_{ij}} \int_{K_{min}}^{K_{max}} f_{ij}(K)\ d\ \ln K \tag{21}$$

where $f_{ij}(K)$ is the generalized oscillator strength[21] given by

$$f_{ij}(K) = \frac{\Delta E_{ij}}{K^2} \sum_{k=1}^{N} \int \Phi_i^N\ e^{iK\cdot r_k}\ \Phi_j^N\ dr_1 \ldots . dr_N \tag{22}$$

and $\underline{K} = \underline{k}_i - \underline{k}_j$ is the momentum transfer involved in the collision.

The advantage of the PWBA is that a partial wave decomposition of the scattered electron wave is not required and a closed expression for the collision strength is obtained.

g. Semi-Empirical Formulae

Most radiation models of plasmas only account for the strong dipole-allowed transitions. Semi-empirical formulae are often used to estimate the excitation collision strengths for these transitions, one of the most noteable being due to Van Regemorter and Seaton.[22] Their formula for the collision strength is

$$\Omega_{ij}(X) = \frac{\omega_i}{\Delta E_{ij}} \cdot f_{ij} \cdot g(X) \tag{23}$$

where f_{ij} is the optical oscillator strength for the transition, and g(X) is called the Gaunt factor which has the form

$$g(X) = a\ (1 + \ln(1 + c(X-1))) \tag{24}$$

Van Regemorter replaced g by an effective Gaunt factor $\bar{g}$, which from analysis of experimental and available theoretical results he deduced should be $\bar{g} = 0.2$ at X = 1. This applies only to $\Delta n \neq 0$ transitions however, and Bely[23] later gave the more correct value of $\bar{g} = 0.8$ at X = 1 for $\Delta n = 0$ transitions.

In general semi-empirical formulae should be used with caution, consistent agreement with sophisticated calculations to 20% is extremely unlikely.

4. TARGET STATE WAVE FUNCTION APPROXIMATIONS

a. Electron-Electron Correlation Effects

In the previous sections we have been concerned with approximations for the continuum electron wave function. We now turn our attention to the approximations made for the bound target state wavefunctions. For ions in the Be, B, C, N, O, Mg, Al, Si, P, S isoelectronic sequences there are significant correlation effects due to orbital degeneracy, in both the ground and lowest excited states. These correlation effects do not disappear in the limit of infinite Z, and it is well known from bound state calculations that they perform a fundamental role in correctly predicting energy levels and radiative transition probabilities. It is not surprising therefore that they play an equally important role in the calculation of collision strengths. In Fig. IV we show some results for the $1s^2 2s^2 2p\ ^2P^0$ - $1s^2 2s 2p^2\ ^2P$ collision strength of the CII ion. The lettered curves correspond to CC calculations,[24] while the numbered curves correspond to UCBO calculations.[24] The curves 1 and A correspond to using single configuration wave functions in the initial and final states. Curves 2 and B result from adding the configuration $1s^2 2p^3\ ^2P^0$ into the ground state wave function, and curve 3 corresponds to allowing for correlations in both states with configurations having a single electron promoted to the n=3 shell. The target correlation effects clearly dominate differences in the continuum wave functions above X=2. The situation is considerably more serious for the second row ions, where because of the open 3d shell many more correlations occur. An analagous example to the B-like transition just mentioned is shown in Fig. V for the $3s^2 3p\ ^2P^0$ - $3s3p^2\ ^2D$

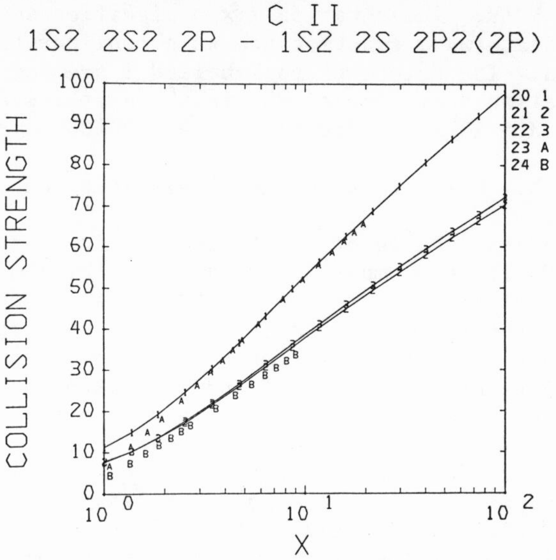

Fig. IV. Target correlation effects on a collision strength of C II.

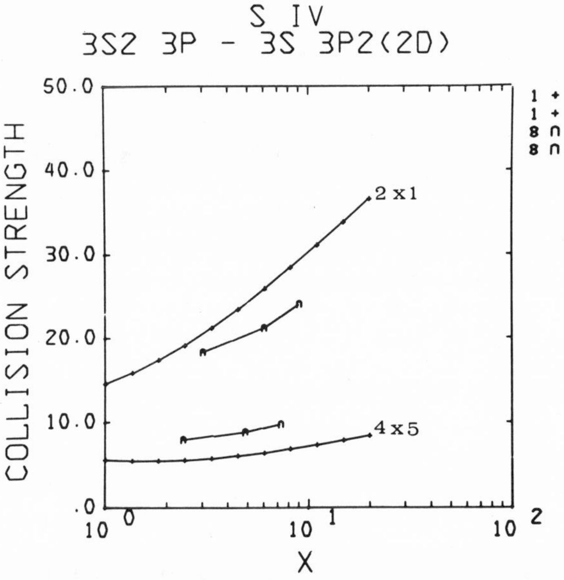

Fig. V. Target correlation effects on a collision strength of S IV.

transition in S IV. The notation m x n signifies an m-configuration initial state wave function and an n-configuration final state function. The calculations labelled 1 are due to Mann[25] and those labelled 8 are due to Bhatia.[26] Here instead of a 30% difference at X = 10 as in Fig. IV we have a 400% difference.

b. Spin-orbit and Relativistic Effects

As the nuclear charge of the target ion increases, the additional Breit-Pauli terms must be included in the Hamiltonian to allow for the relativistic motion of the target electrons.[27] The spin-orbit interaction mixes the orbital and spin angular momenta of the electrons so that configurations with different LS values can mix to represent a given level and we must go to a πSLJ representation of states. In addition the mass-velocity and Darwin terms can significantly alter the threshold energies of $\Delta n = 0$ transitions. Provided Z is not too large we can treat the Breit-Pauli terms as perturbations to the non-relativistic wave functions, proceed with our calculations in πLS coupling and recouple to πSLJ coupling using the procedure detailed by Saraph.[28]

An example of these intermediate coupling effects is shown is Fig. VI for the $1s^2\ {}^1S - 1s2p\ {}^3P^0$ transition of Fe XXV. Here the spin-orbit term mixes the $1s2p\ {}^3P_1$ and $1s2p\ {}^1P_1$ configurations in the ratio 0.96 to 0.28, and so at approximately X = 4 the component of the collision strength from the dipole allowed transition $1s^2\ {}^1S_0 - 1s2p\ {}^1P_1$ begins to dominate.

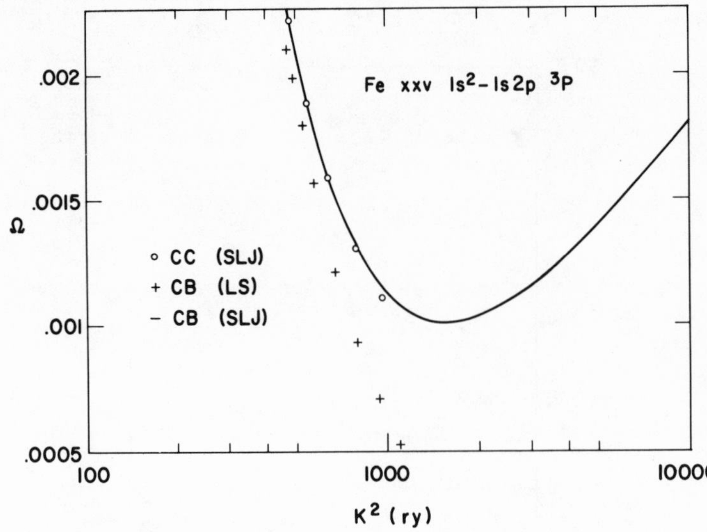

Fig. VI. Intermediate coupling effects in the target wavefunctions.

5. RESONANCES

Until now we have neglected "resonant excitation" processes and have dealt with effects on the non-resonant or background collision strength. Resonant processes are shown in Fig. VII, where we have three levels of the target ion labelled initial, final and upper. The upper level u of the ion may capture the incident electron into an orbital $n\ell'$ of negative energy to form a compound state c of the N+1 electron system. This is not a bound N+1 electron ion state since the electron is free to return to the continuum, and we refer to it as a "doubly-excited", "resonance" or "autoionizing" state.

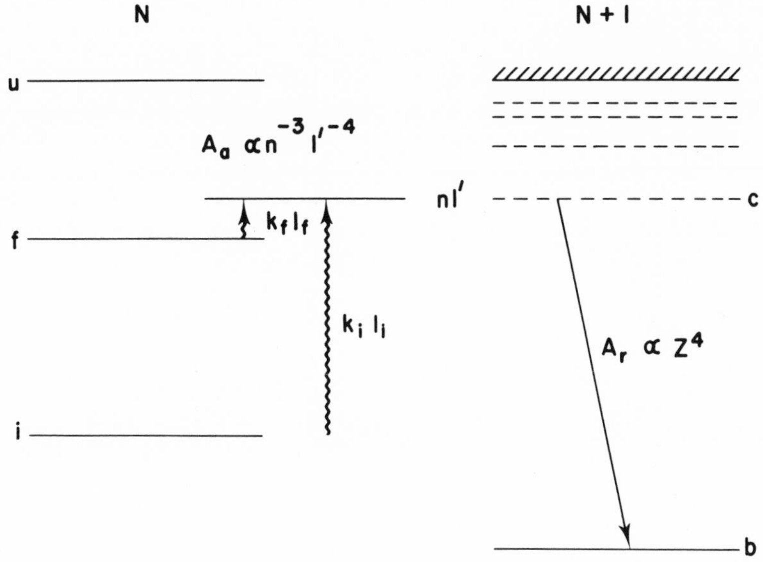

Fig. VII. Energy level diagram for resonant processes.

The process of resonant excitation involves an electron of energy k_i^2 incident on the initial state of the N electron ion, such that $E_c \doteq E_i + k_i^2$. Capture into the resonance state c takes place through the inverse process to autoionization. The state c now has the option of autoionizing into the continua associated with the N-electron ion states i or f or radiatively decaying to a bound state b. If autoionization is into the continuum of state i we have resonant elastic scattering, if autoionization is into the continuum of state f then we have resonant excitation. If radiative decay to the state b occurs then the process of dielectronic recombination[29] has occurred.

A CC calculation involving these three target states will automatically include all of the resonance states c. Such a calculation for the excitation of the $1s^22s^2\ ^1S$ - $1s^22s2p\ ^3P^0$ transition in O V was reported recently by Berrington et al.[30] using the R-matrix method. We show their results in Fig. VIII. Here the dashed resonances are coupled to the $1s^22s2p\ ^1P^0$ and $1s^22p^2\ ^3P$, ^1D, ^1S "upper" states, while the solid curve resonances are coupled to the $1s^22s3\ell^{1,3}\ell$ "upper" states.

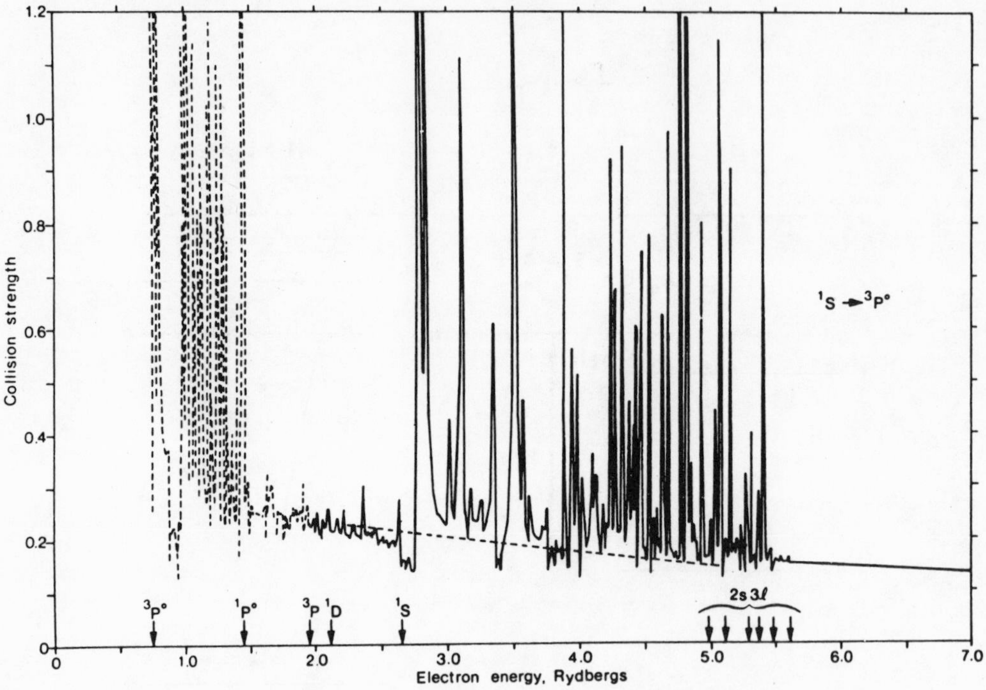

Fig. VIII. Resonance structure in the OV ($2s^2\ ^1S$-$2s2p^3P^0$) collision strength.

Calculations with the DW and CB methods usually involve only the "initial" and "final" states of the transition, and so resonance contributions must be calculated separately. A procedure for computing the resonant contribution to the excitation rate coefficient has been examined by Cowan,[31] which is conceptually and computationally easier to understand than the CC approach. It can be outlined as follows. The rate coefficient for the capture of the incident electron into state c is related, through the principle of detailed balance to the autoionization rate A^a_{ci} by

$$\alpha(i \rightarrow c) = 4\pi^{3/2}\, T^{-3/2}\, e^{-E_{ci}/T}\, \frac{\omega_c}{\omega_i}\, A^a_{ci} \tag{25}$$

The branching ratio which gives the probability of the state c autoionizing into the continuum of the state f is given by:

$$b_{cf} = \frac{A^a_{cf}}{\sum_{i'} A^a_{ci'} + \sum_b A^r_{cb}} \tag{26}$$

where A^r_{cb} is the radiative transition probability, which can be calculated using standard bound state techniques. The resonance excitation rate from state i to state f is then

$$\alpha_{rx}(i \rightarrow f) = \sum_c \alpha(i \rightarrow c) b_{cf} \tag{27}$$

The only remaining problem is the computation of the autoionization probabilities $A^a_{ci'}$, and this is accomplished using the "Golden-Rule" formula:[32]

$$A^a_{ci'} = \frac{2\pi}{\hbar} |\langle c|H^{N+1}|i'\rangle|^2 \tag{28}$$

The state c can be represented by a bound state configuration and the continuum i' by a DW state as in Eq. (14a). These computations can be performed extremely rapidly using standard bound state computer codes with the states c and i' represented in either LS, SLJ or jj coupling.

Results obtained by Cowan[31] using this procedure, together with non-resonant DW calculations, for the OV $1s^22s^2\ ^1S$ - $1s^22s2p\ ^3P^0$ transition described above are shown in Fig. IX and compared with the rates obtained by Berrington et al.[30] The DWXII curve represents the non-resonant contribution to the rate coefficient while the n'=2,m curves represent the full rate coefficient with resonant excitation contributions associated with the upper ionic states n'=2,m. The procedure appears capable of reproducing the CC results to 10-20% accuracy.

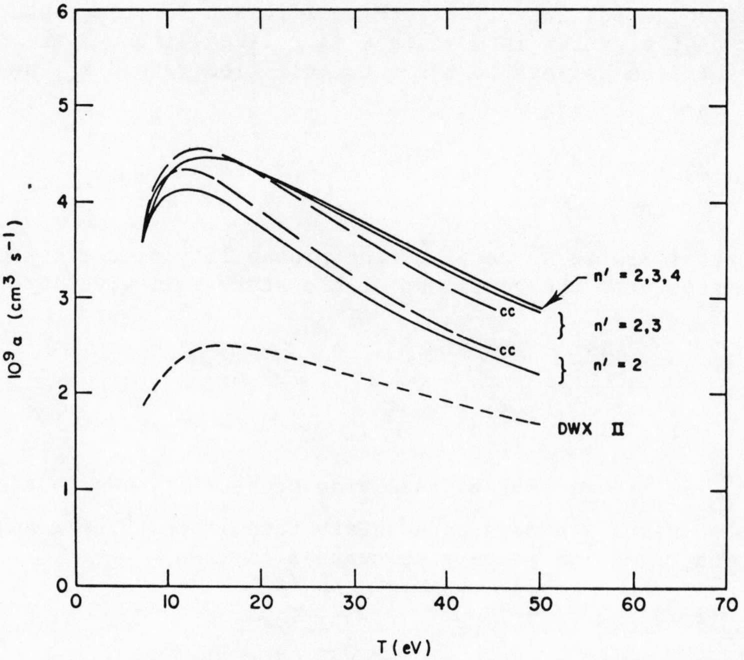

Fig. IX. Comparison of the OV ($2s^2\ {}^1S$-$2s2p\ {}^3P^0$) rate coefficient using CC and DW + resonant excitation methods.

6. FITTING AND SCALING OF COLLISION STRENGTHS

It is useful to be able to fit the non-resonant collision strength to an analytic function of X, and then fit an isoelectronic sequence of such collision strengths as a function of Z. To fit individual collision strengths we use the formulae for spin-allowed transitions:

$$\Omega(X) = C_o + \frac{C_1}{X} + \frac{C_2}{X^2} + C_3 \ln X \tag{29}$$

and for spin-forbidden transitions:

$$\Omega(X) = \frac{C_o}{X^2} + \sum_{n=1}^{4} C_n e^{-n\alpha X} \tag{30}$$

For non-dipole transitions in Eq. (29) we have $C_3 = 0$, and in the case of relativistically mixed collision strengths (as shown in Fig. VI) we use a combined form of Eqs. (29) and (30). Fits over a wide variety of transitions and ions using these formulae typically yield better than 5% agreement with the data for spin-allowed transitions, and on the order of 10% agreement with the data for spin-forbidden transitions.

The traditional Z-scaling of the collision strength arising from the scaling of the wave functions and Hamiltonian in Eq. (13) gives

$$Z^2\ \Omega(X) \rightarrow \text{constant as } Z \rightarrow \infty \tag{31}$$

However, this is only true in the limit and it is known[15b,33] that for finite Z two factors must be taken into account, 1) screening of the full nuclear charge by the target electrons, 2) different scaling for the low and high energy parts of the collision strength. For $\Delta n=0$ dipole allowed transitions, Magee and Merts[33] have used the formula

$$\Omega_Z(X) = \frac{1}{(Z-s_1)^2}\left[C_o + \frac{C_1}{X} + \frac{C_2}{X^2}\right] + \frac{1}{(Z-s_2)^2} C_3 \ln X \tag{32}$$

and obtained a maximum error of 8% in fitting Type (i) collision strengths from X = 1 to 100 for ions with Z = 5 to 42.

7. SUMMARY

For highly stripped ions appearing in CTR-plasmas it appears possible that 20% accuracy in the rate coefficients is possible provided sufficient care is taken. The Coulomb-Born, Distorted Wave and Close-Coupling techniques all have general application, whereas the Plane-Wave Born and Semi-empirical methods have rather limited usefulness. Essential ingredients to be considered for all calculations are exchange, unitarization, distortion, target state correlations, relativistic mixing of target states and resonances.

ACKNOWLEDGMENTS

I would like to thank my colleagues R. D. Cowan, N. H. Magee, J. B. Mann and A. L. Merts of Group T-4 at LASL for many useful conversations and contributions to this article. This work was supported by the United States Department of Energy.

REFERENCES

1. A. L. Merts, R. D. Cowan and N. H. Magee, Jr., Los Alamos Scientific Laboratory Informal report, LA-6220-MS (1976): J. Davis, V. L. Jacobs, P. C. Kepple and M. Blaha, JQRST 17, 139 (1977); D. Post, R. V. Jensen, C. B. Tarter, W. H. Grasberger and W. A. Lokke, At. Data and Nucl. Data Tables, 20, 397 (1978).

2. E. U. Condon and G. H. Shortley, The Theory of Atomic Spectra (Cambridge, 1935).

3. J. Reader and J. Sugar, J. Phys. and Chem. Ref. Data, 4, 353 (1975).

4. P. G. Burke and M. J. Seaton, Methods Comp. Phys. 10, 1 (1971).

5. M. J. Seaton, Proc. Roy. Soc. A218, 400 (1953).

6. H. E. Saraph and M. J. Seaton, Proc. Phys. Soc. 80, 1057 (1962).

7. R. Marriott, Proc. Phys. Soc. 72, 121 (1958).

8. W. N. Sams and D. J. Kouri, J. Chem. Phys. 51, 5809 (1969).

9. P. G. Burke and W. D. Robb, Advances in Atomic and Molecuar Physics 11, 143 (1975).

10. M. J. Seaton, J. Phys. B 7, 1817 (1974).

11. K. A. Berrington, P. G. Burke, M. LeDourneuf, W. D. Robb, K. T. Taylor and Vo Ky Lan Comp. Phys. Commun. 14, 367 (1978).

12. M. A. Crees, M. J. Seaton and P.M.H. Wilson, Comp. Phys. Commun. 15, 23 (1978).

13. N. F. Mott and H.S.W. Massey, The Theory of Atomic Collisions §XIII.2 (Oxford 1965); J. M. Peek and J. B. Mann, Phys. Rev. A16, 2315 (1977); J. Davis, P. C. Kepple and M. Blaha, JQRST, 16, 1043 (1976).

14. W. Eissner and M. J. Seaton, J. Phys. B. 5, 2187 (1972).

15a. A. Burgess, D. G. Hummer and J. A. Tully, Phil. Trans. Roy. Soc. London, A266, 225 (1970).

15b. D. H. Sampson and A. D. Parks, Ap. J. Supplement 28, 309 (1974); A. D. Parks and D. H. Sampson, Phys. Rev. A.

16. D. R. Bates, A. Fundaminsky, J. W. Leech and H.S.W. Massey Phil. Trans. Roy. Soc. London A243, 93 (1950).

17. M. A. Hayes, D. W. Norcross, J. B. Mann and W. D. Robb, J. Phys. B. 10, L429 (1977).

18. A. Burgess and J. A. Tully, J. Phys. B. 11, 4271 (1978).

19. A. Burgess, J. Phys. B. 7, L364 (1974).

20. K. L. Bell and A. E. Kingston, Advances in Atomic and Molecular Physics 10, 53 (1974).

21. M. Inokuti, Rev. Mod. Phys. 43, 297 (1971).

22. H. Van Regemorter, Astrophys. J. 136, 906 (1962); M. J. Seaton, Atomic and Molecular Processes (Academic (1962)).

23. O. Bely, Proc. Phys. Soc. London 88, 587 (1966).

24. W. D. Robb and J. B. Mann quoted in N. H. Magee, Jr., J. B. Mann, A. L. Merts and W. D. Robb, Los Alamos Scientific Laboratory report LA-6691-MS April (1977).

25. J. B. Mann, Jr. (private communication).

26. A. K. Bhatia, G. A. Doschek and U. Feldman, Astron. and Astrophys. (in press).

27. A. M. Ermolaev and M. Jones, J. Phys. B. 6, 1 (1973).

28. H. E. Saraph, Comp. Phys. Commun. 15, 247 (1978); ibid 3, 256 (1972).

29. D. R. Bates and A. Dalgarno, Atomic and Molecular Processes, (Academic 1962).

30. K. A. Berrington, P. G. Burke, P. L. Dufton, A. E. Kingston and A. L. Sinfailam, J. Phys. B. (in press).

31. R. D. Cowan, J. Phys. B. (submitted to).

32. U. Fano, Phys. Rev. 124, 1866 (1961).

33. N. H. Magee, Jr., and A. L. Merts (private communication).

THEORY OF RECOMBINATION PROCESSES

Christopher Bottcher

Oak Ridge National Laboratory*

Oak Ridge, Tennessee 37830

§ 1. RADIATIVE RECOMBINATION

Recombination is the removal of charged particles by the association of positive and negative charges.[1] Such processes can take place in liquids and solids or on surfaces, but we confine ourselves to the gas phase. If x^+, y^- are charged bodies (small letters stand for atoms, molecules, electrons, or photons),

$$x^+ + y^- \rightarrow c \tag{1}$$

is impossible, since by time-reversal c is unstable. However,

$$x^+ + y^- \rightarrow a + b \tag{2}$$

or

$$x^+ + y^- + z \rightarrow x + y + z \tag{3}$$

are possible. The rate of disappearance of x^+ or y^- is described by the rate equations

$$\frac{dn(x^+)}{dt} = \frac{dn(y^-)}{dt} = - \alpha n(x^+)n(y^-) \tag{4}$$

---

*Research sponsored by the Division of Basic Energy Sciences, U.S. Department of Energy, under contract W-7405-eng-26 with the Union Carbide Corporation.

where n (cm^{-3}) is a number density, and α ($cm^3\ sec^{-1}$) is the recombination rate coefficient. In microscopic terms

$$\alpha(T) = \langle v\sigma \rangle \tag{5}$$

where σ is the recombination cross section, v the x+y relative velocity, and the brackets imply a Maxwellian average at temperature T.

Examples of (2) are *radiative recombination*, to which Chapters 1 and 2 are devoted, and *dissociative recombination*, to which Chapter 3 is devoted. In Chapter 4 we take up *three-body recombination* defined by (3). The topics of Chapters 1 and 2 are of overwhelming importance in confined plasmas where bare or highly stripped ions predominate. Yet the topics of Chapters 3 and 4 are also important to fusion science in the design of ion sources and the production of intense neutral beams. We shall not discuss mutual neutralization,

$$x^+ + y^- \rightarrow x + y \tag{6}$$

which is sometimes classed as a recombination process.

Unless otherwise stated, we usually employ atomic units. It is convenient to list the conversions:

a.u. length $a_o = 5.2917 \times 10^{-9}$ cm

a.u. time $= 2.4189 \times 10^{-17}$ sec

a.u. velocity $= 2.1877 \times 10^{8}$ cm sec^{-1}

speed of light in a.u. $= \alpha_{fs}^{-1}$ (α_{fs} is the fine structure constant)

$= 137.037$

1 eV $= 11{,}605.7$ deg K

1 a.u. energy $= 27.21$ eV $= 315{,}790$ deg K

a.u. of cross section $\pi a_o^2 = 0.8797 \times 10^{-16}$ cm^2

a.u. of rate $a_o^3 t_o^{-1} = 6.1259 \times 10^{-9}$ cm^3 sec^{-1}

1.1. Formulae for the Radiative Rate

The simplest, and perhaps most important, recombination mechanism is single electron radiative recombination into a hydrogenic (Rydberg) orbit,

$$e + H^{+} \rightarrow H(n\ell m) + h\nu. \qquad (7)$$

Scaling to a hydrogenic ion is trivial (c.f. just below (49)). The cross section for (7) is readily derived from the Einstein A-coefficient for a transition from a continuum state, labelled by the wave-vector $\underline{k}$, to a bound state $n\ell m$,

$$\sigma_R(\underline{k},n\ell m) = \frac{4}{3}(2\pi\alpha_{fs})^3 \frac{\omega^3}{k^2}\left|\langle \underset{\sim}{k}|\underset{\sim}{r}|n\ell m\rangle\right|^2. \qquad (8)$$

The electron initially has energy $\varepsilon = k^2/2$, while the angular frequency of the emitted photon $\omega = \varepsilon + 1/(2n^2)$; $\underline{r}$ is the position vector (dipole moment) of the electron. We now replace the Coulomb wave by an expansion in eigenstates of angular momentum, average over all directions of $\underline{k}$ and sum over all states m to find

$$\sigma_R(k,n\ell) = \frac{8\pi^2\alpha_{fs}^3\omega^3}{3(2\ell+1)k^2}\left[\ell\langle k\ell-1|r|n\ell\rangle + (\ell+1)\langle k\ell+1|r|n\ell\rangle^2\right]. \qquad (9)$$

The kets $|n\ell\rangle$, $|k\ell\rangle$ are associated with the radial bound and energy-normalized continuum orbitals. The rate of recombination into $n\ell$ is then from (5), (9)

$$\alpha_R(n\ell,T) = \int_0^\infty v\,\sigma_R(v,n\ell)f(\varepsilon,T)d\varepsilon \qquad (10)$$

where $v^2 = 2\varepsilon$ and the Maxwellian distribution

$$f(\varepsilon,T) = 2\left(\frac{\varepsilon}{\pi T^3}\right)^{1/2} e^{-\varepsilon/T}. \qquad (11)$$

To evaluate (9), we need matrix elements of r between hydrogenic wavefunctions. Most textbooks express the wavefunctions of bound hydrogenic states as Laguerre polynomials, and of continuum states as Hypergeometric functions. These representations are not very useful for the present problem, and we turn instead to semi-classical methods which are most simply introduced through the JWKB approximation.

1.2. Radial Matrix Elements

For the n state of hydrogen the JWKB approximation[2] to the radial wavefunction is

$$g_{n\ell} = A_{n\ell}\, r^{-1}\, P^{-1/2} \sin\left(\frac{\pi}{4} + \int_{r_1}^{r} P dr\right) \tag{12}$$

where

$$P^2 = 2E_n + \frac{2}{r} - \frac{\lambda^2}{r^2},\ E_n = \frac{-1}{2n^2},\ \lambda = \ell + \frac{1}{2}, \tag{13}$$

$A_{n\ell}$ is a normalization constant and r_1, r_2 are the inner, outer classical turning points. We recast P in the form

$$(nrP)^2 = n^4\varepsilon^2 - (r-n^2)^2,\ n^2\varepsilon^2 = n^2 - \lambda^2 \tag{14}$$

so that ε is the eccentricity of the classical orbit. The substitution

$$n^2 - r = n^2\varepsilon\cos\tau \tag{15}$$

is suggested by (14). Then the time to move on the classical orbit from perigee (τ = 0) is given by

$$t = \int_{r_1}^{r} \frac{dr}{P} = n^3\,(\tau - \varepsilon\sin\tau). \tag{16}$$

The azimuthal angle swept out from perigee can be obtained from conservation of angular momentum

$$r^2\,\frac{d\phi}{dt} = \lambda \tag{17}$$

whence

$$\phi = \lambda \int_{r_1}^{r} \frac{dr}{r\,P} = \eta \int_{0}^{\tau} \frac{dz}{(1-\varepsilon\cos z)},\ \eta = \lambda/n. \tag{18}$$

Replacing z by Z = tan(z/2), we find that

$$\tan(\phi/2) = \frac{\eta}{(1-\varepsilon)}\tan(\tau/2), \tag{19}$$

or equivalently

$$x = r\cos\phi = n^2 (\cos\tau-\varepsilon), \quad y = r\sin\phi = n^2\eta\sin\tau. \tag{20}$$

In classical mechanics[3] τ is the eccentric anomaly, and (16), (19), and (20) define the classical motion in the (r,ϕ) or (x,y) plane.

Returning to quantal mechanics, the radial matrix elements of an operator $F(r)$ are denoted by

$$M(n\ell,n'\ell') = \langle n\ell|F|n'\ell'\rangle. \tag{21}$$

From (12) this is an integral over the product of two rapidly oscillating functions $\sin a \sin a'$ which we replace by $\frac{1}{2}\cos(a-a')$ so that

$$M(n\ell,n'\ell') = \frac{1}{2} A_{n\ell}A_{n'\ell'} \int_r^r \frac{dr}{P} \cos(a_n\Delta n + a_\ell\Delta\ell)F(r), \tag{22}$$

$$\Delta n = n-n' \quad , \quad \Delta\ell = \ell-\ell'. \tag{23}$$

The coefficients a_n, a_ℓ are given by

$$a_n = n^{-3} \int_r^r \frac{dr}{P} = n^{-3}\, t(\tau), \quad a_\ell = -\lambda \int_r^r \frac{dr}{r^2P} = -\phi(\tau) \tag{24}$$

from (16) and (18). As a special case of (22)

$$A_{n\ell} = \left(\frac{2}{\pi n^3}\right)^{1/2}, \tag{25}$$

so that for $n >> \Delta n$, $\ell >> \Delta\lambda$, (22) becomes

$$M(\Delta n,\Delta\ell) = \frac{\omega_0}{\pi} \int_0^\pi dt(\tau)F[r(\tau)]\cos[\omega_0 t(\tau)\Delta n-\phi(\tau)\Delta\ell] \tag{26}$$

where ω_0 is 2π/orbital period (classical) or the separation of adjacent levels (quantal).

We have derived one of a large class of results known as correspondence principles. For $\Delta n = \Delta\ell = 0$, M reduces to the expectation value of F, while (26) gives the time average of F over a classical orbit. When off-diagonal (transition) matrix elements are considered, the argument of the cosine in (26) is the difference between the classical actions in the initial and final orbits. In general, this quantity is of the form

$$\Delta(\text{action}) = \sum Q_j(t)\Delta P_j \tag{27}$$

where Q_j is a canonical coordinate and ΔP_j is the change in the conjugate momentum during the transition; the sum includes a term for the pair $Q = t$, $P = E$. For a dipole transition $F = r$ and $\Delta\ell = \pm 1$, so that we only have to calculate

$$M(c,\pm 1) = M_x(c) \pm M_y(c) \tag{28}$$

where

$$M_x = \frac{\omega_0}{\pi}\int_0^\pi x\cos\theta dt,\ M_y = \frac{\omega_0}{\pi}\int_0^\pi y\sin\theta dt,\ \theta = c\omega_0 t \tag{29}$$

and x, y were defined in (20). Integrating (29) by parts (c.f. (73)), we find

$$M_x = \frac{-1}{\pi c}\int_0^\pi \sin\theta(\tau)dx(\tau),\ M_y = \frac{1}{\pi c}\int_0^\pi \cos\theta(\tau)dy(\tau). \tag{30}$$

Inserting (20) in (30), we obtain integrals which can be expressed in terms of Bessel functions, so that[3,4,5]

$$M_x(c) = \frac{n^2}{c} J_c'(c\varepsilon),\ M_y(c) = \frac{n^2\eta}{c\varepsilon} J_c(c\varepsilon). \tag{31}$$

We usually require the radial matrix elements squared and averaged over ℓ,

$$\begin{aligned}\mathcal{D}(n,c) &= \frac{1}{n^2}\sum_{\ell=0}^{n-1} [\ell M(c,-1)^2 + (\ell+1)M(c,+1)^2] \\ &\simeq 2\int_0^1 \varepsilon d\varepsilon\ [M_x(c)^2 + M_y(c)^2].\end{aligned} \tag{32}$$

The integral over ε can be evaluated using the second-order differential equation satisfied by J_c, with the result that

$$\mathcal{D}(n,c) = \frac{2n^4}{c^3} J_c(c)J_c'(c). \tag{33}$$

For $c \gg 1$ but still $\ll n$, we have

$$J_c(c) \simeq \frac{0.44730}{c^{1/3}}, \quad J_c'(c) \simeq \frac{0.41085}{c^{2/3}} \tag{34}$$

whence

$$\mathcal{D}(n,c) \simeq \frac{2C_1 n^4}{c^4} + \mathcal{O}(c^{-6}), \; C_1 = 0.18377. \tag{35}$$

The asymptotic formula (35) is essentially the much-quoted Gaunt-Kramers result.[6] Even for small c, the accuracy of (34) is reasonable, as Table 1 shows. The $n \to n + c$ oscillator strength is

$$f(n \to n + c) = \frac{2c}{3n^3}\mathcal{D}(n,c) \simeq \frac{C_2}{2n^2}\left(\frac{n}{c}\right)^3, \; C_2 = 0.49007. \tag{36}$$

Notice that the statistical weight of the initial level, $2n^2$, is explicitly displayed.

We now rewrite (36) in such a form that the final state n' can be extrapolated into the continuum. This will lead to the recombination cross section into n, summed over ℓ. From (9) this total cross section is

$$\sigma_R(k,n) = 4\pi^2 \alpha_{fc}^3 \left(\frac{n\omega}{k}\right)^2 \frac{df(n,\varepsilon)}{d\varepsilon} \tag{37}$$

where $df/d\varepsilon$ is the usual oscillator strength density *out* of the bound state,

$$\frac{df(n,\varepsilon)}{d\varepsilon} = \frac{2\omega}{3n^2}\sum_{\ell=0}^{\infty} [\ell \langle n\ell|r|k\ell-1\rangle^2 + (\ell+1)\langle n\ell|r|k\ell+1\rangle^2]. \tag{38}$$

By definition the quantity $2n^2 f(n,n')$ should be symmetric in n,n' so that (36) can be written as

$$2n^2 f(n,n') = C_2(nn'\omega)^{-3} \tag{39}$$

Table 1

c	$J_c(c)J_c'(c)$	C_1/c
1	0.14308	0.18377
2	0.07900	0.09189
3	0.05471	0.06126
4	0.04190	0.04594

Each Rydberg state n' is associated with an energy interval $d\varepsilon = (n')^{-3}$ so that (38) is equivalent to

$$2n^2 \frac{df(n,\varepsilon)}{d\varepsilon} = C_2(n\omega)^{-3}. \tag{40}$$

The recombination cross section

$$\sigma_R(k,n) = \frac{2\pi^2 \alpha_{fs}^3 C_2}{n^3 \omega k^2}. \tag{41}$$

It is worth noting that the cross section for photoionizing a high n state, the inverse process to recombination, is given by[7]

$$\sigma_{PI}(n) = 4\pi^2 \alpha_{fs} \frac{df(n,\varepsilon)}{d\varepsilon} \simeq \frac{2\pi^2 \alpha_{fs} C_2}{n^5 \omega^3}. \tag{42}$$

Thus at threshold $\sigma_{PI}(n) \simeq 16n$ Mb. However, the band width is proportional to the binding energy so that the total oscillator strength in the continuum

$$f_C \simeq C_2 n^{-1}. \tag{43}$$

Most of the oscillator strength sum rule is exhausted by transitions to adjacent bound states.

1.3. Total Rates

The rate of recombination into n is obtained by inserting (41) in (10),

$$\alpha_R(n,T) = \frac{3.759 \times 10^{-6}}{n} \left(\frac{8}{\pi T^3}\right)^{1/2} \int_0^\infty \frac{k \exp(-k^2/2T)dk}{(1 + n^2k^2)}. \tag{44}$$

Introducing the new variable $x = n^2k^2$, we find

$$\alpha_R(n,T) = \frac{3.759 \times 10^{-6}}{n} \left(\frac{8}{\pi T}\right)^{1/2} \phi(X), \quad X = 1/(2n^2T) \tag{45}$$

where

$$\phi(X) = Xe^X E_1(X) \tag{46}$$

and E_1 is the exponential integral.[4] As $X \to \infty$, $\phi \to 1$. In $cm^3\ sec^{-1}$, degK units the coefficient in (45) is replaced by 1.294×10^{-11}. In practice, the most needed quantities are the total rates[8]

$$\alpha_R^{(n)}(T) = \sum_{m=n}^{\infty} \alpha_R(m,T) \tag{47}$$

where n = 1,2 for plasmas which are optically thin, thick in the Lyman-α line. Thus, we write

$$\alpha_R^{(n)}(T) = \frac{2.065 \times 10^{-11}\ cm^3\ sec^{-1}}{(T\ degK)^{1/2}}\ \phi_n(X_1) \tag{48}$$

where

$$\phi_n(X_1) = \sum_{m=n}^{\infty} \frac{1}{m}\, \phi\left(\frac{X_1}{m^2}\right),\ X_1 = \frac{1.5789 \times 10^5}{T\ degK}. \tag{49}$$

Figure 1 shows the variation of ϕ_1, ϕ_2 with X_1. For a hydrogenic ion of nuclear charge q, (48) is multiplied by q^2. It appears from (45) that the favored values of n are those for which $2n^2T \sim 1$. The asymptotic formulae (41), etc. should not be used for $T \gg 3 \times 10^5$ q^2 degK.

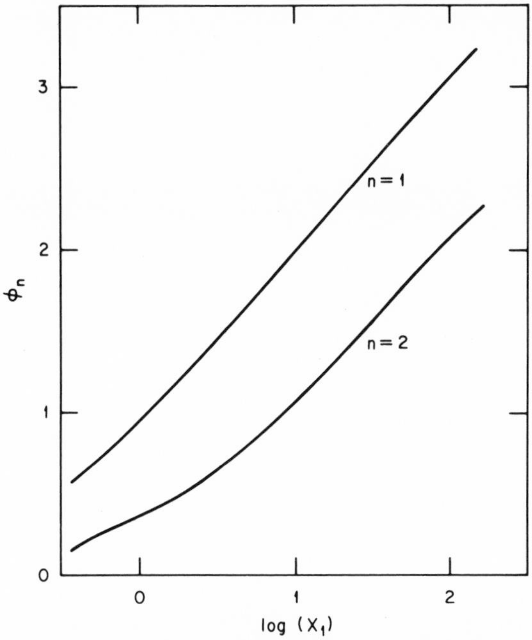

Fig. 1. Variation of functions ϕ_1, ϕ_2 defined in (49).

1.4. Partial Rates

We now turn to the more recondite question of the distribution among ℓ-substates. This is often not of practical importance since the ℓ-s will be statistically redistributed following a single collision with a charged particle. However, in an intense radiation field, low-ℓ states are preferentially photoionized, and since recombination favors just these states, the assumption of redistribution would overestimate the number of neutrals. The dependence of the rates on ℓ was first discussed in detail by Burgess.[9] A convenient table for n-1, $\ell \leq 11$ and T = 1 eV is given in Ref. 1.

First, consider how the bound-bound matrix elements (28) - (31) depend on ℓ. In Fig. 2 we plot $\mu_\pm$ vs. ε for $c \leq 4$, where

$$M(c,\pm 1) = n^2 \; \mu_\pm(c). \tag{50}$$

These numbers were generated by evaluating (26) numerically. Using asymptotic formulae for the Bessel functions (31), we find that

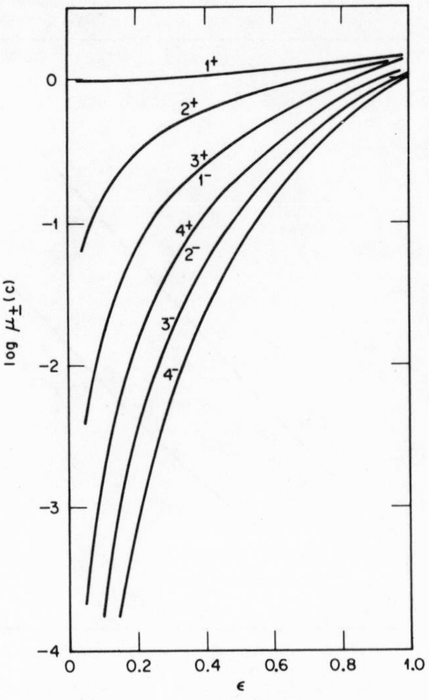

Fig. 2. Variation of $\mu_\pm(c)$, defined by (28) and (50), with ε. Notice that $\mu_+(c) = \mu_-(c-2)$.

$$|\mu_+|^2 + |\mu_-|^2 = \frac{\eta S^{2c}}{\pi c^3 \varepsilon^2}, \quad S = \frac{\varepsilon e^{\eta}}{1+\eta} \; (1-\varepsilon > 1/8 \; c^{2/3}). \tag{51}$$

The accurate calculations and asymptotic formulae agree that, except for $\mu_+(1)$, all the matrix elements → 0 rapidly as $\varepsilon \to 0$, i.e., as the classical orbits become circular. It is not possible to extrapolate (51) into the continuum, though it correctly suggests that very low ℓ-s are favored.

We can rewrite (9) in terms of either velocity or acceleration matrix elements. For exact eigenstates,

$$\begin{aligned} \langle i|\underset{\sim}{r}|f\rangle &= (i\omega)^{-1} \langle i|\underset{\sim}{p}|f\rangle \text{ (velocity)} \\ &= (i\omega)^{-2} \langle i|\underset{\sim}{r}/r^3|f\rangle \text{ (acceleration).} \end{aligned} \tag{52}$$

Then the dipole matrix elements in (9) can be replaced by

$$I_{\pm}(n,k\ell) = \langle n\ell|r^{-2}|k\ell\pm1\rangle \tag{53}$$

times ω^{-2}. Since the acceleration matrix element weighs the region of space near the nucleus, it is possible to replace the high-n orbital by a continuum orbital of zero energy

$$\lim_{n\to\infty} n^{3/2} \langle r|n\ell\rangle = \lim_{k\to o} \langle r|k\ell\rangle \tag{54}$$

when $|k\ell\rangle$ is energy normalized. Then (53) becomes

$$I_{\pm}(n,k\ell) \simeq n^{-3/2} J_{\pm}(k\ell), \quad J_{\pm}(k\ell) = \langle n\ell|r^{-2}|o\ell\pm1\rangle \tag{55}$$

while the recombination cross section

$$\sigma_R(k,n\ell) = \frac{8\pi^2 \alpha_{fs}^2}{3n^3 \omega k^2} S_{\ell}(k) \tag{56}$$

$$S_{\ell}(k) = \frac{1}{(2\ell+1)} [\ell J_-(k\ell)^2 + (\ell+1)J_+(k\ell)^2]. \tag{57}$$

This is equivalent to (41) if

$$\sum (2\ell+1)S_{\ell}(k) = S(k) \tag{58}$$

is constant and close to $2C_1$ over a reasonable range of k. This is verified to be so by direct numerical calculation (Fig. 3). In Fig. 4 we show histograms of

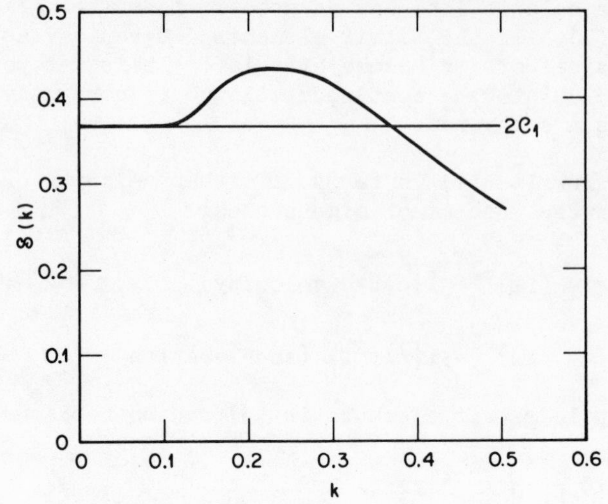

Fig. 3. Variation of $S(k)$ defined by (57), (58) with k.

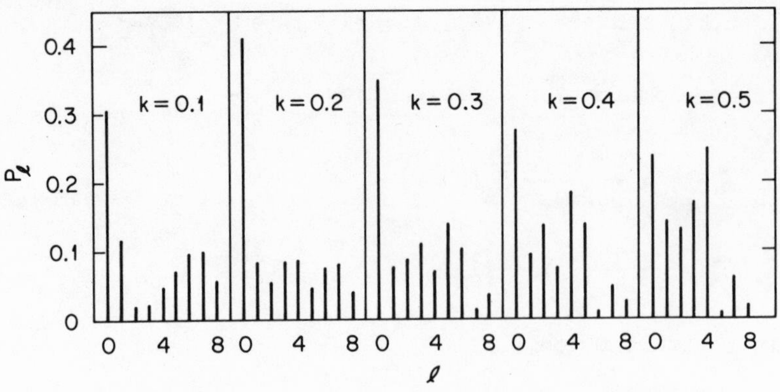

Fig. 4. Histograms of P_ℓ, the distributions defined in (59).

$$P_\ell = (2\ell+1)S_\ell(k)/S(k) \tag{59}$$

which represent the distribution over ℓ for different initial energies. As is well known, $S_\ell(k)$ oscillates as a function of ℓ and k, but this structure does not persist in the rates. Most recombination is into states $\ell \sim 4$. In Fig. 5 we show the rates into all n of a given ℓ,

$$\alpha_R^*(\ell,T) = \sum_{n=\ell+1}^{\infty} \alpha_R(n,T) \tag{60}$$

for several ℓ, compared with the total, $\alpha_R^{(1)}$ of (48).

Burgess used the different approach of extracting the approximate analytic dependence on k from the length matrix elements, so that

$$\omega^{-2}\, I_\pm(n,k\ell) = n^2\, B_\pm(n\ell) \left(\frac{\omega_n}{\omega}\right)^{\beta_\pm(\ell)}, \tag{61}$$

$$\omega_n = 1/(2n^2),\ n^2\, B_\pm(n\ell) = \langle n\ell|r|0\ell\pm1\rangle .$$

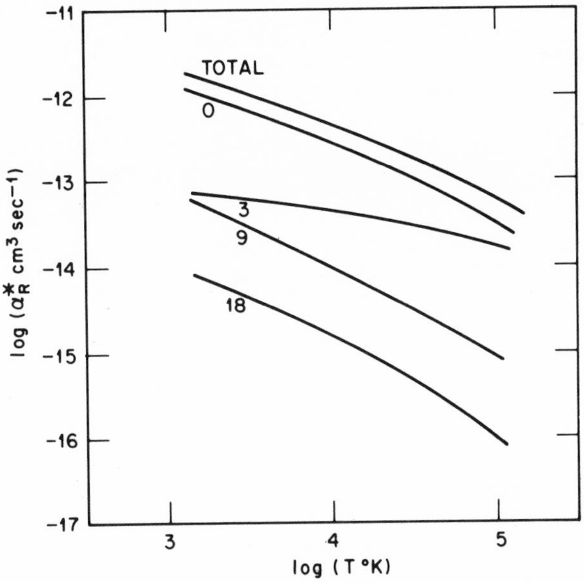

Fig. 5. $\alpha_R^*(\ell,T)$ as defined in (60), vs. T. The total rate into all bound states is also shown.

The functions $\gamma_\pm = 2\beta_\pm - 1$ and $B_\pm = \sigma_\pm/n^2$ are tabulated[9] for $\ell \leq n-1 \leq 11$. This representation is valid for $\omega < 3\omega_n$. Typical values of $\beta_\pm$ are ~ 2.0, while $B_+ \sim 1.3$, $B_- \sim 0.3$. However, $\beta_\pm$ increases with ℓ, and $B_\pm$ have a slow variation with n,ℓ such that

$$\sum_{\ell=0}^{n-1} [\ell B_-(n\ell)^2 + (\ell+1)B_+(n\ell)^2] \simeq 8.8n. \tag{62}$$

We now rewrite (61) as

$$\mathcal{I}_\pm(n,k\ell)^2 = \left(\frac{\omega_n}{\omega}\right)^{2\beta_\pm(\ell)-4} \frac{B_\pm(n\ell)^2}{16n^4} \tag{63}$$

and insert in (56). Taking $\beta_\pm = 2$, which overestimates the contribution of large ℓ, (62) leads back to (41) with $2C_1$ replaced by 0.55. The variation of $\beta_\pm$ with ℓ is such that low ℓ-s are highly favored, as we should expect.

§ 2. DIELECTRONIC RECOMBINATION

2.1. Introduction

In Chapter 1 we saw that single electron radiative recombination falls off reasonably rapidly with the energy of the free electron, simply because the matrix element connecting the free and bound states must decrease as the deBroglie wavelength $2\pi/k$ decreases, (41). If, however, the recombining ion is not a bare nucleus, the incident electron can be slowed by exciting some of the bound electrons. Following this line of thought, one is led to consider *dielectronic recombination* (DIR),

$$e(k\ell) + A^{+q}(\alpha) \rightleftarrows A^{+q-1}(\beta,n\ell') \tag{64a}$$

$$\rightarrow A^{+q-1}(\gamma,n\ell') + h\nu \tag{64b}$$

where $\beta \rightarrow \gamma$ is an allowed dipole transition of A^{+q}. Often $\gamma = \alpha =$ ground state, but γ,α may be distinct excited states. In the first stage of (84) the original ion is excited sufficiently that the electron is captured into a high Rydberg state (Fig. 6). This state is unstable in that it may autoionize back to the initial, or some other, channel. Recombination is achieved if the excited core†

---

†By "core" we always mean A^{+q}, the core of A^{+q-1}. It is not implied that inner shells of A^{+q} are excited, though they might be.

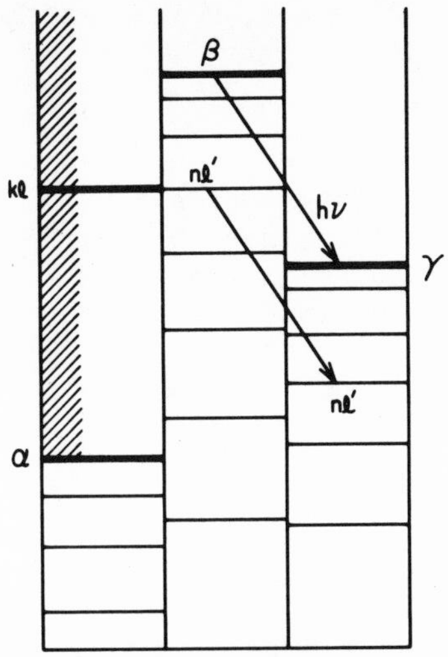

Fig. 6. Scheme of dielectronic recombination (64).

decays radiatively to produce a stable state (we shall not consider the complication that it could produce another unstable state). Typically, the radiative lifetime ~ 10^{-10} sec, while the radiationless lifetime ~ $3 \times 10^{-16}\ n^3$ sec, so that radiative decay (stabilization) is more likely for $n > n_o \sim 70$. For a given electron in the bandwidth of the Rydberg series, the probability of being in a resonance† is ~ 50%, while the probability that the resonance stabilizes is ~ 100% ($n > n_o$) or the ratio of the lifetimes ($n < n_o$), whence the probability of stabilization

$$P_n \simeq 0.5\ \mathrm{MIN}\ [1, (n/n_o)^3] \tag{65}$$

where $n\ell'$ is the Rydberg state nearest in energy. Averaging (65) over energy (multiply P_n by q^2/n^3, sum over n and divide by $q^2/2$) we arrive at a mean probability

$$P \simeq 1/n_o^2. \tag{66}$$

†The autoionizing states are resonances in the electron scattering channel.

If the bulk of the Maxwellian overlaps the Rydberg series and the electronic temperature ~ 10^5 deg K, P is just the rate in a.u.

$$\alpha_{DIR} \simeq 10^{-12}\, N\, \mathrm{cm}^3\, \mathrm{sec}^{-1} \tag{67}$$

where N is the number of Rydberg series (β, ℓ'). In the most favorable cases, α_{DIR} reaches a few times 10^{-11} cm^3 sec^{-1}. Since the radiative rate at T = 3 × 10^5 degK is $\alpha_R = 3 \times 10^{-14}\, q^2$ cm^3 sec^{-1}, dielectronic recombination dominates wherever core transitions $\alpha \to \beta$ are readily excited (at least for $q < 20$). We shall see that α_{DIR} has an intricate dependence on charge state, temperature, and species, though the range of values spanned is usually limited to 10^{-12} - 10^{-11} cm^3 sec^{-1}.

The process (64) was first suggested in the 1930's but early discussions only considered a single autoionizing state thus underestimating the rate by a factor ~ n_o. In the early 1960's discrepancies in the temperature of the solar corona derived from ionization equilibrium theory and that derived from direct measurements of Doppler widths led to the search for a missing recombination process. By considering an entire Rydberg series, Burgess[11] arrived at DIR rates large enough to explain the coronal equilibrium. Notable contributions were also made by Shore[12] and McCaroll.[13] In these lectures we can only present a rather simplified introduction to the problem. For more information, the recent excellent review by Seaton and Storey[14] should be consulted.

2.2. Formal Theory

Since we are not here concerned with scattering theory as such, we shall simply quote the standard result of resonant scattering theory with many channels.[10] For a series of well-separated resonances n, decaying into channels i, the cross section i → j is given by

$$\sigma_{ij} = \sum (2\ell+1)\sigma_{ij}(\ell), \quad \sigma_{ij}(\ell) = \frac{\pi}{k_i^2} \sum_n \frac{\Gamma_{ni}\,\Gamma_{nj}}{(\varepsilon-\varepsilon_n)^2 + \Gamma_n^2/4} \tag{68}$$

where ℓ is the total angular momentum and k_i the initial wavenumber. Each resonance has position ε_n, partial widths Γ_{ni} and total width (also rate of decay in a.u.)

$$\Gamma_n = \sum_j \Gamma_{nj} \tag{69}$$

in general depending on ℓ. The meaning of (68) is clear: each resonance has the overall Breit-Wigner energy dependence, while the branching ratios for different processes are proportional to the products of the entrance and exit partial widths.

It is straightforward to apply (68) to (64). For a single pair of angular momenta ℓ, ℓ'

$$\sigma_{DIR}(\ell,\ell') = \frac{\pi}{k^2} \sum_n \frac{\Gamma_a(\beta,n\ell';\alpha,\ell)\Gamma_r(\beta,n\ell';\gamma)}{(\varepsilon-\varepsilon_n)^2 + \Gamma(\beta,n\ell')^2/4} \tag{70}$$

where the subscripts r,a refer to radiative and autoionizing decays. Though angular momentum couplings are not explicitly shown in (70), it is necessary to *sum* over all initial, intermediate, and final substates, finally dividing by the statistical weight of α. The total width

$$\Gamma(\beta,n\ell') = \sum_{\alpha'\ell} \Gamma_a(\beta,n\ell';\alpha',\ell) + \sum_{\gamma'} \Gamma_r(\beta,n\ell';\gamma'). \tag{71}$$

Certain simplifying assumptions are made:

(i) the resonances are narrow compared with their spacing;

(ii) the energy levels ε_n do not depend on ℓ', i.e. quantum defects are small;

(iii) only the decay channels $\alpha' = \alpha$, $\gamma' = \gamma$ are significant;

(iv) the radiative decay rate is independent of $n\ell'$ and is simply the A-coefficient for $\beta \to \gamma$ in A^{+q}.

If we write

$$\Gamma_a(n\ell') = \sum_\ell (2\ell+1)\Gamma_a(\beta,n\ell';\alpha,\ell), \quad \Gamma(n\ell') = \Gamma_r + \Gamma_a(n\ell') \tag{72}$$

for short, and sum (70) over ℓ, the contribution from a single Rydberg series (β,ℓ') becomes

$$\sigma_{DIR}(\ell') = \frac{\pi}{k^2} \sum_n \frac{\Gamma_a(n\ell')\Gamma_r}{(\varepsilon-\varepsilon_n)^2 + \Gamma(n\ell')^2/4}. \tag{73}$$

To obtain a rate from (73), we must evaluate

$$\alpha_{DIR}(\ell') = \left(\frac{2}{\pi T^3}\right)^{1/2} \int_0^\infty k^2 \, \sigma_{DIR}(\ell') \exp\left(-\frac{\varepsilon}{T}\right) d\varepsilon. \tag{74}$$

Replacing

$$\frac{\Gamma_a \Gamma_r}{(\varepsilon-\varepsilon_n)^2 + \Gamma^2/4} \rightarrow \frac{2\pi \Gamma_a \Gamma_r}{\Gamma} \delta(\varepsilon-\varepsilon_n),$$

we find from (73) and (74) that

$$\sigma_{DIR}(\ell') = \left(\frac{2\pi}{T}\right)^{3/2} \sum_n \frac{\Gamma_a(n\ell')\Gamma_r}{\Gamma_a(n\ell') + \Gamma_r} \exp\left(-\frac{\varepsilon_n}{T}\right). \tag{75}$$

The energy levels must be referred to the initial state of A^{+q}. All discussions of DIR are based on (76).

2.3. Calculation of Rates

The decay process (64a) is closely related to the excitation process

$$c(k\ell) + A^{+q}(\alpha) \rightarrow e(k'\ell') + A^{+q}(\beta). \tag{76}$$

The threshold cross section ($k' \rightarrow o$) and widths ($n \rightarrow \infty$) can be written in terms of the same functions $G_a(k\ell,k'\ell')$,

$$\sigma_{ex}(k) = \frac{\pi}{k^2} \sum_{\ell\ell'} (2\ell+1) G_a(k\ell,k'\ell'), \tag{77a}$$

$$\Gamma_a(n\ell') = n^{-3} \sum_{\ell} (2\ell+1) G_a(k_o\ell,0\ell'), \tag{77b}$$

where $k_o^2/2$ is the threshold energy. Henceforth we write $G_a(k_o\ell,0\ell') = G_a(\ell')$. We shall presently calculate G_a in the Coulomb-Born-Dipole approximation, but first (75) can be developed further.

It will turn out that G_a is insensitive to q, while Γ_r varies strongly,

$$\Gamma_r = \frac{4}{3} \alpha_{fs}^3 \omega^3 |\langle\beta|\underset{\sim}{D}|\gamma\rangle|^2. \tag{78}$$

The dipole operator of A^{+q} is $\underset{\sim}{D}$; (78) implies summation over the substates of γ and averaging over the substate of β. The dipole matrix element ~ q^{-1}, while

$$\omega = E_\beta - E_\gamma \simeq G\, q^{\mu+1} \tag{79}$$

where $\mu = 0,1$ if the principle quantum number changes by

$$|N_\beta - N_\gamma| = 0, \geq 1 \text{ so that}$$

$$\Gamma_r = G_r \, q^{3\mu+1}. \tag{80}$$

Inserting (77b), (80) in (75) we see that a key role is played by the principal quantum number n_o for which $\Gamma_r = \Gamma_a$,

$$n_o(\ell') = \left(\frac{G_a}{\Gamma_r}\right)^{1/3} = \left(\frac{G_a}{G_r}\right)^{1/3} q^{-\mu-\frac{1}{3}}. \tag{81}$$

Then the rate is given by

$$\alpha_{DIR}(\ell') = \Gamma_r \sum_n \frac{F(\varepsilon_n,T)}{[1+(n/n_o)^3]}, \quad F(\varepsilon,T) = \left(\frac{2\pi}{T}\right)^{3/2} e^{-\varepsilon/T}. \tag{82}$$

Most of the sum comes from high values of n for which $\varepsilon_n \simeq \omega$ (if $\gamma = \alpha$ for simplicity). Then (82) becomes

$$\alpha_{DIR}(\ell') = 1.21 \, \Gamma_r \, n_o(\ell') F(\omega,T), \tag{83}$$

and on using (80),

$$\alpha_{DIR}(\ell') = 1.21 \, G_{ra}(\ell') q^{2\mu+\frac{2}{3}} F(\omega,T), \quad G_{ra}(\ell') = [G_r^2 G_a(\ell')]^{1/3}. \tag{84}$$

The total rate is obtained on summing over ℓ' (the sum over substates of ℓ' is already contained in G_a),

$$\alpha_{DIR} = 1.21 \, G_{ra} \, q^{2\mu+\frac{2}{3}} F(\omega,T), \quad G_{ra} = \sum G_{ra}(\ell'). \tag{85}$$

This function peaks at $T = \frac{2}{3}\omega$, as one might have expected, the maximum value being

$$\alpha_{DIR-MAX} = 6.37 \, G_{ra} \, G^{-3/2} \, q^{\frac{1}{2}\mu-\frac{5}{6}}. \tag{86}$$

i.e. for $\mu = 0$, $\alpha_{DIR-MAX} \sim q^{-5/6}$ and for $\mu = 1$, $\sim q^{-1/3}$. Thus, in Saha equilibrium $\mu = 1$ transitions are more important.

Before calculating G_a, we must digress on angular momentum coupling. The state β is usually specified by Russell-Sanders quantum numbers $S_\beta L_\beta J_\beta$, while the outer electron couples to J, giving a total angular momentum K. The full set of quantum numbers is $(\beta S_\beta L_\beta J, n\ell'; KM_K)$. If we average over JKM_K and ignore the small differences in energy between different JK states, the result is as if we had used an uncoupled representation and averaged over the magnetic quantum numbers $(\beta L_\beta M, n\ell' m')$. This procedure should be good except in some small energy ranges which do not significantly affect the rate. Then the autoionization width is given by the "golden rule",

$$G_a(\ell') = 2\pi q^2 |\langle \beta L_\beta M', 0\ell' m' | V | \alpha L_\alpha M, k_0 \ell m\rangle|^2, \tag{87}$$

where we sum over Mm and average over M'm'. The interaction

$$V = \sum |\underset{\sim}{r} - \underset{\sim}{r}_j|^{-1} \tag{88a}$$

where $\underset{\sim}{r}$ is the position of the outer electron and j runs over the electrons of A^{+q}. To evaluate (87) we retain only the dipole term

$$V \simeq \underset{\sim}{D} \cdot \underset{\sim}{r} / r^3 \tag{88b}$$

whence

$$G_a(\ell') = 2\pi q^2 \sum_\ell c(\ell\ell') c(L_\alpha L_\beta) R(\alpha,\beta)^2 \; J(\ell,\ell')^2. \tag{89}$$

The angular integral

$$c(\ell\ell') = \mathrm{MAX}(\ell,\ell') \tag{90}$$

J is the same integral introduced in (55),

$$J(\ell,\ell') = \langle k_0 \ell | r^{-2} | 0\ell' \rangle \tag{91}$$

and R is the radial integral (for a one-electron transition -- more generally it is a reduced matrix element),

$$R(\alpha,\beta) = \langle \alpha | r | \beta \rangle. \tag{92}$$

Since Γ_r is also proportional to R^2, we can express (89) in the form

$$\frac{G_a(\ell')}{\Gamma_r} = \frac{1.217 \times 10^7}{\omega^3 \, q^{-2}} \left[\ell' J(\ell'-1,\ell')^2 + (\ell'+1) J(\ell'+1,\ell')^2\right]. \tag{93}$$

It is interesting that according to (93) the radiationless process (64a) is related to radiative recombination via the acceleration matrix elements. The integral (91) can be scaled to q = 1,

$$J(\ell,\ell'|k_o,q) = J(\ell,\ell'|k_o/q,1). \tag{94}$$

Since $R \sim q^{-1}$, an immediate consequence is that $G_a \sim q^o$, as stated above. Then (93) becomes

$$G_a(\ell') = 1.217 \times 10^7 \frac{\Gamma_r q^2}{\omega^3} T(\ell') \tag{95}$$

where $T(\ell) \simeq (2\ell+1)S_\ell(k_o/q)$, and S_ℓ was defined in (57). From Fig. 4 we see that G_a must fall off rapidly when $\ell' > 5$. From (79), (80), (84), (85), and (95),

$$G_{ra} = 230\ C_3\ G^{-1}\ G_r \quad , \quad C_3 = \sum_\ell T(\ell)^{1/3} \simeq 2.0(k_o < q). \tag{96}$$

Then the total rate

$$\alpha_{DIR} = 278\ C_3 q^{2\mu + \frac{2}{3}} G^{-1}\ G_r F(\omega,T). \tag{97}$$

Finally, the peak rate

$$\alpha_{DIR-MAX} = 1470\ C_3 q^{\frac{1}{2}\mu - \frac{5}{6}} G^{-5/2}\ G_r. \tag{98}$$

To illustrate the zoology of autoionizing states, we look in detail at a series in four-electron ions

$$A^{+q-1}(1s^2\ 2pnp) \to A^{+q}(1s^2\ 2s) + e(ks,kd) \tag{99}$$

which is strongly excited in ion-atom collisions.[15] Thus $\alpha = 2s$, $\beta = 2p$, $\ell' = 1$, $\ell = 0,2$; in (79), $\mu = 0$. Table 2 shows the variation with q of ω/q, qR, $J(p\to s)$ and $J(p\to d)$: ω, $R(2s\to 2p)$ are the best experimental and theoretical values[16] and J was calculated with Coulomb wavefunctions. The regularity of scaling with q is pleasing. From (89) we obtain the reduced widths

$$G_a = 2\pi\ (qR)^2\ [J(p\to s)^2 + 2J(p\to d)^2]. \tag{100}$$

Table 2 shows the *average* width of one multiplet (2pnp)SL, $\frac{1}{9}G_a$. The perturbation theory is valid[10] if $\frac{1}{9}G_a < 2\pi$ which is just about satisfied. Because of the resonance oscillator strength, these widths are unusually large, e.g. in his study of He-like systems, Weisheit[17] used the universal formula

$$\begin{aligned} G_a(\ell') &= 1.6 \times 10^{-5}\ (2\ell'+1) \quad (\ell' \le 7) \\ &= 0 \qquad\qquad\qquad\qquad\quad (\ell' > 7). \end{aligned} \tag{101}$$

Table 2. Variation with q of matrix elements entering DIR

q	1	2	3	5	8	16
$10\ \omega/q$	0.680	0.728	0.735	0.736	0.740	0.779
qR	7.06	7.90	8.20	8.48	8.67	8.76
$J(p\to s)$	0.0829	0.132	0.154	0.169	0.172	0.166
$J(p\to d)$	0.0205	0.0411	0.0514	0.0580	0.0622	0.0674
$\frac{1}{9}\ G_a$	0.268	0.906	1.36	1.77	1.96	1.96

All quantities refer to the process (99) - (103).

However, G_r is compensatingly small because of the ω^3 factor. Using the values at q = 8, G = 0.074 and $G_r = 5.26 \times 10^{-9}$, so that $G_{ra}(1) = 7.87 \times 10^{-6}$ and the total $G_{ra} = 3.27 \times 10^{-5}$. From (97) the total rate is

$$\alpha_{DIR} = (2.42 \times 10^{-11}\ \text{cm}^3\ \text{sec}^{-1})q^{0.67}\ F(0.074q,T) \tag{102}$$

of which $\ell' = 1$ contributes 24%. The maximum rate (at T = 15,600 q degK) is

$$\alpha_{DIR\text{-}MAX} = (6.34 \times 10^{-11}\ \text{cm}^3\ \text{sec}^{-1})q^{-0.83}. \tag{103}$$

The chief defect of (89) - (98), which are intended as illustrations rather than working formulae, is the Coulomb approximation in evaluating J. If distorted waves are used for q > 3, the results would probably be as good as one could desire; for $q \leq 3$ close-coupling calculations might be needed. Referring to (84), α_{DIR} depends on G_a, which must be calculated, only through the one-third power; G_r is usually available in the large body of critically assessed oscillator strengths. If (98) is applied to the He^+ sequence, $G_{ra} = 7.44 \times 10^{-5}$ and $\alpha_{DIR\text{-}MAX} = (1.59 \times 10^{-12}\ \text{cm}^3\ \text{sec}^{-1})q^{-0.33}$, in good agreement with the best calculation;[18] 20% comes from the 2pnp series. For most applications, the semi-empirical universal formula of Burgess[11] is recommended.

In Fig. 7 we show the best available calculations[14,18] of α_{DIR} for He^+, C^{+2}, and Fe^{+21} compared in each case with the radiative rate α_R. The core transition in He^+ has $\mu = 1$ (1s→2p) and that in C^{+2} has $\mu = 0$ (2s→2p); in Fe^{+21} two transitions contribute, one $\mu = 0$ (2s→2p) and one $\mu = 1$ (2p→3d). Since α_{DIR} falls off with q, albeit slowly, and $\alpha_R \sim q^2$, the latter eventually catches up, as can be seen in Fe^{+21}.

Though outside the scope of these lectures, collisions with third bodies may have an important effect on DIR. In dense plasmas

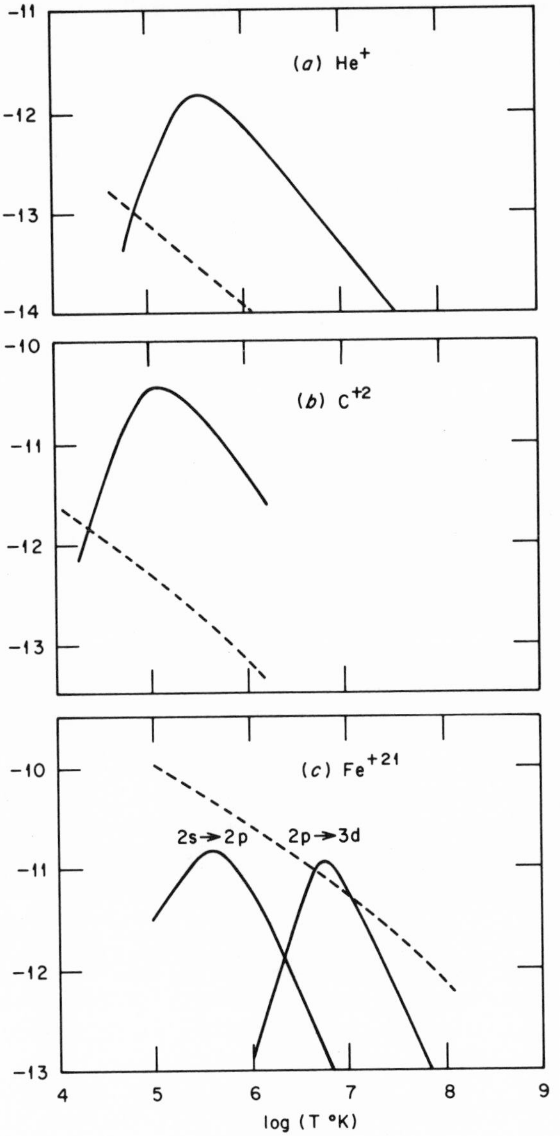

Fig. 7. Variation of DIR rate with temperature. The full lines are α_{DIR} and the dotted lines are α_R. Two transitions contribute in Fe^{+21} as indicated. The results for He^+ are from close-coupling[18] and for C^{+2}, Fe^{+21} from distorted-wave calculations.[14]

α_{DIR} is decreased by the lowering of the continuum.[17] The distribution of ℓ' produced by DIR is similar to that in radiative recombination, being determined by the same matrix elements (95), i.e. $\ell' < 8$ is favored. Thus, if a collision takes place before autoionization, large values of ℓ' are populated, thereby stabilizing the state.

§ 3. DISSOCIATIVE RECOMBINATION

3.1. Introduction

Dissociative recombination (DR) is the process suggested by Bates and Massey in 1947 to account for the removal of O_2^+ in the ionosphere,

$$e + AB^+ \rightarrow A(m) + B(n) \tag{104}$$

where one or both products is usually excited. The process is important at temperatures below 1 eV where molecular ions are abundant. Its rapid rate is explicable if the dissociating state of AB crosses the ionic state in the Franck-Condon (FC) region of the ground vibrational state (Fig. 8). For most systems with more than four electrons, one can almost always find an electronically doubly excited surface AB^{**} satisfying this requirement. The few electron systems of special interest in neutral beam technology provide exceptions to this rule. Accurate potential energy curves[19] for $H_2^+(1\sigma_g)$ and $H_2(1\sigma_u^2\ {}^1\Sigma_g^+)$ are shown in Fig. 9. The crossing point R_x lies within the FC region for initial vibrational states $v \geq 2$. In H_3^+ a similar situation prevails;[20] at least 0.5 eV of excitation is required to reach the lowest dissociating surface from the ground vibrational state. The only molecular ion for which dissociative recombination is never significant may be He_2^+. The molecular orbitals $1\sigma_g$, $1\sigma_u$ based on 1s atomic orbitals are fully occupied in ground state He_2; thus, the lowest doubly excited state involves two orbitals based on 2s,2p atomic states and lies ~ 16 eV above the ionic ground state (relative to $2He^+$, $He + He^+$ ~ -26 eV, while $He_2^{**} \rightarrow 2He^*$ ~ -10 eV).

3.2. Formalism

The formal theory of (104) is an interesting problem on which a large literature has accumulated.[19,20] We only present a simplified account here, confined to diatomics for the most part. Rather than consider the crossing state as the final (adiabatic) channel, we define it to be a diabatic state which crosses the true final state, as illustrated in Fig. 8,

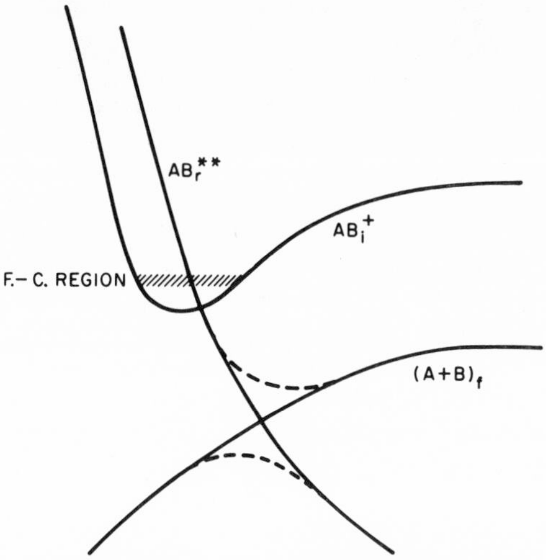

Fig. 8. Scheme of dissociative recombination (104), (105).

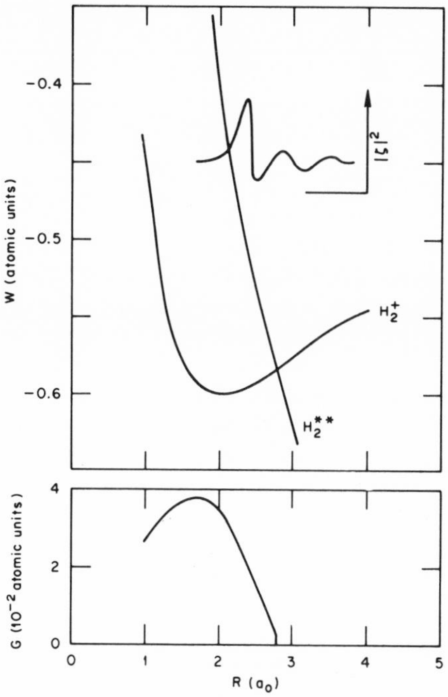

Fig. 9. Accurate potential energy curves[19] for $H_2^+(1\sigma_g)$ and $H_2^{**}(1\sigma_u^2)$. Inset are the width G of the H_2^{**} data and the squared nuclear wavefunction $|\zeta|^2$ of the dissociating state H + H* (c.f. (119)).

$$e(\varepsilon) + AB_i^+ \rightleftarrows AB_r^{**} \rightarrow (A+B)_f. \tag{105}$$

The Born-Oppenheimer electronic wavefunctions at given internuclear separation R of the three states in (105) are denoted by $\Phi_i(\varepsilon)$, Φ_r, Φ_f. In the first place, we suppose that the incoming electron has energy $\varepsilon_i = k_i^2/2$, total angular momentum ℓ_i and projected angular momentum on the internuclear axis λ_i. The rotation of the ion is ignored as is permissible at energies ε_i >> rotational separations. It is vital to realize that AB^{**} can decay by ejecting electrons of energy $\varepsilon \neq \varepsilon_i$, the slack being taken up by the nuclear (vibrational) motion. Much of the literature is concerned with the definition of the resonance state Φ_r which we take as intuitively known. The electronic-vibrational wavefunction of the system (105) can be expanded as

$$\Psi(\ell_i\lambda_i) = F_f(R)\Phi_f + F_r(R)\Phi_r + \int d\varepsilon\ F_i(\varepsilon,R)\Phi_i(\varepsilon). \tag{106}$$

If the potentials are $\varepsilon + W_i$, W_r, W_f and the diabatic couplings are $V_{ri}(\varepsilon)$, V_{rf}, we obtain the following set of coupled equations,

$$(E-K-W_f)F_f = V_{fr}F_r \tag{107}$$

$$(E-\varepsilon-K-W_i)F_i(\varepsilon) = V_{ir}(\varepsilon)F_r \tag{108}$$

$$(E-K-W_r)F_r = V_{rf}F_f + \int d\varepsilon\ V_{ri}(\varepsilon)F_i(\varepsilon) \tag{109}$$

where E is the total (electronic plus vibrational) energy and K the nuclear kinetic energy operator (M is the reduced mass)

$$K = -\frac{1}{2MR^2}\frac{\partial}{\partial R}\left(R^2\frac{\partial}{\partial R}\right). \tag{110}$$

If $V_{ri}(\varepsilon)$ varies slowly with R and F_r is approximately an eigenstate of $K+W_r$, i.e. if the right side of (109) is neglected to a first approximation, (108) has a particular integral equal to F_r times a slowly varying function of R. To this integral must be added a solution of the homogeneous equation

$$(E-\varepsilon_i-K-W_i)\zeta_i = 0 \tag{111}$$

to provide an ingoing wave in (106). Thus

$$F_i(\varepsilon) \simeq \zeta_i\delta(\varepsilon-\varepsilon_i) + (\varepsilon_{ri} - \varepsilon + io)^{-1}\ V_{ri}(\varepsilon)F_r \tag{112}$$

where

$$\varepsilon_{ri}(R) = W_r - W_i \tag{113}$$

is the "vertical ejection energy". If AB was in the state r with the nuclei fixed at R, it would autoionize to AB^+ by ejecting an electron of energy $\varepsilon_{ri}(R)$. The positive imaginary term in the denominator of (112) insures that only outgoing waves appear in the final channel. Substituting (112) in (109), we find that

$$(E-K-W_r)F_r = V_{ri}(\varepsilon_i)\zeta_i + V_{rf}F_f + \mathcal{D}F_r \tag{114}$$

where

$$\mathcal{D} = \int \frac{V_{ri}(\varepsilon)^2 \, d\varepsilon}{\varepsilon_{ri}-\varepsilon+io} \simeq \frac{i}{2} G_r, \quad G_r = 2\pi V_{ri}(\varepsilon_{ri})^2 \tag{115}$$

neglecting a principal value integral. The "vertical" or "local" width G_r is now incorporated in a complex potential

$$\mathcal{W}_r = W_r - \frac{i}{2} G_r \tag{116}$$

so that

$$(E-K-\mathcal{W}_r)F_r = V_{ri}(\varepsilon_i)\zeta_i + V_{rf}F_f. \tag{117}$$

Solution of (107) and (117) leads to the amplitude of the outgoing wave in F_f and hence to the DR cross section. Generalization to many final states is obvious. Back ionization into the initial state is described by the complex potential $\mathcal{W}_r$.

To calculate the cross section, we return to the representation in which W_r crosses W_f so that r is itself the final channel, and V_{rf} is dropped from the right side of (117). It is readily shown from Green's theorem that as $R \to \infty$,

$$F_r \sim A(\ell_i\lambda_i,\varepsilon_i) \left(\frac{2\pi M}{k_f}\right)^{1/2} \frac{e^{i\Omega}}{iR},$$
$$A(\ell_i\lambda_i,\varepsilon_i) = \langle\zeta_r|V_{ri}(\varepsilon_i)|\zeta_i(\varepsilon_i)\rangle \tag{118}$$

where ζ_r is a solution of the homogeneous equation

$$(E-K-\mathcal{W}_r)\zeta_r = 0 \tag{119}$$

so normalized that

$$\delta_r \cdot \left(\frac{2M}{\pi k_f}\right)^{1/2} \frac{\sin\Omega}{R} \quad , \quad \Omega = k_f R + \delta_r \tag{120}$$

where δ_r is a complex phase shift. To obtain the DR amplitude F, we have to combine solutions like (112) to form an ingoing plane wave and then look at the outgoing waves. The total wavefunction will satisfy

$$\Psi(\underset{\sim}{k}_i) \sim k_i^{-1/2} \exp(ik_i z)\zeta_i Y_i + F(\theta) \left(\frac{k_f}{M}\right)^{-1/2} \frac{e^{i\Omega}}{R} \Phi_f \tag{121}$$

where the asymptotic form of Φ_i is

$$\Phi_i \sim \left(\frac{2k_i}{\pi}\right)^{1/2} j_{\ell_i}(k_i r) Y_i + \text{outgoing waves.} \tag{122}$$

Expanding the plane wave in (121) into functions like (122) which connect to outgoing waves (118), we find that

$$F(\theta) = \frac{i\pi}{k_i} \sum_{\ell\lambda} A(\ell\lambda, \varepsilon_i) i^{\ell} P_{\ell}(\cos\theta) \tag{123}$$

whence

$$\sigma_{DR}(\varepsilon_i) = \frac{\pi}{k_i^2} \sum_{\ell\lambda} \left|2\pi A(\ell\lambda, \varepsilon)\right|^2. \tag{124}$$

Usually a single value of $\ell\lambda$ predominates, e.g. the $1\sigma_u^2$ state of H_2 is coupled to a $d\sigma$ wave, $\ell\lambda = 20$.

The total cross section (124) thus depends on the matrix elements (118). Notice that the argument of V_{ri} is $\varepsilon = \varepsilon_i$, *not* ε_{ri}. Some idea of the cross section is obtained by inserting JWKB approximations to ζ_r, ζ_i in (118) and evaluating the integral by the method of stationary phase, with the result that[1]

$$\sigma_{DR} = \frac{4\pi^2}{k_i^2} \frac{\omega_e \, e^{-\rho}}{v(R_x)|W'_{ri}(R_x)|} \, , \; \rho = \int_{R_x}^{R_s} \frac{G_r dR}{v}. \tag{125}$$

The curves are supposed to cross at R_s, beyond which r is stable, $v(R)$ is the local velocity of the separating atoms, $W'_{ri} = (d/dR)(W_r - W_i)$, ω_e is the vibrational frequency of AB^+ and $R_x(\varepsilon) < R_s$ is the point at which

$$W_r - W_i = \varepsilon. \tag{126}$$

The "survival factor" $e^{-\rho}$ is the probability that AB^{**} does not autoionize between R_x and R_s. The reduced mass enters (125) only in $\rho \sim M^{1/2}$, so if $\rho \gtrsim 1$ an isotope effect that σ_{DR} is larger for light species is expected. The semi-classical formula is only valid if R_s lies in the FC region of ζ_i. All too often R_s is on the edge of the FC region and a fully quantal description of the nuclear motion is required. In Fig. 9 we sketch the nuclear wavefunction squared $|\zeta_r(R)|^2$ at a given total energy. As ε increases, the classical turning point moves to smaller R, sweeping across the FC region. Thus below some pseudo-threshold σ_{DR} is small; then it rises rapidly varying with increasing ε as $|\zeta|^2$ varies with increasing R.

The general features just outlined are well illustrated by the case of H_2^+ for which experiments and detailed calculations are available. Figure 9 shows the potentials W_r, W_i and width G_r. We have plotted in Fig. 10 theoretical $\sigma_{DR}(v)$ against energy for H_2^+ initially in the v = 0, 1, and 2 vibrational states. The FC oscillations described in the preceding paragraph are clearly seen. The dip in $\sigma_{DR}(o)$ at ε = 0.5 eV appears in recent experiments.[21] The cross section averaged over a known distribution of many initial vibrational states is compared with the measurements of Peart and Dolder[22] in Fig. 11. The cross section for producing H(n = 2) is also shown in Fig. 11. The distribution of final Rydberg states is predicted[19] over a wide range of energy and initial vibrational states to be

$$P(n) = 25.8\, n^{-3} \left[1 + \left(\frac{n_1}{n}\right)^4\right]^{-1} \tag{127}$$

where n = 4.5, in harmony with the scant experimental evidence.[23] The final state atoms are not, as repeatedly stated without supporting arguments, concentrated in a single low-lying level (e.g. n = 2, 3), but are fairly evenly distributed over the levels n < 10. Within one level the population of each ℓ,m substate $\sim 1/(2\ell+1)$, since a Σ resonance can only couple to final states whose angular momentum projection on the internuclear axis is zero.

3.3. Very Low Energies: The Indirect Mechanism and Large Molecular Ions

At thermal energies (<< 1 eV) the matrix element A defined in (118) tends to a constant nonzero limit,

$$A(\ell\lambda,0) = \lim_{\varepsilon\to 0} \langle\zeta_r|V_{ri}(\varepsilon)|\zeta_i\rangle. \tag{128}$$

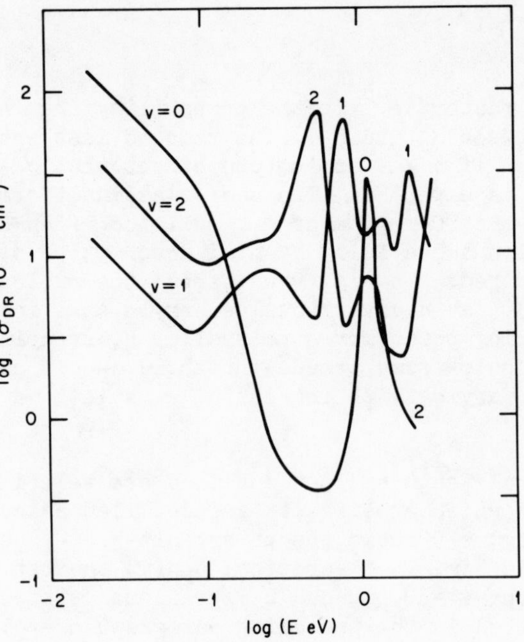

Fig. 10. Cross sections for DR of $H_2^+(v)$ as a function of impact energy.

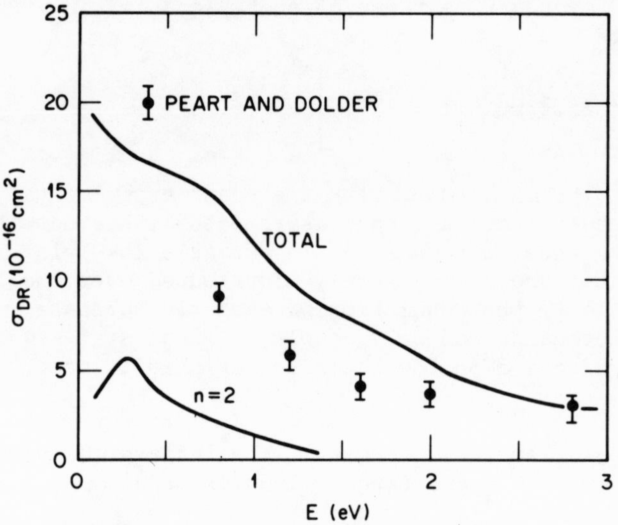

Fig. 11. Cross sections for DR of H_2^+(all v) → H + H(n) as a function of impact energy. Total (all n) compared with experiment.[22] H(n = 2) contribution also shown.

This limit is a consequence of the long-range Coulomb attraction between the ion and electron. Then the cross section (124) becomes

$$\sigma_{DR} = \frac{2\pi^2 G}{k_i^2} \quad , \quad G = 2\pi|A(\ell\lambda,o)|^2. \tag{129}$$

The same form follows from the semiclassical approximation (125) since $R_x \to R_s$ as $\varepsilon \to \infty$. From (129) we get the well-known rate

$$\alpha_{DR} = \frac{(2\pi)^{3/2}}{T^{1/2}} G = (1.717 \times 10^{-10}\ \mathrm{cm}^3\ \mathrm{sec}^{-1}) \frac{G\ \mathrm{degK}}{(T\ \mathrm{degK})^{1/2}}. \tag{130}$$

Since G ~ 2 eV for a broad molecular resonance, $\alpha_{DR} \sim 2 \times 10^{-7}$ $\mathrm{cm}^3\ \mathrm{sec}^{-1}$ at 400 degK, as observed in many species.[20]

The form of (130) is more fundamental than is often supposed as we can see by developing a classical model.[24] Suppose the electron approaches the ion on a trajectory defined by the initial kinetic energy $\varepsilon_i = k_i^2/2$ and impact parameter b; the distance of closest approach is r, the velocity there being v. We assume that within $r < r_o$ recombination takes place with probability P_R. The electron will almost always lose kinetic energy $\Delta\varepsilon$ and angular momentum Δj to the ion, especially to a large ion. The conservation of energy and angular momentum require

$$k_i b = vr + \Delta j \tag{131}$$

$$k_i^2 = v^2 + 2\Delta\varepsilon - 2/r \tag{132}$$

whence recombination is possible if $b < b_o$ given by

$$k_i b_o = \Delta j + (2r_o)^{1/2}\,[1 + r_o(\varepsilon_i - \Delta\varepsilon)]^{1/2}. \tag{133}$$

We are only interested in $\varepsilon_i << 1$ eV while $r_o < 4$ for even the largest ionic cluster, so the square brackets in (133) can be replaced by unity, leading to a cross section

$$\sigma_{DR} = \frac{\pi P_R}{k_i^2} \sum P_T(\Delta j)[\Delta j + (2r_o)^{1/2}]^2. \tag{134}$$

The probability of angular momentum transfer Δj is denoted by $P_T(\Delta j)$, while the sum is over all $\Delta j > -(2r_o)^{1/2}$. If P_T is ~ 1 for $\Delta j = 1$, and $r_o \sim 2$, (134) is consistent with (129) if $9P_R \sim 2\pi G$, i.e. $P_R \sim 0.04$. Thus, unless $\Delta j >> 1$, a situation discussed below, the temperature dependence (130) is inescapable.

In the *indirect* mechanism of DR, first discussed in detail by Bardsley, the electron gives up its energy in vibrational excitation, settling in a Rydberg state until dissociation or reverse ionization takes place,

$$e(k_i \ell_i \lambda_i) + AB^+(v_i) \rightleftarrows [e(n\ell_s \lambda_s)AB^+(v_s)] \to AB^{**} \to A+B. \tag{135}$$

Rotational excitation is possible but does not seem to be of practical importance. The theory of (135) is similar to that of DIR.[19] The indirect rate is of the form (130) with G replaced by a sum like (75),

$$G_{IDR} = T^{-1} \sum_{s,n} \frac{\Gamma_d(s,n)\Gamma_a(s,n)}{\Gamma_d(s,n) + \Gamma_a(s,n)} \exp\left[-\frac{\varepsilon(s,n)}{T}\right]. \tag{136}$$

Relative to the initial state, the resonance energies are given by

$$\varepsilon(s,n) = U_{si} - \frac{1}{2n^2}. \tag{137}$$

The dissociation width Γ_d is analogous to Γ_r in (75), except that it depends on n,

$$\Gamma(s,n) = n^{-3}\, G(v_s). \tag{138}$$

The present autoionization width Γ_a is entirely analogous to the earlier Γ_a, and can be related to the threshold vibrational excitation cross section in the manner of (77),

$$\Gamma(s,n) = n^{-3}\, G_{ex}(v_s) \tag{139a}$$

$$\sigma_{ex}(v_i \to v_s) = \frac{\pi}{k_i^2}\, G_{ex}(v_s) \tag{139b}$$

Inserting (139) and (139) in (136), each term varies as n^{-3}. The sum over n can be done analytically so that

$$G_{IDR} = \sum_s \frac{G(v_s)G_{ex}(v_s)}{G(v_s) + G_{ex}(v_s)} \left[1 - \exp\left(-\frac{U_{si}}{T}\right)\right]. \tag{140}$$

Early discussions of IDR fell down, like early discussions of DIR, in not summing over an entire Rydberg series; thus, they incorrectly found that as $T \to 0$, $G \sim T^{-1}$, a result which is still quoted. If $G \gg G_{ex}$,

$$G_{IDR} \simeq \sum G_{ex}(v_s) \quad , \quad \alpha_{IDR} \simeq \alpha_{TX}, \tag{141}$$

the total rate for vibrational excitation, which *ex hypothesi* $<< \alpha_{DR}$. However, if $G_{ex} >> G$ (independent of v_s),

$$G_{IDR} \simeq NG, \; \alpha \simeq (1 + N)\alpha_{DR} \tag{142}$$

where N is the number of levels s for which $G_{ex}(v_s) >> G$, $U_{si} \gtrsim T$. Thus in a diatomic α_{DR} is usually multiplied by two; this factor has been included in the cross sections in Fig. 10 below 0.1 eV. In polyatomic ions N may be very large[24] e.g. in clusters $Y^+ \cdot X_n$, $N \sim n$. However, (140) still predicts a rate $\sim T^{-c}$, $c \geq \frac{1}{2}$.

Recent experiments of Biondi[25] strongly suggest that $\alpha_{DR} \sim T^0$ for large clusters. In the regime where Δj is small (c.f. (131) - (134)) no reasonable assumption will lead to such a variation since neither r_0 nor P_R can $\sim T$. Nor can we postulate that $G_{ex} \sim k_i$ (which is in any case impossible) since if $\alpha_{IDR} \sim G_{ex}$, $\alpha_{DR} >> \alpha_{IDR}$, as explained above. However, a large cluster with rotational constants $B \sim 10^{-7}$ can absorb a great deal of angular momentum from the incident electron. For a cluster $Y^+ \cdot X_n$ of internal energy E_R, the mean vibrational dipole moment[24]

$$\mu = \frac{K\rho_e}{\omega_e}\left(\frac{BE_R}{n}\right)^{1/2} \tag{143}$$

where $K \sim 1$ and ρ_e, ω_e are the length and frequency of an XY^+ bond. Since $B \sim n^{-1}$, μ should vary little with n. At an impact parameter b_0, the electron dipole interaction can couple states within a band of quantum numbers Δj such that

$$\Delta E_R = 2(BE_R)^{1/2} \; \Delta j \simeq \frac{\mu}{b_0^2}. \tag{144}$$

Inserting (144) in (133) and assuming that $\Delta j >> (2r_0)^{1/2}$, i.e., that the electron loses most of its angular momentum, we obtain

$$\sigma_{DR} = \frac{\pi P_R \mu^{2/3}}{(8BE_R\varepsilon_i)^{1/3}} = \frac{\pi P_R}{(n\varepsilon_i)^{1/3}}\left(\frac{K\rho_e}{\omega_e}\right)^{2/3} . \tag{145}$$

We must emphasize that this is *not* indirect DR since $\Delta E_R \sim 0.01\ \varepsilon_i$ only. If P_R is the probability of exciting a vibrational mode which $\sim n$,

$$\alpha_{DR} = \alpha_o n^{2/3} T^{1/6}, \tag{146}$$

fairly close to the observed behavior ~ nT^o. Trying some numerical values, (145) gives a maximum $\sigma_{DR} \sim 10^4 \pi$ in accord with the observed rates for large clusters ~ 10^{-5} cm^3 sec^{-1}. For T = 5000 deg K, E_R = 400 deg K, (144) gives $b_o \sim 100$, $\Delta j \sim k_i b_o \sim 10$, $\Delta E_R \sim 6$ deg K so that the model is internally consistent.

3.4. Other Topics

The reader is referred to review articles[20,26] and recent papers[27] for more information. A number of theoretical studies have been made of CH^+; while the potentials are fairly well understood, the dynamics of the process are not. The crossing state is the fourth state of $^2\Pi$ symmetry so that a series of curve-hopping transitions are needed to reach the final dissociation products. Such mechanisms are presumably common in atmospheric and astrophysical species and merit investigation. We have already mentioned the difficulty of dissociating H_3^+ from the ground vibrational state at thermal energies. All extant measurements on H_3^+ have used unrelaxed ions (internal energy 2-4 eV) produced by the $H_2^+ + H_2$ reaction. While these experiments are mutually consistent they say nothing about the ground state ion. As far as experimental techniques go, the greatest obstacle to progress is the incapacity to select the initial vibrational state.

§ 4. THREE-BODY RECOMBINATION

4.1. Introduction

We now turn to three-body (or to be erudite, ternery) processes which dominate at high densities. The rate of such a process as (3) is described by a quantity β of dimension L^6T^{-1} such that

$$\frac{dn(x^+)}{dt} = \frac{dn(y^-)}{dt} = -\beta n(x^+)n(y^-)n(z). \tag{147}$$

We can equivalently use an effective ternery rate

$$\alpha_{TR} = n(z)\beta. \tag{148}$$

Many three-body reactions are of great practical importance, e.g. recombination in air,

$$O_{2m}^{+} + O_{2n}^{-} + (N_2,O_2) \rightarrow (m+n)O_2 + (N_2,O_2) \qquad (149)$$

where m,n < 4. But we begin with the electronic process $y^{-} = z = e$ which, e.g., predominates in a helium afterglow X = He,

$$e + e + X^{+} \rightarrow e + X(p). \qquad (150)$$

4.2. Electronic Three-Body Recombination

To arrive at a "rate" for (150), one must consider all the processes illustrated in Fig. 12. The quantity one really wants to know is the net increase in ground state neutral atoms per unit time. We first describe the rate equation approach of Bates *et al.*[28,29,1] A plasma is considered with predominantly singly charged ions X^{+}. The number density of free electrons is n(*) and of neutrals in the Rydberg level p is n(p), all number densities being scaled to $n(X^{+}) = 1$. We suppose (call this assumption A-1) that the free electrons are in Saha equilibrium and that all rates can be averaged accordingly. The processes considered in an optically thin medium, with their rate coefficients, are as follows:

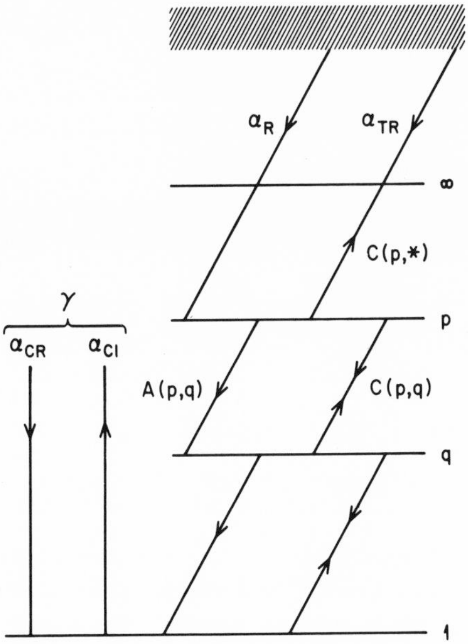

Fig. 12. Scheme of electronic three-body recombination: quantities occurring in (152).

(i) radiative recombination $\alpha_R(p)$,

(ii) spontaneous emission $A(p,q)$, $p > q$,

(iii) collisional excitation and de-excitation, $C(p,q)$,

(iv) collisional ionization $C(p,*)$ and its inverse, ternery recombination.

The last-mentioned rate can be written as in (148),

$$C(*,p) = \beta(p)n(*) = \alpha_{TR}(p). \tag{151}$$

The rate equation governing the relaxation of n(p) is

$$\begin{aligned}\frac{dn(p)}{dt} &= n(*)[\alpha_R(p) + n(*)\beta(p)] + \sum n(q)[C(q,p) + A(q,p)] \\ &\quad - n(p)[\tilde{C}(p) + \tilde{A}(p)]\end{aligned} \tag{152}$$

$$\tilde{C}(p) = C(p,*) + \sum C(p,q) \quad , \quad \tilde{A}(p) = \sum A(p,q). \tag{153}$$

We make a second assumption (A-2) that levels $p \geq 2$ are in a steady state

$$\frac{dn(p)}{dt} = 0 \quad , \quad p \geq 2. \tag{154}$$

If p_{max} levels are explicitly considered, and those with $p > p_{max}$ are in Saha equilibrium (A-3), (154) provides $p_{max}-1$ equations for $n(2) \ldots n(p_{max})$ whose solution is written

$$n(p) = R(p) + n(*)R_*(p). \tag{155}$$

Substituting (155) in (152) for p = 1, we find that

$$\frac{dn(1)}{dt} = -\gamma n(*)n(1) \quad , \quad \gamma = \alpha_{CR} - \alpha_{CI}n(1)/n(*) \tag{156}$$

where γ is a ground recombination coefficient, made up of collisional-radiative and collisional-ionization parts, given by

$$\alpha_{CR} = \alpha_R(1) + \beta(1)n(*) + \sum R_*(q)[C(q,1) + A(q,1)] \tag{157}$$

$$\alpha_{CI} = \tilde{C}(1) - \sum R(q)[C(q,1) + A(q,1)]. \tag{158}$$

The meaning of each term in (156)-(158) is readily grasped.

At low n(*), roughly < 400 (T deg K)^2, $\alpha_{CR} \simeq \alpha_R(1)$. In the other limit one reaches a purely collisional régime where

$$\gamma \simeq \alpha_{CR}^{\infty} = \beta^{\infty}\, n(*). \tag{159}$$

The solutions at moderate and high densities are largely determined by the choice of β(p) and C(p,*). Bates *et al.*[28] used binary encounter expressions. The energy transfer cross section for e(ε) + e'(ε') → e(ε−U) + e' is

$$\sigma(\varepsilon,U)dU = \frac{4\pi}{\varepsilon}\frac{dU}{U^2}. \tag{160}$$

For ionization of a state with binding energy U_p, (160) leads to the cross section

$$\sigma_{BEI}(p) = \frac{2\pi}{U_p \varepsilon} \tag{161}$$

and the rate

$$C(p,*) = \frac{(2\pi)^{1/2}}{U_p T^{1/2}}. \tag{162}$$

The condition for recombination into p is that U lies between $\varepsilon' + U_p$, $\varepsilon' + U_{p-1}$, while e is within the radius of the p state. The cross section for e fixed is

$$\sigma_{BER}(\varepsilon' \rightarrow p) = \frac{4\pi}{\varepsilon}\frac{(2U_p)^{3/2}}{(\varepsilon'+U_p)^2}. \tag{163}$$

To get a ternery rate, one must integrate over the velocity distributions of e,e', the impact parameters of e (already done in (163)) and the configuration space of e', to obtain,

$$\beta(p) = \frac{4\pi^{3/2}}{U_p^{3/2}\; T^{1/2}(T+U_p)^2}. \tag{164}$$

Summing (164) over p such that $U_p \gtrsim T$, i.e. over levels which are unlikely to be re-ionized, we arrive at an estimate of γ,

$$\gamma_{BE} \simeq \frac{0.98}{T^{9/2}} \tag{165}$$

which is surprisingly close to the results of elaborate calculations. Numerical results will be discussed after considering the alternative approach of Keck *et al.*[30]

This second approach is closer to the spirit of statistical mechanics. The process by which a given electron trickles down the ladder of excited states into the ground state while interacting with the third-body electron is treated as a single dynamical process. Then the rate is found by averaging over a ternery hyper-cross-section. A very simple argument of this type was used to derive (164) and (165). The ternary approximation should be valid for densities $< 10^{19}\ \mathrm{cm}^{-3}$.

The earliest statistical model[31] was that of J. J. Thomson who introduced a characteristic energy and associated Bohr radius,

$$E_T = \tau T \quad , \quad r_T = E_T^{-1} \tag{166}$$

where $\tau \sim 1$ (the original guess was $\tau = 3/2$). Electrons which acquire more potential energy than E_T, i.e. which come within r_T of the nucleus, are unlikely to be knocked back into the continuum. Thomson further assumed that if two electrons come within the sphere $r < r_T$, one is likely to be stabilized. The rate

$$\gamma = K \langle v r_T{}^5 \rangle = 1.73\ K\ \tau^{-5}\ T^{-9/2} \tag{167}$$

where $K \sim 1$. A more precise derivation was developed by Keck[30] from Wigner's formulation of chemical transition state theory. He argued that the rate is exactly given by

$$\gamma = \int n\ v_n (1-Q) dS \tag{168}$$

where S is an eleven-dimensional hypersurface in the phase space of the two electrons, dividing the region of two free electrons from that of one bound electron; n is the number density, v_n the velocity normal to S and Q the probability that the trajectory crossing S at a given point later doubles back across S. An upper bound (the "variational estimate" γ_{Var}) is obtained if Q = 0. Thomson's criteria provide a good choice of surface

$$H_1 = -E_T \quad , \quad H_2 \geq -E_T \tag{169}$$

where H_1, H_2 are the total energies of the recombinant and spectator electrons. As so defined γ_{Var} diverges as $r_2 \to \infty$. However, <u>very distant elect</u>rons do not transfer energy on average† if the

†This is not a quantal condition. Mansbach and Keck[30] point out that the same criterion is required in stellar dynamics.

interaction time r_2/v_2 times the classical orbital frequency exceeds a number $\delta \sim 1$, i.e. unless

$$\frac{r_2}{r_T} < \delta\tau^{1/2}. \tag{170}$$

The resulting

$$\gamma_{Var} = 0.470\ T^{1/2}\ r_T^5\ f(\tau),\ f(\tau) \simeq \delta^2\ \tau e^{\tau}(1+8e^{3\tau/5}) \tag{171}$$

where $\tau^{-5} f(\tau)$ has a minimum value $11.5\delta^2$ at $\tau = 2.52$. Thus the best estimate is

$$\gamma_{opt} = 5.40\delta^2\ T^{-9/2} = \frac{(2.44 \times 10^{-8}\ cm^3\ sec^{-1})\delta^2}{(T\ deg\ K)^{9/2}}. \tag{172}$$

The optimum surface is often called the "bottleneck".

To obtain a better estimate, Mansbach and Keck sampled trajectories starting on S to see what fraction doubled back. With a choice of $\delta = 1/2$, this fraction $Q \simeq 0.36$ so that

$$\gamma_{MCarlo} = 0.87\ T^{-9/2} = \frac{2.0 \times 10^{-8}\ cm^3 sec^{-1}}{(T\ deg\ K)^{9/2}} \tag{173}$$

which is close to (165). In Fig. 13 we plot β^{∞} (c.f. (159)) from Bates *et al.*[28,29] against temperature for comparison with γ_{MCarlo}. The agreement is remarkable. The rate equation approach selects the important physical processes, viz. those with fairly large energy transfers, which are not sensitive to the cutoff (170). Hence no such adjustable parameters are required.

4.3. Ionic Three-Body Recombination

We now consider processes of type (3) where x^+, y^- are ions and z a neutral,

$$X^+ + Y^- + Z \rightarrow [X^+Y^-] + Z \rightarrow XY+Z. \tag{174}$$

Harking back to the Thomson model, we suppose that X^+, Y^- must first approach within r_T and then one or the other has to collide with a Z to stabilize the complex $[X^+Y^-]$. The final step, whose details do not affect the rate, is usually that $[X^+Y^-]$ undergoes a curve crossing transition to a covalent state XY, either bound or

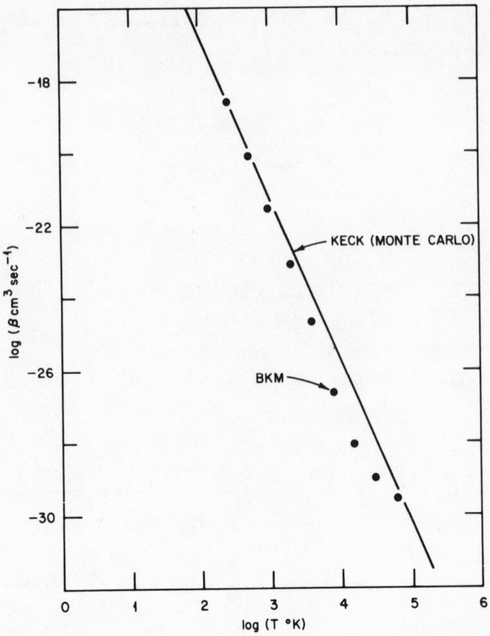

Fig. 13. β^{∞} calculated by Bates, Kingston, and McWhirter[28,29] compared with γ_{MCarlo} calculated by Mansbach and Keck[30] as functions of temperature.

dissociating. The crossing separation is typically ~ 5 a_0 corresponding to a Coulomb potential energy ~ 5 eV. The collision probability is determined by the X + Z, Y^- + Z diffusion cross sections, denoted by σ_X, σ_Y. If Z has a dipole polarizability P_Z, the gas kinetic cross sections are closely given by the Langevin model

$$\sigma_X, \sigma_Y \simeq \pi \left(\frac{P_Z}{T}\right)^{1/2} . \tag{175}$$

A typical collision radius ~ 10 a_0 << r_T ~ 10^3 a_0. The probability of a collision within the Thomson sphere is[31,32]

$$Q = \frac{4}{3} r_T \, (L_X^{-1} + L_Y^{-1}), \; L_X^{-1} = n(Z)\sigma_X, \text{ etc.} \tag{176}$$

where L_X,L_Y are mean free paths for X,Y in Z. The ternery recombination rate is

$$\beta = \frac{\pi r_T^2}{n(Z)} \langle vQ \rangle = \frac{4\pi}{(3M_{XY})^{1/2}} \frac{(\sigma_X + \sigma_Y)}{\tau^3 \; T^{5/2}} \tag{177}$$

where M_{XY} is a reduced mass; r_T and τ were defined by (166). From (175), $\beta \sim T^{-3}$. Although the steep dependence on τ is disturbing, (177) is easy to apply.

The approach of Bates and his collaborators[31,32] is analogous to that used in the electron problem. Rate equations similar to (152) are set up (radiative processes are insignificant here) and solved. The Rydberg states p are now so close together that they might as well be treated as continuous. We have to introduce three semi-arbitrary negative energies:

(i) E_o (> T) is analogous to p_{max} in defining the edge of the continuum. Higher levels are in Saha equilibrium.

(ii) E_d (< T) is the level above which populations are stationary, analogously to p = 2 earlier.

(iii) E_s (< E_d) is the stabilization level at which X^+,Y^- neutralize by charge transfer. As stated above, $-E_s \sim 5$ eV.

The equations analogous to (154) are

$$n_i \int_{E_d}^{\infty} C_{if}\, dE_f = \int_{E_s}^{\infty} n_f C_{if} dE_f \tag{178}$$

where the kernel $C_{if} = C(E_i \rightarrow E_f)$ gives the rate of energy transfer to $X^+ + Y^-$ by collisions with Z; as before C_{if} is calculated in the binary encounter model. Once (178) has been solved for n(E), E
$E > E_d$, the recombination rate

$$\beta = \left\langle \int_{E_d}^{E_o} (n_i - n_f) C_{if}\, dE_f \right\rangle \tag{179}$$

where the triangular brackets denote a Maxwellian average over E_i. Rate coefficients have been extensively tabulated as functions of the mass ratios and ion-neutral cross sections.[31,32]

Bates and Flannery have assessed the validity of the Thomson model.[32] In Fig. 14 we show their calculation of the probability $P(\tau)$ that a system which reaches $E = -\tau T$ does *not* return to the continuum. As $\tau \rightarrow 0$, $P \rightarrow 0$ rapidly, but $P \rightarrow 1$ when $\tau > 2$. For equal masses (177) is very accurate if $\tau = 1.89$. The results of calculations by Bates and Flannery[32] on $O_4^+ + O_2^-$, $O_4^+ + O_4^- + O_2$ are compared in Fig. 15 with recent experiments[33] at densities below

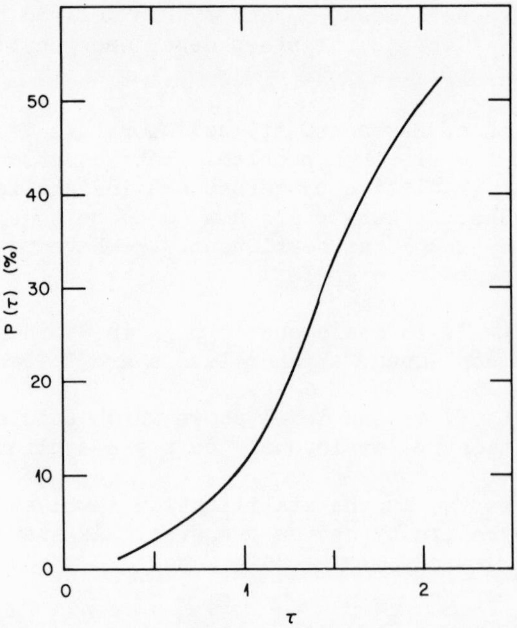

Fig. 14. Probability that a system which reaches $-\tau T$ does not return to the continuum.[32] Parameters (c.f. (177)) are $T = 400$ deg K, $\sigma_X = \sigma_Y = 4 \times 10^{-14}$ cm^2, $M_X = M_Y = 1.17\ M_Z$.

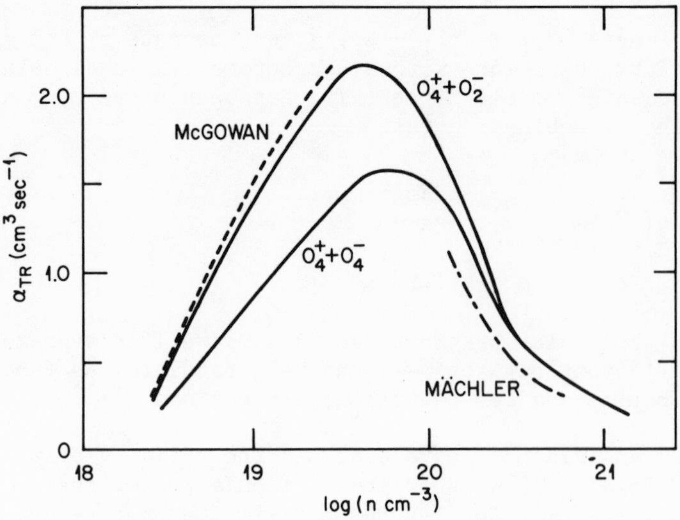

Fig. 15. Recombination in oxygen over a wide range of densities. At $n < 3 \times 10^{19}$ cm^{-3} the Bates-Flannery theory[32] for two mechanisms is compared with the measurements of McGowan.[33] At higher densities the transition to Langevin theory is described by a semi-empirical prescription and compared with the measurements of Mächler on air.[34]

$n = 3 \times 10^{19}$ cm^3. The former mechanism is theoretically favored.

At densities $n > 3 \times 10^{19}$ cm^{-3} the assumption of ternery collisions breaks down. At very high densities $n > 3 \times 10^{20}$ cm^{-3} recombination is described by a theory due to Langevin. He considered the ions diffusing towards each other under the influence of their Coulomb attraction. If the ionic mobilities of X^+,Y^- in Z are K_+,K_-, the mean velocity of approach at a distance r is

$$v_d = \frac{Ke}{r^2}, \quad K = K_+ + K_- \text{ (c.g.s. units).} \tag{180}$$

Then the effective two-body recombination coefficient is

$$\alpha_{TR} = 4\pi r^2 v_d = 4\pi Ke \qquad \text{(c.g.s. units).} \tag{181}$$

Thus $\alpha_{TR} \sim [n(Z)(\sigma_X + \sigma_Y)]^{-1}$ instead of $[\ldots]^{+1}$ as at low densities. Many semi-empirical prescriptions have been suggested for joining (177) to (181) at intermediate densities, none very satisfactory. In Fig. 15 such a theory has been used at high densities, i.e. beyond the linear régime,and compared with very old experiments on air.[34] A true unified theory, valid at all densities, would be a considerable achievement.

References

1. D. R. BATES and A. DALGARNO (1962) in "Atomic and Molecular Processes" (ed. D. R. Bates) p. 245, Academic Press, New York.

2. H. A. BETHE and E. E. SALPETER (1957) "Quantum Mechanics of One- and Two-electron Atoms", Springer, Berlin.

3. L. D. LANDAU and E. M. LIFSHITZ (1971) "Classical Theory of Fields", p. 181, Pergamon Press, New York.

4. M. ABRAMOWITZ and I. A. STEGUN (1964) "Handbook of Mathematical Functions", NBS, Washington.

5. G. N. WATSON (1966) "Treatise on the Theory of Bessel Functions", University Press, Cambridge.

6. D. H. MENZEL and C. L. PEKERIS (1935) MNRAS 96, 77.

7. R. W. DITCHBURN and U. OPIK (1962) in "Atomic and Molecular Processes" (ed. D. R. Bates) p. 79, Academic Press, New York.

8. L. SPITZER (1968) "Diffuse Matter in Space", p. 117, Interscience, New York.

9. A. BURGESS (1958) MNRAS 118, 477.

10. D. R. BATES (1961) "Quantum Theory I. Elements", Academic Press, New York; N. F. MOTT and H. S. W. MASSEY (1966) "Theory of Atomic Collisions", University Press, Oxford.

11. A. BURGESS (1965) 141, 1588; (1964) 139, 776; (1960) Ap. J. 132, 503.

12. B. SHORE (1969) Ap. J. 158, 1205.

13. R. GAYET, D. HOANG BINH, F. JOLY, and R. McCARROLL (1969) Astron. and Astrophys. 1, 365.

14. M. J. SEATON and P. J. STOREY (1976) in "Atomic Processes and Applications", p. 133 (ed. P. G. Burke and B. L. Moiseiwitsch) North-Holland, Amsterdam.

15. M. SUTER, C. R. VANE, S. B. ELSTON, G. D. ALTON, P. M. GRIFFIN, R. S. THOE, L. WILLIAMS, and I. A. SELLIN (1979) Z. f. Physik A289, 433.

16. G. A. MARTIN and W. L. WIESE (1976) J. Phys. Chem. Ref. Data 5, 540.

17. J. WEISHEIT (1975) J. Phys. B 8, 2556.

18. J. DUBAU (1973) Ph.D. Thesis, University of London.

19. C. BOTTCHER (1976) J. Phys. B 9, 2899; (1974) Proc. Roy. Soc. A340, 301.

20. J. N. BARDSLEY and M. A. BIONDI (1970) Adv. Atom. Molec. Phys. 6, 1.

21. D. AUERBACH, R. CACAK, R. CAUDANO, T. D. GAILY, C. J. KEYSER, J. Wm. McGOWAN, J. B. A. MITCHELL, and S. F. J. WILK (1977) J. Phys. B 10, 3797.

22. B. PEART and K. T. DOLDER (1973) J. Phys. B 6, 2409; (1973) 6, L359; (1974) 7, 236.

23. R. A. PHANEUF, D. H. CRANDALL, and G. H. DUNN (1975) Phys. Rev. A11, 528.

24. C. BOTTCHER (1978) J. Phys. B 11, 3887; B. M. SMIRNOV (1977) Sov. Phys. - Usp. 20, 119.

25. C. M. HUANG, M. WHITAKER, M. A. BIONDI, and R. JOHNSEN (1978) Phys. Rev. A18, 64; C. M. HUANG, M. A. BIONDI, and R. JOHNSEN (1976) Phys. Rev. 14, 984; M. T. LEU, M. A. BIONDI, and R. JOHNSEN (1973) Phys. Rev. A7, 292; A8, 413.

26. K. T. DOLDER and B. PEART (1976) Rep. Prog. Phys. 39, 693.

27. J. Wm. McGOWAN, P. M. MUL, V. S. D'ANGELO, J. B. A. MITCHELL, P. DEFRANCE, and H. R. FROELICH (1979) Phys. Rev. Letts. 42, 373; J. B. A. MITCHELL and J. Wm. McGOWAN (1978) Ap. J. Letts. 222, 77; R. D. DUBOIS, J. B. JEFFRIES, and G. H. DUNN (1978) Phys. Rev. A17, 1314; F. L. WALLS and G. H. DUNN (1974) J. Geophys. Res. 79, 1911.

28. D. R. BATES, A. E. KINGSTON, and R. P. W. McWHIRTER (1962) Proc. Roy. Soc. A267, 297.

29. E. W. McDANIEL (1964) "Collision Phenomena in Ionized Gases", John Wiley, New York.

30. J. C. KECK (1972) Adv. At. Molec. Phys. 8, 39; P. MANSBACH and J. C. KECK (1969) Phys. Rev. 181, 275; B. MAKIN and J. C. KECK (1964) in "Atomic Collision Processes", p. 510 (ed. M. R. C. McDowell) North-Holland, Amsterdam.

31. M. R. FLANNERY (1972) Case Studies in Atomic Collision Physics 2, 1.

32. M. R. FLANNERY (1976) in "Atomic Processes and Applications", p. 407 (ed. P. G. Burke and B. L. Moiseiwitsch) Academic Press, New York; D. R. BATES and R. J. MOFFETT (1966) Proc. Roy. Soc. A291, 1.

33. S. McGOWAN (1967) Can. J. Phys. 45, 439.

34. W. MÄCHLER (1936) Z. Phys. 104, 1.

SOME EXPERIMENTAL ASPECTS OF INELASTIC ELECTRON-ATOM COLLISIONS AND COLLISIONS BETWEEN CHARGED PARTICLES

K.T. Dolder

Professor of Atomic Physics

The University, Newcastle upon Tyne NE1 7RU, England

1. INTRODUCTION

This is a very broad field so we must be very selective. The discussion will therefore be restricted to experiments with particular fundamental interest or relevance to fusion. Recent developments will be emphasized and papers or reviews will be cited from which the various topics can be explored more deeply.

Sections 2 and 3 will be devoted to the ionization and excitation of neutral atoms by electron impact whilst Section 4 will discuss inelastic collisions between charged particles.

2. MEASUREMENTS OF THE IONIZATION OF ATOMS BY ELECTRON IMPACT

Seven topics will be considered:

(a) The condenser method for measuring ionization cross sections.

(b) Experiments with thermal beams.

(c) Experiments with fast neutral beams.

(d) (e, 2e) Experiments.

(e) Measurements of near threshold ionization.

(f) Experiments with polarized electron and atom beams.

(g) Inner shell ionization and autoionization.

2.1 The Condenser Method

If we attempt to determine cross sections for the ionization of a neutral atom (A),

$$e + A \rightarrow A^{+} + e + e \tag{1}$$

by measuring the ion current produced, a complication arises because the projectile may eject more than one electron, e.g.

$$e + A \rightarrow A^{++} + e + e + e. \tag{2}$$

It is therefore usual to define an apparent ionization cross section,

$$Q_i^a = Q_i^{(1)} + 2Q_i^{(2)} + 3Q_i^{(3)} + \ldots \tag{3}$$

where $Q_i^{(n)}$ represents the partial cross section for n-fold ionization. Experimentalists were initially content to measure Q_i^a, but techniques to determine partial cross sections were soon developed.

The early measurements (e.g. Compton and van Voorhis (1)) encountered severe difficulties but it was Smith and his collaborators (e.g. Tate and Smith (2)) who built on this experience and obtained results which in some cases compare well with contemporary measurements.

It is instructive to look briefly at Smith's apparatus and discuss two experimental problems which still concern us. Electrons from the cathode (K) were accelerated and collimated through two apertures (S) as illustrated by Figure 1. The beam passed between two plates (P_1 and P_2) to a screened, positively biased collector (C). The whole apparatus was filled with gas and a potential gradient of 5 V cm^{-1} was maintained between P_1 and P_2 so that positive ions, formed by collisions between electrons and the gas, were collected at P_1. An axial magnetic field prevented deflection of the electron beam and suppressed secondary emission from P_1. From a knowledge of the ion and electron currents, the gas density and the dimensions of P_1, the cross section could be deduced.

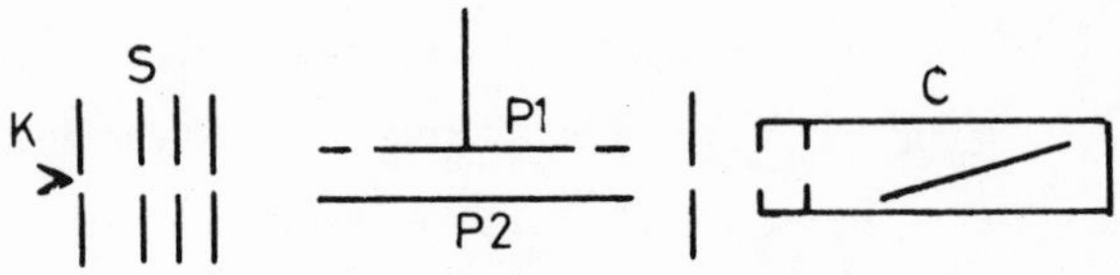

Figure 1. Apparatus used by Tate and Smith to measure total ionization cross sections.

Figure 2 compares these early results for He with more recent measurements. The agreement is good but results for heavier gases (e.g. Ne and Ar) show Smith's results to be 10 - 20% larger than modern values. It is interesting to speculate why this is so.

McLeod gauges were usually used to measure low pressures and a cold trap was interposed between the system and gauge to prevent mercury contamination. But some mercury vapour could stream from the gauge to the trap to form a primitive diffusion pump which maintained a pressure differential between the gauge and the system. The effect is most marked for heavier gases which are less able to back-stream against the vapour. This has become known as the "Ishii effect" and measurements by Schram et al (3) showed errors of 1% for H_2 and 14% for Xe, although the magnitude of errors will depend upon apparatus geometry. This would account for the disparity between Smith's work and more recent experiments. Nowadays many experimentalists use gauges which depend upon the deformation of a membrane by gas pressure.

A second point of experimental interest is the spiralling of electron beam trajectories around the magnetic field lines.

If electrons pass freely through the apparatus it follows that the spiral diameter cannot exceed the diameter (d) of the beam defining apertures. The maximum length of the electron path (ℓ_{max})

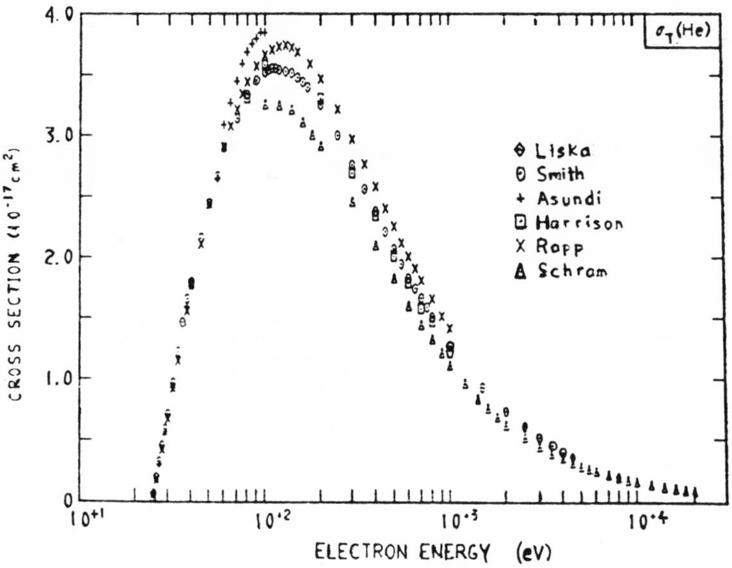

Figure 2. Tate and Smith's measurements for the total ionization of He compared with more recent results.

is therefore related to the distance between apertures (ℓ) by,

$$\ell_{max} \leqslant \ell(1 + 1.1 \times 10^{-4} \frac{d^2H^2}{V}) \qquad (4)$$

where H represents the magnetic field (gauss) and V is the electron energy (eV).

Asundi (4) pointed out that unrealistically large transverse energies would be needed for the spiral diameter to approach d, so equation (3) is likely to overestimate the effect grossly. Taylor et al (5) performed a systematic investigation of spiralling in an electron gun used for crossed beam experiments. They found that, by a judicious choice of electrode potentials (to minimise transverse energy components), the correction for spiralling was only $4 \pm 2\%$ at 3 eV and fell to $0.25 \pm 0.2\%$ at 200 eV. This consideration arises in a variety of experiments but it need not be serious.

Many developments have been made to the condenser technique. Bleakney (6) introduced magnetic analysis of the product ions to permit the measurement of partial cross sections $Q_i^{(1)}$, etc. The extensive series of measurements by Rapp and Englander-Golden (7) deserves mention. These topics were reviewed by Massey et al (8,9) and a compilation of data and two bibliographies have been prepared by Kieffer (10), McDaniel (11), and Kieffer and Dunn (12) respectively. This information, in common with all other topics we shall discuss, is continuously updated in the ORNL bimonthly bulletins on "Atomic data for fusion", edited by Barnett and Wiese.

2.2 Experiments with Thermal Beams

A major advance was the introduction of crossed beam experiments. This enabled measurements to be made with atoms or molecular fragments which do not exist in free states at room temperature.

A very early experiment was performed by Funk (13) who directed electrons at an atomic beam of sodium in an attempt to measure cross sections for,

$$e + Na \rightarrow Na^+ + e + e. \qquad (5)$$

The experiment failed partly because of interaction between the atomic beam and residual gas in the apparatus. A solution of this problem was demonstrated by Boyd and Green (14) who mechanically chopped atomic beams and detected ions formed only at the appropriate frequency. The development and exploitation of this technique was reviewed by Fite (15). An experiment which is particularly relevant to fusion was the measurement of cross sections for,

$$e + H \rightarrow H^{+} + e + e \qquad (6)$$

by Fite and Brackmann (16). Notable experiments with alkali atoms were performed by McFarland and Kinney (17).

2.3 Experiments with Fast Neutral Beams

Free hydrogen or alkali atoms can easily be prepared by thermal dissociation, but other atoms which are found as impurities in thermonuclear plasmas cannot be formed in this way. Examples include C, N and O.

An alternative to the thermal beam approach was therefore suggested by Peterson (18) who prepared fast N beams by neutralizing 2 keV N^+ ions in an N_2 gas target. The ensuing atomic beam was bombarded with electrons to study,

$$e + N \rightarrow N^{+} + e + e. \qquad (7)$$

Apart from providing a fully dissociated nitrogen beam, the method enjoys another advantage. The N atoms are so energetic that their flux can be measured and absolute cross sections obtained without recourse to normalization.

Peterson did not entirely overcome the considerable experimental difficulties but the method was revived by Brook et al (19) who used it to obtain absolute ionization cross sections for He, C, N and O. The result for He is particularly interesting because it enables the fast beam technique to be compared with established methods.

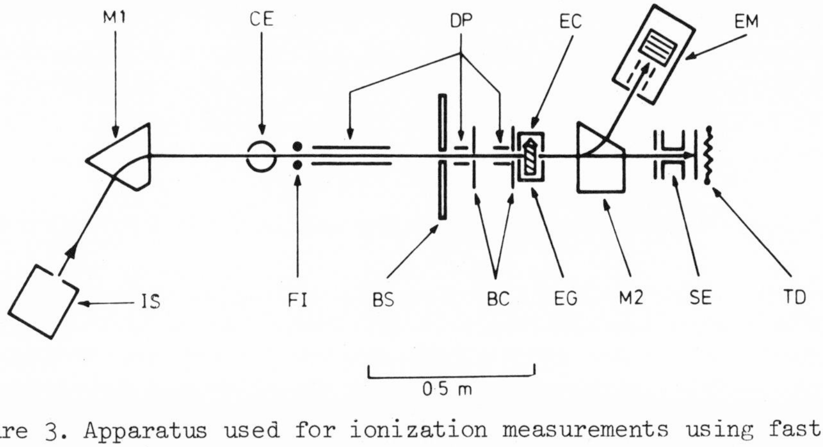

Figure 3. Apparatus used for ionization measurements using fast atomic beams.

The apparatus of Brook et al is sketched in Figure 3. Ions from the source (IS) were selected in the field of a magnet (M1). They then passed through a charge exchange cell (CE) filled with a gas which was chosen to neutralize the ions with the minimum energy defect (χ). This preferentially formed the neutral atoms in their ground state. For example, He^+ was neutralized in He ($\chi = 0$) whilst O^+ was neutralized by Kr ($\chi = 0.38$ eV).

Any atoms in low quantum states ($n \lesssim 8$) could decay before reaching the electron gun (EG) whilst those in higher states ($n \gtrsim 13$) were field ionized (FI) and swept from the beam in the electric field between the deflector plates, DP.

Ions formed by electron bombardment of the atomic beam were individually detected by a particle multiplier (EM) whilst the flux of the neutral beam was measured with a thermopile (TD). Beam modulation techniques were used to separate the required ion signal from backgrounds.

Figure 4 illustrates results for atomic oxygen and compares them with thermal beam measurements by Fite and Brackmann (20), and theoretical predictions by Peach, Omidvar, McGuire and Seaton.

The method has three important attractions:

(a) The results are absolute, without normalization.

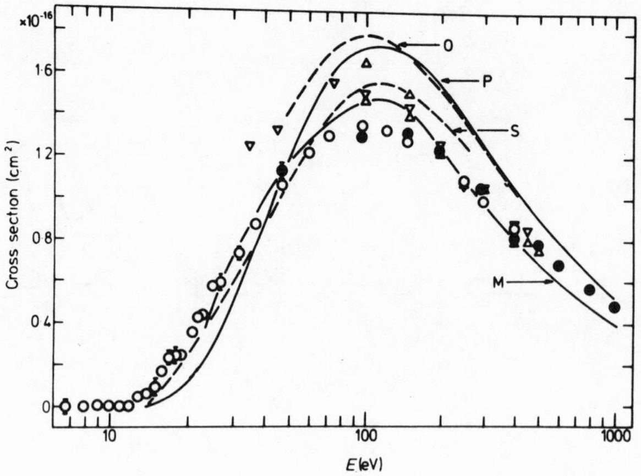

Figure 4. Cross sections for the ionization of atomic oxygen. Measurements by Brook et al (● O) compared with results of thermal beam experiments by Fite & Brackmann (∇) and Rothe (Δ). Theoretical predictions by Peach (P), Omidvar (O), McGuire (M) and Seaton (S) are also illustrated.

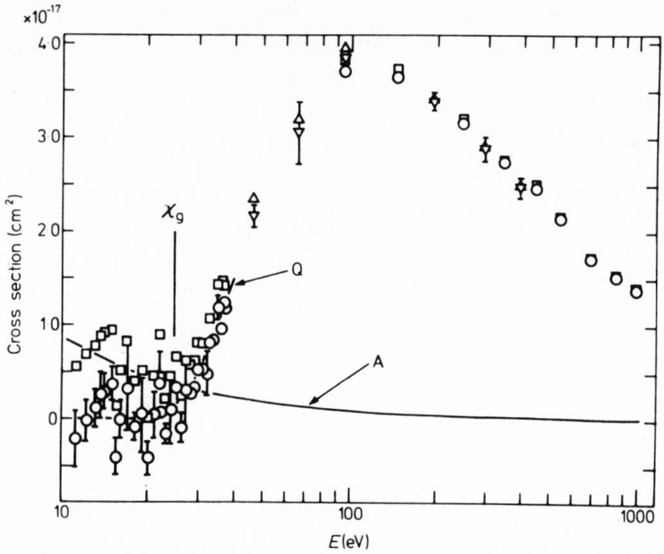

Figure 5. Measurements of the ionization of helium in a fast beam experiment showing non zero apparent cross sections below threshold.

(b) The results are likely to have the accuracy associated with fast crossed beam experiments.

(c) It provides fully dissociated beams of atoms or molecular fragments which cannot, in general, be obtained by thermal dissociation.

But one aspect of these experiments remains to be resolved. Figure 5 shows the results for He and it is seen that the measured cross section was not zero below threshold. This was attributed to the presence of the beam of a small proportion of He atoms in $8 \lesssim n \lesssim 12$ states. A correction was therefore devised and applied to the above-threshold measurements but, when this was done, the results at energies below 70 eV did not agree with the static gas measurements by Rapp and Englander-Golden. de Heer (21) suggested that the two results can be fully reconciled if quite small changes are made to the below-threshold correction and the energy scale of the fast beam measurements. Harrison et al (104) have replied.

Fast beam techniques ave also been adapted to study ionization of metastable atoms. Ionization in plasmas often proceeds via intermediate steps. One collision may excite an atom and a successive collision ionizes it. The lifetimes of excited states are usually so short that it is difficult to devise experiments to study the ionization of excited atoms but the problem becomes simpler if metastable atoms are used and the group at Culham has succeeded in measuring ionization cross sections of metastable atoms with fast beam techniques.

Dixon et al (22) directed protons through a Cs charge exchange cell to form a beam of hydrogen atoms. Cs was chosen because the reaction,

$$H^+ + Cs \rightarrow H(2s) + Cs^+ \quad (8)$$

is nearly resonant ($\chi = 0.51$ eV). The neutral beam was then bombarded by electrons and the flux of protons produced was measured. Allowance was made for the ionization of unexcited hydrogen and the cross section for ionization of H(2s) was deduced.

The fraction (f) of protons converted to H(2s) by single collisions in a Cs cell had been measured by three other groups who obtained $f = 0.25 \pm 0.01$. This confirms the expectation that 2P and 2S states are populated in proportion to their statistical weights, i.e. 3:1.

Dixon et al (23,24) subsequently extended the technique to metastable He (predominantly in the 2^3S state), Ne($^3P_2$, $^3P_2$) and Ar($^3P_2$, $^3P_0$).

The apparatus used in these experiments is sketched in Figure 6 whilst Figure 7 illustrates results obtained for metastable He and compares them with theory and earlier experiments.

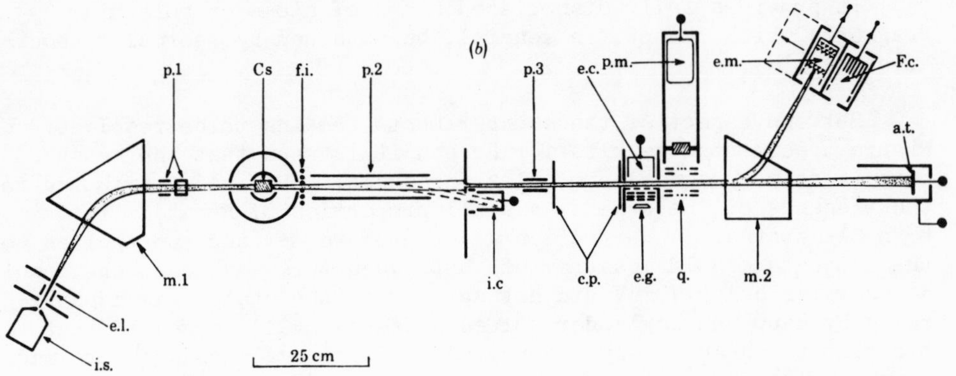

Figure 6. Apparatus used to measure the ionization of H(2s) atoms. Protons from a source (i.s.) were neutralized in a Cs cell (Cs). Highly excited atoms were field ionized (f.i.) and ions were deflected from the beam by an electric field between plates, p.2. Electrons from the gun (e.g.) ionized H(2s) atoms and the protons formed were detected at F.c. The H(2s) component in the beam was monitored using quench plates (q) and a photomultiplier (p.m.).

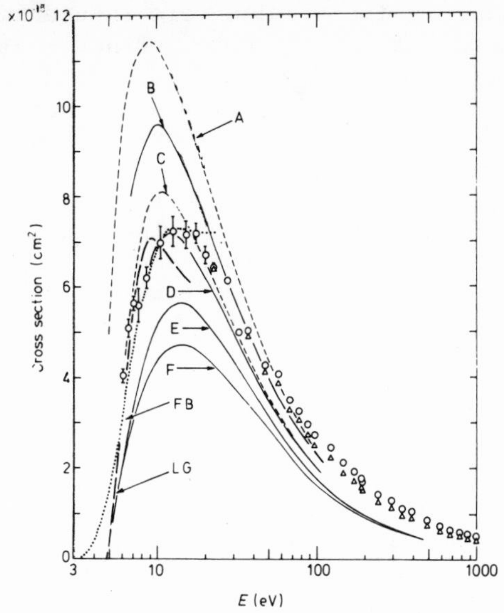

Figure 7. Measured cross sections for ionization of metastable He (O Δ) compared with results of thermal beam experiments (⋯) and various theoretical predictions.

Metastable He had previously been studied with thermal beam techniques by Fite and Brackmann (25), Long and Geballe (26) and others but in these experiments only a small fraction of the neutral

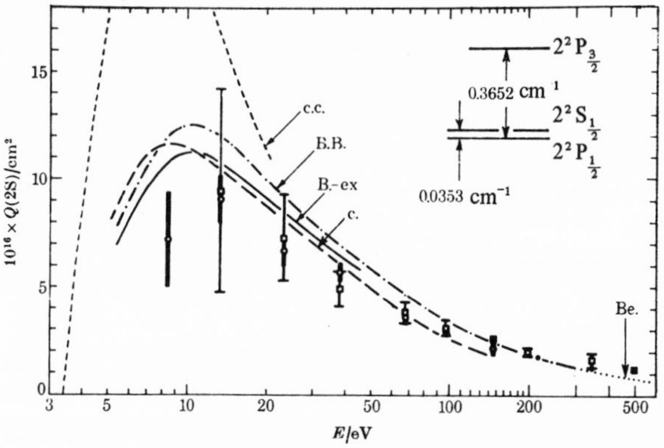

Figure 8. Measured cross sections for the ionization of H(2s) compared with predictions of close coupling (c.c.), Born (BB), Born exchange (B-ex) and classical (c) approximations.

beam was excited and so the results were inaccurate above the threshold for the ionization of unexcited He.

The results for H(2s) are particularly interesting. Classical scaling suggests that the ratio of cross sections for H(1s) and H(2s) should be 16 when the incident electron energy is expressed in units of the relevant threshold energy. The measured ratio was between 10 and 15. Moreover, although the measurements were in fair agreement with the Born and Born-Exchange approximations, they deviated very greatly from the close coupling calculations of Burke and Taylor (27). This is illustrated by Figure 8.

2.4 (e, 2e) Experiments

Modern measurements of ionization cross sections are frequently accurate to better than ±10% but this is insufficient to discriminate between rival theoretical methods which are being developed. A vastly more searching and detailed test of theory is provided by "(e, 2e) experiments". This technique measures the probability that an incident electron with energy E_o produces two final state electrons with energies E_a and E_b directed at angles Θ_a and Θ_b into differential elements of solid angle Ω_a and Ω_b.

Much of the early work was pioneered and reviewed by Ehrhardt's group (28) and a more recent review was given by Paul et al (29).

Considerable attention has been paid to the ionization of He and a recent example is the work of Schubert et al (30). Beaty et al (31) extended the technique to collisions in which the incident and scattered electrons were not coplanar whilst Dixon et al (32) described experiments in which the He^+ ion was left in the $n = 2$ state. The latter experiment strikingly demonstrated the need for theory to use accurate, correlated wavefunctions.

An apparatus used by Ehrhardt's group is illustrated by Figure 9. It shows a 127° electrostatic monochromator (s) with its associated lenses and accelerator. The electron beam is collected at FC but scattered electrons are detected in coincidence by particle multipliers M1 and M2 after they have passed through energy filters (A1 and A2). Both detectors can be rotated to vary the scattering angles Θ_a and Θ_b.

The wealth of detail which these experiments produced revealed severe inadequacies in Born's approximation and other theories of inelastic scattering even by such a simple atom as He.

Improvements in theory have ensued and an example is the recent results of Balashov et al (33) for Ne where accurate account was taken of the distortion of ejected-electron wavefunction. Figure 10

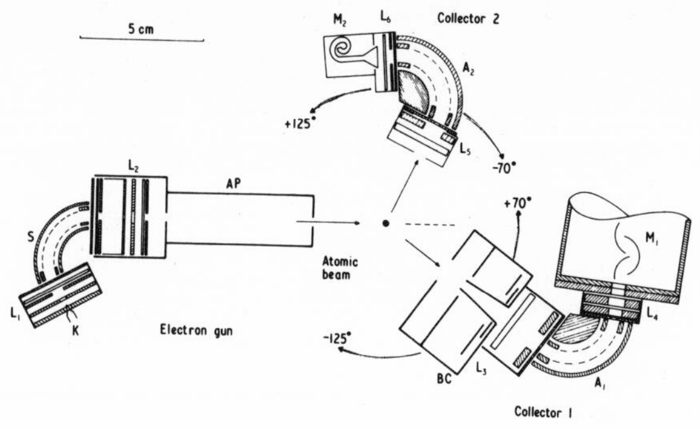

Figure 9. Apparatus used by Ehrhardt's group for the coincident detection of two electrons emerging from an atomic beam after ionization.

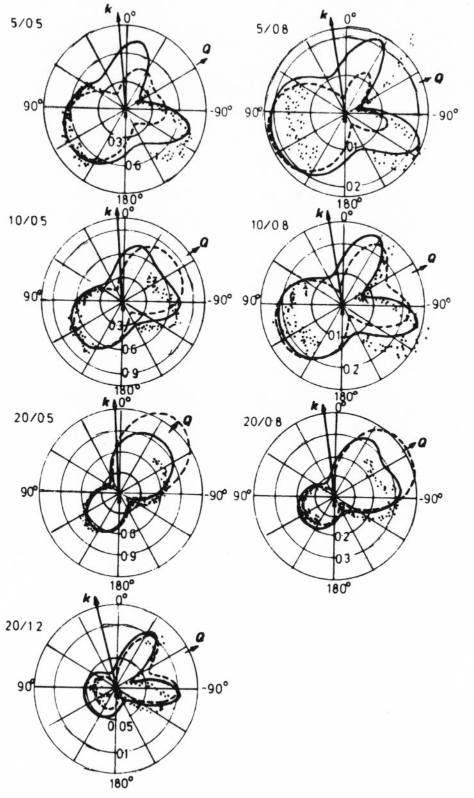

Figure 10. Measured triple differential cross sections for Ne compared with theory. The full curve uses Herman-Skillman wavefunctions for the ejected electron.

illustrates a comparison of this theory with measurements by Ehrhardt's group.

A significant recent advance has been the application of these techniques to atomic hydrogen by Weigold et al (34). This is a difficult experiment and it would be very advantageous if the accuracy of the measurements could be further improved. The papers by Schubert et al (30), Weigold et al (34) and Lal et al (35) are useful starting points from which to explore the literature.

2.5 Measurements of Near Threshold Ionization

The near threshold region of ionization and excitation functions is usually of greatest relevance to plasma physics. It is also very interesting to theorists because strong correlations exist between the two, slowly-moving, escaping electrons. A convenient summary of experimental and theoretical aspects was presented by Read (1976). We merely point out that theory predicts a relation of the form,

$$Q_i^{(1)} \propto E^n \tag{9}$$

between the ionization cross section and energy (E) which the projectile electron possesses in excess of the ionization threshold.

The value $n = 1.127$ has been obtained classically by Wannier (36) and semiclassically by Peterkop (37) although quantum treatment by Temkin's group (38) and Kang and Kerch (39) give values of n closer to unity. None of the theories predict the energy range over which the law is valid.

It is extremely difficult for experimentalists to test this law because cross sections are very small near threshold. Moreover, the measured currents are not very sensitive to variations of the exponent, n.

Nevertheless, experiments indicate that, for He, $n = 1.31 \pm 0.019$ when $E \lesssim 1.5$ eV. Considerable ingenuity was needed to achieve this result and it involves some appeal to theory.

Theory predicts energy distributions $P(E_a)$ and $P(E_b)$ of the two escaping electrons and the distribution of angles between them, $P(\Theta_{ab})$. Quantum and classical theories agree that $P(\Theta_{ab})$ is a maximum when $\Theta = 180^\circ$ and that the width of this distribution is proportional to $E^{0.25}$. They also predict that $P(E_a)$ is independent of E_a and is constant over all possible values in the range $0 < E_a < E$.

This provided the basis for an experiment by Cvejanovic and Read (40) who used the apparatus illustrated by Figure 11 in which an electron beam (not shown) collided with an atomic He beam at the

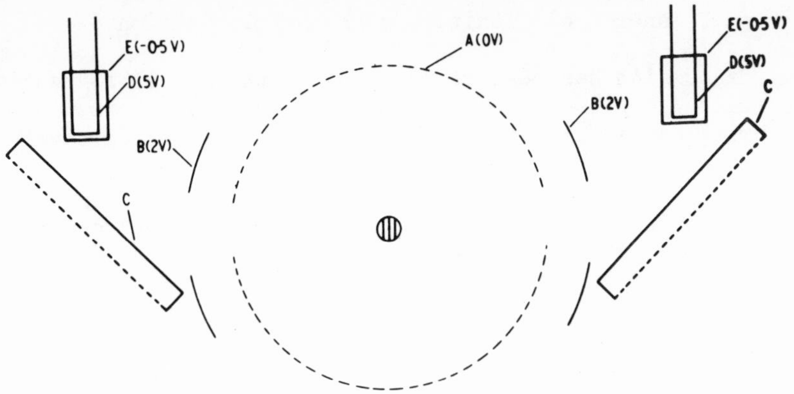

Figure 11. Apparatus used to investigate threshold ionization.

centre of a cylindrical cage. Electrons escaping at 180° were deflected by energy selectors (c) and detected individually in delayed coincidence.

Theoretical predictions of the energy distribution and angular correlation of these electrons was verified and a second experiment was performed to determine the value of n. Only one electron analyser was used and it was tuned to have high transmission efficiency for very slow electrons. The effect of field penetration from electrode B through apertures in cage A also discriminated further against faster electrons and the system was effectively a narrow band filter with a bandwidth (ΔE) of about 15 meV. Since theory and experiment agree that $P(E_a)$ is independent of E_a, it follows that the current of electrons detected will be proportional to $Q_i^{(1)} \cdot \frac{\Delta E}{E}$.

The method is therefore very ingenious because $\Delta E/E$ is proportional to E^{n-1} and a measurement of n-1 leads to a much smaller error in n than would be obtained from a direct measurement. Figure 12 shows the variation with incident energy of the yield of slow electrons from He. Above threshold ($\simeq$24.5 eV) a curve has been drawn corresponding to $E^{0.127}$ as predicted by Wannier's law.

2.6 Experiments with Polarised Electron and Atom Beams

In an effort to extend our understanding of ionization, Hils and Kleinpoppen (41) performed an experiment in which a beam of polarised electrons bombarded a beam of polarised potassium atoms.

The experimental errors were large and one might ask whether the errors of even a refined experiment might not exceed the magnitude of effects due to polarization. Presumably this consideration will determine the usefulness of the technique.

2.7 Inner Shell Ionization and Autoionization

Our discussion has been restricted to outer shell ionization but two related topics are of considerable current interest and deserve brief mention. Both involve strong electron correlation effects.

An energetic projectile or photon may either eject or excite an inner shell atomic electron. In the latter case the excited electron may decay with the emission of radiation or by transferring its energy and ejecting a more weakly bound electron. This is called autoionization. A convenient starting point from which to explore the literature is the paper by Tronc et al (42).

A development which will prove important over the next few years will be the extension of (e, 2e) techniques for inner shell ionization. The theory of the method has been discussed by Berezhko et al (43).

Particular attention has been paid to the study of autoionization thresholds. Just above threshold two electrons move slowly away from the residual ion and correlation effects are strong. The magnitude of this "post collision interaction" (PCI) can easily be estimated. If, for example, a projectile electron recedes with an energy of 1 eV from an atom in an autoionizing state with lifetime 5×10^{-15} s, it would travel only 3×10^{-9} m before the faster electron was ejected. The potential between the two electrons would

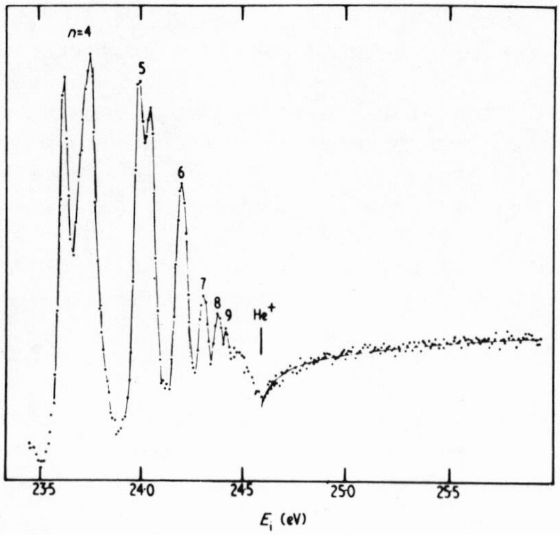

Figure 12. Variation with incident electron energy of slow electrons scattered from He. The smooth curve corresponding to Wannier's law fits the points above the ionization threshold.

be about 0.5 eV and so the energy of the scattered electron would be enhanced.

Hicks et al (44) measured the numbers and energies of electrons ejected at 70° from autoionizing states of He. Figure 13 illustrates their results. Each curve shows four peaks which correspond, from left to right, to the ejection of electrons from the $(2s^2)^1S$, $(2s2p)^3P$, $(2p^2)^1D$ and $(2s2p)^1P$ states. The numbers above each curve show by which the incident electron energy exceeded the autoionization threshold. It can be seen that the energetic position of the ^1S peak shows the expected increase as the projectile energy decreases, whilst the ^3P remain almost unchanged.

This is to be expected because the lifetime of the ^3P state is an order of magnitude longer than that of ^1S so the PCI will be correspondingly less.

An introductory review of this intriguing topic has been given by Read (45) whilst some very refined measurements of PCI effects in autoionization were recently reported by Baxter et al (46) who observed oscillatory structure in the ejected electron spectra of He.

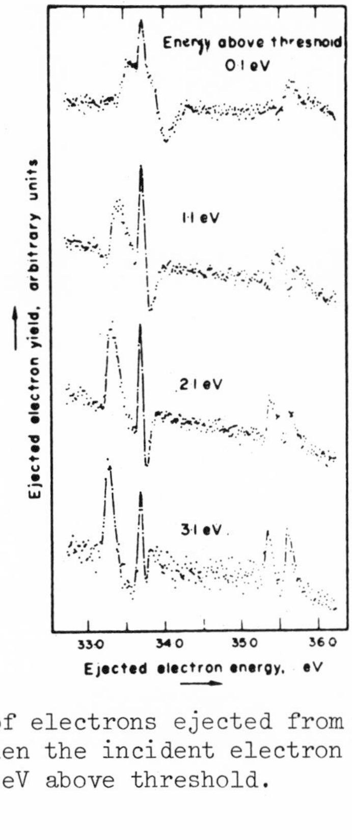

Figure 13. Energy of electrons ejected from four autoionizing states of He when the incident electron energy was 0.1, 1.1, 2.1 and 3.1 eV above threshold.

3. EXCITATION OF ATOMS BY ELECTRON IMPACT

This subject has been studied intensively for more than fifty years and has probably encountered more subtle experimental difficulties than any other branch of collision physics.

In principle it may seem easy to measure excitation cross sections. One merely directs an electron beam at a target of known density and measures the flux of monochromatic radiation produced. But the absolute determination of the radiation flux poses formidable problems and there are further difficulties associated with polarization of the radiation, radiation transfer between target atoms and resonance trapping. Inevitably our discussion must be almost cursory, but we will try to outline some of the experimental difficulties and take note of a technique for the calibration of optical detectors developed by Dunn's group, which sets a very exacting standard.

Our discussion will be confined to the following topics:

(a) Experiments with gas targets.
(b) Experiments with thermal beams.
(c) Calibration of optical detectors.
(d) Electron/photon coincidence experiments.

Relevant reviews, bibliography and tabulated data can be found in references cited at the end of §2.1 whilst Seaton (47) has reviewed the theory of excitation.

3.1 Experiments with Gas Targets

In many experiments an electron beam is directed at a static gas target or atomic beam. The intensity of monochromatic radiation emitted/unit solid angle in a direction θ, to the electron beam, is then measured. For dipole transitions this is related to the excitation cross section, Q_{jk} by

$$I(\theta) = \frac{Q_{jk}(1 - P\cos^2\theta)}{4\pi(1 - P/3)} \qquad (10)$$

The polarization,

$$P = \frac{I_{\parallel} - I_{\perp}}{I_{\parallel} + I_{\perp}} \qquad (11)$$

can be deduced from the radiated intensities parallel ($I_{\parallel}$) and perpendicular ($I_{\perp}$) to the electron beam.

A measurement of Q_{jk} therefore requires a knowledge of the target density and geometry, the electron current and the angular acceptance and efficiency of the radiation detector. Clout and Heddle (48) pointed out that it is not necessary to know the polarization if observations are made at the "magic angle", $\Theta = 54.5^{\circ}$. Then, $\cos^2 \Theta = 1/3$ and equation (10) reduces to,

$$I(\Theta) = \frac{Q_{jk}}{4\pi} \qquad (12)$$

The determination of target density follows the same procedures used in measurements of ionization but, when measuring excitation cross sections, the target density must be low (typically $<10^{12}\text{cm}^{-3}$). Even at moderate densities an excited atom may transfer its energy by collisions before it has time to radiate freely. A more stringent requirement arises when observing resonant radiation because this is very efficiently trapped within a gas, even at low density. Resonant trapping was extensively discussed by Heddle and Samuel (49).

There is insufficient time to discuss these matters in detail but we may have induced wariness in those who use measured cross sections. To underline the point we list results of six independent measurements of the maximum value of the $3^1S - 2^1S$ cross section in He. In units of 10^{-19} cm^2 the experiments gave $Q_{jk}^{max} = 5.3, 3.5, 41.3, 43.6$, and 36.6. These cross sections tended to shrink as the experimental difficulties became better understood.

The last decade has seen spurious effects become eliminated from the better measurements and optical calibration techniques improved. Concurrently, measurements were refined by employing energy resolved electron beams in computer controlled apparatus.

A further check of theory is provided by measuring the polarization of emitted radiation as a function of incident electron energy and provision for these measurements is usually included in excitation experiments. Early measurements obtained values for threshold polarization which were greatly at variance with theoretical predictions but this was largely due to the effects of gas collisions and limited energy resolution in the electron beam. The work on He by Heddle and Keesing (65) and MacFarland (66) overcame these difficulties. The results they obtained by data logging techniques are illustrated by Figure 14.

A potentially useful theoretical technique is being developed by Bransden, McDowell and others which should enable more reliable excitation cross sections to be defined. For example, Bransden and Dewangan (50) analysed experimental data for the excitation of 2^1S and 2^3S states of He in terms of partial waves. Phase shifts for lower order partial waves were than computed to fit best the results of five independent experiments. It was necessary to use theoretical values for the higher order phase shifts but their contribution was relatively small. This technique provides a sophisticated average of a large amount of experimental data and so the cross sections obtained are likely to be reliable. It also offers a probing comparison of the experiments.

3.2 Experiments with Thermal Atomic Beams

Many atoms of particular theoretical or technical interest do not occur in their free state at normal temperatures. Examples include atomic hydrogen and most metals. Free atomic beams can, however, often be obtained by vaporizing or dissociating material in a furnace. Thermal beams can then be prepared and used for targets in excitation experiments. Pioneering work in this field was performed by Fite and his colleagues (Fite and Brackmann (51); Stebbings et al (52)) who studied 1s-2s and 1s-2p excitation of atomic hydrogen. In both experiments a modulated beam of atomic hydrogen was bombarded by an electron beam. A notable feature was the ingenious detector of λ1216 Å photons. With the aid of this counter the excitation function for 1s-2p transitions were measured and the results were made absolute by normalising them to Born's approximation at high energies.

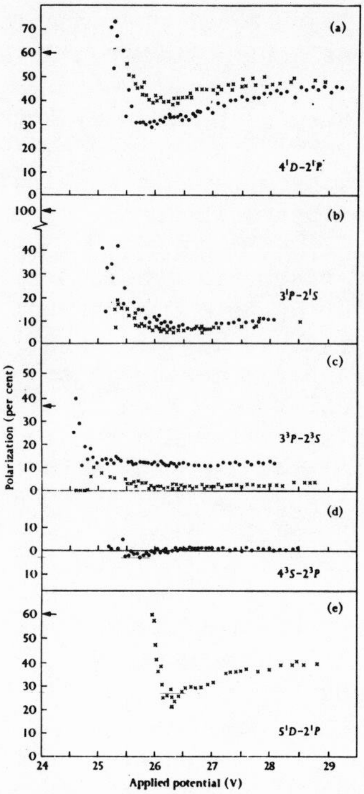

Figure 14. Near threshold measurements of the polarization of four He lines by Heddle and Keesing (●) and MacFarland (x).

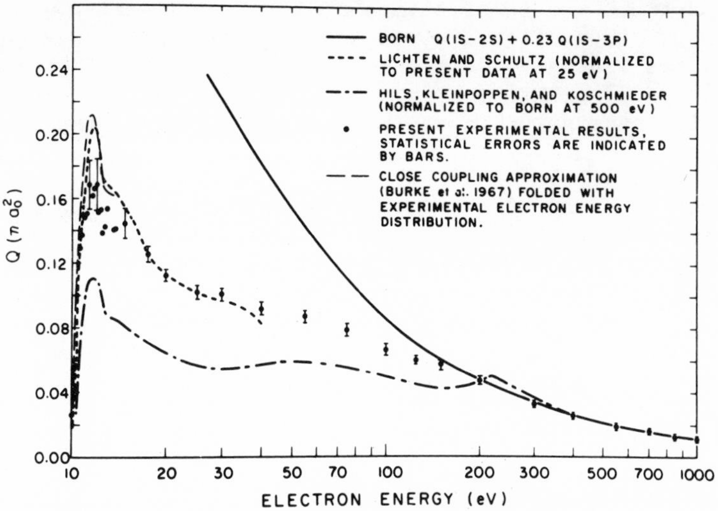

Figure 15. Measurements of total cross sections for the population of H(2s) compared with previous experiments and theory.

In a second experiment the detector was placed downstream from the intersection of the electron and hydrogen beams. Permitted transitions could therefore decay before the atomic hydrogen passed beneath the counter. Only ground state (1s) and metastable (2s) atoms were then present in the beam which was then subjected to an electric field which "quenched" the H(2s) atoms by mixing 2s and 2p states. The emission of λ1216 Å radiation was then proportional to the cross section for 1s-2s excitation.

Unfortunately, the population of H(2s), by cascading from higher levels, is very appreciable so theoretical corrections must be applied to obtain the cross section for direct excitation of the 2s state. But since cascading is relatively unimportant for resonant 1s-2p excitation, measurements of this cross section are more likely to be reliable. Kauppila et al (53) therefore measured the ratio of Q_{1s-2s}/Q_{1s-2p} and deduced values of Q_{1s-2s} which are likely to be more accurate than those given by direct measurement. The results agree with close coupling calculations by Burke et al (54) to better than 15%. A comparison of the measurements of Kaupila et al with theory and other experiments is illustrated by Figure 15.

Ott et al (55) measured the polarization of Ly α radiation produced by quenching H(2s) atoms in an electric field. They obtained $P = -0.30 \pm 0.02$, which compares well with the theoretical prediction of -0.323.

Thermal beam techniques have recently been extended to the excitation of metals. A recent example is the measurement of excitation functions and polarization of the Na-D lines by Stumpf et al (56). Excitation of several states of Mg has been studied by Williams and Trajmar (57) who measured the flux of scattered electrons with appropriate energy. The same technique was previously used by the same group to study other metals including Li (58).

Autoionizing levels in Ca and several other metals have been observed by ejected electron spectroscopy by the group led by Ross (e.g. Pejčev et al (59)).

3.3 Calibration of Optical Detectors

The absolute calibration of an optical detector used in excitation experiments involves a comparison of the radiation flux from an experiment with a radiometric standard (e.g. a "black body"). This is difficult because, in a crossed beam experiment, the required photon flux might only be $\sim 10^3\ s^{-1}$ whereas 1 cm^2 of a black body at 3000 K radiates $\sim 10^{21}$ photons s^{-1}. There are further problems associated with the angular distribution and polarization of the radiation, transmission efficiencies of monochromators, geometry of the emitting volume, the motion of excited particles during the interval before they decay, etc.

A detailed and instructive study of calibration was made by Taylor (60) and an abbreviated account was published by Taylor and Dunn (61).

Briefly, Taylor and Dunn constructed a "uniform, non polarized, isotropic, monochromatic light source" of the same area as the cross section through the interaction region in their experiment. This source was a 1 cm dia. glass sphere connected by a flexible light pipe to a monchromator ($\Delta\lambda \simeq 2.5$ A), illuminated by a quartz iodine lamp. The sphere was masked so that it emitted from the required small rectangular area.

This source was then placed in their apparatus and moved so that the efficiency of their optical detection system could be measured as functions of source position and wavelength. Absolute calibration of the source was accomplished in two stages. First, it was compared with a standard tungsten lamp and the lamp was then checked against a black body operating at the melting point of copper (1357.8 K).

The discussion by Taylor is especially valuable because it considers numerous diverse and subtle sources of error and indicates corrections which are required.

Even with these precautions, the errors in excitation cross sections were typically $\sim 10\%$. Less meticulous experiments may be much less accurate.

3.4 Electron/Photon Coincidence Experiments

We have seen that it is more difficult to make accurate, absolute measurements of cross sections for excitation than for ionization. It was also explained that (e, 2e) experiments provide an extremely searching test of ionization theory. Experiments have therefore been devised to detect, in delayed coincidence, scattered electrons and photons emanating from excited atoms. This provides an exacting test for theory without the need for absolute optical calibration. The method has the further advantage that cross sections for excitation of the various degenerate magnetic substates of the target can be deduced from measured angular correlations between photons and scattered electrons.

The theory of the method was developed by Macek and Jaecks (62) whilst typical experiments have been described by Eminyan et al (63) and Ugbabe et al (64).

In the case of He($1^1S - 2^1P$), if spin is neglected, the excitation at a given incident energy is completely described by three parameters, σ, λ and χ. These are defined in terms of the excitation amplitudes a_0, a_1 and a_{-1} of the magnetic sublevels by,

$$\sigma = 2|a_1|^2 + |a_0|^2 \tag{13}$$

$$\lambda = \frac{|a_0|^2}{2|a_1|^2 + |a_0|^2} \tag{14}$$

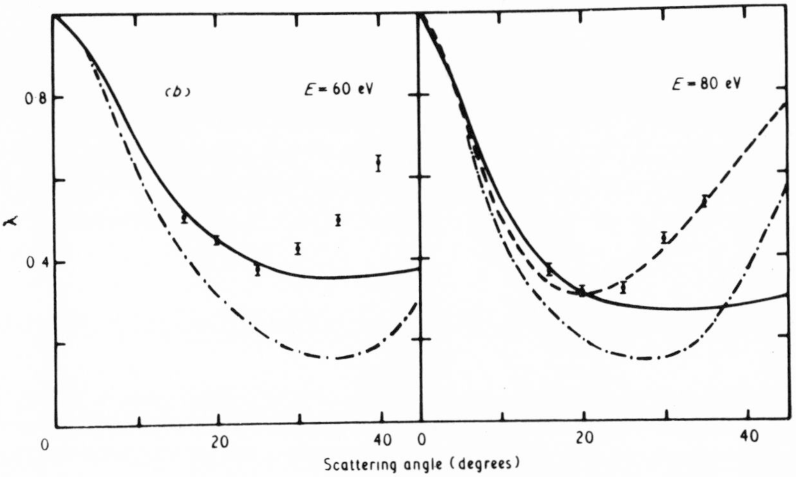

Figure 16. Measured values of the parameter λ for He 2^1P excitation compared with theory. The continuous, dashed and chain-dashed curves illustrate results of Born approximation, Distorted Wave and Many Body Green's function calculations.

whilst χ is the relative phase of a_0 and a_1. In experimental apparatus the incident electrons and both detectors are coplanar and the parameters λ and χ can be deduced from measurements of the angular variation of the coincidence count rate.

Figure 16 illustrates measured values of λ for He(1^1S - 2^1P) and compares them with three theoretical predictions.

This technique is certainly capable of much further application and development.

4. COLLISIONS BETWEEN CHARGED PARTICLES

This subject has been reviewed several times (67,68,69,70) whilst a new account of electron ion collisions is in preparation (71). We will therefore deal briefly with the subject and stress recent trends.

The following four methods have been developed to study charged particle collisions:

(a) Intersecting beams.
(b) "Static" ion traps.
(c) Spectroscopic observations of plasmas.
(d) "Sequential mass spectrometry".

Results have been reported for the following types of reaction:

(a) Single and double ionization of ions by electron impact,

e.g. $$e + He^{+} \rightarrow He^{2+} + 2e \qquad (15)$$

$$e + Li^{+} \rightarrow Li^{3+} + 3e \qquad (16)$$

(b) Single and double detachment from negative ions by electron impact,

e.g. $$e + H^{-} \rightarrow H + 2e \qquad (17)$$

$$e + H^{-} \rightarrow H^{+} + 3e \qquad (18)$$

(c) Excitation of ions by electron impact,

e.g. $$e + Ca^{+}(4^2S_{\frac{1}{2}}) \rightarrow Ca^{+}(4^2P_{\frac{3}{2}}) + e \qquad (19)$$

(d) Dissociative ionization, excitation and "pair production",

e.g.
$$e + H_2^{+} \rightarrow H^{+} + H^{+} + 2e \qquad (20)$$

$$e + H_2^{+} \rightarrow H^{+} + H + e \qquad (21)$$

$$e + H_2^{+} \rightarrow H^{+} + H^{-} \qquad (22)$$

(e) Dissociative recombination,

e.g. $e + H_2^+ \rightarrow H + H$ (23)

(f) Single and double detachment from negative ions by proton impact,

e.g. $H^+ + H^- \rightarrow H^+ + H + e$ (24)

$ \rightarrow H^+ + H^+ + 2e$ (25)

(g) Mutual neutralization,

e.g. $H^+ + H^- \rightarrow H + H$ (26)

(h) Double charge transfer between positive and negative ions,

e.g. $H^+ + H^- \rightarrow H^- + H^+$ (27)

(j) Charge transfer between positive ions,

e.g. $He^+ + H^+ \rightarrow He^{++} + H.$ (28)

All of these reactions have been investigated with intersecting beams (this is the most versatile and prolific of the experimental techniques), but the other three experimental approaches have usefully extended or supplemented this information.

4.1 General Principles of Experiments

These call for quite elaborate apparatus and they encounter several subtle experimental problems which are fully described in the reviews cited. In passing, we note that the particle density in charged beams rarely exceeds $10^6 - 10^7$ cm^{-3} and in this respect the experiments differ greatly from those with neutral beams where the densities are typically $10^{10} - 10^{12}$ cm^{-3}. Moreover, when two charged beams intersect there are space charge forces between them which, unless precautions are taken, may lead to error.

An example of an apparatus which could be used to study the ionization of positive ions is illustrated schematically by Figure 17. Ions from the source S are accelerated and focused by the lens L, deflected in the field of magnet M1 and collimated into a beam by slits (CS). This produces a well collimated beam of ions of the required type. Typically the current and energy are, respectively, $10^{-8} - 10^{-7}$ A at 2 - 10 keV. The ions are then bombarded by an electron beam ($\sim 10^{-4} - 10^{-3}$ A) passing between the electron gun (G) and collector (Ce). A small fraction ($\sim 10^{-8}$) of the ions are ionized and the parent and product ions are separated by the field of M2 so that singly- and doubly-charged ions are collected at C1 and C2. A shutter (Sh) bearing a narrow horizontal slit can be driven through the beams so that their current density distributions can be moni-

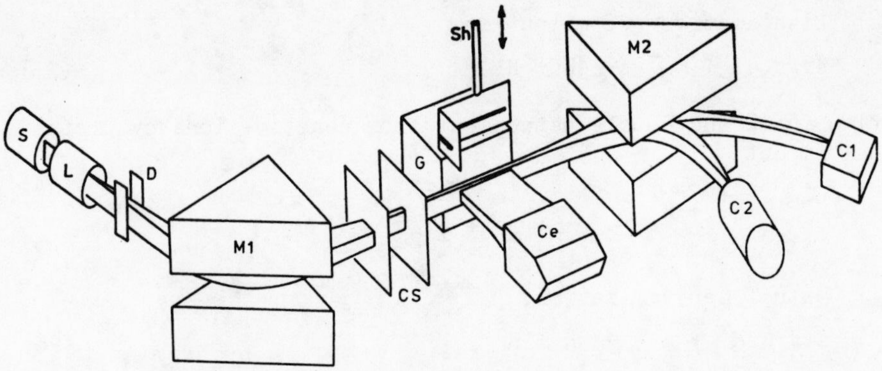

Figure 17. Schematic diagram of apparatus used to study the ionization of positive ions.

tored. The cross section is then deduced from a knowledge of the ion and electron currents and velocities and the beam geometries.

Figure 17 illustrated an arrangement in which beams intersected at 90° but it is sometimes advantageous to use merged ($\theta = 0$) or inclined beams ($\theta \sim 10^{\circ}$).

4.2 Experiments with Merged or Inclined Beams

Figure 18 illustrates four alternative arrangements which could be used in experiments with colliding beams. We will briefly discuss their respective merits.

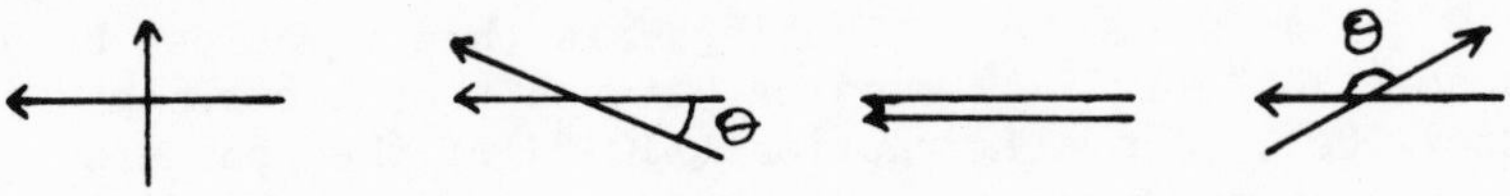

Figure 18. Four alternative arrangements for experiments with intersecting beams.

Perpendicular beams (Figure 18a) have been most widely used, because the required apparatus is easiest to set up, the geometry is well defined, and the electron beam path is short so that space charge spreading of the beam is minimized. Moreover, if electron beams with a rectangular cross section are used, the electrons interact with the ions over an appreciable (∿2 cm) length of the ion path.

But it is sometimes required to study reactions at low interaction energy and it is obvious that if two beams are merged (Figure 18c), and their velocities (V_1 and V_2) are similar, the interaction energy (E) will be small. For beams intersecting at an angle Θ it is given by,

$$E = \frac{m_1 m_2}{m_1 + m_2}\left[\frac{E_1}{m_1} + \frac{E_2}{m_2} - 2\left(\frac{E_1 E_2}{m_1 m_2}\right)^{\frac{1}{2}} \cos\Theta\right] \quad (29)$$

where E_1, E_2, m_1 and m_2 are the laboratory energies and masses of the particles.

Less obvious is the fact that the energy resolution in the centre of mass frame can be much better than in the laboratory frame (this can easily be shown by differentiating equation (29) to obtain dE/dE_1).

For example, for particles of equal mass ($m_1 = m_2$) and $\Theta = 0$ equation (29) reduces to,

$$E = \tfrac{1}{2}(E_2^{\frac{1}{2}} - E_1^{\frac{1}{2}})^2 \quad (30)$$

whilst the energy spread (δE) in the centre of mass frame is given by,

$$\frac{\delta E}{E} = \frac{2\delta E_1}{E_2 - E_1} . \quad (31)$$

Consider the case in which $E_2 = 5.1$ keV and $E_1 = 5.0$ keV. It follows that $E \simeq 0.25$ eV and, if one of the beams is not monoenergetic (e.g. $\delta E_1 = \pm 20$ eV) then δE would be only 0.16 eV.

Merged beams therefore give access to very low interaction energies coupled with greatly enhanced resolution. They are, however, more difficult to set up and there may be difficulties due to space charge forces acting over a long interaction length and ambiguities may arise from uncertainties associated with the extent of the region in which the beams are merged and demerged. These problems have, however, been overcome and merged beams have been used very effectively to study ion-ion (72) and electron-ion (73) collisions.

Inclined beams ($\theta < 90^{\circ}$) also provide enhanced resolution and low interaction energies although a minimum attainable interaction energy is set by the transverse component of the beam velocities. They do, however, have a sharply defined collision geometry and this makes them suitable for accurate, absolute measurements.

Sometimes it may be necessary to extend the range of a measurement beyond those energies which could be reached merely by enhancing the beam energies. This could be achieved with the arrangement illustrated by Figure 18d although energy resolution is then sacrificed.

4.3 Ionization of Ions by Electron Impact

This process is clearly important in determining the equilibrium ionization in hot plasmas. The review by Dolder and Peart (69) includes a tabular list and references to 26 measurements of the ionization of positive ions which had been reported prior to June 1976. Single ionization of He^+, Li^+, Na^+, K^+, Rb^+, Cs^+, Mg^+, Mg^{2+}, Ca^+, Sr^+, Ba^+, Tl^+, C^+, N^+, N^{2+}, O^+, O^{2+}, and Ne^+ had been studied, as was double ionization of Li^+.

Four points of particular interest emerge from these measurements. First, the rise of the ionization function from threshold is more rapid for a positive ion than for a neutral atom. This is presumably due to the attraction which the ionic Coulomb field exerts on the negative projectile.

Second, the classical scaling law based on Thomson's elementary theory is of some value when comparing or predicting ionization cross sections. This law is widely embodied in empirical formulae used notably by astrophysicists and plasma physicists. It states that cross sections (Q) for two isoelectronic systems are related to their ionization energies (χ) by,

$$\frac{Q_1}{Q_2} = \left(\frac{\chi_2}{\chi_1}\right)^2 \qquad (32)$$

if the projectile energy is expressed in units of the relevant threshold. Some tests of this law are illustrated by Figure 19 which compares measurements of the following isolectronic systems, H/He^+; He/Li^+· Ar/K^+ and $Ne/Na^+/Mg^{2+}$. Except in the case of Ne, the simple law works well and the results also illustrate the enhancement of cross sections near threshold for positive ions.

Third, at very high energies the Bethe approximation should be valid but the range of its validity can only be determined by experiment. This approximation predicts that ionization cross sections (Q) and interaction energies (E) are related by,

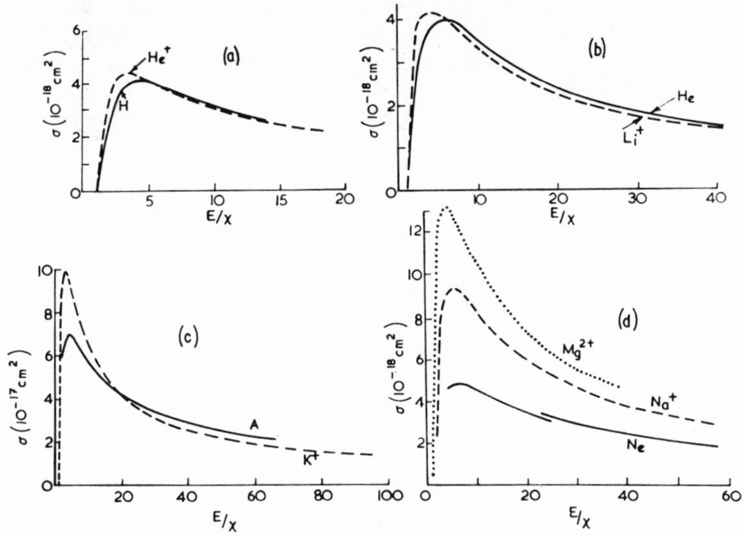

Figure 19. Comparison of classically scaled cross sections measured for the isoelectronic systems H/He$^+$; He/Li$^+$; Ne/Na$^+$/Mg^{2+}; Ar/K$^+$.

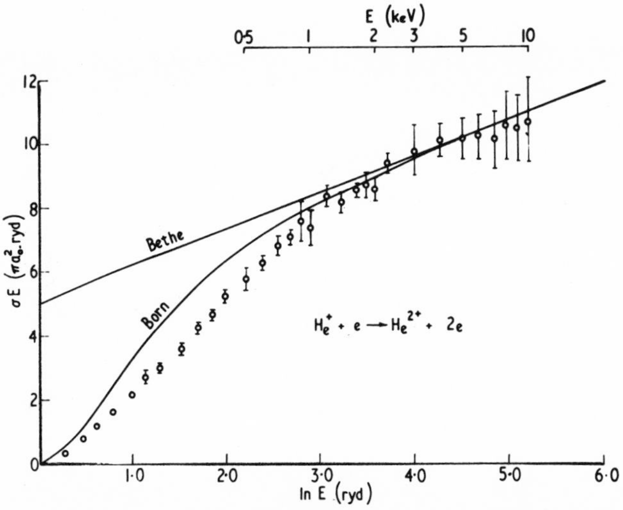

Figure 20. Measured cross sections for the ionization of He$^+$ compared with Born and Bethe predictions.

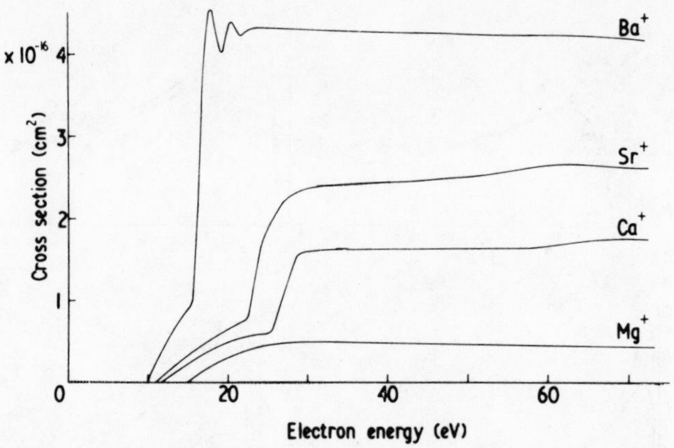

Figure 21. Measured ionization cross sections for Mg^+, Ca^+, Sr^+, and Ba^+ ions showing large autoionization contributions for all but Mg^+.

$$QE = A \lg E + B \tag{33}$$

where A and B are constants. Results for He^+ are compared with Born and Bethe predictions in Figure 20 and it is seen that the Bethe region was attained only when $E \gtrsim 2.5$ keV, which is about 50 times threshold.

Fourth, in some cases autoionization may make a large, even dominant, contribution to the ionization cross section. This was particularly marked for Ca^+, Sr^+, and Ba^+ as illustrated by Figure 21.

The main inadequacy of the experiments we have mentioned is that they are confined to ions in low states of initial ionization. In thermonuclear plasmas impurities may be very highly ionized but it is difficult to prepare adequate beams of, say, C^{5+} which could be used in crossed beam experiments. Substantial progress is, however, being made, notably at ORNL, and reports of the ionization of C^{3+}, N^{4+} and O^{5+} have recently appeared (72). These important advances were made possible by a massive ion source at ORNL which provides adequate beams ($\sim 10^{-7}$ A) of highly charged ions. Perhaps the most spectacular of these results were those for O^{5+} which revealed a large autoionization contribution above 500 eV. This is illustrated by Figure 22. Autoionization was also observed for C^{3+} and N^{4+}. In this isoelectronic series (as in the case of the Ca^+ series) autoionization became relatively larger as the ionic charge increased. This may have important consequences for plasma physics and astrophysics.

Other recent measurements with crossed beam experiments have been reported for Rb^+ (73), Ar^+ (74), and C^{2+} (75) ions.

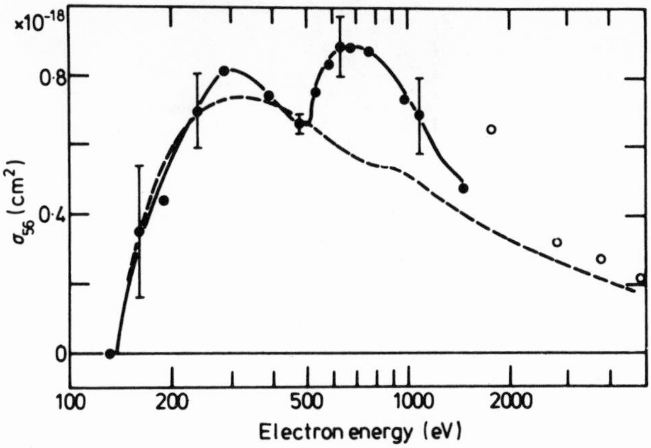

Figure 22. Measured ionization cross section of O^{5+} showing the onset of autoionization at about 500 eV.

4.4 Excitation of Positive Ions by Electron Impact

The most significant work in this field has come from Dunn's group who developed the techniques for absolute optical calibration outlined in §3.3. Intersecting electron and ion beams were viewed with a photomultiplier equipped with a filter and polarizer to isolate the required radiation and measure its polarization. The review by Dolder and Peart (69) includes a tabular list of 18 crossed beam measurements of excitation published prior to June 1976. An example of particular relevance to astrophysics was a measurement of cross sections for the excitation and polarization of the Ca^+ K line (76) (λ 3934 Å). The measurement is compared with 3-state close coupling (CC), Coulomb distorted wave (CDW) and classical binary encounter (CL) approximations in Figure 23. The finite threshold characteristic of the excitation of positive ions is apparent.

Subsequently, this group reported measurements of the excitation of $Hg^+(6p\ ^2P^0_{3/2})$, $Hg^+(7s\ ^2S_{1/2})$ and of λ4797 Å radiation from Hg^{++} (77, 78) excited by collisions of electrons with Hg^+ and Hg^{++}. Cross sections for the spin-changing excitation $Li^+(1^1S - 2^3P)$ have also been reported (79) but the most exciting advance was a measurement of the resonant excitation of C^{3+} (80). This carried cross beam techniques towards the regime of direct relevance to fusion physics and it also involved the absolute calibration of optical detectors in the far UV (λ 1550 Å).

A perplexing problem arises from measurements of the excitation of $He^+(1s - 2s)$ by electron impact. Two crossed beam measurements have been reported (81, 82). In both experiments a beam of $He^+(1s)$ was excited by electrons. The $He^+(2s)$ formed was quenched down-

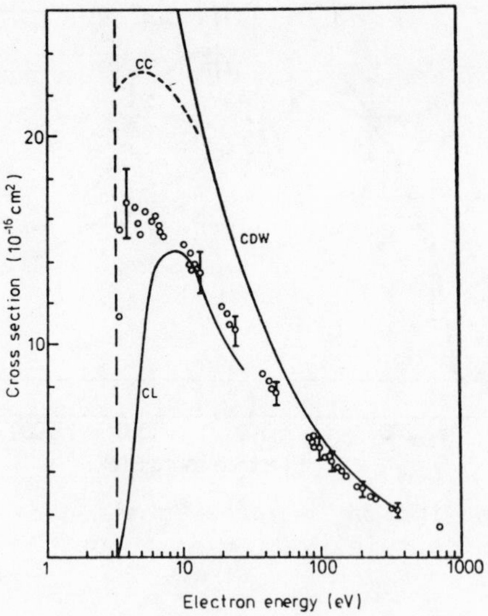

Figure 23. Measured cross sections for excitation of the Ca^+ K line compared with close coupling (CC), Coulomb distored wave (CDW) and classical binary encounter (CL) approximations.

stream so that the emitted radiation could be isolated, detected and counted. To obtain the cross sections for direct excitation, the contribution from cascading was estimated from theory and the results were then made absolute by normalizing to Born's approximation. When this was done, the measured threshold cross section was less than half that predicted by close coupling theory (83). Seaton (47) has summarized the present position. He concluded "that there are errors in low-energy theories, in high-energy theories, or in experiments. Or there could be some combination of errors". A new experiment is in progress.

4.5 Other Crossed Beam Experiments on Electron-Ion Collisions

Several of the reactions mentioned in §4 are particularly interesting. For example, measurements of detachment from H^- and O^- reveal resonances which may be associated with short-lived excited states of H^{--} and O^{--} but we will concentrate here on reactions which are more obviously relevant to fusion.

Reactions involving molecular ions do not occur in hot plasmas but they may be significant in ancilliary devices. For example, ion sources are needed to produce intense beams of D^- ($\sim 10^2$ A). These are stripped to form energetic D atoms which bombard a plasma and can provide supplementary heating. Consequently, there is keen interest in processes leading to the formation and destruction of H^- (84).

It has been demonstrated (85) that cross sections for,

$$e + H_2^+ \rightarrow H^+ + H^- \qquad (34)$$

are of order 3×10^{-18} cm^2. This is one or two orders larger than for dissociative attachment,

$$e + H_2 \rightarrow H^- + H \qquad (35)$$

which is usually considered to be chiefly responsible for the production of H^- in ion sources. It must be remembered that in a source there will be much less H_2^+ than H_2 so reaction (34) is unlikely to dominate.

It has been suggested (84) that significant quantities of H^- might be produced from H^+ by,

$$e + H_3^+ \rightarrow H^- + H_2^+. \qquad (36)$$

However, the cross section has recently been shown to be less than 1.6×10^{-18} cm^2 (86) which implies that reaction (36) is unlikely to be important in ion sources.

Dissociative recombination is important in many environments and measurements for H_2^+ are particularly interesting to theorists because this (with the possible exception of He_2^+) is the only ion for which ab initio calculations are feasible. The first measurements for H_2^+ and D_2^+ were made with inclined beams (87, 88) but these have been supplemented by merged beam experiments (89, 103) which reveal more structure in the cross section and extend the range of the measurements to considerably lower energies.

A brief summary of experiments on collisions between electrons and H_2^+ has been published (90).

4.6 Charge Transfer Between Positive Ions

These reactions may be relevant to fusion in two quite separate ways. Consider intereactions between D^+ and an ionized impurity (X^{n+}) in a fusion plasma,

$$D^+ + X^{n+} \rightarrow D + X^{(n+1)+}. \qquad (37)$$

This neutralization of D^+ results in loss of fuel from the reactor and the enhancement in change of the impurity with the consequent increase in bremsstrahlung radiation. Moreover, the fast D atom will collide with the walls of the containment vessel and sputter more impurities into the plasma.

Measurements of,

$$H^+ + He^+ \rightarrow H + He^{2+} \quad (38)$$

$$H^+ + Mg^+ \rightarrow H + Mg^{2+} \quad (39)$$

$$H^+ + Ti^+ \rightarrow H + Ti^{2+} \quad (40)$$

$$H^+ + Fe^+ \rightarrow H + Fe^{2+} \quad (41)$$

$$He^+ + He^{2+} \rightarrow He^{2+} + He^+ \quad (42)$$

have been reported and there have been several measurements of the inverse type of reaction, e.g.

$$C^{2+} + H \rightarrow C^+ + H^+ \quad (43)$$

or the near resonant reaction,

$$Ti^{2+} + H \rightarrow Ti^+ + H^+ \ (\Delta E = -0.03\ eV). \quad (44)$$

An excellent, extensive review of these processes and of other heavy particle collisions relevant to fusion has recently been prepared by Gilbody (91).

Figure 24 illustrates measurements for collisions between H^+ and He^+. This may form He^{2+} by charge transfer, or by ionization,

$$H^+ + He^+ \rightarrow H + He^{2+} \quad (45)$$

$$H^+ + He^+ \rightarrow H^+ + He^{2+} + e. \quad (46)$$

The open and closed circles show results of two independent measurements (92, 93) of cross sections for the formation of He^{2+} (i.e. the sum of reactions (45) and (46)) whilst the triangles denote measurements (94) solely for charge transfer. The dashed curve is a theoretical estimate (95) of the charge transfer contribution. It is seen that ionization was significantly only above 100 keV.

A second application to fusion arose during design studies for "heavy ion fusion" experiments. These involve the bombardment of a small deuterium pellet with very intense, energetic pulses of heavy ions. It is similar to laser fusion except that ions replace the laser pulse and it is hoped that the production of the necessary ion pulses will consume less energy.

Long acceleration times are needed and the particle density in the accelerating beams will be so high that the cross sections for charge transfer, e.g.

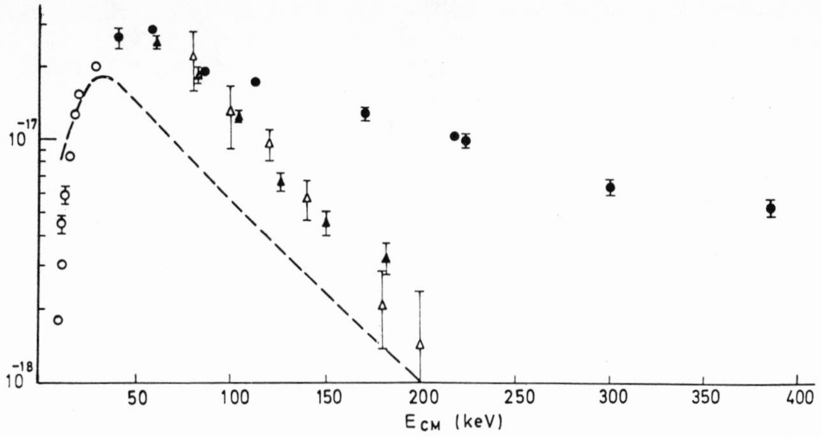

Figure 24. The production of He^{++} by collisions between H^+ and He^+. The circles show two measurements for He^{++} production. The triangles and dashed line represent measured and calculated cross sections for charge transfer.

$$Cs^+ + Cs^+ \rightarrow Cs + Cs^{2+} \qquad (47)$$

vitally affect the design of the accelerators. These are so costly (£2 × 10^8) that firm values of charge transfer cross sections are essential.

4.7 Techniques Employing Trapped Ions

Problems arise when experimenting with molecular ions because their reaction cross sections frequently depend very sensitively upon their initial state of vibrational excitation. It follows that unless the vibrational population is known the results may be uninterpretable. An elegant and effective approach has been developed (96) in which ions are trapped by steady electric and magnetic fields until they have cooled to their lowest vibrational state. Trapping times of 28h have been reported for NH_4^+.

Noise currents induced in the trap electrodes indicate the number of trapped ions and by monitoring the decay of ions when energy-resolved electrons are ejected into the trap, it is possible to measure the variation of dissociative recombination with energy.

Another approach involves trapping ions in the space charge of an electron beam. The ions and electrons then interact and, by measuring the flux of collision products, it is possible to study ionization (97), excitation (98), and recombination (99). Neither of these methods is well suited to absolute measurements and the latter technique does not provide molecular ions in a well defined initial state.

4.8 The Deduction of Ionization and Excitation Reaction Rates from Observations of Plasmas

Ionization and excitation rates clearly determine the equilibrium properties of plasmas. Conversely, if the temperature and density of a plasma is known, it is possible to deduce ionization and excitation rates from spectroscopic observations of a plasma. The method is particularly valuable for highly-charged species which are beyond the reach of beam experiments. An example is the measurement (100) of rate coefficients for Fe VIII, IX and X. Rate coefficients for resonant excitation of H, Li, Na, Ca, Ca^+ and Ba^+ have also been reported (101) and the topic was reviewed by Gabriel and Jordan (102).

REFERENCES

(1) K.T. Compton and C.C. van Voorhis, Phys. Rev. 26, 436 (1925).
(2) J.T. Tate and P.T. Smith, ibid, 39, 270 (1932).
(3) B.L. Schram, F.S. de Heer, M.J. van der Wiel and J. Kistemaker, Physica 31, 94 (1965).
(4) R.K. Asundi, Proc. Phys. Soc. 82, 372 (1963).
(5) P.O. Taylor, K.T. Dolder, W.E. Kauppila and G.H. Dunn, Rev. Sci. Instr. 45, 538 (1974).
(6) W. Bleakney, Phys. Rev. 36, 1303 (1930).
(7) D. Rapp and P. Englander-Golden, J. Chem. Phys. 43, 1464 (1965).
(8) H.S.W. Massey and E.H.S. Burhop, Electronic and Ionic Impact Phenomena, 1, Clarendon Press (Oxford 1969).
(9) H.S.W. Massey, E.H.S. Burhop and H.B. Gilbody, ibid, 5 (1974).
(10) L.J. Kieffer, JILA Information Center Rept. 13, Univ. Colorado, (1973).
(11) E.W. McDaniel, U.S. Army Missile Command, Redstone Arsenal, Alabama 35809, Tech. Rept. RH-77-1 (1977).
(12) L.J. Kieffer and G.H. Dunn, Rev. Mod. Phys. 38, 1 (1966).
(13) H. Funk, Ann. Phys. 4, 49 (1930).
(14) R.L.F. Boyd and G.W. Green, Proc. Phys. Soc. 71, 351 (1958).
(15) W.L. Fite, Atomic & Molecular Processes, Ed. D.R. Bates (Academic Press: New York: 1962).
(16) W.L. Fite and R.T. Brackmann, Phys. Rev. 113, A1141 (1958).
(17) R.H. McFarland and J.D. Kinney, ibid, 137, A1058 (1965).
(18) J.R. Peterson, Atomic Colln. Processes, Ed. M.R.C. McDowell (North Holland, Amsterdam: 1964).
(19) E. Brook, M.F.A. Harrison and A.C.H. Smith, J. Phys. B 11, 3115 (1978)

(20) W.L. Fite and R.T. Brackmann, Phys. Rev. 113, 815 (1959).
(21) F.S. de Heer, J. Phys. B 12, L429 (1979).
(22) A.J. Dixon, A. von Engel and M.F.A. Harrison, Proc. Roy. Soc. A 343, 333 (1975).
(23) A.J. Dixon, M.F.A. Harrison and A.C.H. Smith, J. Phys. B 9, 2617 (1976).
(24) A.J. Dixon, M.F.A. Harrison and A.C.H. Smith, Proc. VIII ICPEAC, Belgrade, 405 (1973).
(25) W.L. Fite and R.T. Brackmann, Proc. VI Int. Conf. Ionization Phenomena in Gases, 21 (1963).
(26) D.R. Long and R. Geballe, Phys. Rev. A 1, 260 (1970).
(27) P.G. Burke and A.J. Taylor, Proc. Roy. Soc. A 287, 105 (1965).
(28) H. Ehrhardt, K. Hesselbacher, K. Jung and K. Willmann, Case Studies in Atomic Colln. Phys. (North Holland: Amsterdam) 2, 2107 (1972).
(29) D. Paul, K. Jung, E. Schubert and H. Ehrhardt, Proc. IX ICPEAC, Seattle, 194 (1975) Invited paper.
(30) E. Schubert, A. Schuck, K. Jung and S. Geltman, J. Phys. B 12, 967 (1979).
(31) E.C. Beaty, K.H. Hesselbacher, S.P. Hong and J.H. Moore, ibid, 10, 611 (1977).
(32) A.J. Dixon, I.E. McCarthy and E. Weigold, ibid, 8, L195 (1976).
(33) W.E. Balashov, A.N. Grum-Grzhimailo and M.N. Kabachnik, ibid, 12, L27 (1979).
(34) E. Weigold, C.J. Noble and S.T. Hood, ibid, 12, 291 (1979).
(35) M. Lal, A.N. Tripathi and M.K. Srivastava, ibid, 12, 945 (1979).
(36) G.H. Wannier, Phys. Rev. 90, 817 (1953).
(37) R. Peterkop, J. Phys. B 4, 513 (1971).
(38) A. Temkin, ibid. 7, L450 (1974).
(39) I.J. Kang and R.L. Kerch, Phys. Lett. 31A, 172 (1970).
(40) S. Cvejanovic and F.H. Read, J. Phys. B 7, 1841 (1974).
(41) D. Hils and H. Kleinpoppen, ibid, 11, L283 (1978).
(42) M. Tronc, G.C. King and F.H. Read, ibid, 12, 137 (1979).
(43) E.G. Berezhko, M.N. Kabachnik and V.V. Sizov, ibid, 11, 1819 (1978).
(44) P.J. Hicks, S. Cvejanovic, J. Comer, F.H. Read and J.M. Sharp, Vacuum 24, 573 (1974).
(45) F.H. Read, Proc. IX ICPEAC, Seattle, 176 (1975) Invited paper.
(46) J.A. Baxter, J. Comer, P.J. Hicks and J. McConkey, J. Phys. B 12, 2031 (1979).
(47) M.J. Seaton, Advances in Atomic & Molec. Phys. 11, 83 (1975) (Academic Press: New York).
(48) P.N. Clout and D.W.O. Heddle, J. Opt. Soc. Am. 59, 715 (1969).
(49) D.W.O. Heddle and M.J. Samuel, J. Phys. B 3, 1593 (1970).
(50) B.H. Bransden and D.P. Dewangan, ibid, 11, 3425 (1978).
(51) W.L. Fite and R.T. Brackmann, Phys. Rev. 113, 1151 (1958).
(52) R.F. Stebbings, W.L. Fite, D.G. Hummer and R.T. Brackmann, Phys. Rev. 119, 1939 (1960).
(53) W.E. Kauppila, W.R. Ott and W.L. Fite, Phys. Rev. A 1, 1099 (1970).

(54) P.G. Burke, S. Ormonde and W. Whitaker, Proc. Phys. Soc. 92, 319 (1967).
(55) W.R. Ott, W.E. Kauppila and W.L. Fite, Phys. Rev. A 1, 1089 (1970).
(56) B. Stumpf, K. Becker and G. Schulz, J. Phys. B 11, L639 (1978).
(57) W. Williams and S. Trajmar, ibid, 11, 2021 (1978).
(58) W. Williams and S. Trajmar, ibid, 9, 1529 (1978).
(59) V. Pejčev, T. Ottley and K. Ross, ibid, 11, 531 (1978).
(60) P.O. Taylor, Ph.D. Thesis, Univ. Colorado 1972 (Univ. Microfilms Inc., Ann Arbor, Mich.).
(61) P.O. Taylor and G.H. Dunn, Phys. Rev. A 8, 2304 (1973).
(62) J. Macek and D.H. Jaecks, ibid, 4, 2288 (1971).
(63) M. Eminyan, K.B. MacAdam, J. Slevin and H. Kleinpoppen, J. Phys. B 7, 1519 (1974).
(64) A. Ugbabe, P.J.O. Teubner, E. Weigold and H. Arriola, ibid, 10, 71 (1977).
(65) D.W.O. Heddle and R.G. Keesing, Proc. III ICPEAC, London (1963).
(66) R.H. MacFarland, Phys. Rev. A 136, 1240 (1964).
(67) M.F.A. Harrison, Methods in Experimental Physics 7B, Eds. W.L. Fite and B. Bederson (Academic Press: New York: 1968).
(68) G.H. Dunn, Atomic Physics 1, Ed. V.W. Hughes (Plenum Press: New York: 1969).
(69) K. Dolder and B. Peart, Rep. Prog. Phys. 39, 693 (1976).
(70) K. Dolder, Proc. X ICPEAC, Paris (1977), Invited paper.
(71) J.W. McGowan, Proc. XI ICPEAC, Osaka (1979), Invited paper.
(72) D.H. Crandall, R.A. Phaneuf and P.O. Taylor, Phys. Rev. A 18, 1191 (1978).
(73) R.K. Feeney, W.E. Sayle and T.F. Divine, ibid, 18, 82 (1978).
(74) P.R. Woodruff, M-C. Hublet and M.F.A. Harrison, J. Phys. B 11, L305 (1978).
(75) P.R. Woodruff, M-C. Hublet, M.F.A. Harrison and E. Brook, ibid, 11, L679 (1978).
(76) P.O. Taylor and G.H. Dunn, Phys. Rev A 8, 2304 (1973).
(77) D.H. Crandall and G.H. Dunn, ibid, 11, 1223 (1975).
(78) R.A. Phaneuf, P.O. Taylor and G.H. Dunn, ibid, 14, 2021 (1976).
(79) W.T. Rogers, J.Ø. Olsen and G.H. Dunn, ibid, 18, 1353 (1978).
(80) P.O. Taylor, R.A. Phaneuf, D. Gregory and G.H. Dunn, Proc X ICPEAC, Paris (1977).
(81) D.F. Dance, M.F.A. Harrison and A.C.H. Smith, Proc. Roy. Soc. A 290, 73 (1966).
(82) K. Dolder and B. Peart, J. Phys. B 6, 2415 (1973).
(83) P.G. Burke and A.J. Taylor, ibid, 2, 44 (1969).
(84) J.R. Hiskes, M. Bacal and G.W. Hamilton, Lawrence Livermore Lab. Report UCID-18031 (1979).
(85) B. Peart and K. Dolder, J. Phys. B 8, 1570 (1975).
(86) B. Peart, R.A. Forrest and K. Dolder, ibid, (in course of publication).
(87) B. Peart and K. Dolder, ibid, 7, 236 (1974).
(88) B. Peart and K. Dolder, ibid, 6, L359 (1973).
(89) D. Auerbach, R. Cacak, R. Caudano and J.W. McGowan, ibid, 10,

3797 (1977).
(90) K. Dolder, Comm. on At. and Mol. Phys. 5, 97 (1976).
(91) H.B. Gilbody, Adv. At. and Mol. Phys. 15 (in course of publication).
(92) B. Peart, R. Grey and K. Dolder, J. Phys. B 10, 2675 (1977).
(93) J.B. Mitchell, K.F. Dunn, G.C. Angel, R. Browning and H.B. Gilbody, ibid, 10, 1897 (1977).
(94) G.C. Angel, E.C. Sewell, K.F. Dunn and H.B. Gilbody, ibid, 11, L297 (1978).
(95) R.E. Olson and A. Salop, Phys. Rev. A 16, 531 (1977).
(96) F.L. Walls and G.H. Dunn, Physics Today 27, 30 (1974).
(97) M. Hamdan, K. Birkinshaw and J.B. Hasted, J. Phys. B 11, 331 (1978).
(98) N.R. Daly and R.E. Powell, Phys. Rev. Lett. 19, 1165 (1967).
(99) D. Mathur, S. Khan and J.B. Hasted, J. Phys. B 11, 3615 (1978).
(100) R. Datla, M. Blaha and H-J. Kunze, Phys. Rev. A 12, 1076 (1975).
(101) D.H. Crandall, G.H. Dunn and A. Gallagher, Astrophys. J. 191, 789 (1974).
(102) A.H. Gabriel and C. Jordan, Case Studies in At. Colln. Phys. 2, 211 (1972).
(103) P.M. Mul and J.W. McGowan, J. Phys. B 12, 1591 (1979).
(104) M.F.A. Harrison, A.C.H. Smith and E. Brook, ibid, 12, L433 (1979).

EXPERIMENTS ON ELECTRON CAPTURE AND IONIZATION BY IONS

F.J. DE HEER

FOM-Institute for Atomic and Molecular Physics

Kruislaan 407, 1098 SJ Amsterdam, The Netherlands

1. INTRODUCTION

In controlled thermonuclear fusion, electron capture and ionization by ions play an important role. Several articles have been written on this topic (for references see further on) and therefore it is not our intention to give a complete review. Gilbody[1] has classified the relevant heavy particle collision processes playing a role in plasma heating, in the production of the fast beam of hydrogen atoms injected for plasma heating and in plasma diagnostics (see also other papers in this book).

For the heating of the plasma a fast neutral beam of an appropriate isotope of hydrogen can be used in the keV energy region. The neutrals undergo electron loss by collisions with protons of the plasma via electron capture and ionization:

$$H^+ + H \rightarrow H + H^+ \qquad \text{(electron capture)} \tag{1}$$

$$H^+ + H \rightarrow H^+ + H^+ + e \qquad \text{(ionization)} \tag{2}$$

In these equations H represents some isotope of hydrogen (i.e. deuterium). It is known that the relevant cross sections are independent of the isotope of hydrogen when the velocity in the center of mass system is the same. The resulting fast protons are trapped in the magnetic confining field of the fusion reactor and can give up their energy in collisions with the plasma constituents.

In practice the plasma heating is disturbed by the presence of small fractions of partially and fully ionized impurities of C, N and O and partially ionized atoms of high atomic Z number such as

Fe, Mo, Ti, Nb, V and W arising from the interaction at the chamber walls. These ions can easily react with the beam of injected hydrogen particles according to

$$X^{q+} + H \rightarrow X^{(q-1)+} + H^{+} \quad \text{(electron capture)} \tag{3}$$

$$X^{q+} + H \rightarrow X^{q+} + H^{+} + e \quad \text{(ionization)} \tag{4}$$

where X^{q+} stands for the impurity ion of charge q. As a consequence the neutrals do not reach the center of the plasma and the protons are trapped in the plasma boundary as a consequence of the deflection action of the magnetic field. Both processes (3) and (4) may lead to short lived excited states or to metastable states of the impurity ion.

Particle loss from the plasma can occur via ion-ion collision processes:

$$H^{+} + X^{q+} \rightarrow H + X^{(q+1)+} \quad \text{(electron capture)} \tag{5}$$

Fast H atoms formed in this way escape from the magnetic confinement.

The fast beam of hydrogen atoms used for heating of the plasma has to be produced by an injection system. In many cases H^+, H_2^+ or H_3^+ ions are accelerated and pass through a metallic vapour target (often Cs), where partial conversion takes place to fast neutral atoms. Electron capture processes combined with ionization and dissociation play an important role. Because at higher energies the electron capture cross sections become relatively small, one has also considered passing H^- ions through metallic vapours, leading to H neutrals by electron loss with much larger probability at these energies.

Finally electron capture and ionization processes are important in plasma diagnostics. Kislyakov and Petrov[2] have used 4 - 14 keV H atom beams to probe a plasma. Such a beam is attenuated in the plasma by processes (1) and (2) and in addition by ionization by plasma electrons. The attenuation is dependent on the plasma density and temperature. In process (1) excited hydrogen atoms can be formed as well, and the Doppler shift of the resulting radiation is related to the temperature of the plasma.

In this paper we shall discuss the different experimental methods which have been used for the determination of electron capture and ionization cross sections (section 2). A short discussion will be given about ion sources used for the production of multiply charged ions (section 3). Experimental results for electron capture and ionization are given in section 4 and compared with theory.

In section 5 we consider the formation of excited states in electron capture processes.

It is useful to give some general references about the topic discussed in this paper. The most recent activities in the field are summarized in the Book of Invited Papers of the XI ICPEAC conference in Kyoto[3], the Book of Abstracts of the same conference[4], the Proceedings of the conference in Nagoya on Atomic Processes in Fusion Plasmas[5] and by Gilbody[1]. Tawara and Russek[6] have reviewed charge exchange processes by H^+. A very good review of experimental methods and analysis of results on charge changing and ionizing collisions is given in the book of Massey and Gilbody[7]. Atomic data for fusion are collected by Barnett and Wiese[8] and the IAEA in Vienna[9].

2. EXPERIMENTAL METHODS TO DETERMINE ELECTRON CAPTURE AND IONIZATION CROSS SECTIONS

One of the most extended reviews on this subject has been given by Massey and Gilbody[7] and here we shall confine ourselves to a short summary (see also Gilbody[1]).

A. Condensor Plates Method

We consider the processes:

$$A^{q+} + B \rightarrow A^{(q-1)+} + B^{+} \qquad \text{(electron capture)} \tag{6}$$

$$A^{q+} + B \rightarrow A^{q+} + B^{+} + e \qquad \text{(ionization)} \tag{7}$$

A beam of ions A^{q+} enters a chamber filled with a target gas of atoms B. The slow ions B^+ and electrons e, formed by reactions (6) and (7) are collected on the condensor plates by means of an elec-

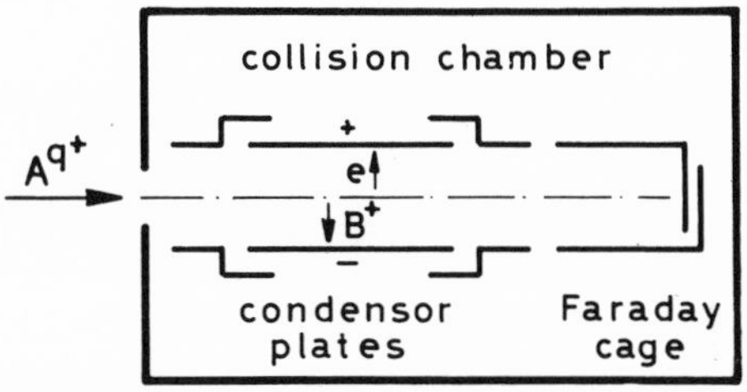

Fig. 1 - Condensor plates system

tric voltage across these plates. Most ions B^+ indeed are slow (in the eV energy region), because glancing collisions (large impact parameter) are dominating in processes (6) and (7). In these collisions little transfer of kinetic energy from A^{q+} to B takes place. We evaluate the cross sections for production of slow ions and electrons according to

$$\sigma_{i;e} = q\, I_{i;e}/L\, I_q\, N \tag{8}$$

I_i respectively I_e are the currents of slow ions and electrons formed in the target gas of density N, determined by a membrane manometer. I_q is the current of projectile ions of charge q measured at a Faraday cage. L is the length of the condensor plates. From equations (6) and (7) it follows directly that

$$\sigma_e = \sigma_I \quad \text{and} \quad \sigma_{q,q-1} = \sigma_i - \sigma_e \tag{9}$$

where σ_I is the ionization cross section and $\sigma_{q,q-1}$ the electron capture cross section. The equations (9) are a good approximation if electron stripping from the projectile, multiple ionization and multiple electron capture are negligible, so in particular for atomic hydrogen as a target at low impact velocities.

B. Charge State Selection of the Projectile.

When A^{q+} collides with atom B a variety of processes are possible, presented by

$$A^{q+} + B \rightarrow A^{(q-n)+} + B^{m+} + (m-n)e \tag{10}$$

where $m-n \geq 0$.
A set-up used by Crandall et al.[10] is presented in Fig. 2 to illustrate the charge-state selection of the projectile. A beam of multiply charged ions is passed through an oven of high temperature ($\sim$2800 K) filled with H_2. At this temperature about 90% of the gas in the oven is dissociated into atomic hydrogen. The charge of the beam particles, after having passed the hydrogen oven, can be determined by means of electrostatic deflection (see the plates in Fig. 2 connected with power supply 2). Then we find the cross section for capture of n electrons according to

$$\sigma_{q,q-n} = (I_{q-n}/I_q\, L\, N) \times q/(q-n) \tag{11}$$

where the symbols have a similar meaning as under A. Only counting techniques instead of current measurements are applied. When the target is not H, but contains many electrons, $\sigma_{q,q-n}$ includes all possible electron capture processes with simultaneous changes in the target (see Eq. (10)).

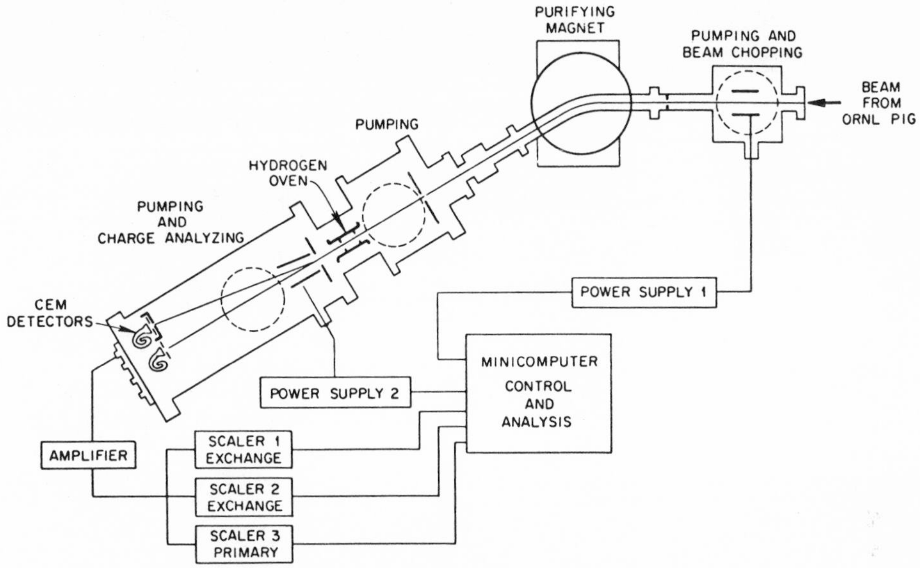

Fig. 2 - Charge state selection (Crandall et al.[10])

C. Charge State, Energy and Angle Selection of Projectile (and Target) Ions, Coincidence.

This type of experiment gives complete information about the kinematics of the collision, including the potential energy change, ΔE, involved in the reaction. So one can also determine whether the particles after the reaction have been formed into an excited state or the ground state. In fact one can determine the cross sections for all processes indicated by equation (10) separately.

In Fig. 3 m_1 with energy E_0 collides with m_2 with energy zero, leading to scattering of m_1 at angle θ and of m_2 at angle ϕ, where m_1 and m_2 may have changed their kinetic and potential energy. The collision is governed by conservation of energy and momentum according to:

$$E_0 = E_1 + E_2 + \Delta E \tag{12}$$

$$m_1 v_0 = m_1 v_1 \cos\theta + m_2 v_2 \cos\phi \tag{13}$$

$$0 = m_1 v_1 \sin\theta - m_2 v_2 \sin\phi \tag{14}$$

These are 3 equations with 6 unknown quantities, E_0, E_1, E_2, θ, ϕ, ΔE. If we measure three of these, for instance E_0, E_1 and Θ, we get all information about the reaction. In that case one measures $d\sigma/d\theta$

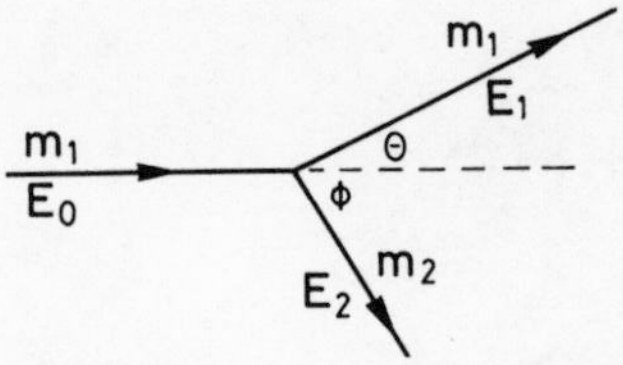

Fig. 3 - Kinematics of the collision

for one of the reactions represented by equation (10). By integration of $d\sigma/d\theta \times \sin\theta$ over angle θ, the total cross section for that reaction can be obtained. Coincidence techniques are required when one measures E_0, θ and ϕ or E_0, E_1 and E_2. For more details the reader is referred to Kessel[11] and to Afrosimov et al.[12]. In Fig. 4 we show the experimental set up of Afrosimov et al.[13] who investigated the reaction

$$He^{++} + He \rightarrow He^{+}(n_1) + He^{+}(n_2) \qquad (15)$$

where n_1 and n_2 are the principal quantum numbers of the He^+ states. The kinetic energies E_0 of the He^{++} projectiles and E_1 of the He^+ particles formed by reaction (15) can be measured by means of the electrostatic parallel-plate analyzer positioned at angle θ. E_2 is approximately equal to $E_0 \sin^2\theta$. Then the energy states of the reaction products can be derived from the equation

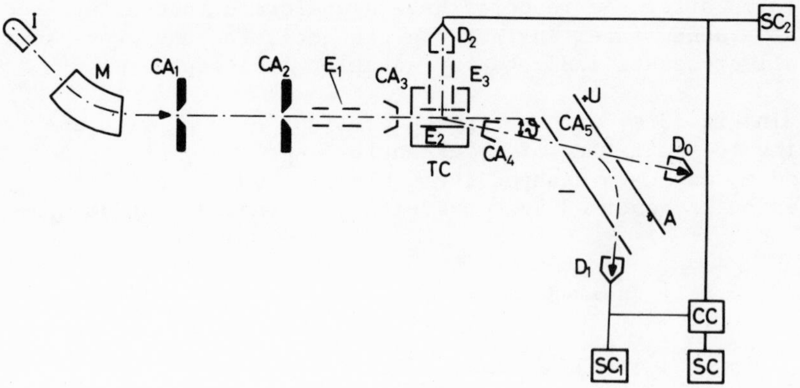

Fig. 4 - Set up of Afrosimov et al.[13] to measure E_0, E_1 and θ in the case of He^{++} incident on He. I - ion source, TC - target chamber, A - electrostatic analyzer, D_0,D_1,D_2 particle detectors, CC - coincidence circuit.

$$\Delta E = E_0 - E_1 - E_2 = I[He(1s^2)] - I[He^+(n_1)] - I[He^+(n_2)] \qquad (16)$$

where I stands for the ionization potential. Measuring the countrate at angle θ of the He^+ particles relative to the current of the He^{2+} projectiles entering the collision chamber, $d\sigma/d\theta$ can be evaluated for process (15) if the gas pressure and the effective length of the collision cell are known.

D. Measurement of Photon Emission.

Photons will be emitted when short living excited states are formed in the reaction, for instance

$$A^{q+} + B \rightarrow A^{(q-1)+*} + B^+ \qquad (17)$$

$$A^{q+} + B \rightarrow A^{q+} + B^* \qquad (18)$$

where the excited states are indicated by an asterisk. Generally the photons are observed by monochromators which collect photons emitted in a direction perpendicular to or at the magic angle (54°44') with respect to the ion beam. These measurements provide σ_{exc}, the cross sections for formation of projectile or target ions and atoms into excited states. So some information is equivalent to that obtained under C where the potential energies of the end products are determined. The cross section for electron capture into excited states of the projectile is part of the cross section $\sigma_{q,q-1}$ as discussed in A and B. The optical method does not give information about electron capture into the ground state and into metastable states of the projectile ion. For more technical details on optical measurements see for instance references 14 and 15.

In all the experiments described before, the target atoms were part of a "static" gas introduced in the collision chamber. As we shall see in section E, in particular cases it is advantageous or necessary to use an atomic beam (or an ion beam) as target, crossed by the projectile ion beam. This technique has been considered by Dolder[16] in a previous paper of this book. We shall limit ourselves to a short discussion about this topic.

E. Crossing and Merging Beams Techniques.

In Fig. 5 we show the set up of Fite et al.[17] in which a proton beam is crossed by a neutral H beam. The processes investigated are

$$H^+ + H \rightarrow H + H^+ \qquad \text{(electron capture)} \qquad (17)$$

$$H^+ + H \rightarrow H^+ + H^+ + e^- \qquad \text{(ionization)} \qquad (18)$$

Later on Stebbings et al.[18] have used the same set-up to observe Ly-α photons from excited states of H. The H beam comes from a

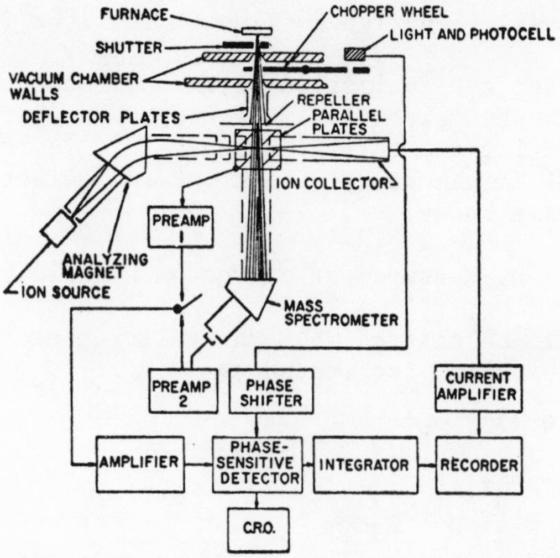

Fig. 5 - Schematic diagram of measurements on H^+ crossed by H[17].

hydrogen furnace and is interrupted with a chopper wheel. The frequency of interruption is used for phase sensitive detection, in order to discriminate processes (17) and (18) from processes of H^+ with the background gas. The measuring methods applied here are: condensor plates for collection of slow ions and electrons (A), analysis of charge of the target beam (B), also to discriminate H_2 from H in the neutral beam (the furnace does not completely dissociate H_2) and detection of photons from excited H atoms (D). Using the hydrogen oven as a target gas cell, as shown under (B) in fig.2, it is difficult or even impossible to apply methods (A) and (D), and no information is obtained about ionization and excitation pro-

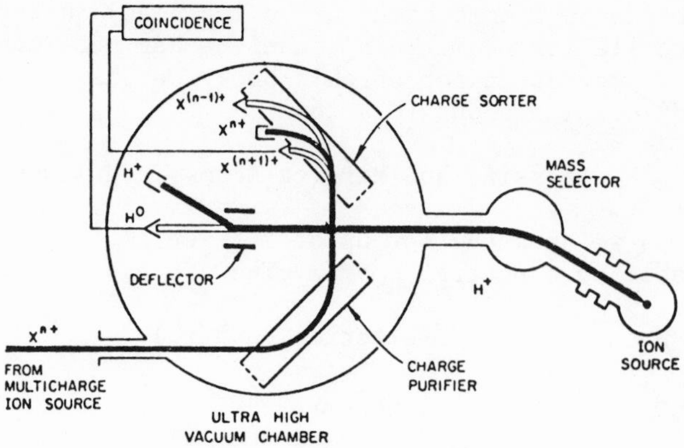

Fig. 6 - Schematic of Cross Beams Experiment for $H^+ + X^{n+} \rightarrow H^0 + X^{(n+1)+}$ (ref. 21).

cesses. Recently, instead of a hydrogen oven, a Wood discharge has been used to produce a high density atomic hydrogen beam[19,20].

In order to investigate reactions between two charged particles crossed beams techniques are unavoidable. In fig. 6 is given the scheme of the set-up for studying reactions of the following kind:

$$H^+ + X^{n+} \rightarrow H + X^{(n+1)+} \qquad \text{(electron capture)} \qquad (19)$$

$$H^+ + X^{n+} \rightarrow H^+ + X^{(n+1)+} + e^- \qquad \text{(ionization)} \qquad (20)$$

The measurements are of the type discussed in (C), i.e. analysis of charge and energy with application of coincidence techniques in order to discriminate processes (19) and (20). Experiments of this kind have been performed in the group of Gilbody[1] and Dolder[16].

In fig. 7 we see the first set up of merging beams, introduced by Trujillo et al.[22] to study low energy reactions of the kind

$$Ar^+ + Ar \rightarrow Ar + Ar^+ \qquad \text{(electron capture)} \qquad (21)$$

The method enables one to study reactions at low energy with merging beams at high energy. The beams merge behind the first collimating aperture. When Ar^+ has an energy, E_1, of 5000 eV and Ar an energy, E_2, of 5100 eV, the interaction energy in the center of mass system is given by

$$(E_1 - E_2)^2/4(E_1 + E_2) = 0.025 \text{ eV}.$$

These measurements are based on methods in (B), analysis of charge, and have been extended to coincidence measurements as discussed in (C). Activities of this kind are going on in the group of Brouillard (see for instance ref. 23).

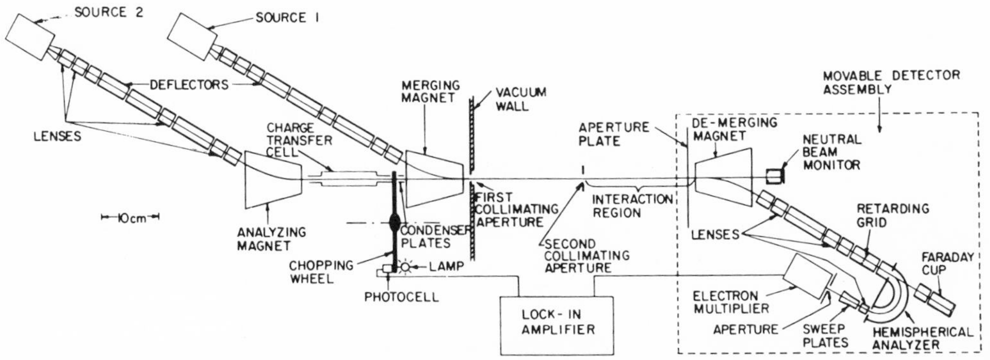

Fig. 7 - Schematic of merging beams apparatus for studying ion-neutral reactions[22].

3. ION SOURCES FOR MULTIPLY CHARGED IONS

In connection with controlled thermonuclear fusion it is important to carry out many of the experiments discussed in section 2 with multiply charged ions. Sources for production of these ions are discussed in a paper of Clark[24] and of Winter[25].

What is needed for these experiments?

a) Beams with charges up to about $q = 25$ and a velocity range up to about 4 a.u. (1 a.u. $\rightarrow 2.2 \times 10^8$ cm/s $\rightarrow$ 24.9 keV/a.m.u.): High charges up to $q \approx 10$ can be obtained either from electron bombardment low arc sources or in dense plasma sources ($n_e\tau$ is large). Application of stripping in beam foils leads to even higher charges.
b) Relatively large intensities in photon emission experiments as discussed in section 2 under (D): Large intensities are obtained in sources with large arc currents (Ampères), which causes a relatively large spread in the energy of the extracted ions.
c) Good energy definition in energy loss or gain experiments as discussed in section 2 under (C): Good energy definition is obtained in sources with low arc currents (milli Ampères), yielding relatively small ion currents. When large arc currents are used, a good energy definition can be obtained by application of electrostatic or magnetic energy dispersing devices.

As an example of a typical low energy spread ion source we mention the source of Menzinger and Wåhlin[26] (see also Huber et al.[27]). The large arc sources summarized by Clark[24] are Penning Ion Gauge, Duoplasmatron, Electron Cyclotron Resonance and Electron Beam Ion Source.

4. EXPERIMENTAL RESULTS AND COMPARISON WITH THEORY

4.1. Introduction

In this section we shall explain the experimental results, for singly charged ions (paragraph 4.2) and for multiply charged ions (paragraph 4.3) in a qualitative way. In paragraph 4.4 we summarize some of the theories which have been used for calculation of electron capture and ionization cross sections. In paragraph 4.5 we treat the characteristic features of the reactions between multiply charged ions and H and H_2 and make a few remarks about reactions with other targets. In paragraph 4.6 we consider the scaling of the data. In section 5 attention will be given to experiments on the formation of excited states in the projectile.

4.2. Singly Charged Ions Colliding with Neutrals.

In Figures 8 and 9 we illustrate results for

$$H^{+} + He \rightarrow H + He^{+} \quad (\Delta E = -11 \text{ eV}) \qquad \text{electron capture} \tag{22}$$

$$H^{+} + He \rightarrow H^{+} + He^{+} + e \quad (\Delta E = -24.58 \text{ eV}) \qquad \text{ionization} \tag{23}$$

These results have been obtained applying the condensor plate method A. The qualitative behaviour of these cross sections as a function of impact energy or velocity can be explained by the adiabatic criterion of Massey[29]. We apply the empirical "adiabatic maximum rule" as formulated by Hasted[30], "a simple but crude method" to predict the general behaviour of cross section versus energy curves. According to this rule a maximum in the cross section is obtained when a king of "resonance" occurs in which the collision time, given by a/v, times the transition frequency, $\Delta E/h$, equals one:

$$\frac{a}{v_{max}} \frac{\Delta E}{h} = 1 \tag{24}$$

a is a parameter of atomic dimensions, v is the velocity and v_{max} the velocity at which the maximum in the cross section is reached, ΔE is the internal energy difference of the reactants and products at infinite separation and h is Planck's constant. At small velocities the collision is adiabatic in character and the cross section may increase roughly as

$$\sigma \approx A \exp(-B\,\Delta E/hv) \tag{25}$$

in which A and B are constants. At higher velocities than v_{max} the perturbation time in the collision becomes small and therefore the cross section becomes smaller with increase of velocity.

This criterion can only be valid if the electronic transition occurs by a single jump from the initial state to the final one; it would not hold if the transition occurs through intermediate states. In general, the internal energy difference ΔE depends upon nuclear separation and the value of $\Delta E(R)$ should be used in the region of R where the electronic transition occurs. However, it is frequently found, as indicated schematically in Fig. 10, that ΔE does not vary much with R, and then we can use its value at infinite separation in the criterion (24).

The cross section versus energy behaviour according to the adiabatic criterion is reflected in Figs. 8 and 9. For the electron capture the maximum cross section is at E = 20 keV or $v_{max} = 0.9$ a.u. (1 a.u. $= 2.2 \times 10^{8}$ cm/s), $\Delta E = -11$ eV $= 0.4$ a.u. and applying eq. (24) we find $a = 14$ a.u. (1 a.u. = 0.529 Å). Similarly for ionization the maximum is at 100 keV and with $\Delta E = 24.58$ eV we again find $a = 14$ a.u.

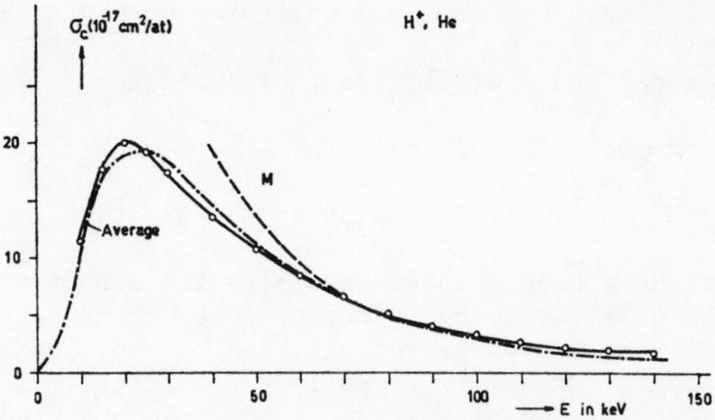

Fig. 8 - Cross sections for electron capture in the case of H^+ on He as presented by de Heer et al.[28]. Open circles are experimental data of ref. 28 connected by a solid line. The dash-dotted curve is the average of cross sections from other experiments mentioned in ref. 28. The dashed curve with symbol M stands for the theory by Mapleton.

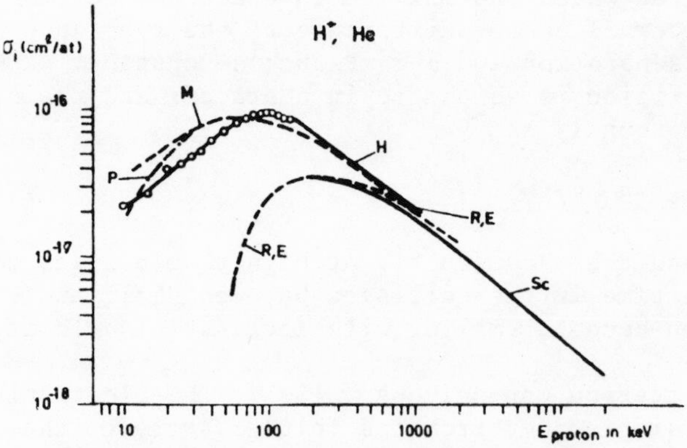

Fig. 9 - Cross sections for ionization of He by H^+ and electrons as presented by de Heer et al.[28]. Open circles are experimental data of ref. 28 for H^+ connected by a solid line and those of Hooper et al. are indicated by symbol H. M and P correspond to the theory of Mapleton and Peach. The curves labelled with R, E (Rapp and Englander-Golden) and Sc (Schram et al.) refer to experimental data for ionization of He by electrons. The energy scales of protons and electrons reduce to the same velocity according to $E_{proton} = E_{electron} \times (m_{proton}/m_{electron})$. All referecenes of the groups are given in ref. 28.

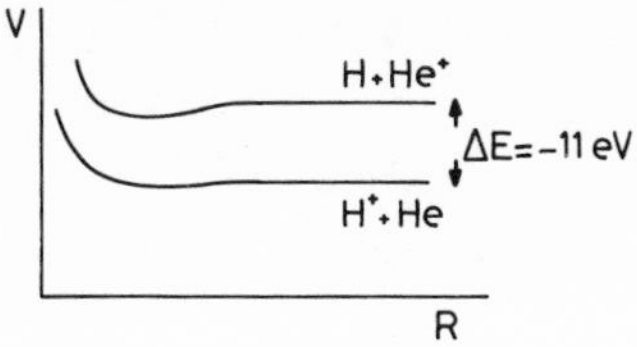

Fig. 10 - Qualitative picture of the potential curves playing a role in the electron capture process (22).

This value is close to that which has been found by Hasted and co-workers[30] for a large number of electron capture processes. This value of a is rather large compared with atomic dimensions. Sometimes the adiabatic criterion is written with $\hbar$ instead of h; this leads to smaller a values which might be more appropriate.

It may not be correct to apply the adiabatic criterion for ionization as we do here. In fact for ionization of He ΔE may have a variety of values dependent on the kinetic energy given to the ionized electron. Further, the relevant potential curves will have $\Delta E(R)$ values which depend on R because of the Coulomb repulsion of H^+ and He^+.

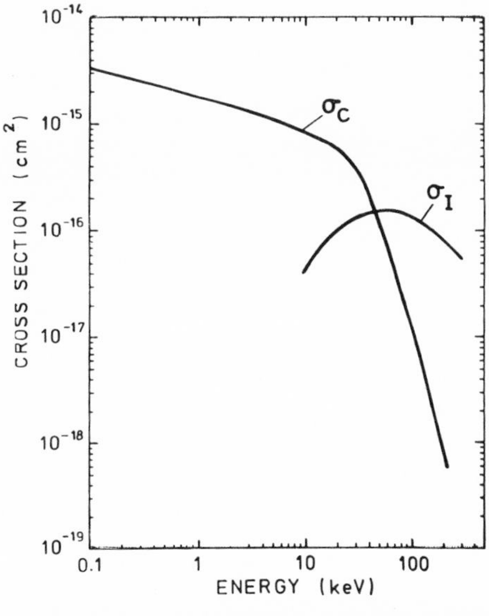

Fig. 11 - Cross sections for electron capture and ionization in $H^+ - H$ collisions (from ref. 1).

In Fig. 11 we show results for

$$H^+ + H \rightarrow H + H^+ \quad (\Delta E = 0) \qquad \text{resonance electron capture} \qquad (26)$$

$$H^+ + H \rightarrow H^+ + H^+ + e \quad (\Delta E = -13.6 \text{ eV}) \text{ ionization} \qquad (27)$$

The data are a combination of those of different groups, applying methods A, B and either oven or crossed beams techniques (further details in ref. 1). We see that the resonance capture cross section has a maximum at E or v equal to zero, in agreement with the adiabatic criterion. At low energies, according to a semiclassical impact parameter treatment (see section 4.4)

$$\sigma_c^{\frac{1}{2}} = a - b \ln v$$

where a and b are constants (see ref. 31). Similarly to exact resonance, cases of accidental resonance show a maximum at zero energy, for instance

$$H^+ + O \rightarrow H + O^+ \quad (\Delta E = 0.02 \text{ eV}) \text{ electron capture} \qquad (28)$$

This is illustrated in Fig. 12. The data have been obtained by using crossed beams and method B, analysing the charge of the target particles.

Next we want to indicate that at small velocity v $\sigma_c > \sigma_I$ and at large v $\sigma_c < \sigma_I$ (see again Fig. 11). The fact that the electron capture cross section becomes very small at high velocities is due

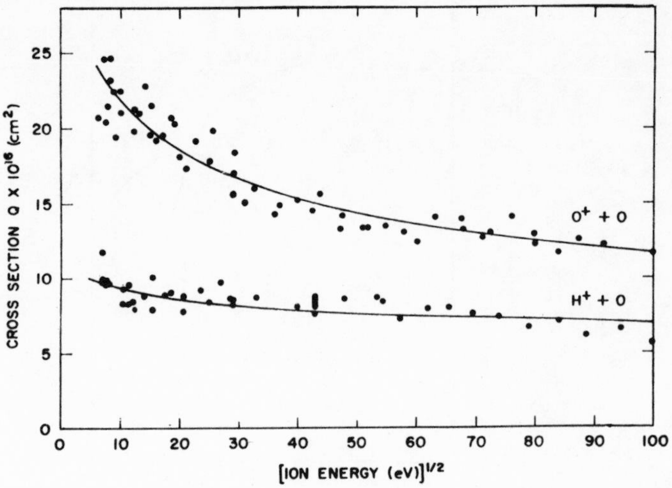

Fig. 12 - Charge transfer between oxygen atoms and O^+ and H^+ ions. The cross sections, in units of 10^{-16} cm^2, are plotted against the square roots of the ion energies (ref. 32).

to the change of the electron momentum when it jumps from the target to the fast projectile. On the other hand, the capture process is favoured at low velocities by the adiabatic criterion, since ΔE is smaller for electron capture than for ionization.

Finally we mention the work of Schlachter[33] and coworkers, who, using method B, investigated charge changing processes for H^+, H and H^- shot into a Cs oven. This work is important in connection with the injection of neutral H atoms into the fusion reactor. If we consider protons as a projectile and electron capture, different reactions are important which all lead to neutral H atoms:

$$H^+ + Cs \rightarrow H(1s) + Cs^+ (\Delta E = 9.7 \text{ eV}) \quad (29)$$

$$H^+ + Cs \rightarrow H(2s,2p) + Cs^+ (\Delta E = -0.49 \text{ eV}) \quad (30)$$

$$H^+ + Cs \rightarrow H(1s) + Cs^+ \ (\Delta E \approx -3.6 \text{ eV}) \quad (31)$$

When one applies the adiabatic criterion of Massey, all these processes have to be taken into account, because the experimental electron capture cross section in method B corresponds to a superposition of these processes.

4.3. Multiply Charged Ions Colliding with Neutrals.

We want to compare the cross sections for electron capture and ionization with those of singly charged ions discussed in the previous section. There we have seen that the general behaviour of the cross section as a function of impact energy for singly charged ions can be explained qualitatively by the adiabatic criterion of Massey both for non-resonance, resonance and accidental resonance reactions. Further σ_c was larger than σ_I at low energies whereas at high energies the reverse is true. How is this behaviour for multiply charged ions, similar or different?

Let us first consider a non-resonance ($\Delta E \neq 0$) electron capture reaction of the kind

$$A^{2+} + B \rightarrow A^+ + B^+ \ (\Delta E \neq 0) \quad (32)$$

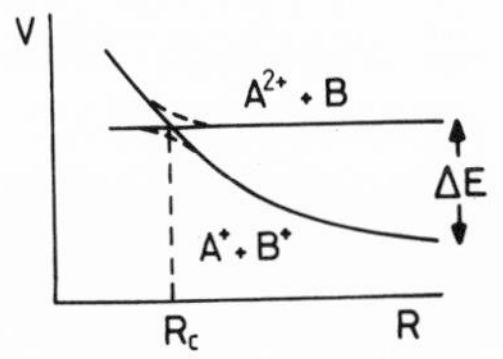

Fig. 13 - Potential diagram related to reaction (32). Dashed curves correspond to adiabatic states, solid curves to diabatic states in the molecule.

and the potential diagram given in Fig. 13. This diagram shows that for electron capture by multiply charged ions the energy difference of the potential curves varies strongly as a function of nuclear distance, mainly as a consequence of the Coulomb repulsion of the reaction products A^+ and B^+. In Fig. 10 we have seen that for electron capture by singly charged ions the energy difference of the potential curves might not vary much with R. However, we have to keep in mind that also for singly charged ions, many cases exist where ΔE varies strongly with R in the relevant interaction region. In any case it is clear that for multiply charged ions we cannot use the asymptotic value of ΔE, when we apply the adiabatic criterion of Massey. Before discussing the modification of the adiabatic criterion as formulated in (24) for singly charged ions, we shall discuss the meaning of the potential diagram in Fig. 13 in more detail.

In Fig. 13 we see that as a consequence of the Coulomb repulsion of the end products, a so-called crossing radius R_c occurs approximated by

$$R_c = (q-1)/\Delta E \text{ in a.u.} \tag{33}$$

where ΔE is the potential energy difference at $R=\infty$, and q is the charge state of the projectile (i.e. $q=2$). In this equation we disregard polarization effects between the collision partners and only consider ΔE and the Coulomb repulsion. If the molecular states corresponding with $A^{2+}+B$ and $A^+ + B^+$ have the same symmetry, then according to the non-crossing rule of Wigner and Witmer (see ref. 34), the curves representing eigenvalues of the electronic Hamiltonian as a function of R cannot cross each other. The dashed curves correspond to these "adiabatic" states of the molecule and these curves are followed in a very slow collision. However, when the heavy particles have a moderate or large velocity, a jump is possible from one adiabatic state to another. The system then follows the solid curves which correspond to "diabatic" states of the molecule (see ref. 35). It is clear from Fig. 13 that applying the adiabatic criterion of Massey, the effective value of ΔE will be much smaller than the asymptotic value. Because of this reduction in ΔE, the maximum in the cross section will be reached at lower velocities. At low velocities, $v < 1$ a.u. the molecular aspect of the collision is important. When the relevant molecular levels have the same symmetry, one can apply the theory of Landau-Zener[35,36,37], according to which the adiabatic criterion of Massey has to be modified into:

$$\frac{a' \, \Delta E(R_c)}{h \, v_{max}} = 1 \tag{34}$$

where

$$a' = \frac{\Delta E(R_c)}{\frac{d}{dR}(V_1 - V_2)_{R=R_c}} \tag{35}$$

$$\Delta E(R_c) = 2H_{12} \tag{36}$$

where H_{12} is the matrix element for coupling of diabatic states. The denominator in (35) is equal to the difference of the slopes of the diabatic potential curves at $R=R_c$. On the basis of geometry it is understandable that the maximum cross section according to Landau-Zener is of the order of πR_c^2.

Hasted and Chong[38] studied a series of reactions with multiply charged ions ($q = 2-4$) using method B. From the position of the maximum in the electron capture cross section and the R_c value evaluated by means of the potential diagram, they derived values for the matrix element H_{12}. On the other hand Olson et al.[39] introduced an empirical formula to derive the H_{12} values. Bates[35,40] considered the limitations of the Landau-Zener theory, where at higher energies the two-state approximation is no longer correct and the coupling is no longer localized near R_c.

The prediction of a maximum in the cross section becomes more complicated when more crossings play a role in the reaction. This is demonstrated in Fig. 14 for the reaction

$$C^{6+} + H \to C^{5+}(n) + H^{+} \tag{37}$$

The electron can be captured in different excited states of C^{5+}, each corresponding to a different R_c value. As we shall see further on, the region around $R_c \sim 7$ a.u. is the most important. For this case Salop and Olson[41] introduced a kind of multi-crossing Landau-Zener-theory. It appears that for higher charges more and more crossings play a role and lead to a flattening of the curve of σ_c vs. E

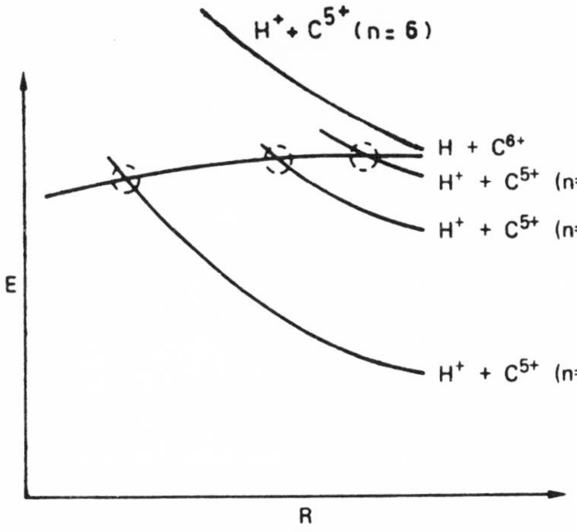

Fig. 14 - Diagram of diabatic potential curves for reactions between C^{6+} and H.

at low energies or a less pronounced maximum in the cross section. This has been verified by experiment (see section 4.5). So far we have only considered coupling between molecular states of the same symmetry or radial coupling of adiabatic states. It has also been shown that rotational coupling, which connects states of different symmetry is increasingly important at higher velocities (see for instance Russek[42]).

After having treated the non-resonance case, we now look to the resonance case:

$$A^{q+} + A \rightarrow A + A^{q+} (\Delta E = 0) \tag{38}$$

It has been found that reactions of this kind with $q > 1$ show a cross section behaviour similar to that for $q = 1$ with a maximum at $v = 0$. For this the reader is referred to the recent work of Okuno et al.[43] They used a so-called injected-ion drift tube technique for $Ar^{2+} \rightarrow Ar$, $Kr^{2+,3+} \rightarrow Kr$ and $Xe^{2+} \rightarrow Xe$ between about 0.04 eV to 3.8 keV.

The accidental resonance case as measured for

$$Ti^{2+} + H \rightarrow Ti^{+} + H^{+} \quad (\Delta E = -0.03 \text{ eV}) \tag{39}$$

by Nutt et al.[44] differs strongly from that for singly charged ions. The maximum in the cross section, as shown in Fig. 15, is not at zero energy. This can be understood by considering the potential diagram (see Fig. 16). The reason that the simple adiabatic maximum rule cannot be applied in this case is again due to the Coulomb repulsion of the reaction products Ti^{+} and H^{+}. At R is infinity ΔE is almost zero, but the coupling between the channels is weak because the particles are so far apart that there is no overlap of the atomic

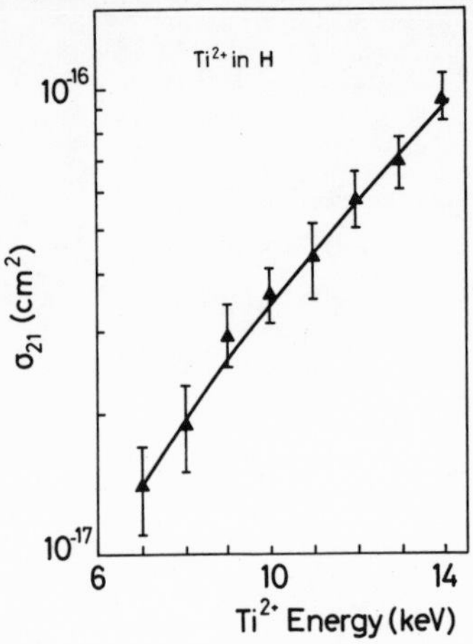

Fig. 15 - One-electron capture measured for Ti^{2+} on H (Nutt et al.[44]).

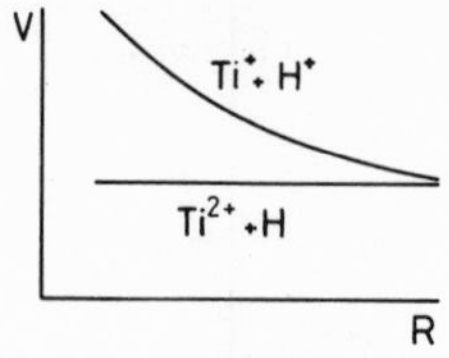

Fig. 16 - Potential diagram for reaction (39).

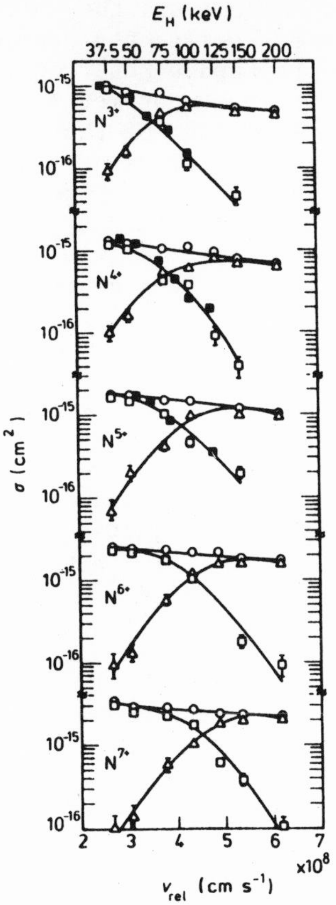

Fig. 17 - Experimental results on electron capture and Monte-Carlo calculations on electron capture and ionization for N^{q+} incident on H (q = 3-7). (Phaneuf et al.[46]).
■ exp. $\sigma_{q,q-1}$, □ theory $\sigma_{q,q-1}$
△ theory σ_I, ○ $\sigma_{q,q-1} + \sigma_I$.

orbitals. At the interaction distance where these orbitals start to overlap significantly (~5 a.u.), the potential curves are far apart, causing a maximum in the cross section at non-zero velocity. In fact one has to apply the adiabatic criterion with a ΔE value around R ~ 5 a.u. instead of at infinity (see also ref. 45).

Finally in Fig. 17 we show that also for multiply charged ions at low energies $\sigma_c > \sigma_I$ and at high energies $\sigma_c < \sigma_I$. For electron capture both experimental and theoretical results are shown for N^{q+} on H (q = 3-7) and for ionization only theoretical results.

4.4. Theory of Capture and Ionization.

Before we show the further features of the reactions for multiply charged ions incident in H and H_2 and make comparison with theoretical calculations, we shall mention the different theoretical methods and make a few remarks about them. For a good and more complete review the reader is referred to the progress report of Olson[47] and the articles of Bransden[48] and Joachain[49]. For an introduction to theory reference 35 is very useful.

At low energies $v < 1$ a.u. ($E < 24.9$ keV/a.m.u.) the molecular aspect of the collision is important and the dominant process is electron capture. We already introduced the Landau-Zener method in section 4.3. In general one often uses an impact parameter treatment, a semi-classical method, in which the motion of the particles is described classically and the relevant transitions between the quasi-molecular levels quantum mechanically. In Fig. 18 as an example A^{2+} collides at an impact parameter b with B and one is interested in the chance P(b) that the electron is captured from B by the projectile A^{2+}. The total cross section for electron capture is then given by

$$\sigma = 2\pi \int_0^\infty P(b)\, b\, db \tag{40}$$

Dependent on the number of relevant states involved in the reaction, the solution of the time dependent Schrödinger equation leads to a number of coupled states differential equations, which have to be solved. They lead to P(b) values in the different exit channels relevant for the different states formed. Radial and rotational coupling between quasi-molecular states play a role, where radial coupling connects molecular levels of the same symmetry ($\Sigma \to \Sigma$, $\Pi \to \Pi$ etc.) and rotational coupling connects levels of different symmetry ($\Sigma \to \Pi$, $\Pi \to \Delta$ etc.). To describe properly the electron capture process, one has to take into account the change of translational motion of the electron, which is often neglected. Its influence on the cross section will increase at higher velocities. Recently, Ryufuku and Watanabe[50] used a full quantum mechanical theory, the so-called unitarized distorted wave approximation (UDWA) based on travelling atomic orbitals, taking this change of translational motion into account. Therefore this method can also be applied at higher velocities.

In the region $1 < v/v_0 < q$, both electron capture and ionization processes are important ($v_0 = 1$ a.u. $= 2.2 \times 10^8$ cm/s). Here Salop and Olson[51] have successfully applied a full classical theory by means of their Monte Carlo calculations. In this method Hamilton's equations of motion for a three-body system are numerically solved for numerous trajectories. When A^{q+} collides with B, the three bodies are A^{q+}, B^+ and the electron around B^+. The forces between these

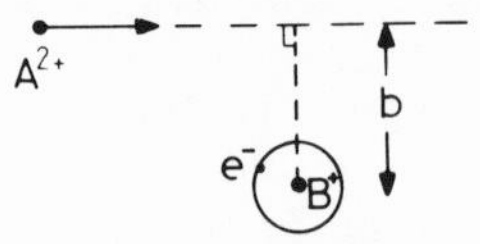

Fig. 18 – A^{2+} collides with atom B at impact parameter b.

bodies are taken Coulombic. At the end of the collision it has to be determined whether the electron is still in its original position around B^+ (no reaction), bound to A^{q+} (electron capture) or free (ionization). The calculation has to be done for 1000 to 2000 trajectories to reduce the statistical error.

For $v/v_0 > q$ ionization is most important and the cross section for electron capture becomes relatively small. In this region the first order Born approximation is used to calculate σ_I, a full quantum mechanical perturbation approach. Oppenheimer and Brinkman and Kramers (see ref. 35) also used a first order Born approximation (O.B.K. approximation) to calculate the electron capture cross section at high energies, neglecting the nuclear-nuclear term in the interaction potential. The O.B.K. approach generally leads to electron-capture cross sections that are too large. Chan and Eichler[52] have used an eikonal treatment to derive a simple formula to show that capture cross sections scale to the O.B.K. values. They used a Fock density matrix expression for the target electron momentum distribution.

Some theoretical developments have not been considered here. We mention the work of Presnyakov and Ulantsev[53], based on a generalized Landau-Zener method, Olson and Salop[54] using an absorbing-sphere model based on the Landau-Zener method introduced in section 4.3, so-called tunnelling models of Chibisov[55] and Grozdanov and Janev[56], and Bottcher[57] applying the so-called Magnus approach.

In general one is often interested in the dependence of the cross section on the projectile charge q (see ref. 47). At low velocities by a simple classical argument Bohr and Linhard[58] have shown that $\sigma_{q,q-1}$ is proportional to q. In their model one determines the internuclear separation where the force on the electron to be captured from target B by projectile A^{q+} is equal for A^{q+} and B^+. Then it is assumed that the electron will be removed from the target atom at all impact parameters smaller than this separation. At high velocities, according to the O.B.K. approximation $\sigma_{q,q-1}$ is proportional to q^3. In this velocity region σ_I is proportional to q^2 according to the Born approximation. We shall consider these dependencies in more detail for the experiment in the next sections.

4.5. Further Results for Multiply Charged Ions Incident on H and H_2.

It is not our intention to present the large amount of experimental results available now, but to illustrate the typical features of the reactions by making use of a few experimental data.

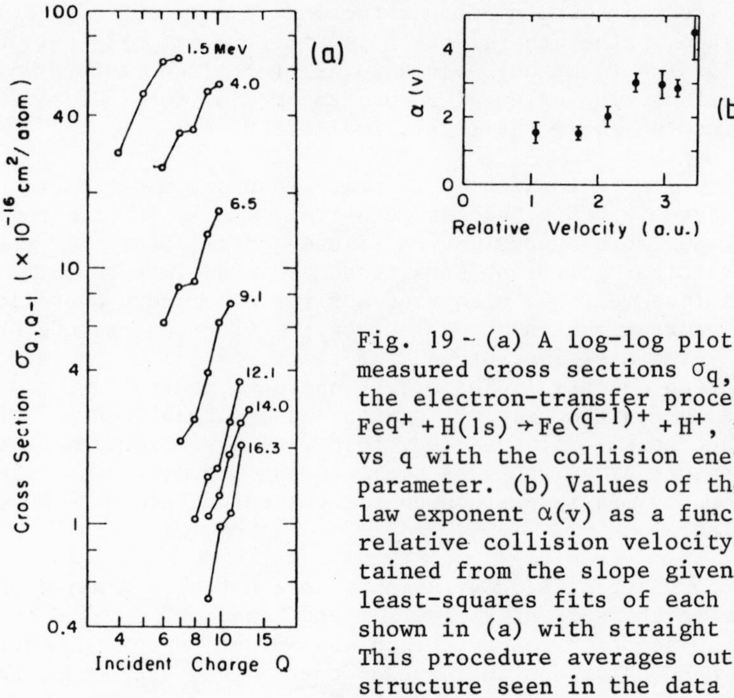

Fig. 19 - (a) A log-log plot of the measured cross sections $\sigma_{q,q-1}$ for the electron-transfer processes $Fe^{q+} + H(1s) \rightarrow Fe^{(q-1)+} + H^+$, plotted vs q with the collision energy as parameter. (b) Values of the power-law exponent $\alpha(v)$ as a function of relative collision velocity v, obtained from the slope given by least-squares fits of each curve shown in (a) with straight lines. This procedure averages out shell structure seen in the data and reflected in the shown error bars generated by the fitting routine (Gardner et al.[59]).

In Figs. 19a and 19b we show the data of Gardner et al.[59] who studied the dependence of $\sigma_{q,q-1}$ as a function of q at different velocities for reactions of the kind

$$Fe^{q+} + H \rightarrow Fe^{(q-1)+} + H^+ \tag{41}$$

The charge was varied between 4+ and 13+ and energies between 1.5 and 16.3 MeV were used. Measurements were performed by charge analysis, method B. In Fig. 19a we see that $\sigma_{q,q-1}$ increases with q and decreases with energy, because we are in the high velocity region far beyond v_{max} (see eq. (24)). In Fig. 19b the q dependence is presented as a function of the relative collision velocity according to

$$\sigma_{q,q-1} \sim q^{\alpha(v)} \tag{42}$$

At relatively small v, $\alpha(v)$ approaches to 1, the value predicted by Bohr and Linhard[58] and at relatively high v, $\alpha(v)$ is in the neighbourhood of 3, the value predicted by the O.B.K. approximation[59].

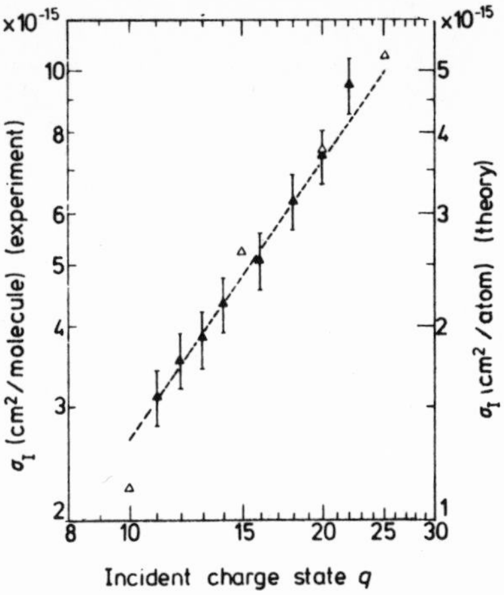

Fig. 20 - Calculated impact-ionization cross sections for Fe^{q+} (q = 10,15,20,25) + H collisions at 1.1 MeV amu^{-1} (open triangles, right ordinate) and experimental impact-ionization cross sections for Fe^{q+} (q = 11-22) + H_2 collisions at 1.1 MeV amu^{-1} (closed triangles, left ordinate). The broken line shows a least-squares fit of the experimental cross sections to eq.(43). (Berkner et al.[60]).

Similarly the impact ionization cross section, σ_I, has been measured by Berkner et al.[60] for Fe^{q+} incident on H_2 as a function of q(11-22), using the condensor plate method A. The experimental values of σ_I have been compared with those calculated for Fe^{q+} incident on H (q = 10,15,20,25) using the Monte Carlo method (see section 4.4) at an energy of 1.1 MeV per amu (v ≈ 6.6 a.u.). The results are presented in Fig. 20, where the σ_I values for H have been multiplied by 2. It is seen that the experimental and theoretical data fall on a line which can be represented by

$$\sigma_I = \sigma_1 \, q^{\alpha} \tag{43}$$

with $\alpha = 1.43 \pm 0.05$ and $\sigma_1 = (9.9 \pm 0.2) \times 10^{-17}$ cm^2.
Apparently the coefficient α is smaller than 2, the value predicted in the asymptotic high velocity region according to the Born approximation (see section 4.4). So far we showed the dependence of $\sigma_{q,q-1}$ and σ_I on q.

In Fig. 21 we illustrate that at low velocities $\sigma_{q,q-1}$ has little dependence on v for sufficiently large q ($\gtrsim$ 4).
As we have explained in section 4.3 this behaviour is caused by the importance of many crossings in the reaction for sufficiently large q, leading to a flattening of the cross section versus velocity curve or a less pronounced maximum. Such a behaviour has been found

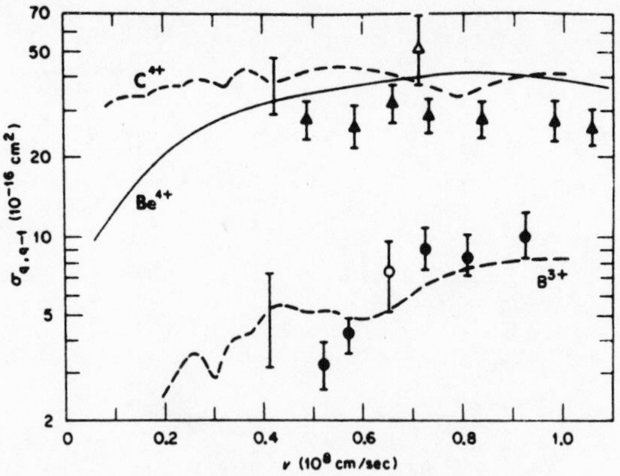

Fig. 21 - Electron capture cross sections for C^{4+} and B^{3+} in atomic hydrogen. ▲ C^{4+} data and ● B^{3+} data from experiment[10], △, ○ similar data of Bayfield et al.[61,62]. Dashed curves are theoretical results of Olson et al.[63] for these ions and the solid curve is the theoretical result of Harel and Salin[64] (Crandall et al.[10]).

experimentally also by Salzborn and coworkers[65] for multiply charged collisions of noble gas ions and atoms. The theoretical results in Fig. 21 have been obtained with the impact parameter treatment and agree with experiment within the combined errors of experiment ($\sim 10\%$) and theory ($\sim 30\%$). Olson et al.[63] evaluated both radial and rotational coupling between the molecular states within the coordinate system centered on the center of mass of the colliding system without electron translational factors. Eight molecular states were included for $(BH)^{3+}$ and seven for $(CH)^{4+}$. Harel and Salin[64] have calculated $Be^{4+} + H$ electron transfer and take a three state calculation in their impact parameter treatment. They suggest a large similarity for $Be^{4+} + H$ and $C^{4+} + H$ on the basis of the relevant potential diagrams and this is indeed the case (see Fig. 21). So far Fig. 21 was especially used to show that $\sigma_{q,q-1}$ is little dependent on v at low velocities for $q \gtrsim 4$, due to the multiplicity of crossings.

Next we show that for $q \gtrsim 4$, $\sigma_{q,q-1}$ is little dependent on the structure of the projectile (in Fig. 21 this was already indicated by theory for C^{4+} and Be^{4+} incident on H) and that for $q < 4$ this is no longer the case. In Fig. 22 we see that the capture cross section for $q = 2$ and 3 changes strongly with variation of the (structure of the) projectile (i.e. the projectiles are ions of O, N, C and B). However for $q = 4$, the cross section appears to be almost independent of the (structure of the) projectile. It is understandable that as q increases partially stripped ions will behave more closely as

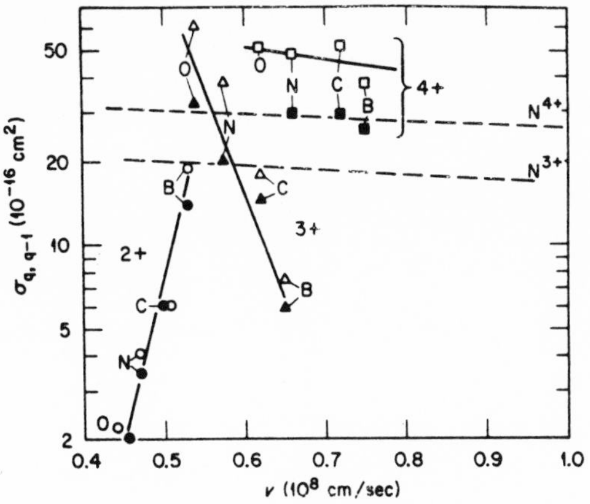

Fig. 22 - Cross sections for different multiply charged ions with H. Black symbols are from Crandall et al.[10] and open symbols from Bayfield et al.[61] corrected by Gardner[62]. Circles, triangles and squares represent data for ions of charge 2+, 3+ and 4+ respectively. Dashed lines represent data for nitrogen ions of charge 3+ and 4+ as a function of collision velocity (Crandall et al.[10]).

fully stripped ions of the same charge and structure influences become smaller.

The next point is the comparison of capture and ionization cross sections for H and H_2 as a target. Let us first start with ionization. In a large energy range

$$\sigma_I(H) \approx \tfrac{1}{2}\,\sigma_I(H_2) \tag{44}$$

so that approximately the additivity rule can be applied, similarly to the case of ionization by electrons (see refs. 66 and 67). The validity of (44) is suggested by the results in Fig. 20 where we compared experimental σ_I values for Fe^{q+} in H_2 with theoretical ones for Fe^{q+} in H.

For electron capture it has been found that

$$\sigma_{q,q-1}(H) > \sigma_{q,q-1}(H_2) \text{ at small v} \tag{45}$$

$$\sigma_{q,q-1}(H) < \tfrac{1}{2}\,\sigma_{q,q-1}(H_2) \text{ at large v} \tag{46}$$

This is demonstrated in Figs. 23a and b. Olson and Salop[54] have considered the differences of $\sigma_{q,q-1}$ for H and H_2 theoretically at low velocities. Two factors lead to a smaller coupling matrix ele-

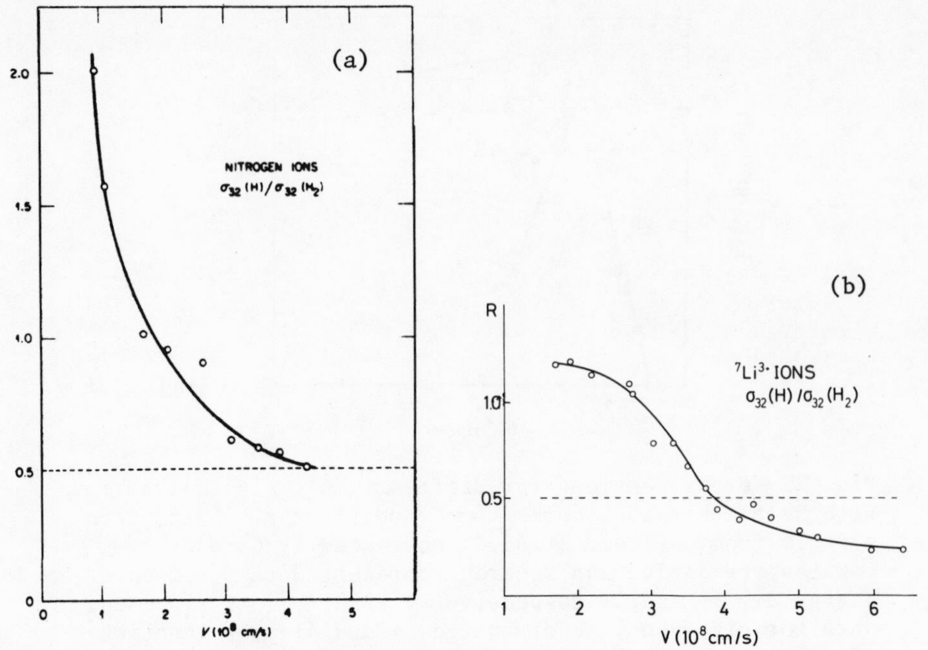

Fig. 23 - The ratio of the electron-capture cross section σ_{32} for (a) N^{3+} and (b) Li^{3+} incident on H and H_2, respectively, as a function of relative velocity.
((a) Phaneuf et al.[68], (b) McCullough et al.[69].)

ment in H_2. The first is the greater ionization potential of H_2 relative to H and the second is the presence of the Franck-Condon factor in the molecular matrix element. Thus, electron capture cannot take place at as large internuclear separation for the molecular target as for the atomic target. At high energies one could expect that $\sigma_{q,q-1}$ (H) is about half of $\sigma_{q,q-1}$ (H_2) as found in Fig. 23a at $v \sim 4.5 \times 10^8$ cm/s for N^{3+} projectiles. In Fig. 23b, however, McCullough et al.[29] find that the ratio even becomes smaller above $v \sim 4 \times 10^8$ cm/s for Li^{3+} projectiles. We are not able to explain this.

In this section we make a few remarks about collisions with other target atoms (see for instance Klinger et al.[70] and Müller et al.[71]). In targets different from H several electrons can be captured simultaneously. Further it has been found generally that

$$\sigma_{q,q-1} > \sigma_{q,q-2} < \sigma_{q,q-3} \text{ etc.} \qquad (47)$$

and that $\sigma_{q,q-1}$ increases for targets with smaller electron binding energy, as would be expected theoretically.

4.6. Scaling of Data.

Olson et al.[72] have used theoretical calculations, confirmed by experiment, to obtain a scaling rule for electron loss from a hydrogen atom in collision with a multiply charged heavy ion. Electron loss from H is determined by the sum of charge exchange and ionization cross sections:

$$\sigma_{loss} = \sigma_{q,q-1} + \sigma_I \qquad (48)$$

They calculated this cross section for the energy range 50 to 5000 keV/amu ($1.4 < v/v_0 < 14$) and for q in the range 1 - 50. The theoretical calculations employ the classical trajectory Monte Carlo method introduced in section 4.4. The results are presented in Fig. 24. The curve of the Monte Carlo calculations can be approximated by

$$\sigma_{loss} = 4.6q \times 10^{-16} \times \{(32q/E)[1 - \exp(-E/32q)]\} \text{ cm}^2 \qquad (49)$$

where E is given in keV/amu and q is the ion charge state. We see that at low energies, where σ_{loss} is almost equal to $\sigma_{q,q-1}$, σ_{loss}

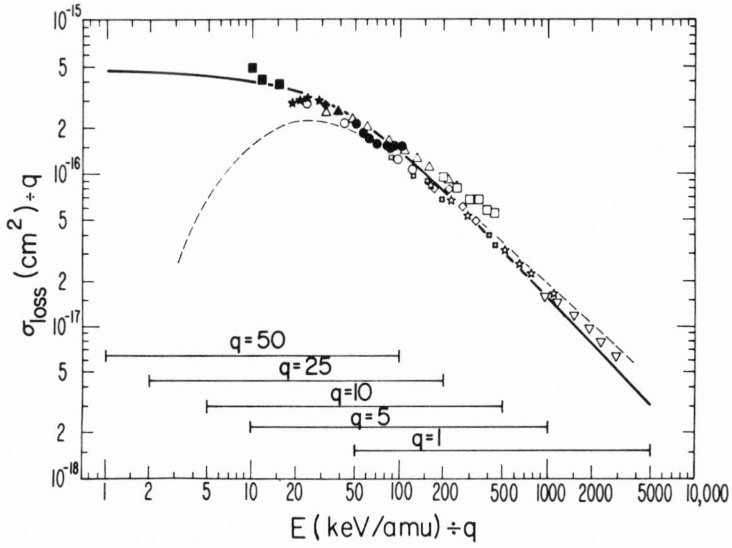

Fig. 24 - Solid line: Calculated cross section σ_{loss} for electron loss by atomic hydrogen in collision with an ion in charge state q; this curve is valid for $1 \leq q \leq 50$ and for energies in the range 50 to 5000 keV/amu. Dashed line: plane-wave-Born approximation cross section for ionization only. The points represent experimental data for a variety of projectiles, from H^+ to Fe^{22+} All ionization cross sections for a H_2 target are divided by two. For the origin of experimental data we refer to references given by Olson et al.[72].

is proportional to q, in agreement with what has been stated in sections 4.4 and 4.5. At high velocities σ_{loss} is almost equal to σ_I and becomes proportional to q^2, in agreement with the Born approximation as stated in section 4.4.

We shall not go into detail about the other work in scaling of data. Ryufuku and Watanabe[73] used their UDWA (see section 4.4) to derive a scaling law for $\sigma_{q,q-1}$ in the case of multiply charged ions incident on atomic hydrogen. Chan and Eichler[52] applied an eikonal treatment (see section 4.4) for calculation of $\sigma_{q,q-1}$ for similar collision systems and scaled the cross sections to the Oppenheimer-Brinkman-Kramer values. Müller and Salzborn[74,75] carried out a systematic experimental study of the capture reactions for multiply charged ions colliding with different atomic and molecular targets. They derived the next empirical formula:

$$\sigma_{q,q-k} = A_k \, q^{\alpha_k} \, I^{\beta_k} \qquad k=1,\ldots.4 \tag{50}$$

where q is used for the initial state of the projectile, k for the number of electrons captured and I for the first ionization potential of the target. A_k, α_k and β_k are empirical parameters. We shall confine ourselves to $k = 1$. In that case at 25 keV/amu they find $\alpha_k = 1.17 \pm 0.09$ and $\beta_k = -2.76 \pm 0.19$, using the best fit to experiment. In this region of relative low energies the value $\alpha_k = 1.17$ is very close to 1.12 predicted by Ryufuku and Watanabe[73] for atomic hydrogen and not too far from the value of 1 predicted by Bohr and Linhard[58] (see section 4.4).

5. FORMATION OF EXCITED STATES

In the previous section 4.5 we have discussed experimental results on electron capture and ionization for multiply charged ions on H and H_2. So far the results presented there are those for total electron capture and ionization cross sections, not giving any information about the excited states involved. Very little experimental work has been carried out dealing with the formation of excited states for multiply charged ions into H and H_2. In the case of atomic H, optical measurements are rather difficult, because one has to use crossed beam techniques (see section 2 at E). When using the high temperature tungsten oven for atomic hydrogen as a target chamber, the photon measurements will be disturbed by the radiation from the heated tungsten. In principle it must be possible with the oven to extend the measurements as carried out by method B (charge state selection of the projectile ions) to method C (charge state, energy and angle selection of projectile ions) as explained in section 2. Such a measurement has been carried out by Park et al.[76] to investigate ionization of atomic hydrogen for H^+ on H.

Using crossed beams some optical work has been performed for ions incident on atomic hydrogen. Stebbings et al.[18] investigated the formation of 2s and 2p states in $H^+ - H$ collisions both by electron capture and excitation of the target. The metastable 2s state was quenched by an electric field. Other optical studies have been made about formation of $He^+(2s)$ states formed after electron capture by He^{2+} on H by Bayfield and Khayrallah[77] and by Shah and Gilbody[78], again quenching these metastable states by an electric field.

For other targets many experiments have been carried out on the formation of excited states using either optical techniques (method D) or the selection method C. As far as multiply charged ions are concerned, for recent progress in this field we refer to the contributions in references 3 and 4. Here we want to limit ourselves to the work which has been carried out at our institute and has been summarized in reference 3.

The experiments we discuss deal with Ne^{q+} ($q = 1$ to 4) and Ar^{q+} ($q = 1$ to 6) in the energy range of 20 to 1200 keV shot into noble gases and H_2. The condensor method has been used to determine $\sigma_{q,q-1}$, the optical method to observe the short-lived radiating excited states by means of monochromators in the wavelength region of 10 - 600 nm, and an electrostatic analyser system to determine the energy spectra of the electrons formed in the reaction. Our experiments are mostly in a velocity region around $v = 0.5$ a.u. so that a molecular model can be used for qualitative interpretation. In Fig.

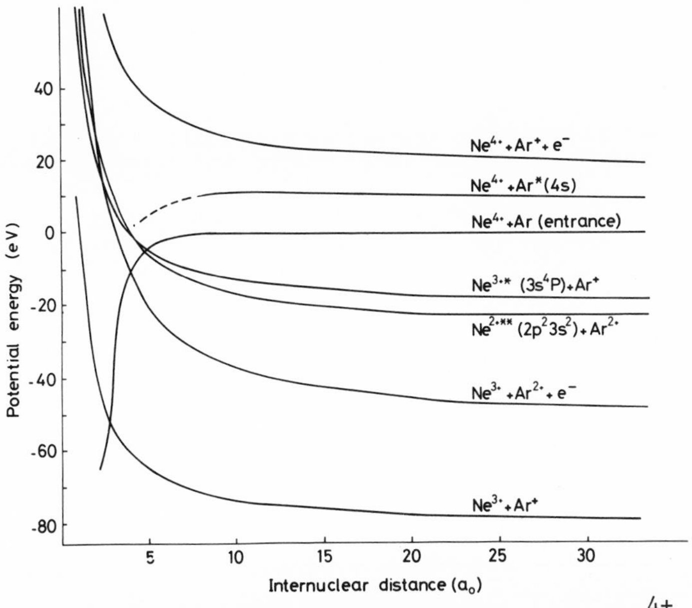

Fig. 25 - Potential-curve diagram for the system $Ne^{4+} - Ar$.

25 we show the potential diagram for Ne^{4+} colliding with Ar. We see a variety of exit channels corresponding to the different possible reactions. It is clear that there are many more processes than in the case of atomic hydrogen as a target. The question now is what reactions are most important. It has been found by Zwally and Koopman[79] and by Winter et al.[80] that moderate exothermic processes dominate that have a crossing radius in the interval of 3 au < R_c < 10 au. This can be qualitatively understood as follows. For the reaction R_c and the coupling matrix element $\langle\psi_{in}|H|\psi_{exit}\rangle$ are the important parameters. Here H is the Hamiltonian of the system and ψ_{in} and ψ_{exit} are the wavefunctions in the entrance and relevant exit channels. Because of the overlap of molecular orbitals, the coupling matrix element becomes larger at smaller R_c values. But since the geometrical cross section, equal to πR_c^2, decreases at smaller R_c, the processes become most important at an intermediate region of R_c values.

In Fig. 26 we show some electron capture results of our group for Ne^{q+} (q = 1-4) shot into Ar in the velocity region of v ≈ 0.5 -

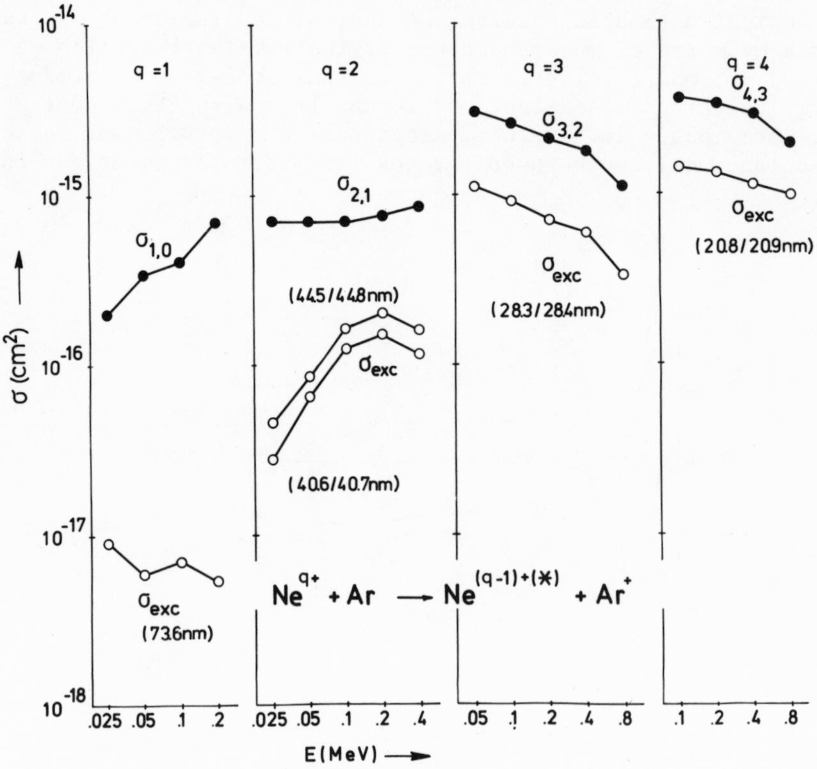

Fig. 26 - Cross sections for capture into excited projectile states, σ_{exc}, and total single electron capture cross sections, $\sigma_{q,q-1}$, for Ne^{q+} incident on Ar (Winter et al.[81]).

$\sim 3 \times 10^8$ cm/s. The curves, indicated with σ_{exc}, correspond with capture of electrons into excited states of the projectile and are determined by photon emission measurements. We limit ourselves to the most important photon emission in the vacuum ultraviolet part of the spectrum. The total capture cross sections are indicated by $\sigma_{q,q-1}$. If in Fig. 26 we move from $q = 1$ to 4, we see that σ_{exc} becomes larger with respect to $\sigma_{q,q-1}$. For $q = 4$ the probability for electron capture into an excited state is so large, that a kind of "excitation inversion" takes place. In this connection the great importance of reactions with highly charged ions lies in the possible application for stimulated emission in the short wavelength region (see I.I. Sobel'man[82]).

What is the reason that for larger q the excited projectile states are relatively important? If in Fig. 25 we enter via $Ne^{4+} + Ar$, then we first cross with $Ne^{3+*}(3s^4P) + Ar^+$ and later with channels having $Ne^{3+}(2p^3)$ in the ground state. Thus it is very well possible that $Ne^{3+*}(3s^4P) + Ar^+$ is preferred as exit channel above the channel with Ne^{3+} in the ground state. We can see this more clearly by using the criterion that processes with a crossing radius $3a_0 < R_c < 10\ a_0$ dominate. In Fig. 25 for Ne^{4+} on Ar we see that formation of $Ne^{3+*}(3s^4P)$ corresponds to $R_c = 5\ a_0$ and of the ground state, $Ne^{3+}(2p^3)$, to $R_c = 2.7\ a_0$. $Ne^{3+*}(3p)$ quartet states with $R_c \approx 7\ a_0$, also formed via electron capture, will contribute by cascade to $Ne^{3+*}(3s^4P)$ (see ref. 80). This example illustrates the importance of excited projectile states for Ne^{4+} on Ar. For Ne^{3+} on Ar similar considerations can be given. For Ne^{2+} and Ne^{1+}, the importance of excited projectile states is less important, because the corresponding electron capture processes are endothermic and electron capture into the ground state is exothermic. Some electron capture processes may give rise to electron production (as discussed further on) and fill the gap between σ_{exc} and $\sigma_{q,q-1}$ in Fig. 26 if electron capture to the ground state is relatively small.

A study was started about the principal and azimuthal quantum number of the excited projectile states formed in collisions of 200 keV ($v \approx 0.5$ a.u.) $Ar^{6+}(3s^2)$ ions incident on noble gases. The results are summarized in Fig. 27.

For a qualitative understanding of the results we give the ΔE and R_c values corresponding to the different excited projectile states in Table 1, where R_c is estimated according to Eq. (33). A few important observations can be made: Firstly, the total capture excitation cross sections tend to increase significantly with the increase in target Z, although the values for Kr and Xe are roughly equal. This is qualitatively in agreement with the predictions of Olson and Salop[54] using an absorbing sphere model and of Grozdanov and Janev[56] using an electron-tunneling theory. Secondly, $\sigma_{exc}(n)$, representing the cross section for electron capture into excited states with principal quantum number n, shifts to larger n for

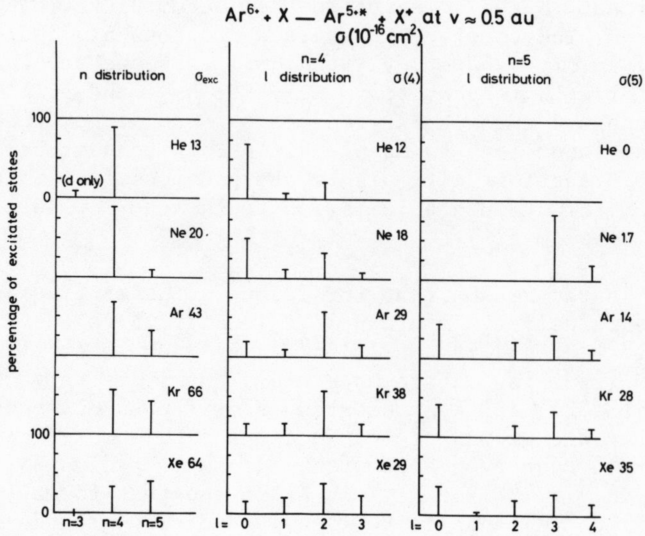

Fig. 27 - Cross sections for capture into excited projectile states in the case of 200 keV Ar^{6+} on noble gases (El-Sherbini et al.[83]). σ_{exc} refers to the sum of all the cross sections for electron capture into excited states. Percentage distribution over n and ℓ levels in a given n-shell are given.

TABLE 1

Level	He		Ne		Ar		Kr		Xe	
	ΔE	R_c	ΔE	R_c	ΔE	R_c	ΔE	R_c	ΔE	R_c
3d	39.6	3.43	42.7	3.19	48.5	2.81	50.2	2.71	52.1	2.61
4s	24.3	5.60	27.3	4.98	33.1	4.11	34.9	3.90	36.8	3.70
4p	17.8	7.66	20.8	6.55	26.6	5.12	28.3	4.80	30.2	4.50
4d	10.4	13.1	13.4	10.2	19.2	7.09	20.9	6.49	22.8	5.96
4f	6.58	20.7	9.60	14.2	15.4	8.83	17.2	7.93	19.0	7.15
5s	3.36	40.5	6.38	21.3	12.2	11.2	13.9	9.76	15.8	8.60
5p	-0.02		3.00	45.3	8.80	15.5	10.6	12.9	12.4	10.9
5d	-2.14		0.88	155	6.68	20.4	8.44	16.1	10.3	13.2
5f	-4.65		-1.63		4.17	32.6	5.93	22.9	7.80	17.4
5g	-4.96		-1.94		3.86	35.2	5.62	24.2	7.49	18.2

Change of interval energy, ΔE(eV), and $R_c(a_o)$ values for potential curve crossings for the various Ar^{5+*} excited states formed in collisions of Ar^{6+} with noble gases.

larger Z. This is consistent with our simple curve crossing model:

$$R_c = (q-1)/\Delta E = (q-1)/(I_{Ar^{5+*}(n)} - I_Z) \qquad (51)$$

where I is the ionization energy of $Ar^{5+*}(n)$ and of target Z. For larger Z, I_Z becomes smaller, and to remain in the same domain of R_c values $I_{Ar^{5+*}(n)}$ has to become smaller. This occurs when n increases, which is in agreement with our experimental result. Thirdly, for n = 4, $\sigma_{exc}(4,\ell)$ shifts to larger ℓ with larger Z. This is understandable for instance with regards to the $Ar^{5+}(4f)$ level which has $R_c = 20.7$ au for He and $R_c = 7.2$ au in Xe (see table 1), where only for Xe the R_c value falls inside the important domain. Fourthly for n = 5, 5d, 5f and 5g levels are found for Ar, Kr and Xe as a target, although R_c is large (13 - 35 au). In general at higher velocities ($v \gtrsim 0.5$ au) rotational coupling becomes important (see ref. 42) and coupling may occur at smaller internuclear distances than these large R_c values.

It is of interest to compare our results with the measurements of charge transfer in Ar^{6+} + He collisions at lower energies by Afrosimov et al.[84] and by Panov[85] between 6 keV and 100 keV, using method C for charge and state selection. They also find (see Fig. 28) formation of Ar^{5+*} 3d, 4s and 4p levels with R_c values between 3 and 10 au.

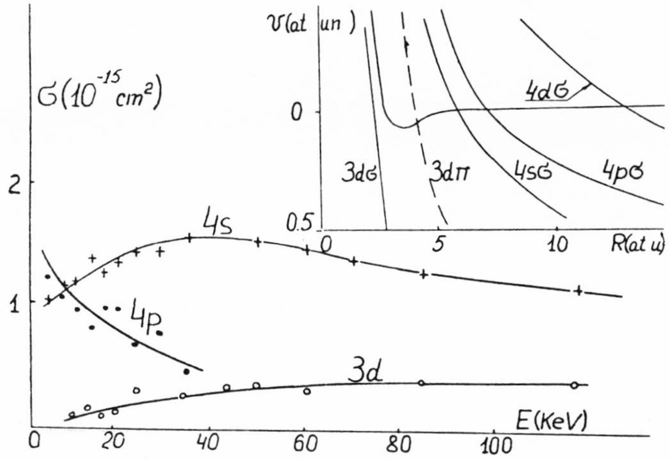

Fig. 28 - Cross section for one-electron capture by Ar^{6+} on He into different electronic states of Ar^{5+}. The relevant potential diagram is included in the figure (Panov[85]).

However, they do not observe the 4d level which is clearly seen in our case at 200 keV (see Fig. 27). This may be due to the fact that for 4d $R_c = 13.1$ a.u. At lower energies the coupling is more localized near R_c and transitions might not occur at this large R_c value.

Qualitatively these results are in agreement with the different theoretical calculations which have been done for different ion-atom combinations (see Ryufuku and Watanable[86], Salop[87] and Opradolce et al.[88]).

The theoretical UDWA results of Ryufuku and Watanabe[86] are summarized by them as follows: At low impact energies, < 10 keV/amu, the most probable n-state of the projectile, n_m, is determined by the effective level crossing point with the largest R_c value. For $n > n_m$ the excitation is vanishing small. At higher energies, up to about 100 keV/amu, the distribution over n-states becomes broad with the same n_m still as the most probable state. Above 100 keV/amu, the distribution over n becomes narrow due to the effect of momentum transfer.

At low impact energies the distribution over ℓ for a given n shows larger values at particular values of ℓ due to the effects of level crossing. At intermediate energies, the distributions become maximum at $\ell = n - 1$ for the higher values of n and $\ell < n - 1$ for the lower values of n. At high impact energy the distributions of ℓ are suppressed to smaller values of ℓ due to momentum transfer effects.

Next we consider some experiments concerned with the production of electrons and their energy spectrum. Winter et al.[81,89] have determined the total cross section, σ_e, for electron production by the condensor method A for Ne^{q+} ($q = 1 - 4$) on He, Ne and Ar and for Ar^{q+} ($q = 1 - 8$) on He, Ne, Ar, Kr and Xe at impact velocities around $v \sim 0.5$ au. The results show that σ_e values become very large when the relevant "transfer" ionization processes become exothermic. As we see further on (also consider Fig. 25), these processes are:

$$A^{q+} + B \rightarrow A^{(q-1)+} + B^{2+} + e^- \tag{52}$$

$$A^{q+} + B \rightarrow A^{(q-2)+**} + B^{2+} \tag{53}$$

where the doubly excited states of $A^{(q-2)+**}$ autoionize and lead to an electron of specific energy determined by the autoionization level. Reaction (52) leads to a continuous energy spectrum of electrons. In Fig. 29 we show the energy spectrum of electrons formed in collisions of Ne^{q+} ($q = 1 - 4$) with Ar at 100 keV ($v \sim 0.5$ au) observed at 90° with respect to the ion beam by Woerlee et al.[90]. We see different peaks, in particular for Ne^{4+}, which are superposed on a continuum. The peaks are due to electrons arising from autoionization of doubly excited states. For instance Ne^{2+**} states are formed by capture of two electrons in the case of Ne^{4+} projectiles. When we

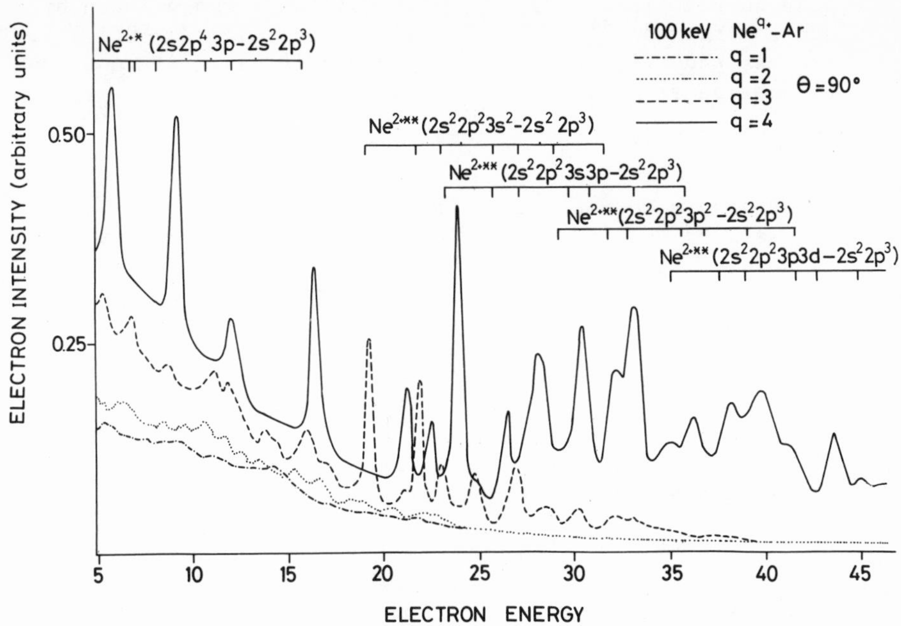

Fig. 29 – Electron energy spectra for 100 keV Ne^{q+} – Ar ($\theta = 90^{o}$). The bars indicate calculated energies of the electrons from autoionization levels corrected for a Doppler shift (Woerlee et al.[90]).

consider the potential diagram in Fig. 25 we see that the exit channel $Ne^{2+**} + Ar^{2+}$ corresponds to an exothermic reaction with a crossing radius of about 5 a.u. So these results strongly suggest that reactions with double electron capture are also very important when they occur in the domain of crossings $3 < R_c < 10$ au. No data exist on the autoionization states of Ne^{2+**} found in the spectrum, but the average energies of the autoionizing configurations, calculated by means of the computer program of Froese Fischer (see ref. 90) correspond roughly with those of the peaks. Similar results were found by Woerlee et al.[90] for Ne^{q+} on Kr and Xe. However, the results obtained for Ne^{q+} on Ne, not shown here, were different. In that case only peaks were observed due to autoionization states of the target, Ne^{+**}, formed according to

$$Ne^{q+} + Ne \rightarrow Ne^{(q-1)+} + Ne^{+**} \quad (54)$$

Considering the potential diagram for the Ne^{4+} – Ne system, it is found that this process is exothermic with $R_c \approx 3$ au. In this system formation of Ne^{2+**} is not likely, as found experimentally, because the relevant process (eq. 53) is endothermic.

It is interesting to compare the results of Woerlee et al.[90] at $v \sim 0.5$ au on reactions (52), (53) and (54) with those taken by Niehaus and coworkers[91,92] at relatively low energies ($v \sim 0.1$ au). The main process they find for electron production is different from that of Woerlee et al.:

$$A^{q+} + B \rightarrow [A^{(q-1)+*} + B^{+}] \rightarrow A^{(q-1)+} + B^{2+} + e \tag{55}$$

The electrons are formed via an intermediate state, which decays spontaneously by Penning or autoionization. Such a process has been clearly observed in the case of He^{2+} incident on Hg. No continuum is present as in the case of reaction (52) observed by Woerlee et al.[90]. Niehaus and Ruf[91] show that reaction (55) only occurs at impact velocities smaller than about 0.1 au. Probably at higher velocities the collision time is too short for autoionization of the intermediate molecular complex. In this velocity region, due to the nuclear motion, the continuum spectrum of electrons is probably formed via dynamic coupling (radial and/or rotational) in the quasimolecule as explained by Watanabe et al.[93].

6. CONCLUSION

In describing different electron capture and ionization experiments we have tried to give a review on the different activities going on in this field, with some emphasis on reactions of multiply charged ions colliding with atomic hydrogen. As has been stated by Olson[47] "this research area is rapidly approaching maturity and is being vigorously pursued" as is illustrated in his table of investigated reactions and in refs. 1, 3, 4 and 5. Recently the interest in these processes has been extended to very low energies in connection with the cool plasma edge of the fusion reactor. More electron capture data are required between about 1 eV and 100 keV, where it becomes more difficult to produce multiply charged ions. Some effort in this direction is going on in the group of Gilbody, Shah and McCullough at Belfast.

As far as atomic hydrogen as a target is concerned, several groups are considering the possibility of measuring the formation of excited projectile states after electron capture using the method of charge and energy state selection discussed in section 2.

The author is indebted to Dr. J. Delos, Drs. R.W. Wagenaar, Drs. E. Bloemen and Mr. D. Dijkkamp for their critical remarks on the manuscript and to Mrs. T. Köke-van der Veer for typing this manuscript.

This work is part of the research program of the Stichting voor Fundamenteel Onderzoek der Materie (Foundation for Fundamental Research on Matter) and was made possible by financial support from the Nederlandse Organisatie voor Zuiver-Wetenschappelijk Onderzoek (Netherlands Organization for the Advancement of Pure Research).

REFERENCES

1. H.B. Gilbody in *Advances in Atomic and Molecular Physics, Volume XV*, eds. D.R. Bates and B. Bederson (Academic Press, New York).
2. A.I. Kislyakov and M.P. Petrov, Soviet Phys.-Techn.Phys. *15*, 1252 (1971).
3. *Symposium on electron capture by multiply charged ions in the Book of the Invited Papers of the XI ICPEAC in Kyoto, 1979*, eds. K. Takayanagi and N. Oda (North-Holland Publishing Company, Amsterdam (in Press).
4. *Abstracts of Contributed Papers, XI ICPEAC, Kyoto, 1979*, eds. K. Takayanagi and N. Oda (North-Holland Publ.Comp., Amsterdam, 1979).
5. *Proceedings of the Conference in Nagoya on Atomic Processes in Fusion Plasmas*, 1979, organizer S. Ohtani (Institute of Plasma Physics, Nagoya University, Nagoya, in press).
6. H. Tawara and A. Russek, Rev.Mod.Phys. *45*, 178 (1973).
7. H.S.W. Massey and H.B. Gilbody in *Electronic and Ionic Impact Phenomena, Volume IV*, eds. H.S.W. Massey, E.H.S. Burhop and H.B. Gilbody (Oxford, Clarendon Press, 1974).
8. C.F. Barnett and W.L. Wiese *Atomic Data for Fusion* (OakRidge National Laboratory and National Bureau of Standards).
9. K. Katsonis, J. Rumble and F.J. Smith, *International Bulletin on Atomic and Molecular Data for Fusion* (IAEA Nuclear Data Selection, Vienna).
10. D.H. Crandall, R.A. Phaneuf and F.W. Meyer, Phys.Rev. *19*, 504 (1979).
11. Q.C. Kessel in *Case Studies in Atomic Collision Physics I*, eds. E.W. McDaniel and M.R.C. McDowell, pages 401-460 (North-Holland Publ.Comp., 1972).
12. V.V. Afrosimov, Yu.S. Gordeev, A.M. Polyanskii and A.P. Shergin, Soviet Physics-Techn.Phys. *17*, 96 (1972).
13. V.V. Afrosimov, A.A. Basalaev, G.A. Leiko and M.N. Panov, Soviet Physics - JETP *47*, 837 (1979).
14. F.J. de Heer in *Advances in Atomic and Molecular Physics, Volume II*, pages 327-384, eds. D.R. Bates and I. Esterman (Academic Press, New York, 1966).
15. J.A.R. Samson, *Techniques of Vacuum Ultraviolet Spectroscopy*, (John Wiley and Sons, New York, 1967).
16. K. Dolder, see previous article of this book.
17. W.L. Fite, R.F. Stebbings, D. Hummer and R.T. Brackmann, Phys. Rev. *119*, 663 (1960).
18. R.F. Stebbings, W.L. Fite, D.G. Hummer and R.T. Brackmann, Phys. Rev. *119*, 1939 (1960).
19. J.D. Walker and R.M. St. John, J.Chem.Phys. *61*, 2394 (1974).
20. S.T. Hood, A.J. Dixon and E. Weigold, FIAS-R-32 (1978) Institute for Atomic Studies, The Flinders University of South Australia.
21. D.H. Crandall in *Proceedings of the Fourth Conference on Scientific and Industrial Applications of Small Accelerators*, p.157, eds. J.L. Duggan and J.A. Martin (IEEE Serv.Centre, Piscataway, 1976).

22. S.M. Trujillo, R.H. Neynaber and E.W. Rothe, Rev.Sci.Instr. *37*, 1655 (1966).
23. M. Burniaux, F. Brouillard, A. Jognaux, T.R. Govers and S. Szucs, J.Phys.B: Atom.Molec.Phys. *10*, 2421 (1977).
24. D.J. Clark, IEEE Trans.on Nucl.Sci. *24*, 1064 (1977).
25. H. Winter in *Experimental Methods in Heavy Ion Physics*, ed. K. Bethge (Springer-Verlag, Heidelberg, 1978).
26. M. Menzinger and L. Wåhlin, Rev.Sci.Instr. *40*, 102 (1969).
27. B.A. Huber and H.J. Kahlert, paper submitted to J.Phys.B: Atom. Molec.Phys. 1979.
28. F.J. de Heer, J. Schutten and H. Moustafa Moussa, Physica *32*, 1766 (1966).
29. H.S.W. Massey, Rep.Progr.Phys. *12*, 248 (1949).
30. J.B. Hasted in *Advances in Electronics and Electron Physics*, Volume XIII, pages 1-78, ed. L. Marton (Academic Press, New York, 1960) and Proc.Roy.Soc. (London) *A212*, 235 (1952).
31. D. Rapp and W.E. Francis, J.Chem.Phys. *37*, 2631 (1962).
32. R.F. Stebbings, A.C.H. Smith and H. Ehrhardt, ICPEAC III, page 814, ed. M.R.C. McDowell (North-Holland Publ.Comp., Amsterdam, 1963).
33. A.S. Schlachter, Report LBL-6838, Lawrence Berkeley Laboratory, University of California (1977).
34. G. Herzberg, *Molecular Spectra and Molecular Structure, I. Spectra of Diatomic Molecules* (D. van Nostrand Company, Inc., Princeton, New Jersey).
35. D.R. Bates in *Atomic and Molecular Processes*, ed. D.R. Bates, pages 549-622 (Academic Press, New York, 1962).
36. L. Landau, Z.Phys.Sowjet *2*, 46 (1932).
37. C. Zener, Proc.Roy.Soc.A *137*, 696 (1932).
38. J.B. Hasted and A.Y.J. Chong, Proc.Phys.Soc. *80*, 441 (1962).
39. R.E. Olson, F.T. Smith and E. Bauer, Applied Optics *10*, 1848 (1971).
40. D.R. Bates, Proc.Roy.Soc. *A257*, 22 (1960).
41. A. Salop and R.E. Olson, Phys.Rev. *A13*, 1312 (1976).
42. A. Russek, Phys.Rev. *A4*, 1918 (1971).
43. K. Okuno, T. Koizumi and Y. Kaneko, ref. 4.
44. W.L. Nutt, R.W. McCullough and H.B. Gilbody, J.Phys.B: Atom. Molec.Phys. *11*, L181 (1978).
45. T.R. Dinterman and J.B. Delos, Phys.Rev. *A15*, 463 (1977).
46. R.A. Phaneuf, F.W. Meyer, R.H. McKnight, R.E. Olson and A. Salop, J.Phys.B: Atom.Molec.Phys. *10*, L425 (1977).
47. R.E. Olson, in reference 3.
48. B.H. Bransden, see other article of this book.
49. C.J. Joachain, see other article of this book.
50. H. Ryufuku and T. Watanabe, Phys.Rev. *A18*, 2005 (1978).
51. R.E. Olson and A. Salop, Phys.Rev. *A16*, 531 (1977).
52. F.T. Chan and J. Eichler, Phys.Rev.Letters *42*, 58 (1979).
53. L.P. Presnyakov and A.D. Ulantsev, Soviet J.Quant.Electron *4*, 1320 (1975).
54. R.E. Olson and A. Salop, Phys.Rev. *A14*, 579 (1976).

55. M.I. Chibisov, JETP Lett. *24*, 46 (1976).
56. T.P. Grozdanov and R.K. Janev, Phys.Rev. *A17*, 880 (1978).
57. C. Bottcher, J.Phys.B: Atom.Molec.Phys. *10*, L213 (1977).
58. N. Bohr and J. Linhard, K.Dan.Vidensk.Selsk.Mat.-Fys.Medd. *28*, 1 (1954).
59. L.D. Gardner, J.E. Bayfield, P.M. Koch, H.J. Kim and P.H. Stelson, Phys.Rev. *A16*, 1415 (1977).
60. K.H. Berkner, W.G. Graham, R.V. Pyle, A.S. Schlachter, J.W. Stearns and R.E. Olson, J.Phys.B: Atom.Molec.Phys. *11*, 875 (1978).
61. J.E. Bayfield, P.M. Koch, L.D. Gardner, I.A. Sellin, D.J. Pegg, R.S. Peterson and D.H. Crandell, *Abstracts of the Fifth International Conference on Atomic Physics, Berkeley, 1976*, eds. R. Marrus, M.H. Prior and H.A. Shugart (University of California, Berkeley, 1976).
62. L.D. Gardner, Thesis (Yale University, 1978).
63. R.E. Olson, E.J. Shipsey and J.C. Browne, J.Phys.B: Atom.Molec. Phys. *11*, 699 (1978).
64. C. Harel and A. Salin, J.Phys.B: Atom.Molec.Phys. *10*, 3511 (1977).
65. G. Salzborn, see references 3 and 5.
66. J.W. Otvos and D.P. Stevenson, J.Am.Chem.Soc. *78*, 546 (1956).
67. L.J. Kieffer, JILA Report no.30, 1965, *A compilation of Critically Evaluated Electron Impact Ionization Cross Section Data for Atoms and Diatomic Molecules* (University of Colorado, Boulder, Colorado, 1965).
68. R.A. Phaneuf, F.W. Meyer and R.H. McKnight, Phys.Rev. *A17*, 534 (1978).
69. R.W. McCullough, M.B. Shah and H.B. Gilbody, private communication.
70. H. Klinger, A. Müller and E. Salzborn, J.Phys.B: Atom.Molec.Phys. *8*, 230 (1975).
71. A. Müller, C. Achenbach and E. Salzborn, Phys.Letters *70A*, 410 (1979).
72. R.E. Olson, K.H. Berkner, W.G. Graham, R.V. Pyle, A.S. Schlachter and J.W. Stearns, Phys.Rev.Letters *41*, 163 (1978).
73. H. Ryufuku and T. Watanabe, Phys.Rev.*A19*, 1538 (1979) and ref. 3.
74. A. Müller and E. Salzborn, Phys.Letters *62A*, 391 (1977).
75. A. Müller and E. Salzborn, Inst.Phys.Conf.Ser. *38*, 169 (1978).
76. J.T. Park, J.E. Aldag, J.M. George, J.L. Peacher and J.H. McGuire, Phys.Rev. *A15*, 508 (1977).
77. J.E. Bayfield and G.A. Khayrallah, Phys.Rev. *A12*, 869 (1975).
78. M.B. Shah and H.B. Gilbody, J.Phys.B: Atom.Molec.Phys. *11*, 121 (1978).
79. J.H. Zwally and D.W. Koopman, Phys.Rev. *A2*, 1851 (1970).
80. H. Winter, E. Bloemen and F.J. de Heer, J.Phys.B: Atom.Molec. Phys. *10*, L1 (1977).
81. H. Winter, E. Bloemen and F.J. de Heer, J.Phys.B: Atom.Molec. Phys. *10*, L599 (1977).
82. I.I. Sobel'man in *The Book of the Invited Papers of the XI ICPEAC in Kyoto, 1979*, eds. K. Takayanagi and N. Oda (North-Holland Publ.Comp., Amsterdam).

83. Th.M. El-Sherbini, A. Salop, E. Bloemen and F.J. de Heer, J.Phys.B: Atom.Molec.Phys. *12*, L579 (1979).
84. V.V. Afrosimov, A.A. Basalaev, M.N. Panov and G.A. Leiko, JETP Lett. *26*, 699 (1977).
85. M.N. Panov, see ref. 3.
86. H. Ryufuku and T. Watanabe, Phys.Rev. submitted and JAERI-memo 8337 (1979).
87. A. Salop, J.Phys.B: Atom.Molec.Phys. *12*, 919 (1979).
88. N. Opradolce, P. Valiron and R. McCarroll in ref. 4.
89. H. Winter, Th.M. El-Sherbini, E. Bloemen, F.J. de Heer and A. Salop, Physics Letters *68A*, 211 (1978).
90. P.H. Woerlee, Th.M. El-Sherbini, F.J. de Heer and F.W. Saris, J.Phys.B: Atom.Molec.Phys. *12*, L235 (1979).
91. A. Niehaus and M.W. Ruf, J.Phys.B: Atom.Molec.Phys. *9*, 1401 (1976).
92. M.W. Ruf, Thesis (University of Freiburg, 1976).
93. T. Watanabe, P.H. Woerlee and Yu.S. Gordeev, ref. 4.

EXPERIMENTAL STUDIES OF ENERGY LEVELS AND OSCILLATOR STRENGTHS OF HIGHLY IONIZED ATOMS

Indrek Martinson

Department of Physics, University of Lund

S-223 62 Lund, Sweden

INTRODUCTION

The structure of highly ionized atoms is being vigorously investigated at present. The physical quantities of interest include wavelengths, excitation and ionization energies, fine- and hyperfine structure, lifetimes of excited levels and probabilities for decay by photon or electron emission. Such data are needed in atomic theory where electron correlation and relativistic effects present challenging problems. Spectroscopic material also facilitates studies of atomic excitation and collision processes. Astrophysical observations of the sun, stars, nebulae and interstellar medium require atomic data such as wavelengths (for the identification of various chemical elements) and transition probabilities (for determination of element abundances). Additional strong motivation for research in experimental atomic spectroscopy is provided by the fact that information about highly ionized atoms is very much needed in plasma physics and fusion research.

The studies of highly ionized atoms were initiated many years before most of these applications were known, however. Already in 1925 Bowen and Millikan[1] observed transitions in Cl VII which belongs to the Na I isoelectronic sequence. A few years later a very important experimental breakthrough was made in M. Siegbahn's laboratory at Uppsala, where a powerful light source (vacuum spark) and an efficient grazing-incidence spectrograph for wavelengths below 200 Å were constructed. Using this equipment Edlén[2] investigated the Na-like spectra of K IX - Cu XIX. Work on heavier systems, e.g. the Co I and Cu I isoelectronic sequences[3], was extended to Sb XXV and In XIII, respectively. By studying optical transitions at very short wavelengths (10 - 25 Å) Tyrén[4] mapped the Ne-like spectra of Cr XV - Co XVIII. Perhaps the most spectacular result of these early, accurate studies was Edlén's[5] identification of several strong lines in the solar corona as transitions between fine-structure levels in 9-15 times ionized Ca, Fe and Ni. A strong corona line at 5303.4 Å was thus explained as the 3p $^2P_{1/2}$ - 3p $^2P_{3/2}$ magnetic dipole (M1) transition in Al-like Fe XIV. The existence of such ionization degrees made clear that the temperature in the corona can reach $2 \cdot 10^6$ K. The pioneering character of the Uppsala work is evident from the fact that it would take more than 20 years before other groups obtained comparable results. The solar spectrum below 3000 Å was first studied in 1946 when an optical spectrograph was flown on a V-2 rocket[6]. Many observations of solar and stellar spectra using rocket- and satellite-borne equipment have followed[7] and wealth of data on highly ionized atoms has thereby been obtained. The presence of light impurities in high-temperature plasmas was established more than 20 years ago. Somewhat later also heavy impurities were identified[8]. A thorough review of this work was given by Drawin[9]. The results from astrophysics and fusion research have stimulated developments of compatible laboratory light sources for the production of highly stripped atoms. The vacuum spark was refined and various plasma light sources (theta pinch, plasma focus etc) have been successfully tried. Exploding wires have been found to produce high ionization degrees. Laser-produced plasmas belong to the most efficient and widely used light sources for studies of highly stripped atoms. All these light sources are mainly being used for time-integrated studies of spectra. When time resolution is needed (lifetime measurements) fast ion beams from particle accelerators can be used.

In this article on highly ionized atoms experimental methods and results are discussed. Because of space limitations only a small fraction of the available material can be mentioned. References will therefore be given to several excellent review articles.

LIGHT SOURCES FOR HIGH IONIZATION DEGREES

Vacuum Spark

The basic idea of a vacuum spark light source - an old device in atomic spectroscopy[10] - is quite simple. Two electrodes in vacuum are connected to a capacitor (C) which is charged to high voltage (U) until electric breakdown occurs. In the work of Edlén and Tyrén[2-4] C was 0.5 μF and U = 50-80 kV. The discharge current was about 50 kA. Different ionization stages were reached by varying the inductance L in the circuit. This open spark configuration can be modified by introducing an insulator between the electrodes. Such a "sliding spark" has been developed by Bockasten[11] and others. The light source is now easier to operate in a controlled and reproducible way but it is more difficult to obtain very high ioni-

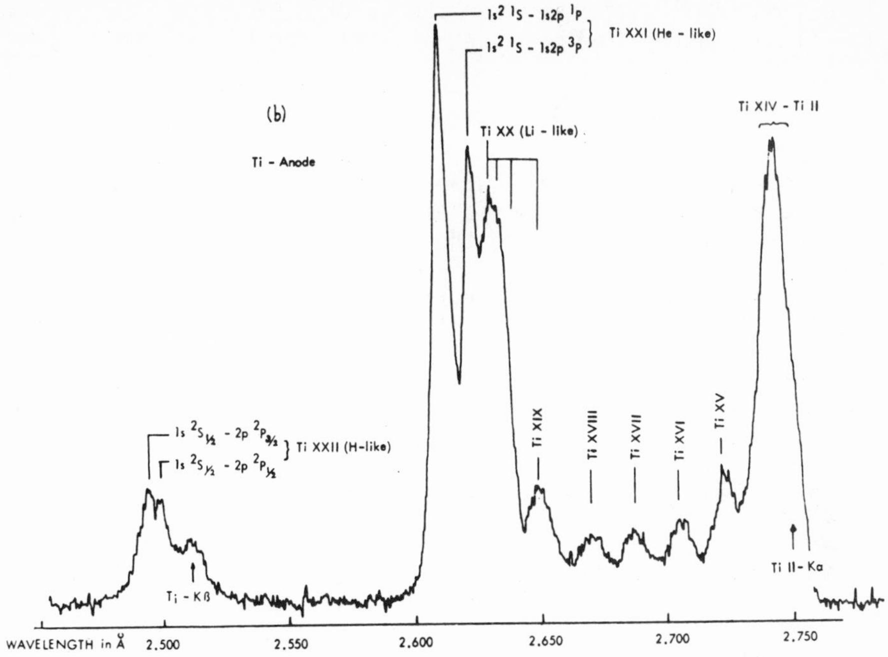

Fig. 1. Spark spectrum in the X-ray region of highly ionized Ti (Lie and Elton[14]).

zation stages. The sliding spark is therefore often used to produce 2-6 times ionized atoms.

Very high charge states are obtained using the so-called "low-inductance vacuum sparks", developed by several authors[12-17]. The light source of Feldman et al.[12] had the following parameters $U \sim 20$ kV, $C = 13$ μF, $L = 160$ nH and $I \sim 150$ kA. The voltage is not sufficient for a direct breakdown, the discharge is therefore usually triggered by a high-voltage prepulse. In the work of Turechek and Kunze[16] the values $U = 10$ kV, $C = 30$ μF, and $L = 2.5$ nH were used. The electron temperature was estimated to 33 - 40 keV and the electron density exceeded 10^{21} cm^{-3}. By studying the X-rays from a vacuum spark Cohen et al.[18] observed the $1s^2$-1snp transitions in Ti XXI - Ni XXVII (He sequence) and the 1s-2p transition in H-like Fe XXVI. Lie and Elton[14] recorded transitions in Ti XV - Ti XXII, Fe XIX - Fe XXVI and Cu XXVII - Cu XXVIII. One of their spectra is shown in Fig. 1. Transitions in He-like Ti, Fe and Cu were also studied by Cilliers et al.[15] while Beier and Kunze[17] observed radiation from He-like and Li-like Mo (Mo XLI and Mo XL).

The mechanisms which create highly ionized species in spark discharges have been discussed by several authors[13-18]. Some interesting similarities between the spectra from spark sources and solar flares (see below) have been investigated by Feldman et al.[19].

Exploding Wires

When a thin (10 - 100 μm) wire undergoes very sudden electric heating electromagnetic light is emitted[20]. If high-energy electrons are used the wire explodes and the radiation is predominantly in the extreme UV and X-ray regions[21]. Burkhalter et al.[22] have recently developed this method for the spectroscopy of highly ionized atoms. Their exploding-wire plasma was produced with 10^{12} W discharges of relativistic electrons in the Gamble II generator[23]. The X-ray spectra for a large number of wire materials (from Al to U) were recorded with a curved-crystal spectrograph. The plasma was found to consist of small pinched regions with high electron density (10^{21} cm^{-3}) and temperature (1-2 keV) separated by cooler (50-200 eV) regions. Transitions in H-, He- and Li-like Al, Ti and

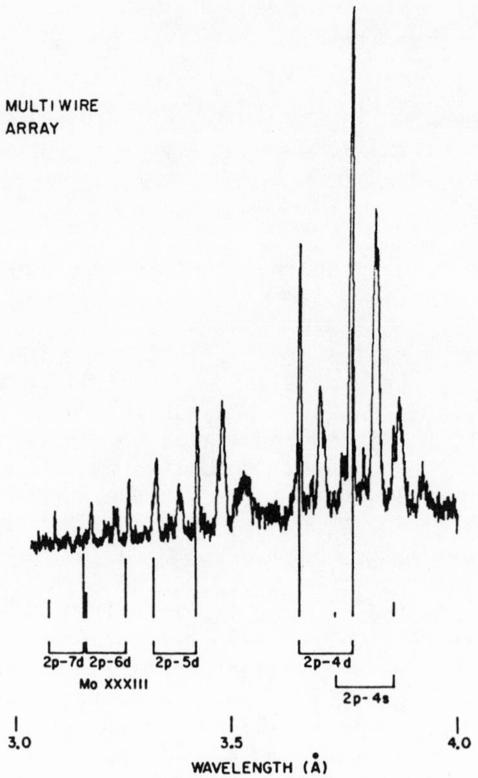

Fig. 2. X-ray spectrum of Mo XXXIII (Ne I sequence), recorded with the exploding-wire method[24].

Fe were emitted from the pinched regions whereas the cooler regions showed lines belonging to 11-13 times ionized Ti and Fe. With heavier wire materials the Ni-like spectra of Dy XXIX, W XLVII, Pt LI and Au LII could be observed. In a later publication[24] spectra of Mo XXXI-Mo XXXIV were investigated. These spectra were found to be quite complex because of overlapping ionization states. In addition there were pronounced satellite structures. Part of a Mo spectrum is shown in Fig. 2. The plasma temperature was about 4 keV in the Mo-experiment[24]. Additional information about the exploding-wire light source - which seems to be very efficient for production of highly stripped atoms - is available in a recent publication[25].

Laser Plasmas

The plasmas produced when a powerful laser irradiates solids have been used as spectroscopic light sources for more than 10 years. In one of the first experiments of this kind Fawcett et al.[26] used a 400 MW ruby laser (0.694 μm wavelength). The power density on the solid was approximately 10^{12} Wcm^{-2}. With Fe and Ni targets new transitions in Fe XV, Fe XVI, Ni XVII and Ni XVIII were observed. One of the interesting results of that early work was that the spectra were quite clean (no impurity lines) and they only showed transitions from a few ionization stages (e.g. Fe XV and Fe XVI).

Much work has since been performed with this light source. In the majority of cases Nd-glass lasers ($\lambda = 1.06$ μm) have been used but some spectroscopic work has also been reported with CO_2 lasers ($\lambda = 10.6$ μm). The laser power has reached several GW and the power densities often exceed 10^{14} Wcm^{-2}. A detailed review of ionization and excitation processes in laser-produced plasmas has been given by Peacock[27]. Relations between the electron temperature T_e and the light flux ϕ are thus discussed in that article. Experimental evidence seems to favor a power-law dependence $T_e \sim \Phi^p$ with p in the range 0.5 - 0.7.

The charge states which are produced can therefore be calculated. The ionization occurs so fast that only a few ionization stages contribute to the radiation (as noted above). Compared to the spark light source a laser plasma thus gives much cleaner spectra. Fawcett[28] points out that in the spark spectra of Fe the important $3s^2 3p^n - 3s^2 3p^{n-1} 3d$ transitions in Fe IX - Fe XIV may be obscured by $3p^6 3d^n - 3p^5 3d^{n+1}$ transitions in Fe V - Fe VIII, the latter being emitted during colder phases of the discharge. The ionization degrees from laser plasmas can be varied by changing the laser energy or defocusing the beam. There are also possibilities for spatial resolution of ionization stages in the plasma[28].

The electron densities in laser plasmas are typically 10^{19} - 10^{20} cm^{-3} and the temperatures may reach several keV. With such high electron densities many levels in a given spectrum can be populated. The spectral lines from high-power lasers often show asymmetries due to mass motion and Stark effects.

Only a few examples of the spectral studies of highly ionized atoms using laser plasmas can be listed here. Detailed information can be found in the paper by Boiko et al.[30] who investigated the H, He, Li, Be and Ne isoelectronic sequences (for $Z = 10 - 42$). The authors also identified many new transitions in the spectra of Fe XVII - Fe XXIV. Highly ionized members of the F and Na isoelectronic sequences[31-33] have also been investigated in this way. In a recent experiment Fawcett et al.[34] worked with 200 ps, 50 J pulses from a powerful Nd-glass laser obtaining 10^{16} Wcm^{-2} on the target. Transitions in Li- and Be-like Fe and Ni were studied. Highly ionized Mo (Mo XIV - Mo XXIV), of substantial relevance to impurity problems in Tokamaks, has been investigated in several laser-plasma experiments [35-39]. Using a 4 GW Nd-glass laser Fawcett and Hayes[40] observed up to 25 times ionized atoms of Cu - Br. Some years ago Aglitskii et al.[41] showed that laser irradiation of heavy elements such as Ta and W yields more than 50 times ionized atoms. In a study of the Cu I isoelectronic sequence Reader et al.[42] have extended the observations to W XLVI.

Plasma Light Sources

The Tokamak is a very efficient light source for highly ionized atoms[9,38,43,44]. The ionization stages that can be observed are similar to those in low-inductance sparks or laser-produced plasmas. However, the electron densities are much lower in Tokamaks, the value for PLT being $(3 - 10) \times 10^{13}$ cm^{-3}, while the electron temperature is in the region 1 - 2.5 keV[45]. One of the consequences of this low electron density is that Tokamak spectra are usually limited to transitions between low-lying levels, e.g. $\Delta n = 0$ and $\Delta n = 1$ resonance transitions in various ions. On the other hand, forbidden transitions (M1 or E2) can be observed in emission. In denser light sources collisional de-excitations make radiative decays by M1 or E2 processes less likely. Another interesting advantage of magnetically confined plasma light sources is the possibility of separating lines from different ionization stages by studying the temporal intensity variations during discharges. For detailed information about atomic physics with Tokamaks we refer to the papers by Hinnov[46] and Klapisch[47].

Magnetically confined plasmas have for years played an important role in atomic spectroscopy. As early as 1961 Fawcett et al.[48] studied the radiation emitted from the Harwell ZETA, a toroidal pinch with 3 ms pulse length, 10^{14} cm^{-3} electron density and 20 eV temperature. The spectra showed many new multiplets in highly ionized Ne, Ar, Kr and Xe, small quantities of which had been introduced into the light source. Using the toroidal discharge SCEPTRE IV Bockasten et al.[49] obtained excellent spectra of highly ionized B, C, N, O, F and Ne. Many authors have used theta pinch light sources for studies of highly ionized atoms. Using a 40 kJ theta-pinch Tondello and Paget[50] obtained detailed spectra of Ne VII and Ne VIII. Additional information about atomic physics with these sources can be found in Fawcett's review article[28].

Astrophysical Light Sources

Astrophysical observations of solar and stellar spectra have stimulated much laboratory research in atomic physics. However, the sun is also an excellent light source for atomic spectroscopy and much information about the level structure in highly ionized atoms has been obtained from solar spectra, particularly those in the UV and X-ray regions.

More than 70 chemical elements have been identified in the solar spectrum. The abundances (relative to hydrogen), as compiled by Engvold[51], are shown in Fig. 3.

The spectra of the solar photosphere mostly show absorption lines belonging to neutral and singly ionized species. Above the photosphere the temperature first reaches a minimum and then increases to $2 \cdot 10^6$ K in the corona. The electron density varies between 10^{11} cm^{-3} and 10^8 cm^{-3}. The light elements are completely ionized in the corona. In the case of iron transitions in Fe XV - Fe XVII are pronounced. Much higher ionization stages occur in solar flares where stored magnetic energy suddenly heats the plasma to $2 \cdot 10^7$ K.

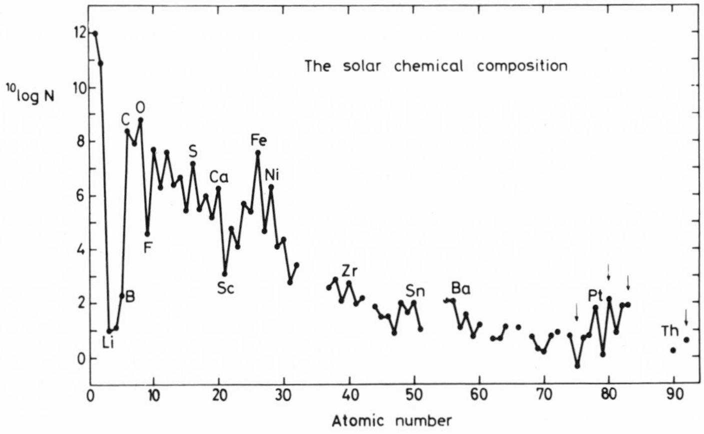

Fig. 3. Solar abundances of elements[51].

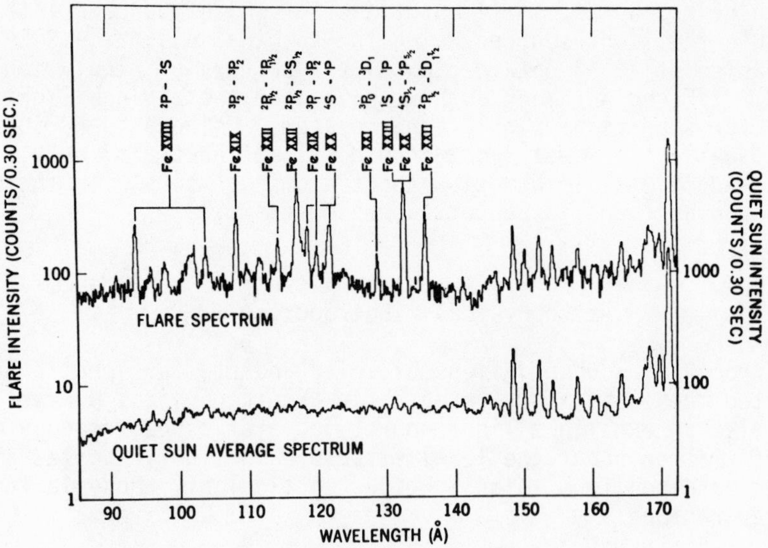

Fig. 4. Solar flare spectrum, obtained by Kastner et al.[54]. The identifications of Fe lines are from Fawcett and Cowan[55].

The electron density is about 10^{13} cm^{-3}. Elements up to Fe and Ni may become totally stripped in the solar flares[52,53]. Two spectra in the same wavelength region, one from the quiet sun and the other from a flare are shown in Fig. 4 (from the work of Kastner et al.[54] and Fawcett and Cowan[55]).

Much interesting spectroscopic material has in recent years been obtained from satellite studies of the solar spectrum. The information has been summarized in several reviews[28,52,53]. Excellent spectra were recently also reported from the Skylab mission [56-58].

Stellar UV spectra of high quality have also been recorded in the 1970's and promising developments can here be expected[52,53].

Excited Ion Beams

The spectra of highly ionized atoms can also be studied with the aid of heavy ion accelerators. This particular method, usually called beam-foil spectroscopy (BFS) was introduced by Kay[59] and Bashkin[60] in the early 1960's.

The principle of BFS is illustrated in Fig. 5. A particle accelerator (e.g. isotope separator, single or tandem Van de Graaff,

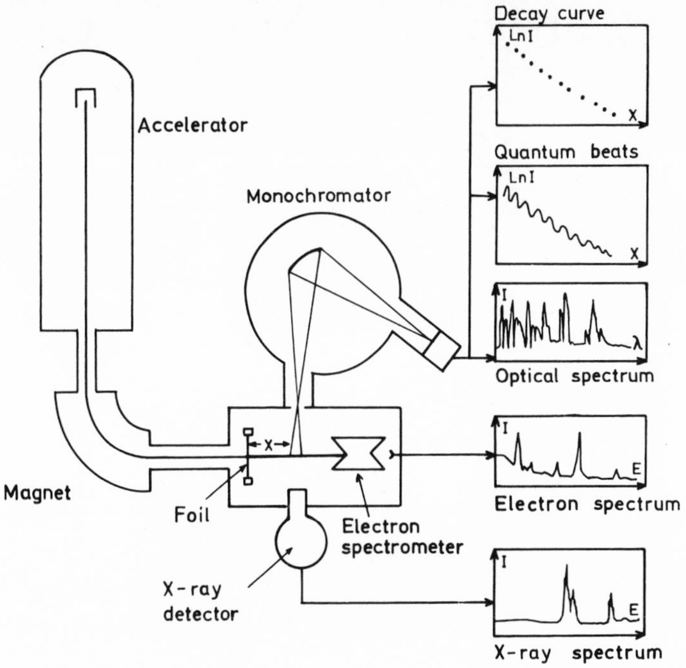

Fig. 5. Experimental arrangement for studies of atomic spectra and transition probabilities with BFS.

heavy ion linear accelerator, cyclotron) is used to produce a beam of fast ions which are directed through a thin carbon foil. While inside the foil the fast ions may undergo ionization and the emergent ions are often also in excited levels, the decay of which can be studied with optical, X-ray and electron spectroscopy (Fig. 5). The main strength of BFS lies in the excellent time resolution (a few ps in favorable cases) which permits studies of lifetimes and other time-dependent processes. However, even as a spectroscopic light source BFS has some interesting properties. Practically all elements can be accelerated in modern heavy-ion machines. Furthermore very many ionization stages can be reached by varying the ion energy. The excitation is unselective which e.g. means that a large number of levels are populated in each spectrum. Because of high vacuum (10^{-6} torr) and very low particle densities (typically 10^5 cm^{-3}) collisional processes are negligible after the foil, instead all excited levels decay spontaneously. This fact facilitates studies of high-lying Rydberg states in highly ionized atoms, multiply excited levels as well as forbidden transitions.

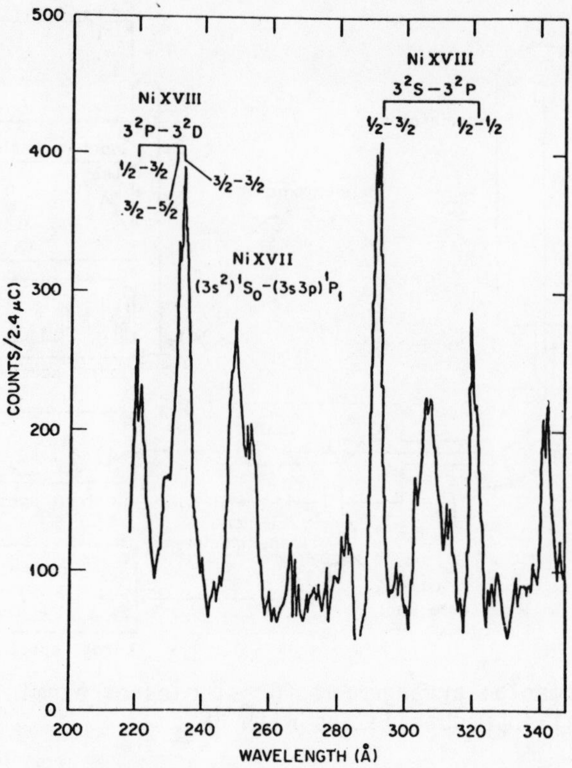

Fig. 6. Beam-foil spectrum of highly ionized Ni in the far UV region[61].

A spectrum observed with a beam of 64 MeV Ni ions from the Brookhaven tandem accelerator is displayed in Fig. 6. Transitions in Na- and Mg-like Ni are indicated.

The wavelength resolution in BFS is inferior to that attainable with many other light sources. The light is emitted by fast ions (v/c may be several per cent) and Doppler effects thus contribute to line widths. Although progress has been made in eliminating such effects[62], the BFS method is barely suited for classification work on highly stripped systems. However, a considerable amount of valuable data have been obtained for multiply excited configurations which are abundantly populated at the ion-foil interaction. The decay of these multiply-excited (or inner-shell excited) levels can be studied from photon or electron spectra[63-65]. Several review articles[66-69] summarize recent BFS experiments.

An interesting variation of BFS consists of bombarding gas targets with very heavy, highly ionized atoms. The collisions with the gas atoms also lead to highly ionized target atoms. Since these are slowly recoiling the spectral resolution is quite high. This very promising method for obtaining highly ionized atoms has been developed by Beyer et al.[70]

COMMENTS ON ATOMIC SPECTROSCOPY

Several additional light sources for the production of highly ionized species are possible[28,71,72]. Some years ago a new idea, based on inner-shell vacany production, was also tried. It is well known that the creation of a vacany in an atomic inner shell is accompanied by Auger - and shake-off processes which lead to highly ionized atoms[73]. Erman and Berry[74] have made a preliminary experiment in which inner-shell vacancies were created with beams of keV electrons. Transitions in e.g. O VI and Ar VII were observed; more work along these lines would be worthwhile.

Discussions of technical kind,concerning e.g. spectrographs, monochromators, detectors, electronics, data handling etc. will be omitted here. Detailed information about these questions can be found elsewhere[71,75-77]. We will instead report - in a rather sketchy way - some recent results in atomic spectroscopy. Thorough discussions of atomic spectra have been given by Edlén[78-80] and Fawcett[28].

In establishing atomic energy level diagrams one usually begins with a measured spectrum. Experimentally determined wavelengths are converted into energy differences (in cm^{-1} or eV) from which the energy level structure is established e.g. by means of the Ritz combination principle. It is well known that very high spectral resolution is necessary in such work. Otherwise blends of spectral lines (particularly serious in line-rich spectra) will cause problems. High wavelength accuracy is also essential because accidental coincidences of energy differences would lead to confusion in the establishment of energy levels. This is especially important in the far UV, as pointed out by Edlén[71]. Because of the relation $\Delta\sigma = \Delta\lambda/\lambda^2$ (where $\Delta\sigma$ and $\Delta\lambda$ are the uncertainties in wavenumber and wavelength, respectively) very high accuracy is needed for work at short wavelengths.

When light atoms and ions are studied it is usually possible to establish a very detailed and complete energy level diagram. The task is simplest for systems with only one valence electron, e.g. the Li I and Na I isoelectronic sequences. Available information about these spectra is summarized and critically analyzed by Edlén [81,82]. One example of such an analysis is shown in Fig. 7 which illustrates the smooth variation of level energies with nuclear charge Z along the Na I isoelectronic sequence. Similar work, for the Be I sequence (two valence electrons) has also appeared[83].

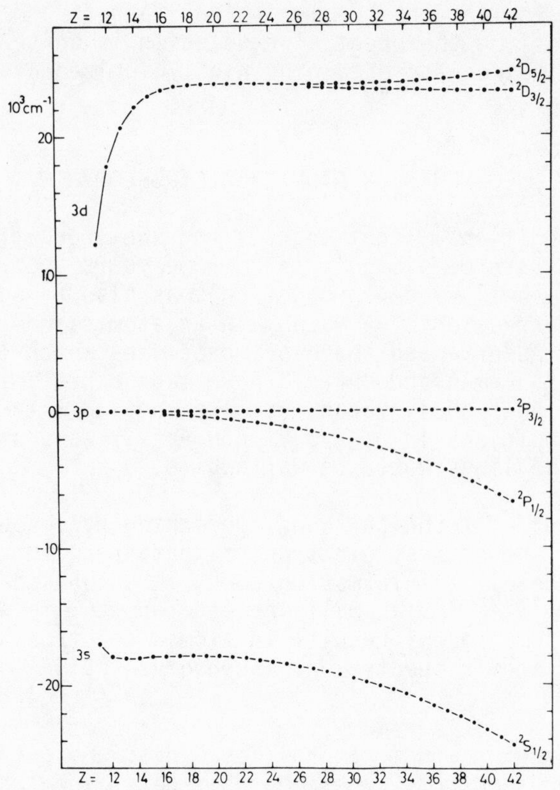

Fig. 7. Relative positions of the 3s ^2S, 3p ^2P and 3d ^2D levels in the Na I isoelectronic sequence, obtained by plotting the redcued energy expression $[E(LJ) - E(^2P_{3/2})]/(Z - 10)$ vs Z. (From Edlén[82]).

The energy level structure is quite satisfactorily known for many light atoms and ions but there are still several fundamental spectra (e.g. O II, O III, F IV - F VI) where much more information is needed[80].

As an example of an analysis which is complete for all "practical" purposes Fig. 8 from Zetterberg and Magnusson[84] shows the energy level diagram for P IV (Mg I isoelectronic sequence). There is much configuration mixing in this spectrum. Because of deviations from LS coupling several intercombination lines, which violate the $\Delta S = 0$ rule are also found. The accuracy of the work[84] can e.g. be judged from the fact that the ionization energy was determined with an uncertainty of $2.4 \cdot 10^{-6}$.

When more complex spectra are studied, e.g. those belonging to the iron group elements, such completeness is not realistic.

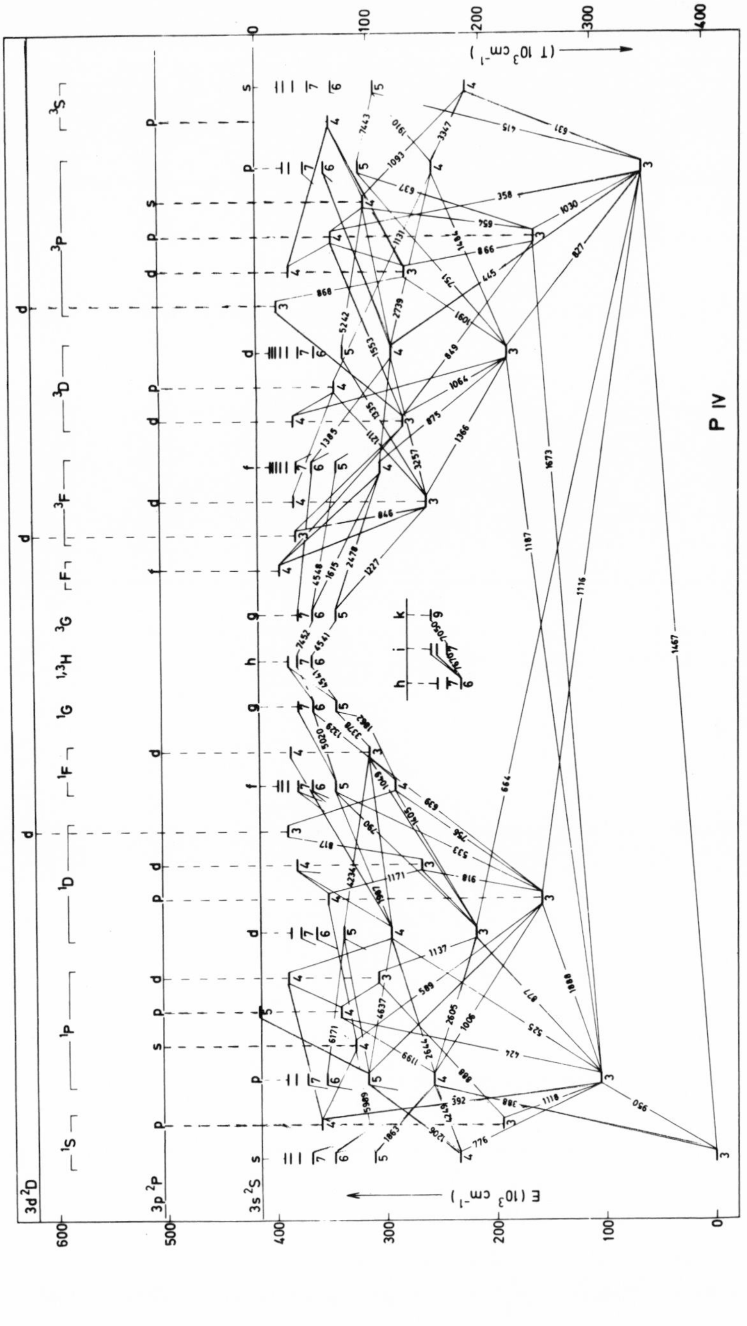

Fig. 8. Energy level diagram for P IV according to Zetterberg and Magnusson[84].

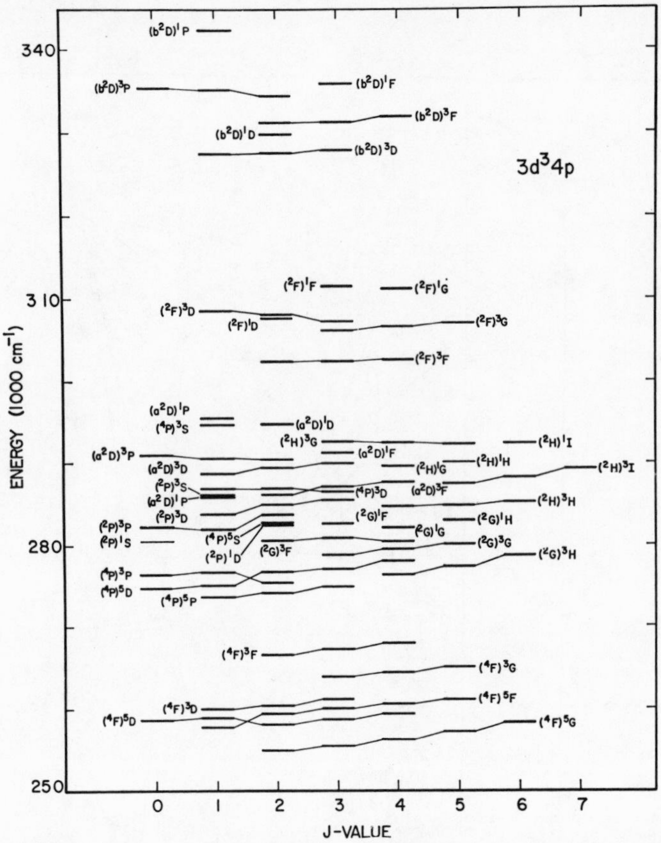

Fig. 9. Structure of the $3d^34p$ configuration in Fe V according to Ekberg.[85]

Edlén[79] has pointed out that in the cases with half-filled 3d shell it is practically impossible to make a complete analysis of more than the lowest configurations such as $3d^k$, $3d^{k-1}4s$ and $3d^{k-1}4p$. The situation for Fe V ($k = 4$) can be studied in a paper by Ekberg[85] from which the structure of the $3d^34p$ configuration is shown in Fig. 9. There are 110 levels within this configuration!

In a study of the Fe II spectrum Johansson and Litzén[86] classified 481 lines as belonging to the $3d^6(^5D)4d - 3d^6(^5D)4f$ transition array.

The complications increase even more when the rare-earth spectra are studied. For example the $4f^7$ configuration has 476 levels!

The isoelectronic comparisons are important also when complex spectra are studied. Interesting effects may here occur when high-lying levels with low principal quantum numbers move through the level system when Z increases. In the Xe I sequence, for example, the ground state is $5p^6\ ^1S$ and the excited levels initially show the typical character well known in noble-gas spectra. From Pr VI on the $5p^6\ ^1S$ ground level is replaced by hundreds of $4f^6$ levels[87] and the isoelectronic sequence ends in practice. A somewhat similar, although less drastic change, for the K I sequence was noted by Mansfield et al.[38]. In K I the ground state is $4s\ ^2S$ while in Mo XXIV it is $3d\ ^2D$. In the latter system there are also low-lying inner-shell excited configurations such as $3p^5 3d^2$, $3s3p^6 3d^2$ and $3p^4 3d^3$ which complicate spectral analyses.

Forbidden transitions are often due to M1 transitions between fine-structure levels within an LS term. When the energy levels and their separations are very accurately known, forbidden lines can be identified. Fig. 10, from a recent study by Edlén[88], shows forbidden lines within the $2s^2 2p^4$ complex in the O I isoelectronic sequence. The lines observed in auroral, nebular, coronal and flare spectra are identificated. Discussions of forbidden lines in the solar corona are found in recent articles by Jordan[53] and Smitt[89]. The latter paper also includes a list of all observed corona lines.

The power of isoelectronic analyses for identifying forbidden

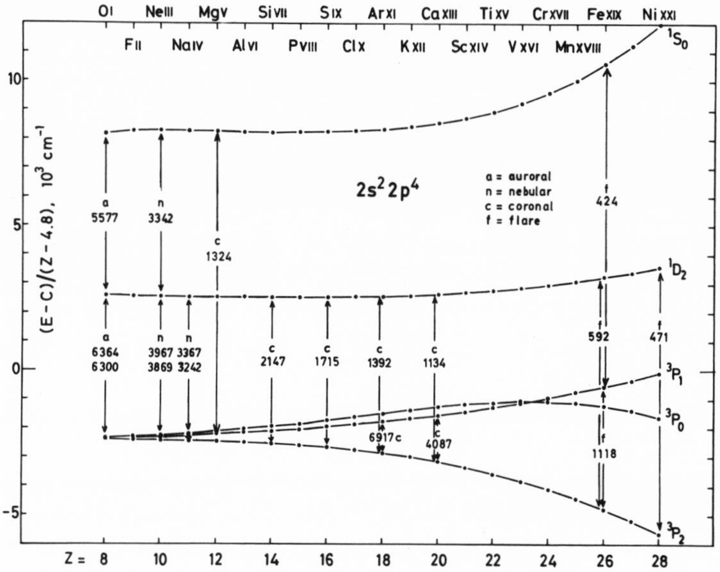

Fig. 10. Fine structure of the $2s^2 2p^4$ configurations according to Edlén[88]. The wavelengths of the forbidden lines are indicated.

corona lines is nicely illustrated in Fig. 10. Additional examples have been provided by Edlén and Smitt[90] who studied the metastable $3p^43d$ and $3p^53d$ configurations in the Cl I and Ar I sequences, respectively. A graphical method for extrapolating wavenumbers of the forbidden lines was thereby used. The difference between the observed and calculated value for the forbidden line wavenumber was found to vary regularly with Z which permitted highly accurate predictions of forbidden-line wavelengths in Fe IX, Fe X, Ni XI and Ni XII.

Forbidden lines play an important role in Tokamak plasmas, as shown by Suckewer and Hinnov[91]. In the PLT Tokamak a line at 2665.1 Å was identified as an M1 transition in Fe XX. The high wavelength made this line convenient for determination of ion temperatures. Similarly, Klapisch et al.[92] have observed the emission of E2 transitions in Mo XV from the TFR Tokamak plasma. Recent news about such forbidden lines will be found in this volume[46,47].

ATOMIC LIFETIMES AND TRANSITION PROBABILITIES

Basic Concepts

Excited states in atoms and ions decay by the emission of photons or electrons. The decay of a level follows the relation

$$N(t) = N(0)\exp(-t/\tau) \qquad (1)$$

Here N(0) is the initial population of the level and N(t) that at the time t. The lifetime τ is defined as the time after which the level population has decreased from N(0) to $N(0)/e = 0.37\cdot N(0)$. Excited states in neutral atoms have typical lifetimes of 10^{-7}-10^{-8}s, while much shorter lifetimes can be found in highly stripped systems. Thus, the 3p ^2P level in Li I has a lifetime of 216 ns while in the isoelectronic ion Si XII the value is 0.0022 ns[93].

For a level with a single decay mode the inverse of τ equals the probability A_{ab} for a spontaneous transition from a to b. In the E1 approximation the transition probability (in s^{-1}) can be expressed as

$$A_{ab} = \frac{4}{3}\,\frac{\omega^3}{\hbar c^3}\,|\langle\phi_a|\,e\underline{r}\,|\phi_b\rangle|^2 \qquad (2)$$

Here $\hbar\omega$ is the energy of the transition, ϕ_a and ϕ_b are the wavefunctions of the levels and $e\underline{r}$ is the dipole moment.

A convenient unit is the (dimensionless) absorption oscillator strength, or f-value, numerically related to the transition probability as follows

$$f = 1.499 \cdot 10^{-16} \cdot \lambda^2 \cdot A_{ab} \cdot g_a/g_b \quad (3)$$

Here λ is the photon wavelength (in Å), while g_a and g_b are the statistical weights of the levels.

Many techniques have been developed for the determination of atomic lifetimes and transition probabilities. The intensities of spectral lines in emission or the equivalent widths in absorption are proportional to transition probabilities. The variation of refractive index of a gas with wavelength in the vicinity of a spectral line also provides information about the f-value. These three techniques (the emission, absorption and hook method, respectively) have been successfully used for many years[94,95]. More direct lifetime methods are based on the excitation of atoms with short pulses (or sinusoidally varying waves) of photons (e.g. laser light) or electrons (sometimes ions) and the subsequent measurement of the fluorescence radiation. There are also level-crossing and resonance methods such as the Hanle effect and optical double resonance techniques. Detailed information is available in several reviews[96-99]. All these lifetime methods work very well for neutral atoms and they can usually also be applied to singly ionized species. Extensions to higher ionization degrees are usually not possible, however. Only BFS can be routinely applied to several times ionized atoms. This article will therefore be limited to lifetime determinations by BFS and related fast-ion methods.

Measurements of Lifetimes by BFS

The principle of BFS lifetime measurements is illustrated in Fig. 5 . The intensity I(x) of a spectral line is determined as the function of the distance x from the foil. Eq.(1) can therefore be rewritten as

$$I(x) = I(0) \exp(-x/v\tau) \quad (4)$$

The velocity v of the excited particles can be determined quite accurately. Because the ion velocity is usually very high short lifetimes can be measured. When ^{12}C ions are accelerated to 10 MeV their velocity is 12.7 mm/ns. A lifetime as short as 10 ps can therefore be conveniently measured. The lifetime range in BFS experiments is approximately 1 ps - 1 µs. Using very high spatial resolution along the beam Knystautas and Drouin[100] obtained the experimental value of 1.2 ± 0.2 ps for the 2p ^1P lifetime in C V. Even shorter lifetimes have been determined by using beams of MeV energies and observing the yield of X-rays from the foil as a function of foil thickness[101]. Lifetimes as short as 10^{-14} s have been measured in this way[101,102]. The accuracy is typically ±10% in BFS lifetimes but much higher precision is sometimes possible.

The BFS method is applicable to many levels in atoms and ions. Lifetime data have been obtained for H I as well as Kr XXXV. Many levels in each spectrum are accessible, the excitation is assumed to approximately follow the $(2\ell+1)n^{-3}$ distribution. However, as the result of electron capture processes high n,ℓ states in multiply ionized atoms are often abundantly populated. The excitation process has been investigated by Veje[103].

The BFS method for lifetime determination is in principle very simple. The decay curves are obtained in a straight-forward way and no corrections are needed for collisional processes, trapping of resonance radiation etc. Examples of typical decay curves are shown in Fig. 11. In the simplest case (a) the decay curve appears as a single exponential. The curves (b) and (c) are more complex, showing the effects of cascading processes.[66]

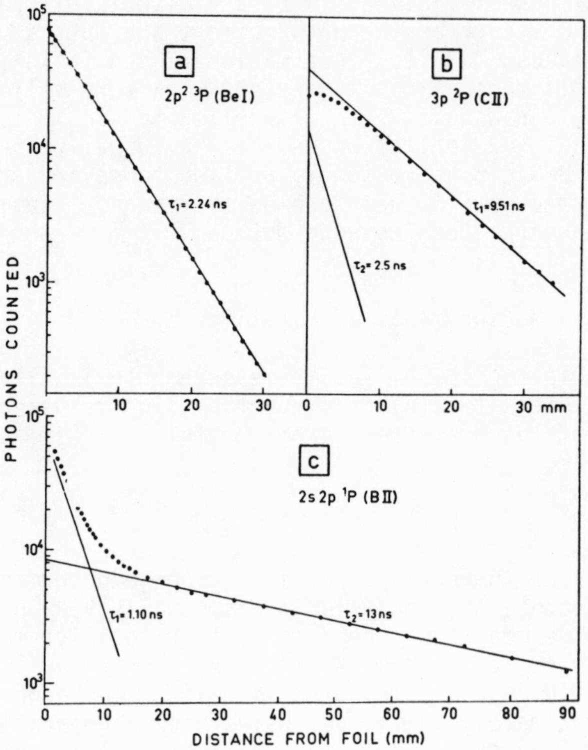

Fig. 11. Three examples of decay curves encountered in BFS experiments. The curve (a) is practically unaffected by cascades while the other curves show the repopulation from a short-lived (b) or a long-lived (c) higher level.[66]

Data Analysis and Experimental Problems

Cascading is one of the main problems in BFS lifetime measurements. Because many levels are populated at the ion-foil interaction the level under study is repopulated from higher-lying states. The observed decay curves (e.g. (b) and (c) in Fig. 11) consist of several exponentials, $I(x) = \Sigma I_i(0) \exp(-x/v\tau_i)$ and must be analyzed by means of a non-linear least-squares fitting procedure. Computer programs have been developed for this[104] but experience shows that it is already very difficult to decompose a curve into 3 (or more) exponentials. The problems are particularly severe when the primary and cascade lifetimes are similar.

When heavy cascading occurs direct curve-fitting methods are often replaced by the "analysis of correlated decay curves" (ACDC). This technique, introduced by Curtis[105,106], is illustrated by the following example.

In a measurement of the 4p level lifetime in the Cu isoelectronic sequence cascading into the 4p level from nd (n = 4,5...) and ns (n = 5,6...) levels must be considered. The population rate equation for the 4p level can therefore be written as

$$dI_{4p}/dt = -I_{4p}(t)/\tau_{4p} + \xi_{4d}\, I_{4d}(t) + \sum_{n=5}^{\infty} [\xi_{ns} I_{ns}(t) + \xi_{nd} I_{nd}(t)] \tag{5}$$

The intensities $I_{n\ell}(t)$ are measured while the normalization constants $\xi_{n\ell}$ are determined using the fact that Eq.(5) holds for all values of t. It is usually sufficient to measure a few cascade contributions. In the Cu I sequence the main cascading into 4p comes from 5s and 4d. Furthermore, the 4p-4d decay curve is complex and it contains information about cascading from the chain 4d-4f-5g-6h... which may be pronounced in beam-foil excitation.

Using this method Curtis[106] and Pinnington et al.[107] have found that the 4p ^2P lifetime in Kr VIII is about 40% lower than the value obtained from multi-exponential curve fits. The cascading processes in the Cu I sequence have also been throughly analyzed by Younger and Wiese[108].

Several methods have been developed to eliminate or reduce cascade effects in BFS. Andrä et al.[109] introduced the beam-laser method in which ions from the accelerator are excited with monochromatic laser light (instead of the foil). The excitation is now selective and there is thus no cascading. Very accurate lifetimes are obtained but the method has only been applied to neutral and singly ionized atoms. All results so far obtained are listed by Andrä.[69] Another interesting approach consists of combining fast-beam and Hanle-effect techniques whereby cascading is reduced[110]. Coincidence methods have also been tried[111] but the counting rates are unfortunately very low. While cascading may create serious problems in BFS experiments, it should be realized that it is either negligible or can be successfully analyzed in a majority of cases.

At low ion energies, when single or doubly ionized atoms are studied, the energy loss in the foil may be substantial[112]. However, this is usually not a serious problem when ions of MeV energies are used and lifetimes in highly ionized atoms are studied. In such a case the energy-loss correction is very small and it can also be accurately estimated.

The problems caused by modest spectral resolution may often be much more serious. To avoid systematic errors the decay curves must be measured at the highest possible spectral resolution. Line blending can be better estimated if the level structures of the ions studied are accurately known. There are thus important linkes between investigations of atomic energy levels and transition probabilities.

Results of Lifetime Measurements

Lifetimes for many levels in highly stripped atoms have already been determined by BFS.[99,113,114] For studies of impurity problems in Tokamaks $\Delta n = 0$ transitions in the Li I, Be I, Na I, Mg I, Cu I and Zn I sequences are particularly important[43].

Atomic f-values have interesting regularities, e.g. within isoelectronic sequences[99,114,115]. The following relation thus holds for oscillator strengths

$$f = f_0 + f_1(1/Z) + f_2(1/Z)^2 + \ldots \qquad (6)$$

where Z is the nuclear charge, f_0 the hydrogenic value and f_1 and f_2 constants. By displaying f versus 1/Z smooth curves are obtained (unless strong perturbations or cancellations occur).

The f-values for the 2s ^2S - 2p ^2P resonance transition in the Li I sequence have been measured (by BFS) for many species from Li I to Kr XXXIV. Some of the data are shown in Fig. 12, together with the results of relativistic calculations[116,117]. The latter predict a steep increase of the $^2S_{1/2} - {}^2P_{3/2}$ f-value in highly ionized atoms. This trend has been nicely confirmed in recent BFS studies of Fe XXIV[118] and Kr XXXIV[119].

Much experimental material exists also for allowed transitions in the Be I isoelectronic sequence. Here the $\Delta n = 0$ transitions within the L-shell are particularly interesting for atomic theory, plasma physics and astrophysics. Because of correlations between the

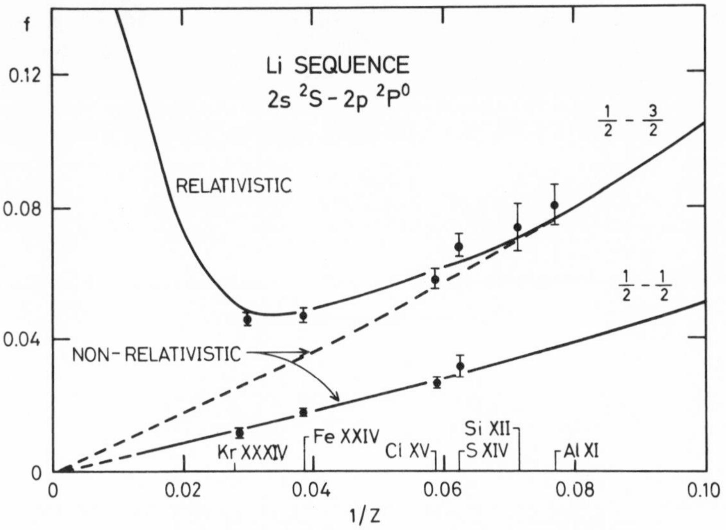

Fig. 12. Oscillator strengths for the 2s $^2S_{1/2}$ - 2p $^2P_{1/2,3/2}$ transitions in the Li I sequence.

two valence electrons the calculations of f-values are much more complicated than for the Li I sequence. Relativistic multiconfiguration calculations have been reported also for highly ionized members of this sequence[116,117,120]. For lower Z (Be I - Ne VII) it has been possible to calculate rigorous error bounds[121] (in the non-relativistic approximation). The experimental f-values for the important $2s^2$ ^1S - 2s2p ^1P (a), 2s2p ^1P - $2p^2$ ^1S (d), 2s2p ^1P - $2p^2$ ^1D (c) and 2s2p ^3P - $2p^2$ ^3P (b) transitions in Be-like Al X - Cl XIV are displayed in Fig. 13. The data originate from a recent paper by Pegg et al.[122] The agreement between theory and experiment is satisfactory but it should be possible to reduce the experimental errors by performing ACDC analyses.

The f-values for the ns ^2S - np ^2P resonance lines in the Na I (n = 3) and Cu I (n = 4) isoelectronic sequences have also been extensively studied in recent years. Substantial discrepancies between theory and experiments have been noted in several cases[99,108,123]. The theoretical problems in calculating these f-values are not overwhelming (the effects of core-polarisation can be rather accurately estimated) and the shortcomings were therefore assumed to be on the experimental side. In constrast to the Li I sequence where there is no $\Delta n = 0$ cascading into the 2p ^2P level (and $\Delta n = 1$ cascades play a minor role for high Z) there is appreciable $\Delta n = 0$ cascading through the whole sequence for Na-like (3p-3d) and Cu-like (4p-4d-4f) ions. Some careful experiments for the Cu I sequence have already been

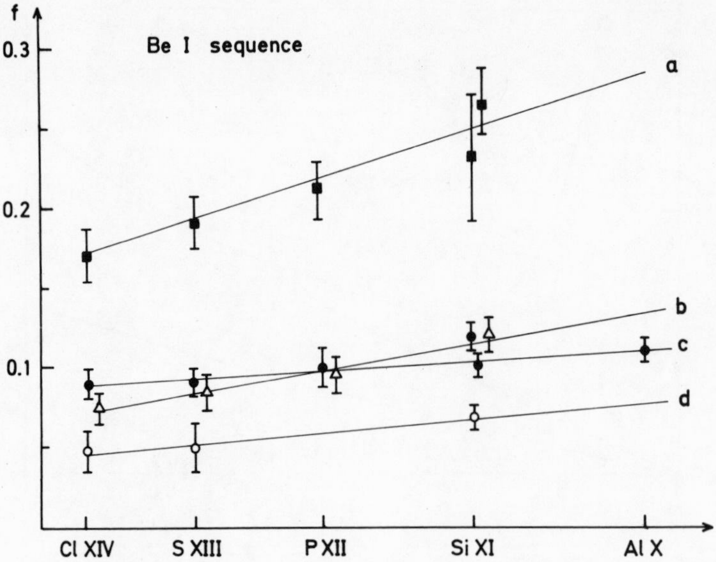

Fig. 13. Experimental oscillator strengths for the $2s^2\ ^1S - 2s2p\ ^1P$ (a), $2s2p\ ^3P - 2p^2\ ^3P$ (b), $2s2p\ ^1P - 2p^2\ ^1D$ (c) and $2s2p\ ^1P - 2p^2\ ^1S$ (d) transitions in the Be I sequence[122].

mentioned[106,107]. Pegg et al.[124] have recently measured the 3p ^2P lifetime in Na-like Cu XIX. These new results essentially bridge the gap between theory and experiment.

Discrepancies (up to 30-40%) have also been noted between theoretical and experimental f-values for the resonance lines in the Mg I and Zn I isoelectronic sequences[125,126]. Here also new, careful experimental work is needed. The problems are complicated by the fact that the level structure in several times ionized members of these sequences is rather fragmentarily known.

Forbidden Transitions

Forbidden transitions (M1, M2, E2, spin-forbidden E1 etc) have very low rates in neutral or mildly ionized atoms and lifetime measurements are therefore quite difficult. However, the transition probabilities scale strongly with Z and the forbidden lines thus become accessible for decay studies in highly stripped systems. Examples of forbidden transitions, in hydrogenlike (a) and heliumlike (b) systems are shown in Fig. 14.

The 2s $^2S_{1/2}$ level in hydrogenlike ions decays by two-photon

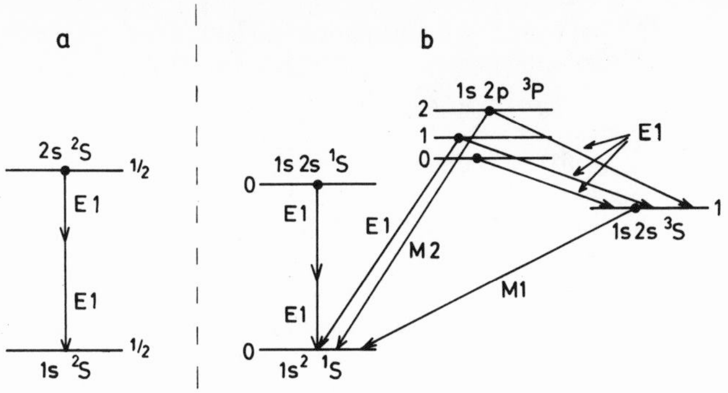

Fig. 14. Forbidden transitions in H-like (a) and He-like (b) systems.

emission. The theoretical decay probability can theoretically[123] be expressed as $8.2294 \cdot Z^6$ (s^{-1}). This has been verified in the BFS experiments of Cocke et al.[129] (for O^{7+} and F^{8+}) and Marrus and Schmieder (S^{15+} and Ar^{17+})[130]. The latter authors also obtained a value for the two-photon decay rate for the 1s2s ^1S state in Ar^{16+}.

The lowest-lying triplet state in He-like systems, 2s $^3S_1$ decays by M1 transitions to the ground level. The corresponding spectral lines have been observed in the solar corona[131]. Using the BFS method the M1 transition probabilitiy has been determined for many ions, from S XV to Kr XXXV[132]. The agreement with theory is excellent. The 2p ^3P term decays by allowed transitions to 2s ^3S. For the $^3P_2$ level also magnetic quadrupole (M2) decays are possible. This rate scales approximately as Z^8 (to be compared with the allowed branch which scales as Z). Theory predicts that the E1 and M2 decay modes have equal probabilities in Ar XVII. This has been experimentally verified by Marrus and Schmieder[130].

The 2p $^3P_1$ level also decays to the $1s^2$ $^1S_0$ ground state (Fig. 14). This intercombination transition is caused by the spin-orbit interaction which mixes the 2p $^3P_1$ level with 2p $^1P_1$ (and - to lesser extent - with other $^1P_1$ levels)[1]. This transition probability is low for He I and Li II but it increases rapidly with Z being proportional to Z^{10}. Beam-foil studies of this probability have been reported for N VI - Ar XVI. In addition the results summarized by Marrus and Mohr[132] a recent study of the 2p $^3P_1$ lifetime has been reported. Using X-ray spectroscopy Träbert et al.[133] thus obtained an accurate value for the intercombination line transition probability in Si XIII. All experimental data are in good agreement with theory[134].

The 2p $^3P_0$ level can only decay to 2s $^3S_1$. In systems with non-zero nuclear spin hyperfine-induced transitions between $^3P_0$ and $^1S_0$

are possible[132]. This effect has been calculated by Mohr[135] and experimental data are already available for V XXII (nuclear spin $I = 7/2$)[136] and Al XII ($I = 5/2$)[137].

In the Li I isoelectronic sequence the doubly excited $1s2pn\ell$ ^4L states have been extensively investigated in BFS experiments [63,64,138,139]. Becuase of the $\Delta S = 0$ selection rule these highly excited states cannot auto-ionize via Coulomb repulsion. The decay mechanisms are either spin-orbit and spin-spin autoionization or radiative transitions. The $1s2s2p$ $^4P_{5/2}$ level is particularly interesting. It mainly de-excites by spin-spin autoionization but also M2 transitions have been observed in X-ray spectra[68]. The $^4P_{5/2}$ lifetime has been measured for very many ions between Li I and Ar XVI using photon or electron spectroscopy[68,138].

Much interest has been focussed on the $2s^2$ $^1S_0$ - $2s2p$ $^3P_1$ intercombination lines in the Be I isoelectronic sequence. These transitions are also of considerable astrophysical importance. Many theoretical investigations of this probability have appeared, one of the most recent being the work of Glass and Hibbert[140]. Recently Dietrich et al.[118] have been able to measure the $2p$ $^3P_1$ lifetime in Be-like Fe XXIII. Unfortunately, such measurements are realistic only for very high ionization stages. For lower Z systems the $^3P_1$ lifetime becomes too long to permit BFS measurements.

However, the $2s^2$ $^1S_0$ - $2s3p$ $^3P_1$ transition probability is much higher in the Be I sequence, as experimentally found by Engström et al.[141]. Using a differential lifetime technique these authors obtained values for this probability for N IV, O V and F VI. It should be interesting to search for these intercombination lines in highly ionized systems.

ACKNOWLEDGEMENTS

Discussions with Dr. U. Litzén are gratefully acknowledged. This work was supported by the Swedish Natural Science Research Council (NFR) and the National Swedish Board for Energy Source Development (NE)

References

1 I.S.Bowen, and R.A. Millikan, Phys. Rev., 25, 295 (1925).

2 B. Edlén, Z. Physik, 100, 621 (1936).

3 B. Edlén, Physica, 13, 545 (1947).

4 F. Tyrén, Z. Physik, 111, 314 (1938).

5 B. Edlén, Z. Astrophysik, 22, 30 (1942).

6 R. Tousey, C.V. Strain, F.S. Johnson, and J.J. Oberly, Astrophys. J., 52, 158 (A)(1947).

7 R. Tousey, Astrophys. J., 149, 239 (1967).

8 E. Hinnov, Phys. Fluids, 7, 130 (1964).

9 H.W. Drawin, Phys. Reports, 37, 125 (1978).

10 R.A. Millikan, and R.A. Sawyer, Phys. Rev., 12, 167 (1918).

11 K. Bockasten, Arkiv Fysik, 9, 457 (1955).

12 U. Feldman, M. Swartz, and L. Cohen, Rev. Sci. Instr., 38, 1372 (1967).

13 U. Feldman, and G.A. Doschek, in Atomic Physics 5, (R. Marrus, M. Prior, and H. Shugart, eds., Plenum, New York, 1977) p.473.

14 T.N. Lie, and R.C. Elton, Phys. Rev., A3, 865 (1971).

15 W.A. Cilliers, R.U. Datla, and H.R. Griem, Phys. Rev., A12, 1408 (1975).

16 J.J. Turechek, and H.-J. Kunze, Z. Physik A 273, 111 (1975).

17 R. Beier, and H.-J. Kunze, Z. Physik A 285, 347 (1978).

18 L. Cohen, U. Feldman, M. Swartz, and J.H. Underwood, J. Opt. Soc. Am., 58, 843 (1968).

19 U. Feldman, S. Goldsmith, J.L. Schwob, and G.A. Doschek, Astrophys. J., 201, 225 (1975).

20 B. Ya'akobi, Phys. Rev., 176, 227 (1968).

21 S.K. Händel, B. Stenerhag, and I. Holmström, Nature, 209, 1227 (1966).

22 P.G. Burkhalter, C.M. Dozier, and D.J. Nagel, Phys. Rev., A15, 700 (1977).

23 I.M. Vitkovitsky, L.S. Levine, D. Mosher, and S.J. Stephanakis, Appl. Phys. Letters, 23, 9 (1973).

24 P.G. Burkhalter, R. Schneider, C.M. Dozier, and R.D. Cowan, Phys. Rev., A18, 718 (1978).

25 P.G. Burkhalter, J. Davis, J. Rauch, W. Clark, G. Dahlbacka, and R. Schneider, J. Appl. Phys., 50, 705 (1979).

26 B.C. Fawcett, A.H. Gabriel, F.E. Irons, N.J. Peacock, and P.A.H. Saunders, Proc. Phys. Soc., 88, 1051 (1966).

27 N.J. Peacock, in Beam-Foil Spectroscopy (I.A. Sellin, and D.J. Pegg, eds., Plenum, New York, 1976) p.925.

28 B.C. Fawcett, Adv. Atom. Molec. Phys., 10, 223 (1974).

29 U. Feldman, Astrophys. Space Sci. 41, 155 (1976).

30 V.A. Boiko, A.Ya. Faenov, and S.A. Pikuz, J. Quant. Spectrosc. Radiat. Transfer, 19, 11 (1978).

31 V.A. Boiko, S.A. Pikuz, A.S. Safronova, A.Ya. Faenov, P.O. Bogdanovich, G.V. Merkelis, Z.B. Rudzikas, and S.D. Sadziuviene, J. Phys. B, 12, 1927 (1979).

32 P.G. Burkhalter, G.A. Doschek, U. Feldman, and R.D. Cowan, J. Opt. Soc. Am., 67, 741 (1977).

33 E.Ya. Kononov, A.N. Ryabtsev, and S.S. Churilov, Physica Scripta, 19, 328 (1979).

34 B.C. Fawcett, A. Ridgeley, and T.P. Hughes, Mon. Not. R. Astr. Soc., 188, 365 (1979).

35 J. Reader, G. Luther, and N. Acquista, J. Opt. Soc. Am., 69, 144 (1979).

36 V.A. Boiko, S.A. Pikuz, A.S. Safronova, and A.Ya. Faenov, J. Phys. B, 11, L503 (1978).

37 P.G. Burkhalter, J. Reader, and R.D. Cowan, J. Opt. Soc. Am., 67, 1521 (1977).

38 M.W.D. Mansfield, N.J. Peacock, C.C. Smith, M.G. Hobby, and R.D. Cowan, J. Phys. B, 11, 1521 (1978).

39 H. Gordon, M.G. Hobby, N.J. Peacock, and R.D. Cowan, J. Phys. B, 12, 881 (1979).

40 B.C. Fawcett, and R.W. Hayes, J. Opt. Soc. Am., 65, 623 (1975).

41 E.V. Aglitskii, V.A. Boiko, O.N. Krokhin, S.A. Pikuz, and A. Ya. Faenov, Sov. J. Quant. Electron., 4, 1152 (1975).

42 J. Reader, G. Luther, and N. Acquista, to be published.

43 E. Hinnov, Phys. Rev., A14, 1533 (1976).

44 J.L. Schwob, M. Klapisch, N. Schweitzer, M. Finkenthal, C. Breton, C. De Michelis, and M. Mattioli, Phys. Letters, 62A, 85 (1977).

45 E. Hinnov, Astrophys. J., 230, L197 (1979).

46 E. Hinnov, these proceedings.

47 M. Klapisch, these proceedings.

48 B.C. Fawcett, B.B. Jones, and R. Wilson, Proc. Phys. Soc., 78, 1223 (1961).

49 K. Bockasten, R. Hallin, and T.P. Hughes, Proc. Phys. Soc., 81, 522 (1963).

50 G. Tondello, and T.M. Paget, J. Phys. B, 3, 1757 (1970).

51 O. Engvold, Physica Scripta, 16, 48 (1977).

52 A.K. Dupree, Adv. Atom. Molec. Phys., 14, 393 (1978).

53 C. Jordan, in Progress in Atomic Spectroscopy, (W. Hanle and H. Kleinpoppen, eds., Plenum, New York, 1979) p.1453.

54 S.O. Kastner, W.M. Neupert, and M. Swartz, Astrophys. J., 191, 261 (1974).

55 B.C. Fawcett, and R.D. Cowan, Mon. Not. R. Astr. Soc., 171, 1 (1975).

56 G.A. Doschek, U. Feldman, and F.D. Rosenberg, Astrophys. J., 215, 329 (1977).

57 U. Feldman, G.A. Doschek, and F.D. Rosenberg, Astrophys. J., 215, 652 (1977).

58 K.P. Dere, Astrophys. J., 221, 1062 (1978).

59 L. Kay, Phys. Letters, 5, 36 (1963).

60 S. Bashkin, Nucl. Instr. Methods, 28, 88 (1964).

61 K.W.Jones, J. Physique, 40, C1-197 (1979).

62 J.A. Leavitt, J.W. Robson, and J.O. Stoner,Jr., Nucl. Instr. Methods, 110, 423 (1973).

63 H.G. Berry, Physica Scripta, 12, 5 (1975).

64 I.A. Sellin, Nucl. Instr. Methods, 110, 477 (1973).

65 R. Bruch, D. Schneider, W.H.E. Schwarz, M. Meinhart, B.M. Johnson, and K. Taulbjerg, Phys. Rev., A19, 587 (1979).

66 I. Martinson, and A. Gaupp, Phys. Reports, 15C, 113 (1974).

67 H.G. Berry, Rep. Prog. Phys., 40, 155 (1977).

68 C.L. Cocke, in Methods of Experimental Physics, 13B, (D. Williams, ed., Academic, New York, 1977) p.213.

69 H.J. Andrä, in Progress in Atomic Spectroscopy, (W. Hanle, and H. Kleinpoppen, eds., Plenum, New York, 1979) p.829.

70 H.F. Beyer, F. Folkmann, and K.-H. Schartner, J. Physique, 40, C1-17 (1979).

71 B. Edlên, Rep. Prog. Phys., 26, 181 (1963).

72 L. Minnhagen, J. Res. NBS, 68C, 237 (1964).

73 T.A. Carlson, W.E. Hunt, and M.O. Krause, Phys. Rev., 151, 41 (1966).

74 P. Erman, and H.G. Berry, Phys. Letters 34A, 1 (1971).

75 J.A.R. Samson, Techniques of Vacuum Ultraviolet Spectroscopy, (Wiley, New York, 1967).

76 J.A.R. Samson, in Methods of Experimental Physics, 13A, (D. Williams, ed., Academic, New York, 1976) p.204.

77 K. Heilig, and A. Steudel, in Progress in Atomic Spectroscopy, (W. Hanle, and H. Kleinpoppen, eds., Plenum, New York, 1978) p.263.

78 B. Edlên, in Handbuch der Physik, Vol 27, (S. Flügge, ed., Springer, Berlin, 1964) p.80.

79 B. Edlén, Optica Pura y Aplicada, 5, 101 (1972).

80 B. Edlén, in Beam-Foil Spectroscopy, (I.A. Sellin, and D.J. Pegg, eds., Plenum, New York, 1976) p.1.

81 B. Edlén, Physica Scripta, 19, 255 (1979).

82 B. Edlén, Physica Scripta, 17, 565 (1978).

83 B. Edlén, Physica Scripta, 20, 129 (1979).

84 P.O. Zetterberg, and C.E. Magnusson, Physica Scripta, 15, 189 (1977).

85 J.O. Ekberg, Physica Scripta, 12, 42 (1975).

86 S. Johansson, and U. Litzén, Physica Scripta, 10, 121 (1974).

87 J.O. Ekberg, and J. Reader, unpublished work.

88 B. Edlén, Mem. Soc. Roy. Sci. Liège, 9, 235 (1976) and unpublished work.

89 R. Smitt, Solar Phys., 51, 113 (1977).

90 B. Edlén, and R. Smitt, Solar Phys., 57, 329 (1978).

91 S. Suckewer, and E. Hinnov, Phys. Rev. Letters, 41, 756 (1978).

92 M. Klapisch, J.L. Schwob, M. Finkenthal, B.S. Fraenkel, S. Egert, A. Bar-Shalom, C. Breton, C. DeMichelis, and M. Mattioli, Phys. Rev. Letters, 41, 403 (1978).

93 A. Lindgård, and S.E. Nielsen, At. Data Nucl. Data Tabl., 19, 534 (1977).

94 E.W. Foster, Rep. Prog. Phys., 27, 469 (1964).

95 M.C.E. Huber, Physica Scripta, 16, 16 (1977).

96 A. Corney, Adv. Electron Electron. Phys. 29, 115 (1969).

97 R.E. Imhof, and F.H. Read, Rep. Prog. Phys., 40, 1 (1977).

98 I. Martinson, in Excited States in Quantum Chemistry, (C.A. Nicolaides, and D.R. Beck, eds., D. Reidel, Dordrecht, 1978) p.1.

99 W.L. Wiese, in Progress in Atomic Spectroscopy, (W. Hanle, and H. Kleinpoppen, eds., Plenum, New York, 1979) p.1101.

100 E.J. Knystautas, and R. Drouin, in Beam-Foil Spectroscopy, (I.A. Sellin, and D.J. Pegg, eds., Plenum, New York, 1976) p.377.

101 H.-D. Betz, F. Bell, H. Panke, G. Kalkoffen, M. Welz, and D. Evers, Phys. Rev. Letters, 33, 807 (1974).

102 S.L. Varghese, C.L. Cocke, B. Curnutte, and G. Seaman, J. Phys. B, 9, L387 (1976).

103 E. Veje, J. Physique, 40, C1-253 (1979).

104 D.J.G. Irwin, and A.E. Livingston, Comput. Phys. Comm., 7, 95 (1974).

105 L.J. Curtis, in Beam-Foil Spectroscopy, (S. Bashkin, ed., Springer, Berlin, 1976) p.63.

106 L.J. Curtis, J. Physique, 40, C1-139 (1979).

107 E.H. Pinnington, R.N. Gosselin, J.A. O'Neill, J.A. Kernahan, K.E. Donnelly, and R.L. Brooks, Physica Scripta, 20, 151 (1979).

108 S.M. Younger, and W.L. Wiese, Phys. Rev., A17, 1944 (1978).

109 H.J. Andrä, A. Gaupp, and W. Wittmann, Phys. Rev. Letters, 31, 501 (1973).

110 O. Poulsen, T. Andersen, and N.J. Skouboe, J. Phys. B, 8, 1393 (1975).

111 K.D. Masterson, and J.O. Stoner,Jr., Nucl. Instr. Methods, 110. 441 (1973).

112 T. Andersen, Nucl. Instr. Methods, 110, 35 (1973).

113 J.B. Fuhr, B.J. Miller, and G.A. Martin, Bibliography on Atomic Transition Probabilities, (NBS Spec. Publ. 505, Washington, D.C., 1978).

114 W.L. Wiese, in Beam-Foil Spectroscopy, (S. Bashkin, ed., Springer, Berlin, 1976) p.147.

115 W.L. Wiese, and A.W. Weiss, Phys. Rev., 175, 50 (1968).

116 Y.-K. Kim, and J.P. Desclaux, Phys. Rev. Letters, 36, 139 (1976).

117 L. Armstrong, W.R. Fielder, and D.L. Lin, Phys. Rev., A14, 1114 (1976).

118 D.D. Dietrich, J.A. Leavitt, S. Bashkin, J.G. Conway, H. Gould, D. MacDonald, R. Marrus, B.M. Johnson, and D.J. Pegg, Phys. Rev., A18, 208 (1978).

119 D. Dietrich, J.A. Leavitt, H. Gould, and R. Marrus, J. Physique, 40, C1-215 (1979).

120 K.T. Cheng, and W.R. Johnson, Phys. Rev., A15, 1326 (1977).

121 J.S. Sims, and R.C. Whitten, Phys. Rev., A8, 2220 (1973).

122 D.J. Pegg, J.P. Forester, P.M. Griffin, G.D. Alton, S.B. Elston, B.M. Johnson, M. Suter, R.S. Thoe, and C.R. Vane, J. Physique, 40, C1-205 (1979).

123 C. Froese Fischer, in Beam-Foil Spectroscopy, (I.A. Sellin, and D.J. Pegg, eds., Plenum, New York, 1976) p.69.

124 D.J. Pegg, P.M. Griffin, B.M. Johnson, K.W. Jones, J.L. Cecchi, and T.H. Kruse, Phys. Rev., A16, 2008 (1977).

125 M. Aymar, and E. Luc-Koenig, Phys. Rev., A15, 821 (1977).

126 P. Shorer, and A. Dalgarno, Phys. Rev., A16, 1502 (1977).

127 C. Froese Fischer, and J.E. Hansen, Phys. Rev., A17, 1956 (1978).

128 S. Klarsfeld, Phys. Letters, 30A, 382 (1969).

129 C.L. Cocke, B. Curnutte, J.R. MacDonald, J.A. Bednar, and R. Marrus, Phys. Rev. A9, 2242 (1974).

130 R. Marrus, and R.W. Schmieder, Phys. Rev., A5, 1160 (1972).

131 A.H. Gabriel, and C. Jordan, Phys. Letters, 32A, 166 (1970).

132 R. Marrus, and P.J. Mohr, Adv. Atom. Molec. Phys., 14, 181 (1978).

133 E. Träbert, I.A. Armour, S. Bashkin, N.A. Jelley, R. O'Brien, and J.D. Silver, J. Phys. B, 12, 1665 (1979).

134 G.W.F. Drake, and A. Dalgarno, Astrophys. J., 157, 459 (1969).

135 P.J. Mohr, in Beam-Foil Spectroscopy, (I.A. Sellin, and D.J. Pegg, eds., Plenum, New York, 1976) p.97.

136 H. Gould, R. Marrus, and P.J. Mohr, Phys. Rev. Letters, 33, 676 (1974).

137 B. Denne, S. Huldt, R. Hallin, J. Pihl, and R. Sjödin, Physica Scripta, (in press).

138 I.A. Sellin, Adv. Atom. Molec. Phys., 12, 215 (1976).

139 H.D. Dohmann, and H. Pfeng, Z. Physik, A288, 29 (1978).

140 R. Glass, and A. Hibbert, J. Phys. B, 11, 2413 (1978).

141 L. Engström, B. Denne, S. Huldt, J.O. Ekberg, L.J. Curtis, E. Veje, and I. Martinson, Physica Scripta, 20, 88 (1979).

THEORETICAL STUDIES OF OSCILLATOR STRENGTHS FOR THE SPECTROSCOPY OF HOT PLASMAS

Marcel Klapisch

Racah Institute of Physics

Hebrew University, Jerusalem, Israel

§ 1. INTRODUCTION

Most of the atomic processes described in this book by the other contributors will eventually lead to the emission of photons by excited atoms. Whether these photons are considered as beneficial - for diagnostic purposes, or detrimental - causing energy losses, is not our concern here. In either case, however, control of the plasma requires the knowledge of the ratio of the number of photons corresponding to a given transition emitted per unit time from an excited state to the number of atoms lying in the excited state, per unit volume. This defines the transition probability (abbreviated below as TP). Let us recall also that oscillator strengths, (OS) defined below, are needed in the estimation of collision excitation cross sections by semi empirical formulas - which will probably continue to be widely used for a long time.

In many instances, as shown by Martinson in the following chapter of this book, it is very difficult to measure these TP, and one has to rely on theoretical estimations. Crossley (1) in the conclusion of his excellent review (1969) states: "With few exceptions - one and two electron atoms - the calculations of TP to an accuracy of 10% is a very difficult task." Much work has been done since, but the statement remains generally true, although it could be softened for the case of ionized atoms. The present chapter, which is by no means an updating of Crossley's work (1), is meant to draw attention to some recent developments of great importance for hot plasma spectroscopy (i.e. ionized atoms). Thus, ignoring relativistic effects may yield OS off by a factor of 5 for atoms 50 times ionized - which are not uncommon, and neglecting correlation (or configuration interactions) may bring errors by a factor of 10 for atoms ionized 2 to 5 times.

Another very exhaustive review of the theoretical methods available ten years ago is due to Layzer and Garstang (2). In the following, many references to these two reviews will be made, either explicitly or implicitly.

For the sake of clarity, we recall here the relationship between the TP, A, the line strengths, S, and the OS, f, for electric dipole transitions: (in a.u.)

$$A_{fi} = \frac{4}{3}\alpha^3 \frac{E_{fi}{}^3}{g_i} S_{fi}, \qquad f_{fi} = -\frac{2}{3} E_{fi} S_{fi}$$

where the subscript f and i refer to the final and initial level of the transition, E_{fi} is the transition energy and g_i is the initial level statistical weight. f is usually denoted as negative for emission and positive for absorption. S is the matrix element:

$$S_{fi} = |\langle\Psi_f|\ e \sum_{i=1}^{N} r_i\ |\Psi_i\rangle|^2$$

Thus the problem of evaluating TP involves the computation of a pair of energy levels and wavefunctions (wf). In section 2, we shall recall some basic concepts and results of the Z expansion model which enable us to classify the ionization stages and to estimate when correlation effects are important or when relativistic effects dominate. In section 3, we shall review the relativistic theory of radiation and the connection between gauge invariance and the "length or velocity formula" dilemma. Section 4 will be devoted to a simplified description of different atomic models - including RPA and alike - from the unifying point of view of perturbation theory. Section 5 will describe the "relativistic effects". Applications to hot plasma spectroscopy will be reviewed in section 6, and we will conclude by section 7.

§ 2. BASIC CONCEPTS

2.1 The Z-Expansion Framework

Layzer and Bahcall (3) have shown that it is possible to write the energy of a given level J in an isoelectronic sequence as a series in inverse powers of Z (the nuclear charge)

$$E_J(Z) = Z^2 E_J^0 + Z E_J^1 + E_J^2 \dots = Z^2 \sum_{k=0}^{\infty} E_J^k Z^{-k} \qquad (2.1)$$

E_J^0 is the purely hydrogenic nonrelativistic energy

$$E_J^0 = -\frac{1}{2}\sum_{i=1}^{N}\frac{1}{n^2} \qquad \text{(in a.u.)}$$

and correlation energy (difference between Hartree-Fock and exact energies) contributes in varied ways to all other terms (1). On the other hand, relativistic corrections to energy levels are obtained by developing the relativistic energy in powers of

$\frac{v}{c} \sim \alpha Z$

$$E_{R,J} = \sum_{k=0}^{\infty} E_{RJ}^{k} (\alpha Z)^k \tag{2.2}$$

More precisely, each of the terms of the non relativistic expansion (eq. 2.1) behaves as a series like eq. 2.2 (4). These simple facts allow us to stress the following points:

2.1.1 The balance between correlation and relativistic effects will change along an isoelectronic sequence, and one can distinguish grossly three zones of ionization stages:

Zone	Z (ionization stage)	Dominating effects
1	Low ($\lesssim 10$)	Correlation
2	Medium ($\lesssim 10 \lesssim 20$)	-
3	High ($\gtrsim 20$)	Relativistic

Of course, the values of limiting ionization states are purely informative and may vary widely from one sequence to another. As one of the interesting fields of present hot plasma spectroscopy is the study of highly ionized impurities, we shall be concerned mainly with Zone 2 and 3. Thus, we shall not describe involved and accurate methods that were implemented only for 2 - or 3-electron atoms (1). On the other hand, we shall deliberately describe the relativistic theory right from the beginning. Actually it turns out that it is much easier to describe correctly the effects of relativity (primarily one-electron effects) than that of correlation (multi-electron effects).

2.1.2 One must distinguish between transitions involving a change in the principal quantum number ($\Delta n \neq 0$) from those in which there is no such a change ($\Delta n = 0$) in multi-electron atoms. For low Z, the transition energy

$$\Delta E \sim Z^2 \text{ for } \Delta n \neq 0$$

while $\Delta E \sim Z$ for $\Delta n = 0$.

For very high Z, this distinction, connected with the ℓ degeneracy in hydrogenic spectra has to be complemented by the change of the j quantum number of the jumping electron

$$\Delta E \sim Z \quad \text{for } \Delta j = 0 \quad \Delta n = 0$$

$$\Delta E \sim Z^4 \quad \text{for } \Delta j \neq 0 \quad \Delta n = 0.$$

2.1.3 A similar expansion exists for wavefunctions and any matrix element, for low Z:

$$(\Psi_a|A|\Psi_b) = (\Psi_a^0|A|\Psi_b^0) + \frac{1}{Z}[(\Psi_a^0|A|\Psi_b^1) + (\Psi_a^1|A|\Psi_b^0)] + \frac{1}{Z^2}\cdots$$

where the first term is just the hydrogenic value. Hence we obtain the well known expansion for the OS:

$$f = f_0 + \frac{1}{Z} f_1 + \ldots$$

For the case $\Delta n = 0$ (and $\Delta j = 0$ for high Z), $f_0 = 0$. This is the origin of the plots of f as a fuction of 1/Z. Figure 1 ($\Delta n = 0$) and 2 ($\Delta n \neq 0$) show typical behaviours of a resonance line ($\Delta j \neq 0$) in the Mg isoelectronic sequence (5). The three zones are quite conspicuous.

2.1.4 Note the inverse behaviour of the $\Delta n = 0$ and $\Delta n \neq 0$ curves, which is quite frequent for resonance lines. It has a simple explanation in terms of the Thomas-Reiche-Kuhn sum rule (1)

$$\sum_a f_{ba} = N$$

where N is the number of electrons in the system (usually, outside closed shells). Numerous examples of these simple isoelectronic behaviours of experimental energy values can be found in the work of Edlen (6) or others (7), and f values computed by different methods (5,8,9) also follow roughly these simple rules.

2.1.5 It is well known that electric dipole radiation can be described by the "length form" of the matrix element

$$f_L = -\frac{2}{3} E_{ab} \; |\langle a|\Sigma\, r_i|b\rangle|^2 \tag{2.1}$$

or the "velocity form"

$$f_v = -\frac{2}{3} \frac{1}{E_{ab}} \; |\langle a|\Sigma\nabla_i|b\rangle|^2$$

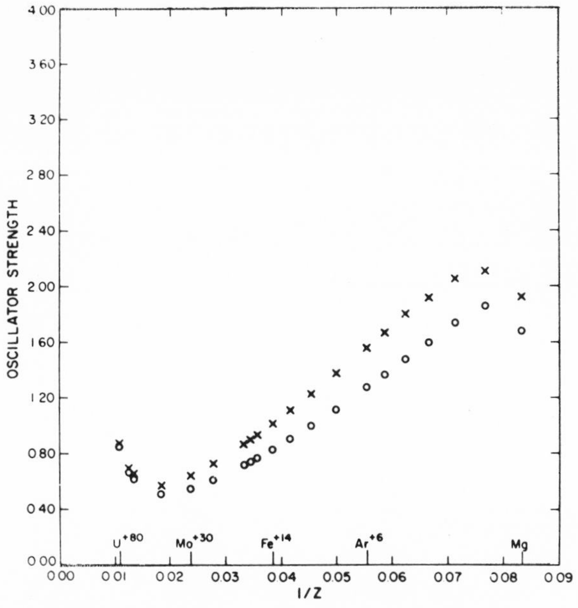

Figure 1. Oscillator strength as a function of Z^{-1} for $3s^2\ ^1S_0$ - $3s3p\ ^1P_1$. O RRPA; X DHF.

These two forms can be shown to be equivalent by using commutation relations (Hypervirial theorem). This is verified in the case of one electron atom in a central field. In actual practice, however, for a multi electron atom, using non exact wfs and energies, different numerical values are generally obtained. This discrepancy may originate in an inaccuracy of the energies, or of the matrix element, or both. Many discussions may be found in the literature about which formula is better and when. It is worth noticing that the two expressions behave inversely one to another with respect to the energy difference E_{ab}. This prompted some authors to consider the geometrical mean of the two values to compensate for this source of errors (1,2). Others use the experimental energy difference. Numerical equality of the two formulas has often been considered as a test of accuracy of the wfs. We shall come back to this question in the sections 2.1.5 and 4.2.6 .

§ 3. RELATIVISTIC THEORY OF RADIATION

Our aim here is not to develop the whole formalism, which can be found in many places (10,11), but to enable the reader to understand the origin of the various "relativistic effects", and also to follow an interesting argument that has developed recently: i.e. the connection between gauge invariance and the equivalence of f_L and f_V.

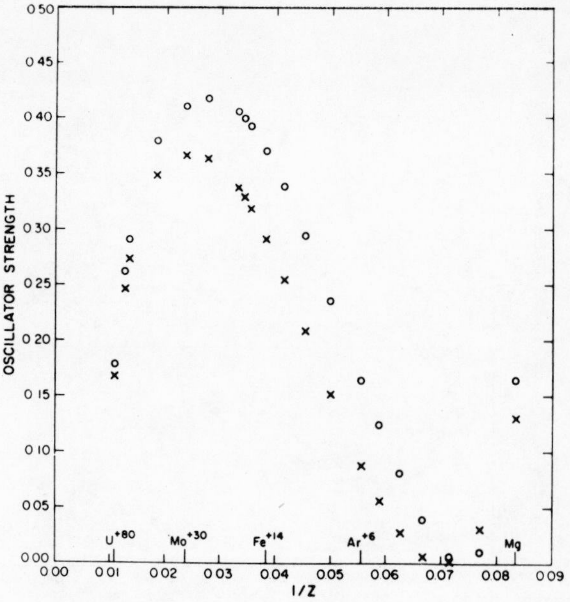

Figure 2. Oscillator strength as a function of Z^{-1} for $3s^2\ ^1S_0$ - $3s4p\ ^1P_1$. O RRPA; X DHF.

3.1 Relativistic wf

For system with one electron in a central field, wfs are solution of the Dirac equation (10,11, 12,13)

$$H_D = c\underline{\alpha}\cdot\underline{p} + \beta mc^2 - eU(r)$$

and can be written as

$$|\Psi_{n\ell j}) = \left| \begin{array}{cc} \dfrac{P_{n\ell j}(r)}{r} & |1/2\ \ell j m_j> \\ i\,\dfrac{Q_{n\ell j}(r)}{r} & |1/2\ \bar{\ell} j m_j> \end{array} \right)$$

$|1/2\ \ell j m_j>$ represents a 2-component spinor. $P_{n\ell j}(r)$ is the "large" radial component associated with the ℓ value which obtains in the non relativistic limit. The small component $Q(r)$ is connected with $\bar{\ell} = 2j - \ell$. So ℓ is no longer a good quantum number, but only j, as well as κ

$$\kappa = (-1)^{\ell+j+\frac{1}{2}} \ (j + 1/2) \ .$$

3.2 Interaction Hamiltonian

Let us consider the system: atom + radiation field, and a transition between two states of the system conserving the total energy. According to the"golden rule", its probability will be

$$|M_{fi}|^2 \sim |(\Psi_f \chi_f | H_{int} | \Psi_i \chi_i)|^2$$

where Ψ_i and χ_i are the initial atomic and field states,while Ψ_f, χ_f are the final states. The interaction H_{int} is obtained from the usual Hamiltonian by replacing the momenta $\underline{p}$ by $\underline{p} + (e/c)\underline{A}$ and electric fields E by E + eΦ ($\underline{A}$ and ϕ are the vector and scalar potentials). The four-vector potential may be written as $\{A^\mu\} = (\Phi,\underline{A})$ and satisfies the wave equation

$$\Box \underline{A} = \partial_u \partial^\mu A^\mu = 0 \qquad \mu = 0,1,2,3 \ . \tag{3.1}$$

Solutions of (3.1) exhibiting harmonic time dependence exp(-iωt) and having specific tensor character can easily be written down. The scalar potential Φ can be expanded in series of functions:

$$\Phi_{LM} = i^L \ (2L + 1)^{\frac{1}{2}} \ j_L\left(\frac{\omega r}{c}\right) \sqrt{4\pi} \ Y_{LM} \ (\Theta,\chi)$$

where $j_L(\frac{\omega r}{c})$ is the usual spherical Bessel function. It comes out that the vector potential is expressed as a linear combination of three types of <u>vector</u> spherical harmonics:

$$\underline{A}^e_{LM} = \frac{1}{\omega[L(L+1)]^{\frac{1}{2}}} \ \underline{\nabla} \times \underline{L} \ \phi_{LM} \qquad \text{(electric type)}$$

$$\underline{A}^\ell{}_{LM} = \frac{1}{i\omega} \underline{\nabla} \ \phi_{LM} \qquad \text{(longitudinal type)}$$

$$\underline{A}^m{}_{LM} = \frac{1}{c[L(L+1)]^{\frac{1}{2}}} \ \underline{L}_{\ LM} \qquad \text{(magnetic type)}$$

$A^\ell{}_{LM}$ is non zero only in a direction parallel to the propagation.

A^μ is defined only up to the gradient of an arbitrary scalar function Λ, with the condition $\Box \Lambda = 0$. Adding such functions to A^μ is called a gauge transformation. It is clear that it should leave invariant any physical result - e.g. the T.P. The "Coulomb gauge" is the choice of Λ that cancels $A^\ell{}_{LM}$, leaving only transverse waves. It annihilates also the scalar potential. To show the effect of other choices of gauge one usually adds a multiple of A^{ℓ}_{LM} to A^{e}_{LM}:

$$A^{\mu}(r) \sim \sum_{LM} c_{LM} \left[A^{e}_{LM}(r) + G_L A^{\ell}_{LM}(r)\right]^{\mu} \qquad \text{(electric type)}$$

$$A^{\mu}(r) \sim \sum_{LM} c_{LM} \left[A^{m}_{LM}\right]^{\mu} \qquad \text{(magnetic type)}$$

The parity of A^{m} is different from A^{ℓ} and A^{e}, so that it is not affected by a gauge transformation. The Coulomb gauge now correspond to $G_L = 0$.

3.3 Matrix Elements

The Einstein coefficients are

$$A_{fi} = \frac{2}{\hbar^2 c} \sum_{\substack{L,M \\ \text{types}}} \left|M^{LM}_{fi}\right|^2$$

Owing to their different parities, electric and magnetic multipole, do not mix, so the transition is actually either electric or magnetic, for a given L and, nearly always, only one L will contribute to a given transition. Using the Wigner-Eckart theorem, one has

$$M^{LM}_{fi} = (-1)^{J_f - M_f} \begin{pmatrix} J_f & L & J_i \\ -M_f & M & M_i \end{pmatrix} (J_f || \Sigma\, h^{\lambda}_{LM} || J_i)$$

the sum runs over the electrons. The mono-electronic operators are

$$h^{\lambda}_{LM} = e\, \underline{\alpha} \cdot \underline{A}^{\lambda}_{LM} - e\, \phi^{\lambda}_{LM}$$

for absorption, and the conjugate for emission. λ stands for "e" or "m". The reduced matrix element, which involves multi-electron wavefunctions, can be decoupled to obtain the result in terms of single particle reduced elements involving radial integrals (11)

$$(\kappa j||h^{e}_{LM}||\kappa j') = a e(\hbar\omega)^{\frac{1}{2}} \begin{pmatrix} j & L & j' \\ \frac{1}{2} & 0 & -\frac{1}{2} \end{pmatrix} \left(\frac{[j][j']}{[L]}\right)^{\frac{1}{2}}$$

$$\times \quad \{[(\frac{L}{L+1})^{\frac{1}{2}} - G_L] \cdot [(\kappa-\kappa')\, I^{+}_{L+1} + (L+1)\, I^{-}_{L+1}]$$

$$- \quad [(\frac{L+1}{L})^{\frac{1}{2}} + G_L] \cdot [(\kappa-\kappa')\, I^{+}_{L-1} - L\, I^{-}_{L-1}]$$

$$+ \quad G_L\, [L] \cdot \frac{(E_f-E_i)}{|E_f-E_i|}\, J_L\}$$

where $a = i^L$ for absorption and $a = (-1)^{L+1}\, i^L$ for emission,

$$I^{\pm}_{L} = \int (P_j Q_{j'} \pm Q_j P_{j'})\, j_L\, (\frac{\omega r}{c})\, dr$$

$$J_L = \int (P_j P_{j'} + Q_j Q_{j'})\, j_L\, (\frac{\omega r}{c})\, dr$$

and for magnetic transitions (11)

$$(\kappa j||h^{m}_{LM}|\kappa' j') = -ae\; i(\hbar\omega)^{\frac{1}{2}}\, (-1)^{L+j+\frac{1}{2}} \begin{pmatrix} j & L & j' \\ \frac{1}{2} & 0 & -\frac{1}{2} \end{pmatrix}$$

$$\times \quad \frac{(\kappa+\kappa')}{[L(L+1)]^{\frac{1}{2}}}\, [j,j',L]^{\frac{1}{2}}\, I^{+}_{L}$$

These matrix elements look much more cumbersome than the corresponding non relativistic matrix elements (13) although their derivation is more straightforward. This prompted many authors to investigate the non relativistic limits (14). However there are several drawbacks to the latter approach: it yields more complicated operators, and it is not sure that terms of successive powers of (αZ) are always smaller. In other words, on the one hand it is important to have in mind the different "contributions" to the matrix elements for the understanding of the physics, but on the other hand when it comes to getting numbers it is just as well to compute the above formulas directly. Anyhow, modern atomic physics cannot do without computers. Actually, a program for relativistic multipoles - especially magnetic, is much simpler than a non relativistic one.

3.4 Gauges and the Length-velocity Dilemma

To make the connection clear, we take the non relativistic limit of the matrix elements, keeping only the first non vanishing term. We have:

$$P_{n\ell j} \to R_{n\ell} \quad ; \quad Q_{n\ell j} \to \frac{-\hbar}{2mc}\left(\frac{d}{dr} + \frac{k}{r}\right) R_{n\ell}$$

$$j_L(x) \sim \frac{x^L}{(2L+1)!!}\left(1 - \frac{\frac{1}{2}x^2}{1!(2L+3)} + \dots\right) \qquad (3.2)$$

So that

$$I_L^+ \to \frac{-\hbar}{2mc}\frac{1}{(2L+1)!!}\left(\frac{\omega}{c}\right)^L (\kappa+\kappa'-L)\int R_{n\ell}R_{n'\ell'}\, r^{L-1}\, dr$$

$$I_L^- \to \frac{-\hbar}{2mc}\frac{1}{(2L+1)!!}\left(\frac{\omega}{c}\right)^L \int R_{n\ell}\left[(\kappa'-\kappa+L)\, r^{L-1} + 2r^L \frac{d}{dr}\right]R_{n'\ell'}\, dr$$

etc.

It is then straightforward to show that for electric dipoles the Coulomb gauge, $G_1=0$, yields the velocity formula, while the choice $G_1 = -\sqrt{2}$ gives the length formula (10,11). This connection is interesting for several reasons: First, it puts the discussion on a more general footing. In the non relativistic theory, the deep reason for the equality of f_L and f_V is somewhat obscure. Here, the principles of Quantum mechanics require any physical result to be gauge independent. Second, it generalizes the problem to multipoles. Finally, several authors used this formalism to re-investigate the dilemma, especially regarding Relativistic Hartree-Fock (10,15,16,17,18). However, the discussion is not closed. Grant (10) states that Coulomb gauge is more natural - leading to velocity form. Kobe (17,18) distinguishes between the gauge invariance of the interaction and of the matrix element. He proposes the form $e\underline{E}\cdot\underline{r}$ instead of $e\,\underline{\alpha}\cdot\underline{A}$ for the interaction. This, not involving the vector potential is "truly gauge invariant". However, his argument is valid only for dipoles. Our personal feeling is that this discussion just re-assesses the theoretical equivalence of the two formulas. To know which one is good for which approximate atomic model requires a method of analyzing the models in this respect. We shall mention this in section 4.2.6.

3.5 Selection Rules and Orders of Magnitude

There are two kinds of selection rules: absolute and approximate.

3.5.1 Absolute rules are always valid in the absence of an external perturbation on the atom.

(i) Parity - To simplify, let us define the parity of a relativistic wavefunction as that of the ℓ associated with the large component. Then, in order for the transition to occur the product of the parities of initial and final states must be $\sim(-1)^L$ for electric multipoles and $\sim(-1)^{L+1}$ for magnetic multipoles. This rule is broken by an external electric field, which mixes atomic parities.

(ii) J value: The total angular momenta J,J' and L must satisfy the triangular inequalities. This rule can be broken by hyperfine interactions, causing J not to be the resultant angular momentum (19).

3.5.2 Approximate rules: if the j quantum numbers of the individual electrons are well defined (i.e. good jj coupling and no configuration interaction), then one finds some additional rules:

(i) Only one j-value can change in a transition (one electron jump)

(ii) j,j' and L must satisfy triangular inequalities.

We do not mention here the selection rules on the spin or orbital angular momenta since these are not good quantum numbers for relativistic wavefunctions. Let us note that the selection rules in jj coupling are formally much simpler than in LS for higher multipoles (20).

Table 1 gives the orders of magnitudes of the numerical factors for the first few multipoles. Owing to the large differences between E1 and the others, these are very often called forbidden transitions. However, some authors prefer to keep the expression "forbidden" for the case when some approximate selection rule is not valid (13), or when an external perturbation enables the "absolute" rule to be violated. The matrix element appearing in the entries is the non relativistic line strength. For instance:

$$S_{E1} = \frac{3c^2}{2\hbar\omega^3} |(J_f||\Sigma\, h_1^e||J_i)|^2 \underset{NR}{\rightarrow} |\langle J_f||\Sigma\, r_i||J_i\rangle|^2 .$$

§ 4. ATOMIC MODELS

4.1 Generalities

In order to obtain numbers for the matrix elements described in the previous section, it is necessary to compute the atomic wf, and energy levels. For many-electron atoms, the relativistic hamiltonian is known only approximately (11,21), and usually

Table 1

The entries are TP in s^{-1}

$$A_L = N_L \, \omega^{2L+1} \, S_L$$

where ω are either wavelengths λ in Å, or energies E in eV. S_L are in atomic units.

	λ(Å)	E (eV)
E1	$2.026\cdot 10^{18}\ \lambda^{-3}\ S_{E1}$	$1.083\ 10^{6}\ E^{3}\ S_{E1}$
E2	$1.680\cdot 10^{18}\ \lambda^{-5}\ S_{E2}$	$5.919\ 10^{-3}\ E^{5}\ S_{E2}$
M1	$2.698\cdot 10^{13}\ \lambda^{-3}\ S_{M1}$	$1.443\cdot 10^{1}\ E^{3}\ S_{M1}$
M2	$6.630\cdot 10^{12}\ \lambda^{-5}\ S_{M2}$	$2.334\ 10^{-8}\ E^{5}\ S_{M2}$

written as:

$$H = \sum_i h_D(i) + \sum_{i<j} \frac{1}{r_{ij}} + \sum_{i<j} B_{ij} \qquad \text{(in a.u.)}$$

B_{ij}, the Breit interaction, is given by

$$B_{ij} = -\frac{1}{2r_{ij}} \left(\underline{\alpha}_i \cdot \underline{\alpha}_j + \frac{(\underline{\alpha}_i \cdot \underline{r}_{ij})(\underline{\alpha}_j \cdot \underline{r}_{ij})}{r_{ij}^2} \right)$$

The approximation made in obtaining B_{ij} discards the electron-positron and positron-positron interactions. Thus, it can only be introduced as a first order perturbation (11,22). The remaining problem is to solve

$$H_R = \sum_i^N h_D(i) + \sum_{i<j}^N \frac{1}{r_{ij}}$$

Now the main difficulty is not solving Dirac equations, but handling electron correlation, which is not a relativistic effect. Thus, we describe first some non relativistic models.

4.2 Non Relativistic Atomic Models

We shall restrict ourselves to the description of some widely used models, from the point of view of perturbation theory. The latter can be applied to isolated atoms thanks to the central field model of Slater (23): The non relativistic hamiltonian can be written as:

$$H = \sum_{i}^{N} - \frac{1}{2}\nabla_i^2 - U_i(r_i) + \sum U_i(r_i) - \frac{Z}{r_i} + \sum_{i<j} \frac{1}{r_{ij}}$$

$$= \qquad H_0 \qquad + \qquad H_1$$

U(r) is a mathematical artefact that is arbitrary, inasmuch as it enables convergence of perturbation expansion (24). If U(r) has spherical symmetry, the zero order wfs come out as determinants of one electron wfs where the coordinates are separable. For open shell atoms, these are highly degenerate. It is useful to recall that this is the definition of "configurations". At that stage (zero order), f_L and f_V are always equal, since the computation of f boils down to a problem with one electron in a central field. When dealing with degenerate zero order wfs, e.g. $|i_r\rangle$, the first order energies are obtained by diagonalizing the matrix of $\langle i_r|H_0+H_1|i_s\rangle$. This usually splits the configuration into terms (pure LS coupling), or levels, if spin orbit interaction is included in H_1. This splitting is responsible for the difference between f_L and f_V in <u>first order</u> models, since the transition energy is no more equal to the difference of one electron energies (which comes out of the commutation relations). Not only does the mere concept of configurations emerge from Slater's idea, but their very position in the spectrum and the importance of first and higher order corrections (configuration interaction) will depend on the particular form of U(r) chosen (25).

<u>4.2.1 The Z-expansion</u> model, already mentioned in section 2.1 stems from the choice U(r) = Z/r. (3,4) As the wfs are hydrogenic, it is possible to compute the first few expansion coefficients exactly, but the following involve a lot of work. As a way of obtaining numbers, its usefulness is limited to systems with a very small number of electrons. However, as a framework for gaining insight into the physics, it is invaluable. It is worth noting that zero order functions are now "aggregate of states characterized by a given set of n quantum numbers and a given parity" - this is called a complex. (2) The Hamiltonian matrix must then be diagonalized on this basis.

<u>4.2.2 The Hartree-Fock model</u> uses a non-local U(r) defined by its matrix elements between one-electron wavefunctions:

$$\langle i|U_{HF}|j\rangle = \langle i|-Z/r|j\rangle + \sum_{n\neq i,j} \langle in|1/r_{12}|jn\rangle - \langle in|1/r_{12}|nj\rangle$$

where the sum runs over all occupied shells. It has the property of cancelling the first order energy correction. For details about numerical methods and other properties, the reader is referred to the book by C. Froese - Fisher (26). Many computations of TP have been performed by this method. (9,26) However, let us state that the Hartree-Fock scheme is not particularly convenient for the calculation of TP. One reason is that the two configurations are obtained independently, and the wavefunctions do not come out orthogonal. This introduces the necessity of dealing with overlap integrals (2), which lengthen the computations. Another reason is that for an unknown transition, the energy difference has to be computed and can be minute compared with total energies, especially for heavy atoms. Hence, there is a very significant loss of accuracy. Finally, there may be some difficulties in obtaining convergence (28). For these and other reasons, many variants of the original Hartree Fock exist. We shall mention only the transitional state model (27) which avoids the necessity of overlap integral, and the statistical-exchange approximation for obtaining a local potential (28), sometimes called Hartree-Fock-Slater.

4.2.3 The Parametric Potential model (25,29,30,31) consists in representing U(r) by an analytical function depending on a small number of parameters (one per closed shell). For any given value of the parameters, the wave functions and the energy can be computed numerically. Then the parameters may be varied in such a way as to minimize the energy of one or a group of levels in intermediate coupling or with configuration interaction. The optimization of the parameters may be achieved also on some other quality criterion. For instance, when some energy levels are known, the potential may be varied to obtain a best fit between experimental and computed energies. TP obtained in this fashion come out very accurate for rare gases (32). This method overcomes the difficulties encountered with Hartree-Fock mentioned above. This method must not be confused with the so-called pseudo-potential model. The latter is not an attempt to solve the Schrodinger equation by perturbation expansion. Rather, it is a semi empirical method, in which only valence electron wfs are computed. The potential used is optimalized to yield the best fit to some experimental energy levels, absorbing in an effective way the interactions (direct and exchange) with core electrons. The wfs so obtained, which do not necessarily have the proper number of nodes are then used to evaluate TP. In this approach, it is impossible to achieve ab initio prediction of energy levels since the hamiltonian itself is approximated - it does not include interelectronic repulsion and is not bounded from below. On the other hand, the parametric potential model involves the computation of all the wfs, even those of core elctrons, and of all exchange integrals. Thus, it enables the minimization of total energy and ab initio predictions. Of course, pseudo-potential calculations are much quicker. (33)

4.2.4 Configuration Interactions. In many instances, especially for the first few ionization stages, the calculations using zero-order wfs, - first order energies - may be quite inaccurate. When second order effects are dominated by one or a few terms in the sum over excited states, it is possible to include these explicitly as if they were degenerate with the zero order energies - that is by including them in the basis of the hamiltonian matrix. The effect on TP may be dramatic. For instance, in the sequence of sulphur, the"mixing" $3s3p^5$ - $3s^23p^33d$ can reduce by a factor of more than 100 the transition to $3s^23p^4$ (34). This matrix diagonalization may be performed with a parametric potential, as above, or within the framework of the Hartree-Fock method, which becomes then "Multi configuration Hartree-Fock" (MCHF) (26,35). Then wf optimalization and matrix diagonalization take place simultaneously. Many results have been published for the isoelectronic sequences of B(36), Zn(37) etc. (26) The dramatic reduction in the OS that may be found (34) is sometimes called "cancellation effects". It must be distinguished from another fact that is unfortunately designated by the same expression. This is the case where the radial integral

$$I = \int R_{n\ell} \, r \, R_{n'\ell'} \, dr$$

changes sign along an isoelectronic sequence (usually for $Z-N \lesssim 7$), causing the OS to become very small, and hence, very sensitive to small changes in the central potential. (2,38)

4.2.5 Computation of Perturbation expansion. With any definite choice of $U(r)$, it is a simple matter to generate numerically a complete set of zero order wfs, and, in principle, this is all that is required for computing any order in the perturbation expansion, including contributions from the continuum. This method, called Many Body Perturbation Theory (MBPT) was described (39) and used by Kelly (40) and others (41) for TP to discrete and continuum states. It is often necessary to use diagrams in order to classify the different contributions (42). The computations can be applied to any atom and any transition, but they are time consuming.

4.2.6 Higher order models. A completely different approach was developed in the past few years, especially efficient when a large number of excited configurations contribute equally. We refer here to the Time Dependent Hartree-Fock scheme (43) or Random Phase Approximation (RPA) (44), and their numerous nearly equivalent variations. (45,46,47) In these models, it is not attempted to obtain good wf in an absolute sense, but rather to include only - but as much as possible of the terms of the infinite sums contributing to the particular transition being computed. Let us illustrate this point in more details. Suppose we chose a given potential generating a zero order basis set of wf $|j\rangle$ (usually HF). We can write:

$$|\Psi_a) = |a\rangle + \sum_{j\neq a} |j\rangle \frac{\langle j|H_1|a\rangle}{E_j^0 - E_a^0} + \ldots$$

and an analogous expression for $|\Psi_b)$. Then the transition matrix element writes:

$$(\Psi_b|D|\Psi_a) = \langle b|D|a\rangle + \sum_j \frac{\langle b|D|j\rangle\langle j|H_1|a\rangle}{E_j - E_a} + \sum_k \frac{\langle b|H_1|k\rangle\langle k|D|a\rangle}{E_k - E_b} + \ldots$$

The electric dipole D is a one electron operator. Consequently, in the sum over all excited configurations $|j\rangle$ in the expansion of $|\Psi_a)$, only those which are one-particle excitations with respect to $|b\rangle$, and satisfy some selection rules, will contribute. Similarly, the sum over $|k\rangle$ can be restricted to some one-particle excitations from $|a\rangle$. This can be looked upon as selective excitations due to the "coupling" between the two states, through the "perturbation" D. One can then look for the pair of determinantal wf which would take this coupling into account. When developing the theory, coupled integrodifferential equations are obtained (43,48):

$$\{H_i^0 - \varepsilon_i^0 \pm \omega\}\, \phi_{i\pm}^1(r_i) + \{V_{i\pm}^1(r_i) + v_i(r_i)$$

$$- \langle\phi_i^0|V_{i\pm}^1|\phi_i^0\rangle \pm \omega\langle\phi_i^0|\phi_{i\pm}^1\rangle\}\, \phi_i^0(r_i) = 0$$

where ε_i^0, $\phi_i^0(r_i)$ are Hartree-Fock orbitals and energies; $V_{i\pm}$ are terms coupling together the positive and negative frequency ($\pm\omega$) components $\phi_{i\pm}^1$ of the "response". $v_i(r_i)$ is the space part of the one-electron perturbation. There are several ways of solving these equations numerically (48). TDHF (or RPA) have been used successfully also for transitions towards a continuum state (photoionization) (46). It is remarkable that f_L and f_V come out nearly identical in these computations. (5,8,49,50). However, these methods are restricted, up to now, to atoms described by determinantal wfs (closed shells ground states). This means that, contrarily to MBPT, they are not applicable when L S coupling breaks down. The domain of usefulness for the non relativistic version is thus zone 1 of ionization stages.

4.2.7 Variationnal principles. The above mentioned agreement is not, by itself, a guarantee of accuracy (see 4.2). This is actually what comes out in some cases from a comparison with experiment (8). Indeed, what one wants is a more reliable quality criterion. A definite variationnal principle involving the matrix element itself was described by Rebane and Braun (51). Weinhold (52) obtained lower and upper bounds for the matrix element and devised a method for the ab initio estimation of error bounds. These were applied to He and Li^+ (52) and to the isoelectronic sequence of Be up to OV (54). In these cases, the authors were able to assess a theoretical precision of 0.05%, with very sophisticated correlated wfs. A very interesting case which has stimulated a number of computations and experiments is that of the M1 transition 1s2s $^3S_1$- $1s^2$ $^1S_0$ in Helium-like ions S XVI, Cl XVII, Ar XVIII. Anderson and Weinhold (56) were able to state an error bound of 0.2% on theoretical values, showing that the experimental results were in error. The source of this error was later found as due to satellite of a lower ionization stage (57).
The most valuable contribution of these variational principles is that they give the possibility of investigating rigorously the relative accuracy of f_L and f_V (55) in terms of other properties of the approximate wfs. In principle, it is now possible to know which of f_L or f_V is nearer to the exact result in each case. However, applications have been restricted to very simple systems.

4.3. Relativistic Atomic Models

The only difference between relativistic and non relativistic models is that zero order wfs are now solutions of Dirac equations with the potential U, and that the perturbing Hamiltonian includes the Breit Operator. Also, one must be aware of the existence of negative energy states (11). The radial functions possess two components, but this can be considered merely as a computational complication. The essential feature - i.e. that mean values of operators are linear combinations of products of angular matrix elements times radial integrals, remains valid. An example of this was given above in section 3.

This is why we shall content ourselves here with refering to computer programs which are the relativistic versions of the methods described in the previous paragraphs. A review of published computer codes for handling relativistic wavefunctions was recently made by Grant (58). These codes include programs for angular coefficients in jj coupling of the Coulomb interaction (59), of the transition operators (60) and of the Breit interaction (61), as well as codes for radial wavefunctions by Relativistic Hartree-Fock-Slater (62) and Multiconfiguration Dirac-Fock (63) (i.e. Multiconfiguration relativistic Hartree-Fock). Tables of matrix elements in jj coupling are available (64). Other works using unpublished programs are based upon Relativistic Z-expansion method (65), Relativistic Analytic Hartree-Fock (66) and Relativistic Hartree-Fock-Slater (67). Last but not least, many accurate results obtained with the Relativistic RPA (RRPA) have been quoted already (5,8). However, as mentioned above, the RPA methods work with determinantal wfs, which are jj coupled in the relativistic version. These may be quite an inadequate description for wf of zone 1 and 2 of § 2.1, so that RRPA is restricted to zone 3 ionization stage (68,69). Empirical corrections may be used to bridge the gap to non relativistic RPA computations for zone 1. One of the practical problems involved in these computations is computer time. Each "nj" wf has 2 components, but there are two values of j for each ℓ - that is 4 time as many wfs as in the non relativistic case. Also, the number of integrals to be computed is much larger than in non relativistic programs. The fact that relativistic configurations are $n\ell j$ wf instead of $n\ell$ means that it is hardly possible to avoid diagonalizing matrices. On the other hand, for highly ionized atoms, correlation effects are not so important. These remarks explain the usefulness of the relativistic version of the parametric potential method (70,71,72). The same arguments justifies the use with fair success of some programs which are only approximately relativistic. For instance, the code of Cowan (73) solves the one electron Schrodinger equation with a statistical approximation to exchange and some relativistic corrections. Numerous results were obtained with this code. (see section 6.1) However, let us conclude this section by recalling again that actual computations may be simpler in a purely relativistic framework than in a non relativistic one with an evaluation of "relativistic effects".

§ 5. RELATIVISTIC EFFECTS

Under this heading, one usually lists the corrections to the Schrodinger equation that are obtained by developing the relativistic equation in successive powers of v/c. Our aim here is not to propose actual computing of these corrections, like other authors did (4,11,14,74). Indeed, our point of view is to advocate the full relativistic computations, which include implicitly all these corrections. We merely wish to point out briefly what are the main differences that are to be expected in the results of relativistic and non relativistic methods.

5.1 Spin-Orbit Interaction

This is related to the difference of binding energies of electrons with different j (but same ℓ) in Dirac equation. It is

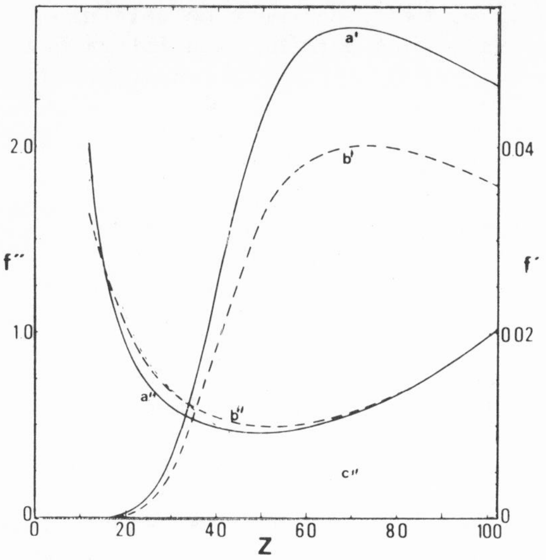

Figure 3. Z-dependence of the O.S. for resonance lines of Mg isoelectronic sequence. a'(f_L) and b'(f_V) correspond to $3s_{\frac{1}{2}} \rightarrow 3p_{\frac{1}{2}}$ ($3s3p\ ^3P_1$ in LS limit) (scale on right). a"(f_L) and b"(f_V) and c" (non relativistic) correspond to $3s_{\frac{1}{2}} - 3p_{3/2}$ ($3s3p\ ^1P_1$ in LS limit) (scale on left).

by far the largest relativistic effect. Its consequences for energy levels are well known - splitting multiplets, breakdown of LS coupling and "intercombination lines" etc. Merely introducing these corrections on the energies in eq. (2.1) gives a substantial j-dependent contribution to O.S. Figure 3 shows this effect for the Mg isoelectronic sequence, obtained with the relativistic parametric potential method (76). It is also shown in the latter work that in the f_L formula, spin orbit splitting gives the bulk of the relativistic j-dependent corrections, while in the f_V formulation, it is more the radial integral that changes with j. These remarks relate specifically to $\Delta n = 0$ transitions. However, as pointed out already in § 2.1.4, the effect of sum rules will be to change also all other transitions.

5.2 Effective Operators

Some effects of the relativistic transition operators acting on the relativistic wfs may be reproduced by effective operators acting on non relativistic wfs (14). These are double tensors, and e $\underline{r}$ is replaced by

$$X = a_{01}\ \underline{w}^{(01)1} + a_{11}\ \underline{w}^{(11)1} + a_{12}\ \underline{w}^{(12)1}$$

where the coefficients a_{01} etc. involve relativistic radial integrals. This shows that even in pure LS coupling, $\Delta S = 1$ and $\Delta L = 2$ transitions may be allowed. This is especially important for intercombination lines (77). However, one must be careful not to include the same effect twice when working with intermediate coupling wfs (78).

5.3 Wavefunction Effects

The above mentioned coefficients a_{01}, a_{11} etc. are usually obtained by formally expanding the radial wfs in powers of v/c and keeping only the first term. However, it has recently been pointed out (79,80) that the second term of the large component (in $(v/c)^2$) may contribute much more to transition operators than the first term of the small component (of order v/c). For $1s$-$2p_{3/2}$ in Hydrogen, this contribution is nearly 4 times larger. However, it does not have an angular behaviour different from the zero order, non relativistic matrix elements, and thus cannot be described by effective operators. This is responsible for the fact that in Rb I, for instance, the ratio $f_{3/2}/f_{1/2}$ in the series 5s-np may be much larger than 2 for high n's. (actually measured as ≃6 for n > 30) (81).

5.4 Indirect Relativistic Effects

All shells tend to contract towards the nucleus because of the relativistic kinetic energy correction, but this effect is more pronounced for smaller j. As a consequence, the inner shells $(s,p_{\frac{1}{2}})$ contract, thereby increasing the screening on the outer shells. Thus the outer shells expand slightly, and this modifies the radial matrix element (82). This effect, can be thought of as a cross term: correlation-relativity.

5.5 Retardation

In the radial integral of the transition operators described in § 3.3 a spherical Bessel function was present. However, the argument of this function is usually small. Indeed, replacing ω and r by hydrogenic mean values, one obtains approximately

$$\frac{\omega r}{c} \sim \frac{1}{c} \cdot \frac{Z^2}{2n^2} \cdot \frac{3n^2}{2Z} \sim \frac{3}{4}\frac{Z}{c} \sim \alpha Z$$

We saw (3.4) that $j_L(x)$ differs from x^L only by a term of order $x^2 \sim (\alpha Z)^2$. This is the so called "retardation effect", and although small, it may improve the agreement with experiment (83). Moreover, for magnetic dipole transitions of the type $1s2s\ ^3S - 1s^2\ ^1S$ (56) the first term contribution vanishes identically, and only the second term allows the transition which is thus a purely relativistic effect.

§ 6. APPLICATION TO HOT PLASMA SPECTROSCOPY

6.1 Extensive Computations for Identification Purposes

One of the problems of the spectroscopy of hot plasmas, is the appearance in the spectra of many hitherto unknown lines. In many cases, one has to rely on extensive computations to identify them. For this purpose, the computations must be quick, reliable and universal - i.e. applicable to any transition of any atoms. The sophisticated and accurate methods like RRPA described in § 4 are inadequate in this context. We report here satisfactory results obtained by simpler methods.

6.1.1 For ionized atoms with a few remaining electrons, the relativistic Z expansion method was used. Good agreement was obtained on the isoelectronic sequences of He (84), Li, Be (85) etc.

6.1.2 A very large number of computations were performed by Cowan, using his HXR method (73) (relativistic corrections in the Schrodinger equations). Here, we just indicate some recent references, including a catalog of lines expected in Tokamaks (86), a description of Fe lines in hot plasma (87), the identification of Mo XV-XX in Tokamaks (88), of Mo XXX- Mo XXXII in Laser produced plasma (89), of Mo XXXI - Mo XXXIV in exploded wires (90). For these highly ionized atoms the agreement between theoretical and experimental wavelengths is around 2% for $\Delta n = 0$ transitions and 0.1% for $\Delta n \neq 0$. These computations include also evaluation of O S which are very useful for identification. However, they serve merely as a guide, since line intensities depend also on plasma conditions and excitation cross sections.

6.1.3 The relativistic parametric potential method (RELAC) was also used extensively for identification purposes on Mo XXIV - Mo XL (91), Mo XV - Mo XXXIII,(92) Fe II - Fe XXV (X rays) (72), for Ta XLIV - Pt L in laser produced plasmas (93,94), etc. The agreement with experiment is good, around 1% for $\Delta n = 0$ transitions and better than 0.1% for $\Delta n \neq 0$. An interesting comparison between the HXR method of Cowan, RELAC, and RRPA (5) is presented in ref. 89., on the resonance transition of the Mg-like Mo XXXI transition $3s^2$ - $3s3p$ $^1P_1$. The discrepancy with the measured wavelength of 115.994 Å is +2.8 Å , -1.1 Å , and + 0.08 Å respectively for HXR, RELAC and RRPA. The O S value is 0.55 both for HXR and RELAC (76), while the RRPA yields 0.549 (f_L) and 0.566 (f_V). Thus the RRPA method plays here the role of a standard, and confirms our earlier statement that for highly ionized atoms, correlation effects are not as important as relativistic effects. As a consequence, simple methods like HXR, RELAC or Dirac-HF (63) are well adapted to extensive computations for identification of highly ionized spectra.

6.2 Transition Arrays

When heavy atoms (Mo,W) become highly ionized, the ground state configurations may comprise open d or f shells. The spectrum then consists of transition arrays, each one including so many lines in a narrow spectral range that they cannot be resolved. For instance in the spectrum of ORMAK, a broad band around 50 Å has been attributed to W XXXI - W XXXV. Several thousands of lines were computed and superimposed by Cowan, in order to try and reproduce the experimental profile (95). A new method has recently been developed for such cases, in which the statistical distribution of wavelengths in an arbitrary array, weighted by line strengths, is obtained directly (96). The results, the mean wavelength and the width of an array, are obtained as sums of products of simple angular coefficients and the Slater integrals pertaining to the initial and final configurations. Comparison with experiment on Mo spectrum obtained in a vacuum spark is very satisfactory (97).

6.3 Identification of Forbidden Transitions

"Forbidden" transitions - we mean here transitions other than E1 - are of great interest in plasma spectroscopy for several reasons. Their intensity is quite sensitive to electron density, and they can thus provide means of diagnosis, Moreover, for M1 or E2 transitions that can occur within a configuration, the wavelengths fall usually in a convenient spectral region (1000-3000 Å) enabling the measurement of ionic temperature by Doppler broadening. Identification of these lines usually requires precise wavelengths computations and careful examination of the intensities in the plasma conditions. The latter usually depend more on excitation cross sections then on T.P. However, when cascades are important - as they often are, it is necessary to compute many excitation cross sections, and one can use well known semiempirical formulae involving the O S (98). This method, used for Mo XV $3d^{10}$ - $3d^{9}4s$ E2 transitions appearing in the TFR, gave very good results (99), as shown on figure 4. In the same way, M2 lines have been identified in Cr XIV, Fe XVII, Ni XIX (isoelectronic to Ne) (100). Recently a M1 line of Fe XX at 2665 Å has been identified in the PLT tokamak, and used for ionic Temperature measurements (101). Computations were made by Cowan (86) and others (102). Other lines are being investigated (103). A review of forbidden transitions of Tokamal interest has been recently written by Feldman (104).

7. SUMMARY AND CONCLUSION

In this brief survey, we tried to describe some of the recent developments in the theory of atomic transition probabilities, in

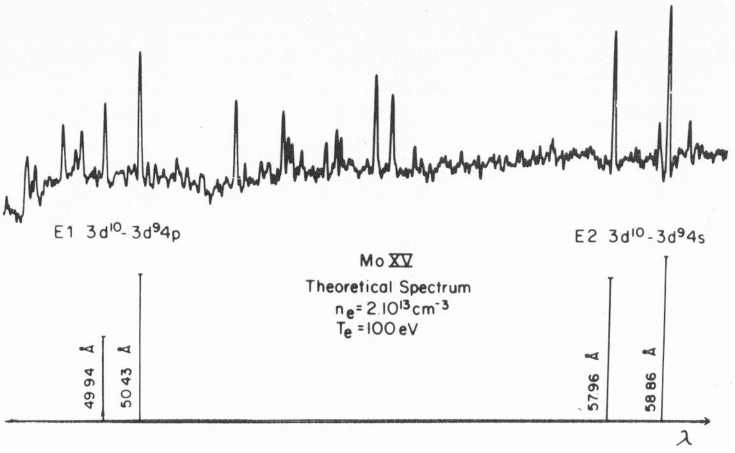

Figure 4. Experimental spectrum from TFR Tokamak and theoretical spectrum, showing forbidden (E2) transitions.

connection with the spectroscopy of highly ionized atoms in hot plasmas. We recalled, with the help of the Z expansion model, that for highly ionized atoms, correlation effects are small while relativistic effects are large. Thus the discussion was deliberately placed within the framework of the relativistic theory. We recalled briefly the main points of the theory of radiation, explaining the connection between gauge transformations and the f_L-f_V dilemma. It was pointed out that methods for estimating the error bounds in either formula were recently proposed by Anderson and Weinhold. Then, we reviewed some atomic models which are of interest, Z expansion model for few electrons ions, RRPA for closed shells ground states, Relativistic Hartree-Fock and Relativistic Parametric Potential for more general cases, with references to published programs. Then analyzing the "relativistic effects", we pointed out that the spin orbit interaction, although by far the most important, is not the only effect. Finally, we reviewed some results relevant to hot plasma spectroscopy. It appears that fairly simple models, such as central field relativistic, or approximately relativistic, are operational for identifying transitions. It is striking that it is easier to compute highly ionized atoms than neutrals. No doubt that the reason lies in the fact that relativistic effects are mainly one-electron operators, while correlation effects stem from electronic repulsion which are a two body operator.

References

1. R.J.S. Crossley, Adv. Atom. Mol. Phys. 5, 237 (1969).
2. D. Layzer, R.H. Garstang, Annual Rev. Astron. Astrophys. 6, 449(1968).
3. D. Layzer, J. Bahcall Ann. Phys. 17, 177 (1962).
4. H.T. Doyle, Adv. Atom. Mol. Phys. 5, 337 (1969).
5. P. Shorer, C.D. Lin, W.R. Johnson, Phys. Rev. A16, 1109 (1977).
6. B. Edlen, Phys. Scripta 7, 93 (1973) and 17, 565 (1978).
7. J. Reader, N. Acquista, Phys. Rev. Lett. 39, 184 (1977).
8. P. Shorer, A. Dalgarno, Phys. Rev. A16, 1502 (1977).
9. E. Biemont, Physica 81C, 158 (1976).
10. I.P. Grant, J. Phys. B7, 1458 (1974).
11. L. Armstrong, Jr., in Structure and Collisions of Ions and Atoms, I.A. Sellin, Editor, Springer-Verlag (1978).
12. A. Messiah, Quantum Mechanics, Vol. II (J. Wiley & Sons, 1962).
13. I.I. Sobelman, Theory of Atomic Spectra, Pergamon 1972.
14. L. Armstrong, Jr., S. Feneuille, Adv. Atom. Mol. Phys. 10, 1 (1974).
15. A.F. Starace, Phys. Rev. A3, 1242 (1971).
16. I.P. Grant, A.F. Starace, J. Phys. B8, 1999 (1975).
17. D.H. Kobe, Int. J. Quant. Chem. 12, 73 (1978).
18. D.H. Kobe, Phys. Rev. A19, 205 (1979).
19. R.H. Garstang, J. Opt. Soc. Am. 52, 845 (1962).
20. R.H. Garstang, Astroph. J. 148, 579 (1967).
21. I.P. Grant, N.C. Pyper, J. Phys. B9, 761 (1976).
22. H.A. Bethe, E.E. Salpeter, Quantum Mechanics of one-and two-electron Atoms, (Springer-Verlag, Berlin, 1957).
23. J.C. Slater, Phys. Rev. 34, 1293 (1929).
24. H.P. Kelly, in Perturbation Theory and Application to Quantum Mechanics, Wilcox ed. Wiley & Sons (1966).
25. M. Klapisch, Ph.D. Thesis, Orsay (France) 1969.
26. C. Froese-Fisher, The Hartree-Fock Method for Atoms, J. Wiley & Sons, 1977.
27. M. Godefroid, J.J.Berger, G. Verhaegen, J. Phys. B 9, 2181 (1976).
28. R.D. Cowan, Phys. Rev. 163, 54 (1967).
29. M. Klapisch, Compt. Rend. Acad. Sci. 265, 914 (1967).
30. M. Klapisch, Comp. Phys. Comm. 2, 239 (1971).
31. M. Aymar, M. Crance, M. Klapisch, J. Physique Coll. C4, 141 (1970).
32. M. Aymar, S. Feneuille, M. Klapisch, Nucl. Inst. Meth. 90, 137 (1970).
33. P. Hafner, W.H.E. Schwarz, J. Phys. B 11, 2975 (1978).
34. M. Aymar, Nucl. Instr. Meth. 110, 211 (1973).
35. C. Froese-Fisher, Comput. Phys. Comm. 4, 107 (1972).
36. W. Dankwort, E. Trefftz, J. Phys. B 10, 2541 (1977).
37. C. Froese-Fisher, J. Hansen, Phys. Rev. A17, 1956 (1978).
38. R.D. Cowan, J. Physique Coll. C4, 31, C4-191 (1970).
39. H.P. Kelly, Adv. Chem. Phys. 14, 129 (1969).
40. E.S. Chang, M.R.C. McDowell, Phys. Rev. 176, 126 (1968).
41. H.P. Kelly, A. Ron, Phys. Rev A5, 168 (1972).

42. B.R. Judd, Second Quantization and Atomic Spectroscopy Johns Hopkins Press, 1967.
43. A. Dalgarno, G.A. Victor, Proc. Roy. Soc. A291, 291 (1966).
44. D.J. Rowe, Rev. Mod. Phys. 40, 153 (1968).
45. T. Shibuya, J. Rose, V. McK y, J. Chem. Phys. 58, 500 (1973).
46. T.N. Chang, U. Fano, Phys. Rev. A13, 263 (1976).
47. A.C. Lasaga, M. Karplus, Phys. Rev. A16, 807 (1971).
48. B.C. Webster, M.J. Jamieson, R.F. Stewart, Adv. Atom. Mol. Phys. 14, 87 (1978).
49. M. Ya. Amusia, N.A. Cherepkov, L.V. Chernysheva, Sov. Phys. JETP 33, 90 (1971).
50. G. Wendin, J. Phys. B 6, 42 (1973).
51. T.K. Rebane, D.A. Braun, Opt. Spectros. 28, 486 (1969).
52. F. Weinhold, J. Chem. Phys. 54, 1874 (1971).
53. M.T. Anderson, F. Weinhold, Phys. Rev. A9, 118 (1974).
54. J.S. Sims, R.C. Whitten, Phys. Rev. A8, 2220 (1973).
55. M.T. Anderson, F. Weinhold, Phys. Rev. A10, 1457 (1974).
56. M.T. Anderson, F. Weinhold, Phys. Rev. A11, 442 (1975).
57. D.L. Lin, L. Armstrong, Jr., Phys. Rev. A16, 791 (1977).
58. I. P. Grant, Comput. Phys. Comm. 17, 149 (1979).
59. I.P. Grant, Comput. Phys. Comm. 14, 312 (1978).
60. N.C. Pyper, I.P. Grant, N. Beatham, Comput. Phys. Comm. 15, 387 (1978).
61. I.P. Grant, N.C. Pyper, N. Beatham, Comput. Phys. Comm. to be published.
62. D.A. Liberman, D.T. Cromer, J.T. Waber, Comput. Phys. Comm. 2, 107 (1971).
63. J.P. Desclaux, Comput. Phys. Comm. 13, 71 (1977).
64. Z.B. Rudzikas, V.I. Sivcev, I.S. Kickin, Atom. Data and Nucl. Data 18, 205 (1976), and ibid, 223 (1976).
65. G.L. Klimchiskaya, L.N. Labzovski, Optika Spectroskopy, 34, 633 (1973).
66. Y.K. Kim, Phys. Rev. 159, 190 (1967).
67. I. Lindgren, A. Rosen, Phys. Rev. 176, 114 (1968).
68. W.R. Johnson, C.D. Lin, Phys. Rev. A14, 565 (1976).
69. G.W.F. Drake, Phys. Rev. A19, 1387 (1979).
70. E. Luc-Koenig, Physica 62, 393 (1972) and Phys. Rev. A13, 2114 (1976).
71. M. Klapisch, J.L. Schwob, B.S. Fraenkel, J. Oreg, J. Opt. Soc. Am. 67, 148 (1977).
72. M. Klapisch, E. Luc-Koenig to appear in Comput. Phys. Comm.
73. R.D. Cowan, D.C. Griffin, J. Opt. Soc. Am. 66, 1010 (1976).
74. D.R. Beck, H. Odabasi, Annals Phys. 67, 274 (1971).
75. C.F. Tull, R.P. McEachran, M. Cohen, Atomic Data 3, 169 (1971).
76. M. Aymar, E. Luc-Koenig, Phys. Rev. A15, 821 (1977).
77. E. Luc-Koenig, J. Phys. B 7, 1052 (1974).
78. G.W.F. Drake, J. Phys. B 9, L169 (1976).
79. S. Feneuille, E. Luc-Koenig, Comments Atom. Mol. Phys. 6, 151 (1977).
80. E. Luc-Koenig, J. Physique Coll. C1, 40, C1-115 (1979).

81. S. Liberman, J. Pinard, Phys. Rev. A20, 507 (1979).
82. S.J. Rose, I.P. Grant, N.C. Pyper, J. Phys. B 11, 1171 (1978).
83. C.P. Bhalla, Nucl. Instr. Meth. 110, 227 (1973).
84. U.I. Safronova, G.L. Khlimchitskaya, L.N. Labzovsky, J. Phys. B 7, 2471 (1974).
85. L.A. Vainshtein, U.I. Safronova, Sov. Astronom. AJ 15, 175 (1971).
86. R.D. Cowan. Los Alamos report LA 6679-MS (1977)(unpublished).
87. A.L. Merts, R.D. Cowan, N.H. Magee, Los Alamos report LA 6620-MS (1976)(unpublished).
88. M.W.D. Mansfield, N.J. Peacock, C.C. Smith, M.G. Hobby, R.D. Cowan, J. Phys. B 11, 1521 (1978).
89. P.G. Burkhalter, J. Reader, R.D. Cowan, J. Opt. Soc. Am. 67, 1521 (1977).
90. P.G. Burkhalter, R. Schneider, C.M. Dozier, R.D. Cowan, Phys. Rev. A18, 718 (1978).
91. M. Klapisch, R. Perel, D. Weill Rapport EUR-CEA-FC-827 (1976) unpublished.
92. J.L. Schwob, M. Klapisch, N. Schweitzer, M. Finkenthal, C. Breton, C.DeMichelis, M. Mattioli, Phys. Letters 62A, 85 (1977).
93. A. Zigler, H. Zmora, N. Spector, M. Klapisch, J.L. Schwob, A. Bar-Shalom, J. Opt. Soc. Am. in the press (1979).
94. A. Zigler, H. Zmora, N. Spector, M. Klapisch, J.L. Schwob, A. Bar-Shalom, Phys. Lett. in the press (1979).
95. R.C. Isler, R.V. Neidigh, R.D. Cowan, Phys. Lett. 63A, 295 (1977).
96. C. Bauche-Arnoult, J. Bauche, M. Klapisch, J. Opt. Soc. Am. 68, 1136 (1978).
97. C. Bauche-Arnoult, J. Bauche, M. Klapisch, Phys. Rev. 1979, in the press.
98. H. Van Regemorter , Astroph . J. 136, 906 (1962).
99. M. Klapisch, J.L. Schwob, M. Finkenthal, B.S. Fraenkel, S. Egert, A. Bar-Shalom, C. Breton, C. DeMichelis, M. Mattioli, Phys. Rev. Lett. 41, 403 (1978).
100. M. Klapisch, A. Bar-Shalom, J.L. Schwob, B.S. Fraenkel,
101. C. Breton, C. DeMichelis, M. Finkenthal, M. Mattioli, Phys. Lett. 69A, 34 (1978).
102. S. Suckewer, E. Hinnov, Phys. Rev. Lett. 41, 756 (1978).
103. S.O. Kastner, A.K. Bhatia and L. Cohen, Phys. Scripta, 15, 259 (1977).
104. S. Suckewer, E. Hinnov, Phys, Rev. A20, 578 (1979).
105. G.A. Doschek, U. Feldmann, J. Appl. Phys. 47, 3083 (1976).

SPECTROSCOPY OF HIGHLY IONIZED ATOMS IN THE INTERIOR OF TOKAMAK PLASMA

Einar Hinnov

Princeton University

Princeton, New Jersey, USA

Introduction

A tokamak is a device to create and confine a high temperature hydrogen isotope plasma, with the ultimate objective to produce thermonuclear fusion energy (see ref. 1 for a comprehensive review). It consists of a toroidal current in an externally applied toroidal magnetic field, and some provision for preventing the current loop from expanding in major radius. Since the plasma current both heats the plasma and produces a magnetic field, and since the local plasma conductivity (and hence current density and power input distribution, and the magnetic field topology) depends on the local electron temperature, the plasma dynamics in tokamaks are extraordinarily complicated, and only vaguely understood in many important respects. In particular, the mechanism of particle transport across the magnetic field has been a mystery of long standing. The principal difficulty in solution of the problem is the scarcity of reliable measurements of local particle transport rates, and other local plasma parameters of interest, in the hot interior of the plasma. It is in such local diagnostics where spectroscopy of highly ionized atoms of e.g. Fe, Ni, Ti, Kr and others offer many interesting possibilities.

Description of Tokamak Discharges

As a representative example, we consider the PLT tokamak[2-4] at Princeton shown schematically in Fig. 1, with some of the principal features described in Fig. 2. The major radius R = 130 cm, the minor radius usually a = 40 cm to the current-aperture limiter (heavy bars of graphite or stainless steel, formerly also of tungsten), and b = 49 cm to the stainless-steel vacuum walls. The

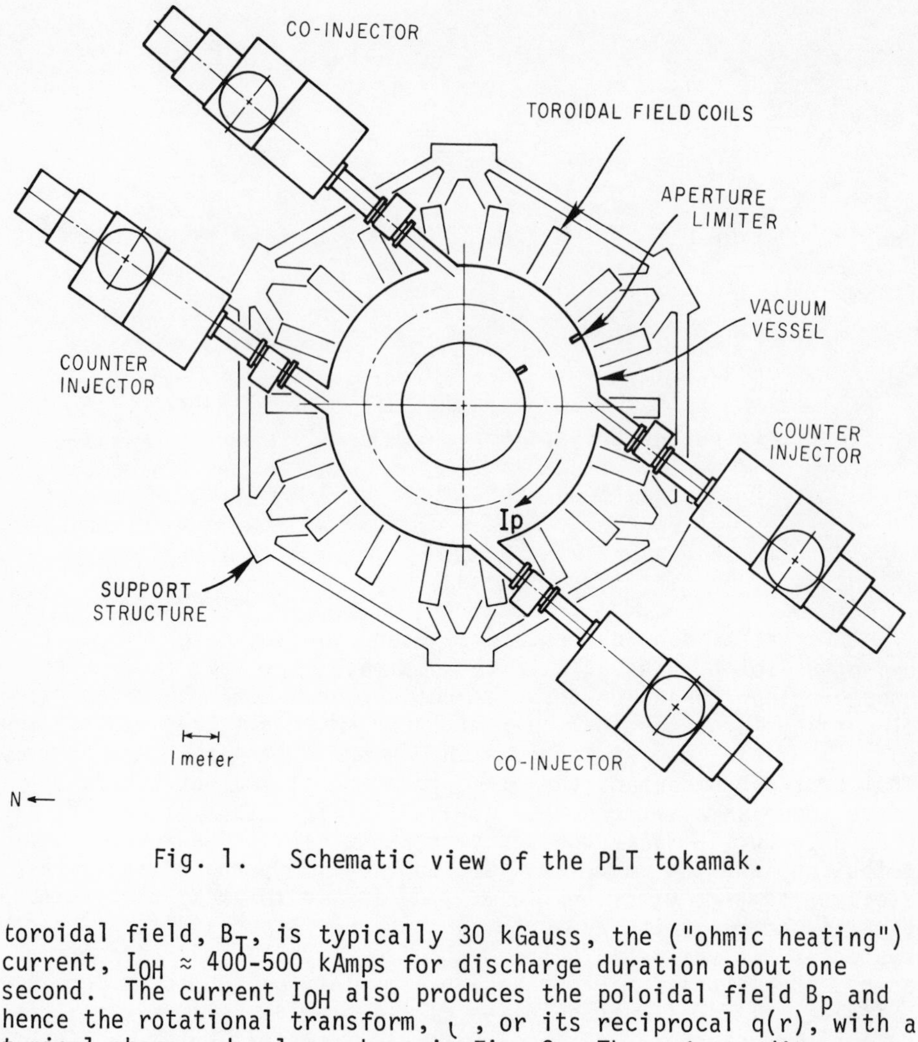

Fig. 1. Schematic view of the PLT tokamak.

toroidal field, B_T, is typically 30 kGauss, the ("ohmic heating") current, $I_{OH} \approx$ 400-500 kAmps for discharge duration about one second. The current I_{OH} also produces the poloidal field B_p and hence the rotational transform, ι, or its reciprocal q(r), with a typical shape and values shown in Fig. 2. The major-radius expansion of current is prevented by a programmable vertical magnetic field (not shown).

After an initial buildup time of about 0.1 sec, the discharge is quasistationary for times much longer than the replacement times of energy (plasma energy/power input) or particles (density/ionization rate). The latter two are comparable to each other and in the range 10-100 msec, depending on plasma conditions. The electron densities, $n_e(r)$, have a roughly parabolic radial profile within the limiter radius, with peak values $n_e(o) \approx (0.3\text{-}1.5) \times 10^{14}/cm^3$; the electron temperature profiles, $T_e(r)$ are somewhat narrower, often exhibiting a flattened top within q = 1 radius, as

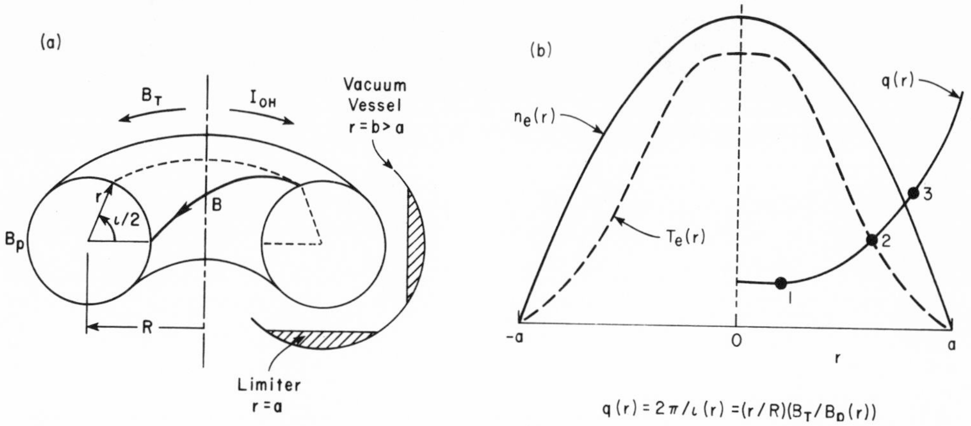

Fig. 2. (a) Principle of the tokamak device and its nomenclature: toroidal current I_{OH}, toroidal (B_T) and poloidal (B_p) magnetic fields, major (R) and minor (r) radii and the rotational transform, ι. (b) Typical radial distributions of electron temperature and density, and the inverse rotational transform, q(r), during a discharge.

shown in Fig. 2. The peak values are $T_e(o) \approx 1.0\text{-}3.0$ keV, with the higher temperature generally corresponding to lower-density discharges.

Two very important features of tokamak discharges are the power input and particle input distributions. In a steady-state straight-cylinder geometry, which gives a fair approximation, electrical conductivity, current density, and power input density all behave as $T_e(r)^{3/2}$, i.e. the ohmic power is deposited predominately near the center (with typical peak values ~ 1 watt/cm^3 in PLT), the peripheral plasma being heated by convective or conductive transport. The particle input, i.e. ionization of gas that is either deliberately added during the discharge, or results from recycling of plasma neutralized on the limiter or vacuum walls, on the other hand occurs predominately near the periphery, in fact at radii where the density is considerably lower than the central peak density. Thus it might be expected that the density distribution should be rather flat within the limiter radius, whereas the current distribution would tend to collaspe to the center. Actually an intermediate qausisteady state obtains, in many cases limited in time only by the ability of the transformer to deliver the current. A sample of measured electron density and temperature distributions in time and space is shown in Fig. 3. In this discharge plasma density was continuously raised by fairly intense gas-feed throughout the discharge. With a less intense gas feed, densities rise less rapidly (or even fall slowly), whereas central temperature would rise considerably higher.

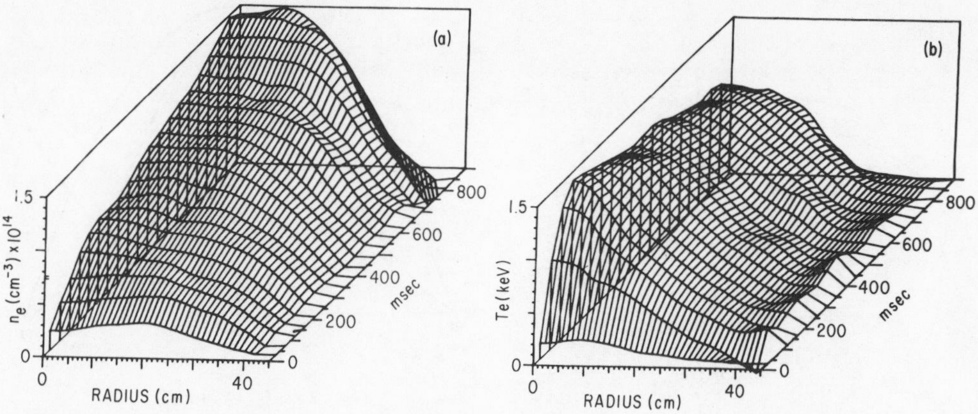

Fig. 3. Time and space evolution of the electron density and temperature in a PLT tokamak discharge.

Besides hydrogen isotopes the plasma contains oxygen and carbon impurities in the 1-10% range, and also smaller quantities (0.1-1%) of wall and limiter materials (Fe, Cr, Ni, formerly also W) and titanium [5-8] The latter is sprayed on the walls between discharges, in order to control the reflectivity of the wall to hydrogen atoms, and thereby allow programming of gas influx rate, and also to control the level of oxygen and carbon concentrations.[3] Like hydrogen, oxygen and carbon are ionized and radiate predominantly near the periphery of the discharge (although per atom they radiate much more strongly than hydrogen), thereby cooling the periphery and forcing the current density (and temperature and power input) distribution to become more centrally peaked. Thus either the gas inflow rate or the level of light-element impurities can be used to sculpture the all-important radial distributions to some extent.[5,9] However, in the hot central region where the confinement characteristics are really determined these atoms are fully stripped and therefore not directly observable. The heavier elements, with their higher-state ionization potentials comparable to the central temperature, on the other hand radiate and are ionized fairly uniformly throughout the discharge, and thus provide the means of localized diagnostics of the interior of the plasma. In this paper we therefore discuss only iron as a typical sample, partly because it is the most abundant of the heavier elements in PLT plasmas, and partly because its spectra are better known than others, largely as a result of past studies of solar flares.

In addition to ohmic heating, PLT plasmas have been heated[4] by injection of 40 keV neutral hydrogen beams, at power levels up to 2.5 MW (3-4 times ohmic heating power), for about 0.1 sec. The neutral beam injection increases primarily the ion temperature (up to 6-7 keV, compared to about 1 keV in ohmic heating), but the electron temperature (up to 3-4 keV) and density are also significantly increased. During the injection the beams also increase the central neutral hydrogen concentration (to about $10^8/cm^3$), and consequently the rate of charge-exchange with plasma ions. The beams furthermore impart momentum to the plasma, and unless balanced by opposing beams result in toroidal rotation with considerable velocity. Localized study of such dynamic effects is another important task for atomic physics in tokamak.

Properties of Iron Ions in the n = 2 Shell

In this section we discuss such features of the iron ions and their spectra that are particularly important for the diagnostics of tokamak type plasmas. It will be useful to bear in mind the principal features of such plasmas, ignoring for simplicity the outer periphery. Thus, we have a quasicylindrical, quasisteady plasma with electron temperature increasing rapidly toward the center to a peak of $\gtrsim$ 1 keV while the electron density is nearly constant [$n_e(r)$ wider than $T_e(r)$] at ~ $10^{14}/cm^3$, and the ion radial motions are retarded by the magnetic field to something of the order of limiter radius/confinement time, or ~ 10^3 cm/sec.

An overview of the relevant energy levels is shown in Fig. 4, with the effective central charge ζ felt by the least-bound electron plotted against the threshold potential for excitation, E_x, or ionization, E_i. The shell structure is clearly evident in the ionization potentials. From FeXVI (3s) to FeXVII ($2p^6$) the ionization potential more than doubles, and from FeXXIV (2s) to FeXXV ($1S^2$) it quadruples, whereas within a shell the changes from state to state are quite small. The states of particular interest, as discussed below, are those where $E_i \sim T_e$, i.e. the n = 2 shell for tokamak plasmas with temperature maxima ~ 1-3 keV.

The excitation potentials shown in Fig. 4 are those of one or a few strongest transitions from the ground state of each ionization stage. Except at the lowest ζ they clearly fall into two distinct groups, depending on whether the transition from ground state involves a change in the principal quantum number or not. Those in the first category remain within about a factor 2 of the ionization potentials (the transition energies from $\Delta n > 1$ fall between those of $\Delta n = 1$ and the ionization potentials). The excitation potentials of the transitions with $\Delta n = 0$ fall progressively below the ionization potentials with increasing ζ, until in the n = 2 shell the gap is rouhgly a factor 10. The radiation emitted by these ions therefore also falls to two distinct wavelength ranges, roughly 5-15 Å for the $\Delta n \neq 0$ and 100-200 Å for the $\Delta n = 0$ transitions (in the iron n = 2 shell). Since the exitations rates scale approximately as $E_x^{-3/2}$ most of the radiation is emitted by the latter group, $\Delta n = 0$. Furthermore, each state of ionization is roughly equal in radiation efficiency, since the $\Delta n = 0$ excitation potentials do not change very markedly with ζ.

This is also true in comparing different elements i.e. a titanium or a nickel atom in the n = 2 shell is not very different from an iron atom, either in the radiation pattern in wavelength range or the radiation efficiency per ion. However, ions in the n = 1 shell (and the $2s^2 2p^6$ configuration, FeXVII) which have no ground-state $\Delta n = 0$ transitions are markedly less efficient as radiators.

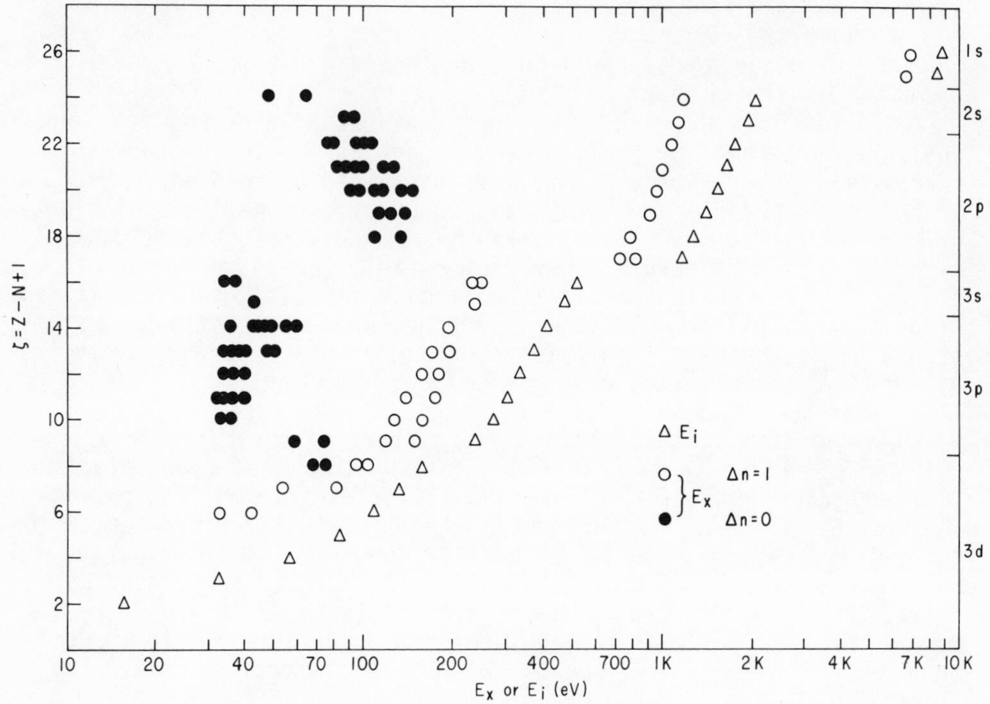

Fig. 4. The potentials of ionization and of excitation of the $\Delta n = 1$ and $\Delta n = 0$ transitions in iron ions.

A more detailed view of FeXXIII and XXIV as representative cases is shown in Fig. 5, together with some important rate coefficients. The ionization rate coefficients,[10-12] S_i, increase rapidly with electron temperature for $T_e \lesssim E_i$ (about 2 keV for these states) but begin to level off at higher temperature. The magnitudes are such that at typical tokamak conditions e.g. $T_e \approx$ 1-2 keV, $n_e \approx 3 \times 10^{13}/cm^3$ the ionization times are a few milliseconds, implying a radial motion of several cm in an ionization time, i.e. distances that are short compared to plasma radius (in PLT), but not very short compared to temperature scale-length ($T_e/\Delta T_e/\Delta r$). The recombination coefficients, α_{r+d} (radiative and dielectronic, the values shown are those of ref. 13), drop with increasing temperature. The points where they cross the corresponding ionization rate curves, marked with circles in Fig. 5, give the temperatures between which the ionization states in question are dominant at steady state ("coronal" ionization equilibrium), although a few neighboring states to either side are also present in appreciable quantities. At higher temperatures the ion is in an "ionizing" regime and at lower temperatures in a "recombining" regime. In tokamak plasmas, with a few very important exceptions noted below, the ions are almost always found in an ionizing regime, i.e. with volume recombination more-or-less negligible. The temperature range of interest for a particular ion is thus somewhat above the equilibrium temperature or, for practical purposes within about factor 2 of the ionization potential, as shown in Fig. 5.

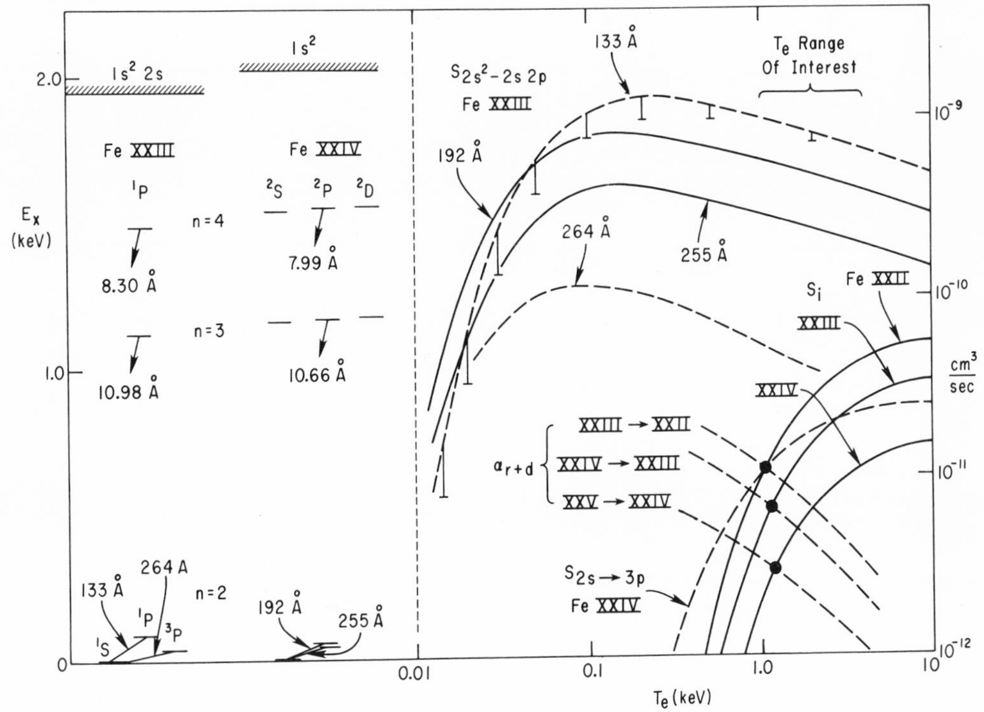

Fig. 5. Simplified energy level diagrams of FeXXIII and XXIV, and some selected ionization, recombination, and excitation rate coefficients.

The remaining curves in Fig. 5 display many important features of the excitation rate coefficients. As mentioned above, the rates for the $\Delta n = 0$ type of transitions (shown at left on Fig. 5) are much larger than than the ionization rates, or the excitation rates for the $\Delta n \neq 0$ transitions (the 2s-3p transition coefficient shown is typical of the strongest of the latter), but the latter two are fairly similar to each other both in magnitude and temperature dependence. Consequently, some 10-100 photons of the $\Delta n = 0$ transitions will be emitted for each of the higher-energy transitions, or of ionization events. Therefore, the line-emission of the $\Delta n = 0$ resonance transitions is completely determined by direct collisional excitation, i.e. radiative cascading caused by recombination or excitation to higher states is negligible. Furthermore, since the excitation rate at tokamak densities is about $S_x n_e \gtrsim 3 \times 10^{-10} \times 3 \times 10^{13} \approx 10^4$/sec and the radiative transition probabilities for allowed (electric dipole) transitions are $> 10^{10}$/sec secondary excitations from excited levels are also negligible, so the local emissivity is simply

$$J(r) = n_i(r)\, n_e(r)\, S_x(T_e) \text{ photons/cm}^3 \text{ sec} \qquad (1)$$

where n_i is the population of the lower level of the transition (generally the ground level of the ion) in question.

It is also evident from Fig. 5 that the excitation rates for the $\Delta n = 0$ transitions are very weakly temperature dependent in the T_e range of interest, a consequence of the average plasma electron energy being much larger than the excitation threshold. Thus, precise knowledge of the local electron temperature is not essential in determining ion densities from measured line intensities. The electron temperature profile enters only indirectly into the problem, by determining the (radial) location of the ion through the ionization (and recombination) rates.

On the other hand, the ratio of a $\Delta n \geq 1$ and a $\Delta n = 0$ transition of the same ion may be sufficiently temperature dependent to allow local electron temperature measurement, provided that (1) an appropriate line ratio can be measured with sufficient accuracy, and (2) the corresponding rate coefficients are adequately known. The second condition restricts the possibilities at present to a very few simple systems, notably the lithiumlike ions such as the FeXXIV shown in Fig. 5. The first condition also represents a formidable problem since measurements at widely different wavelengths are to be compared. Nevertheless, it has been performed successfully in some cases,[14] particularly for the OVI ion, with results that show OVI in tokamak plasmas[15] radiating predominantly at $T_e \approx 90$ eV — somewhat smaller than OVI ionization potential (138 eV) but substantially above coronal equilibrium temperature.

Still another aspect of the $\Delta n = 0$ excitation rates is shown by the vertical bars on the 133 Å ($2s^2$-$2s2p\,^1P$) case in Fig. 5. The rate coefficient is the product of the excitation cross-section and the incident electron velocity averaged over Maxwellian distribution. The cross-sections are directly measured in a few cases (for $\zeta \lesssim 6$), but generally it is necessary to rely on calculations of various degrees of sophistication. The uncertainties of the calculations are usually largest near the excitation threshold and diminish with increasing incident energy, where the incident electron remains more-or-less distinguishable from the target electrons. The vertical bars in Fig. 5 (133 Å line) show the effect on the rate coefficient of a near-threshold factor-of-two uncertainty in the cross-section: in the temperature range of interest the resulting error has diminished to about 20-30%. Thus, these transitions are useable for ion density measurements even if the near-threshold cross-section are not precisely known. The actual rate-coefficients shown in Fig. 5 (and for other Li and Be sequence ions) are probably quite accurate (they are based on as yet unpublished data,[16] but are similar to results deduced from various earlier calculations[17-19]), and are in good agreement with measurements in tokamak plasmas.[20] However in more complicated systems there are still few calculations and large uncertainties remain.

In the beryllium sequence, the 2s2p ^3P term is metastable in LS coupling, and therefore can have appreciable population that has to be taken into account in ion density measurements. This is indeed the case in lighter elements such as CIII and OV, where at tokamak densities the triplet population is typically 1.0-1.3 times the ground state population. However, in heavier elements the radiative lifetimes of the $^3P_1$ level become short[21,22] ($\sim 2 \times 10^{-8}$ sec in FeXXIII) compared to collisional lifetimes and the populations therefore small. The measured line intensity ratio[20] of 16:1 (at T_e = 1.5 keV) for the 133:264 Å lines of FeXXIII is in good agreement with the rate coefficients[16] shown in Fig. 5.

The energy levels[23] and wavelengths of the transitions within the $2s^22p^x$ ground configurations of iron are shown in Fig. 6. The wavelengths enclosed in boxes have been observed in the PLT tokamak and are accurate to about ± 0.1 Å. The wavelengths in parentheses are deduced from observed wavelengths of resonance lines or from calculations, and may be uncertain by several Angstroms, except those identified directly from solar flare spectra.[24] These transitions occur between levels of the same parity (magnetic dipole or electric quadrupole) and the transition probabilities[25-27] are in the range of 10^2-10^5 sec^{-1} or less for the quadrupole transitions. Thus the radiative rates are typically comparable or smaller than collisional rates, and the excited states may have populations comparable to the ground state populations. Because of the relatively high populations, these lines may have photon emissivities comparable to those of allowed (electric dipole) resonance lines, in spite of the low radiative transition probabilities.

The principal interest of such "forbidden" lines in tokamak plasmas lies in their diagnostic applications: here we have fairly

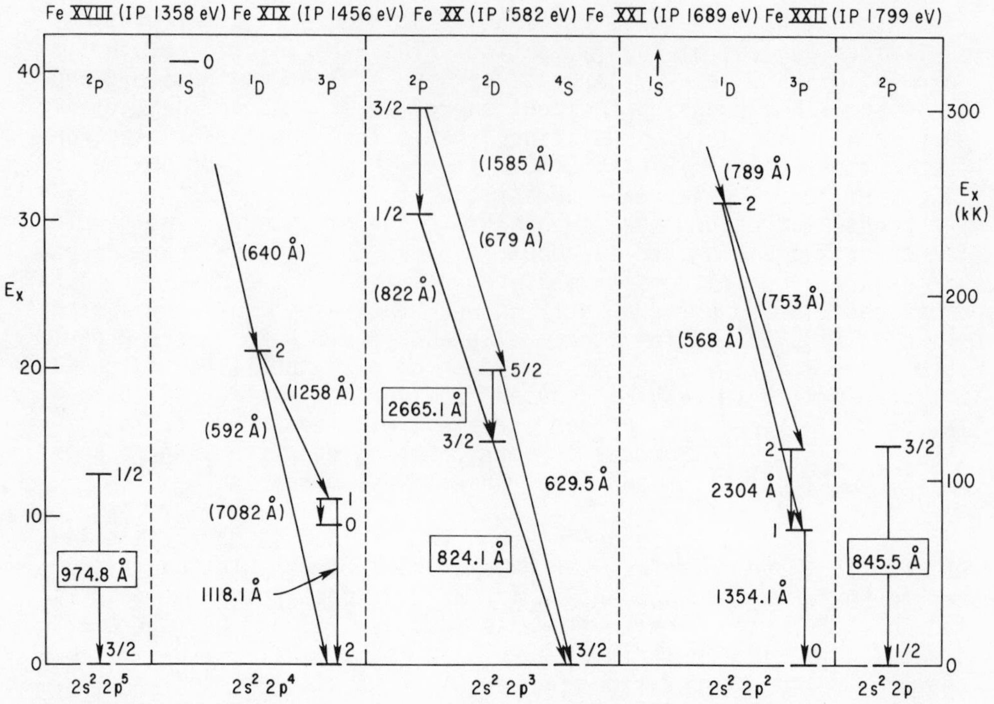

Fig. 6. Energy levels and transitions in the ground configurations of the $2p^X$ subshell of iron. Predicted wavelengths in parentheses. Wavelengths in boxes have been observed in the PLT tokamak.

strong spectrum lines of relatively long wavelengths arising from the high-temperature region (~ 1-3 keV in the case of iron) of the plasma. The long wavelengths allow the use of sophisticated optical techniques involving mirrors and perhaps lenses and windows that are not feasible in the short-wavelength region of grazing-incidence or crystal spectrometers. These forbidden lines of iron, chromium, and titanium atoms have been used extensively in tokamaks, to measure spatial distributions of the ions, ion temperatures and drifts through Doppler widths and shifts of the lines, and the ion densities from absolute intensities of the lines.[28-31] Similar techniques in somewhat heavier elements, especially krypton, would be usable at still higher temperatures, 3-10 keV that may be expected in large tokamaks in the future.

Iron in the PLT Tokamak Discharges

Figure 7 shows the time-behavior of a succession of iron ion lines, measured in a rather low-density PLT tokamak discharge, with steel limiters. The intensities are normalized to their

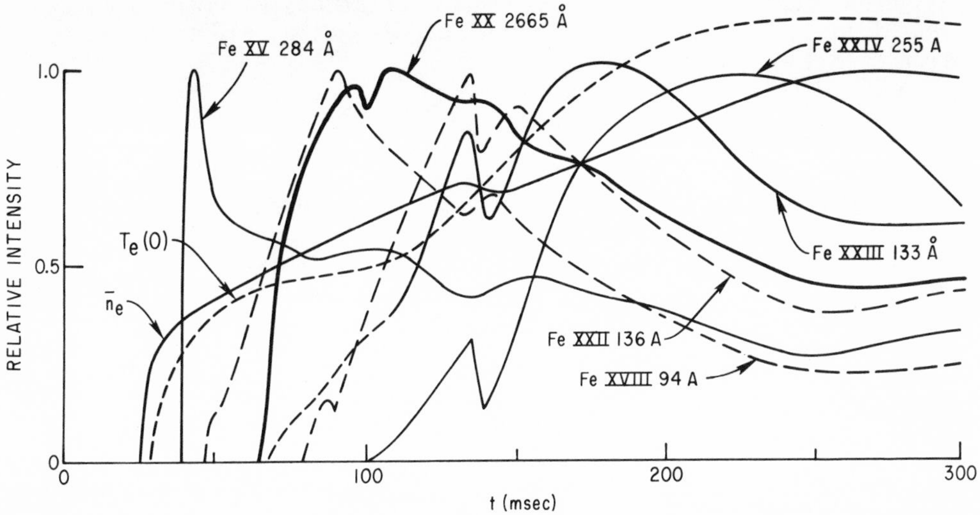

Fig. 7. The line-average electron density (in units of 2 x 10^{13} cm^{-3}), the central electron temperature (in units of 2 keV), and a sequence of iron ion lines observed in a PLT tokamak discharge.

maximum values in the figure. The measurements were made in the equatorial plane, in a series of nominally identical discharges, although some details in the risetimes of the different lines may be due to pulse-to-pulse variations. The central electron temperature was about 0.8 keV at 75 msec, 1.0 keV at 100 msec, 1.7 keV at 150 msec, and 2.3 keV at 200-350 msec. At these temperatures and the indicated electron densities (the central densities are about 1.5 times the line-average $\bar{n}_e$) the ionization times of the iron ions are about 1-5 msec, i.e. quite short on the time-scale of the figure. The observed succession of the ionization states indicateds a significantly slower rate of ionization.

There are several possible explanations for this qualitative observation:

1) The actual ionization rates are smaller than those used in calculations.[10,11] Comparisons with available experimental data[12,32,33] (at lower degrees of ionizations) suggest rather that the experimental rates might be somewhat larger than assumed, as a result of inner-shell ionization and autoionization processess. To be sure, there are some ionization rate measurements in plasmas[34,35] that suggest systematically smaller rates, but the interpretations of these measurements seem to be subject to the same type of uncertainties as our tokamak observations.

2) The electron temperature measurements may be in error. This also seems unlikely, as the temperature is measured by several

different methods with good internal consistency (Thomson scattering of laser light, soft-x-ray continuum spectra, electron cyclotron emission, plasma resistivity).

3) Recombination may slow the net ionization rates. Since the temperatures in question are significantly above the coronal equilibrium temperatures, the radiative and dielectronic recombination rates should have only a very minor effect. However, there is another possible recombination process in tokamaks, the charge-exchange between neutral hydrogen and highly-ionized atoms, that may be sufficiently rapid to be significant. This process will be discussed below in conjunction with neutral beam injection.

4) Radial plasma motion may be too rapid. If an atom moves during an ionization time over a distance where the electron temperature change is significant, the ionization states will lag behind equilibrium values or, in other words ions of a given state appear at higher than equilibrium temperature. [Qualitatively the same effect would be produced by explanation (1) above i.e. if the actual ionization rates were smaller than those used to interpret the observed data.] This is probably the principal reason for the apparently too slow ionization rates in tokamak plasmas. Therefore, if the true ionization and recombination rates were independently known, the differences between observed and expected ion density distributions could be used to deduce radial particle velocities or transport rates.

A particular case, corresponding to t = 200 msec in Fig. 7, is shown in Fig. 8. Here the chord-brightnesses (photon/cm^2-sec-sr) of the various ion resonance lines (measured on an absolute scale by a calibrated spectrometer[36,37] at different distances from the plasma center) have been first Abel-inverted to yield local emissivities (photons/cm^3-sec). Then, from separately measured (e.g. by means of Thomson scattering) electron densities and temperatures, using calculated excitation rate coefficients,[16] the emissivities are converted to local ion densities. The actual electron densities and temperatures at this time were (with steel limiter at r = 43 cm)

r(cm)	0	5	10	15	20	25	30	35	40
$n_e(10^{13}/cm^3)$	2.9	2.7	2.45	2.1	1.85	1.45	1.1	0.8	0.4
T_e(keV)	2.3	2.0	1.5	1.1	0.74	0.52	0.37	0.22	~0.1

The changes in the degree of ionization toward decreasing r, i.e. toward increasing electron temperature is evident. In cases like FeXVIII, with a pronounced minimum in the center the uncertainties in the measurement become amplified by the Abel-inversion process so in such cases the near-central densities cannot be accurately determined. Also, the extreme outer edges of the distributions are somewhat vague, because in almost all cases a substantial background radiation (continuum emission, nearby interfering lines of other origin, and straylight in the spectrometer) must be subtracted from the observed signal. However, the bulk of the distributions, and

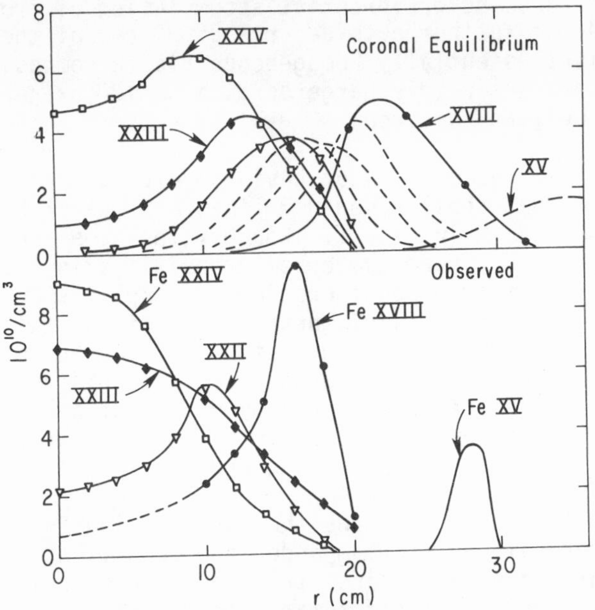

Fig. 8. Radial distributions of several iron ions deduced from space-resolved resonance-line intensities, and the (experimental) electron densities and temperatures given in the Table, compared with distributions expected on the basis of coronal equilibrium.

particularly the location and shape of the maxima is quite accurately determinable in most cases. A special problem arises in the case of FeXXIII. The strong resonance line at 132.9 Å overlaps with a FeXX resonance line[38] ($2s^22p^3\,{}^4S_{3/2}$-$2s2p^4\,{}^4P_{5/2}$) at 132.86 Å. Although the latter is considerably weaker in intensity (depending of course on the actual ionization state distributions and hence on electron temperature) it does contribute noticeably to the outside edge of the intensity distribution (and also to the very early part of the 133 Å radiation, Fig. 7). This is undoubtedly the main reason for the evidently too large density of "FeXXIII" between 15 and 20 cm in Fig. 8. In principle, a correction could be applied for this, e.g. by measuring the chord distribution of the ${}^4S_{3/2}$-${}^4P_{3/2}$ 121.8 Å line of FeXX, calculating the expected ${}^4S_{3/2}$-${}^4P_{5/2}$ intensities from this and substracting from the observed 132.9 Å distribution, although in the present case such a procedure was not attempted.

In spite of such minor uncertainties, the concentrations and distributions of the missing states, e.g. FeXIX, XX, XXI can be fairly reliably interpolated, and hence the total iron density distribution established. This total iron distribution turns out to be homogeneous, i.e. a constant fraction of the electron density. Furthermore, this homogeneity appears to persist throughout the discharge. There is no preferential accumulation of iron either near the center or further outside (r = 25-30 cm) of the discharge. This result, i.e. essentially homogeneous plasma composition has been found under various discharge conditions for oxygen and carbon impurities as well as for iron.

The principal reason for not having measured the distributions of FeXIX-XXI is that their resonance lines are weaker by factors ~ 2-5 than those shown in Fig. 7. Roughly the same amount of radiation is spread over larger number of transitions in these somewhat more complicated electron configurations, $2s^22p^4$, $2s^22p^3$, $2s^22p^2$. As a result the measurements of these states of ionization take more machine-time and are less accurate. This is one of the reasons for the interest in the long-wavelength forbidden transitions in these ions: these are measureable by different and simpler spectrometers, and allow the use of intricate optical systems for light-gathering, spectral resolution, and spatial scanning.

The top half of Fig. 8 shows the ion distributions expected on the basis of coronal equilibrium under the conditions of experimental $T_e(r)$ and total iron concentration. The experimental distributions are clearly shifted toward smaller radii, i.e. higher temperature. The cause of this shift is probably the same as was noted in the discussion of time-dependence: a radial motion of ions sufficiently fast to compete with ionization rates. Estimates of the required inward velocities are 2-4 cm/sec. This would require a total travel time of ions from periphery to center about 10-20 msec. This estimate is consistent with directly observed travel times of argon ions resulting from a sudden puff of argon admitted during the discharge and observing the timelag between emission of near-periperhal ArVII and near-central ArXVIII radiation in PLT tokamak discharges.

Inward ion motion of this magnitude combined with the observed lack of accumulation of ions in the center requires a comparable outward flux. Presumably a large fraction of the outward flux occurs in the heliumlike FeXXV state. This state should therefore occur in overabundance relative to coronal equilibrium, at larger radii (~ 10-30 cm). Although this state is observable[39] in the iron K-line (1.857 Å), there have been as yet no radial scans and in any case it would be difficult to detect at the lower temperatures. Perhaps the best prospect for detecting the recombining FeXXV would be observation of the recombination continua at different radii, in the 3-6 Å region. Again, such a continuum has been observed[40] in argon (recombination of ArXVIII) at temperatures considerably below coronal equilbrium temperatures, indicating a large outward drift velocity of ions. Similar outward drift velocities have been deduced from the behavior of tungsten[41] radiation resulting from experimentally changing the radial temperature profiles in PLT discharges.

Thus, there is a considerable body of circumstantial evidence that large radial ion transport velocities, both inward and outward, exist in tokamak plasmas, and that measurements of ionization state distributions in space and time can be used to measure such transport rates. However, detailed systematic measurements, as well as improvement of present knowledge of ionization and recombination (particularly including recombination through charge-exchange) rates of highly ionized atoms are required for sufficiently quantitative determination of cross-field ion motions.

Another problem of considerable importance in tokamak plasmas is the measurement of local ion temperatures. Because of the fairly well-defined positions of the various ionization states (Fig. 8), the measurement of Doppler profiles of spectrum lines of these ions provides automatic space resolution for the measured ion temperatures. Furthermore, the Doppler profile measures the velocity or energy distribution of the bulk of the ions, i.e. it is not affected by high-energy tails or other minor distortions of the Maxwellians that may affect strongly alternative methods of measurement, like neutron production or the energy spectrum of charge-exchanged neutral hydrogen emitted from the plasma.

At typical tokamak plasma densities and temperatures Stark effect and other collisional line broadening, as well as Zeeman splitting, are usually negligible in comparison with Doppler effect. [In the case of iron K-lines and similar transitions with extremely short radiative lifetimes, the natural broadening may be important.[39]] The measured spectral profile of a line is thus essentially the instrumental profile convoluted with the Doppler profile of the line.

The usual problem in high-resolution measurements is the scarcity of photons in the required spectral and time interval and this is particularly important in the grazing-incidence spectral region ($\lesssim$ 400 Å), where the solid angle of acceptance is small and transmission losses often high. Because of the reciprocal relationship between resolution and luminosity of the spectrometer, the appropriate instrumental width in such cases is comparable to the expected line (Doppler) width.

Figure 9 shows a shot-to-shot scan of the spectral profile of the resonance line of FeXXIV, which originates near the center of the discharge, during a high-power neutral beam injection experiment in PLT tokamak.[4] The profile was measured in the 4th order of a 1 m grazing incidence (85°) spectrometer equipped with a 2400/mm holographic grating. The time-integration was about 2 msec. The solid curve is a Gaussian fit to the experimental data, and the dashed curve is the instrumental profile, measured separately in a low-temperature (~ 50 eV) plasma by means of the 1032 Å resonance line of OVI in the first order, (at which conditions the intrinsic

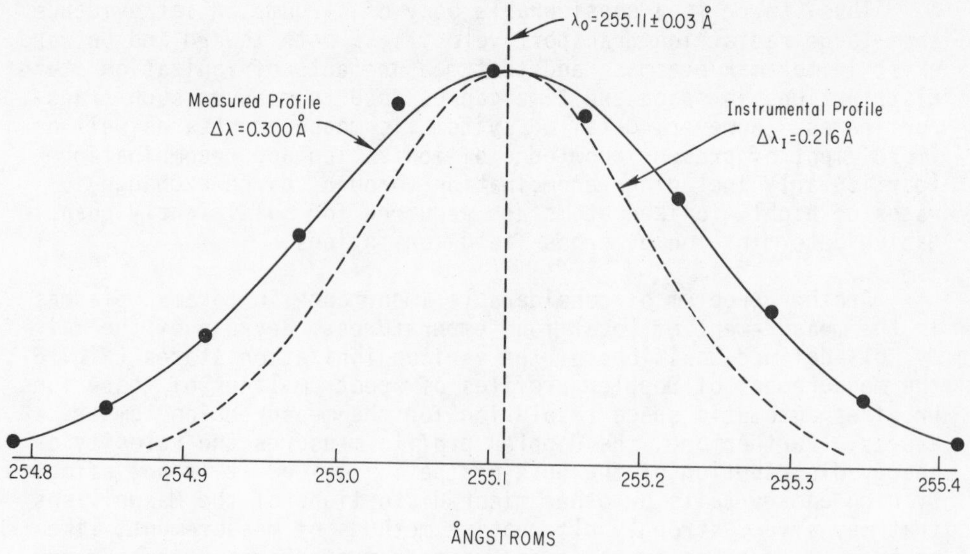

Fig. 9. Spectral profile of the 255 Å line of FeXXIV measured during a 2.1 MW neutral beam injection experiment in PLT, together with experimentally determined instrumental profile.

line-width is negligible). The instrumental width is sufficiently close to a Gaussian to allow deducing the Doppler width, $\Delta\lambda_D$(in Å), as

$$\Delta\lambda_D^2 = \Delta\lambda_M^2 - \Delta\lambda_I^2 = 6.9 \times 10^{-3}\ T_i(\mathrm{keV}) \quad ,$$

for iron, yielding $T_i \approx 6.3$ keV in this case.

The relatively long-wavelength forbidden or intercombination lines arising in transitions between ground-configuration levels are particularly suitable for the Doppler profile measurements. As explained above, such lines may have intensities in tokamak plasmas nearly comparable to resonance-like intensities of the ions while the availability of reflection or transmission optics allows much higher luminosity of the spectrometers. Figure 10 shows the ion Doppler temperatures measured during a neutral beam injection experiment in PLT with the 2665 Å forbidden line of FeXX (see Fig. 6), and with the 255 Å resonance line of FeXXIV. The spectral profile of the FeXX line was scanned repeatedly[42] during the discharge so the entire time behavior could be observed, whereas the shot-to-shot scan of the FeXXIV had too much variation and too low intensity (typical total intensity time-behavior of this line is shown by the lower curve in Fig. 10) to prevent significant measurements before 480 msec.

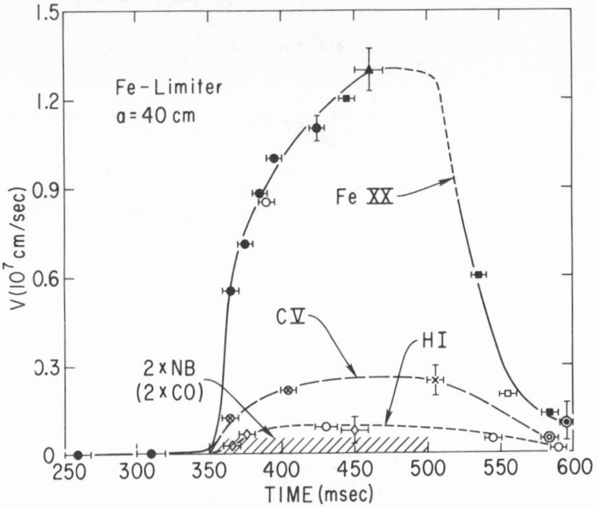

Fig. 11. Plasma toroidal rotation velocities, induced by two neutral beams in the same direction (Co-beams, Fig. 1), measured from Doppler shifts of atomic lines emitted at different radii.

different ionization potentials the rotational velocities could be mapped in considerable detail, both in toroidal and poloidal directions. Again, the long-wavelength lines of forbidden transitions in the ground configurations of the ions are required for these measurements, because of the necessary use of mirrors for viewing the approaching and receeding parts of the rotating plasma.

As a final sample of tokamak plasma diagnostics, we consider the problem of the large (about 30-fold) rise of the FeXXIV light intensity during the high-power neutral beam injection, shown in Fig. 10. Part of this rise is due to the threefold rise in plasma (electron) density mentioned above, but this still leaves about 10-fold increase in the FeXXIV density. The radial distributions of this density before, during, and shortly after the beam injection are shown in Fig. 12. The electron temperature also rose, from about 2.3 keV at 430 msec to about 3.0 keV at 550-600 msec, but this only aggravates the problem as the increased temperature should increase the ionization rate to FeXXV state, and therefore decrease the FeXXIV population (the expected ionization times of FeXXIV decrease from 7-8 msec at 430 msec to 2-3 msec at 550-600 msec.)

The most obvious explanation is of course a general increase of iron concentration, perhaps caused by sputtering by the energetic beam atoms. The rise-time on the time scale of several tens of milliseconds would be consistent with such explanation. However

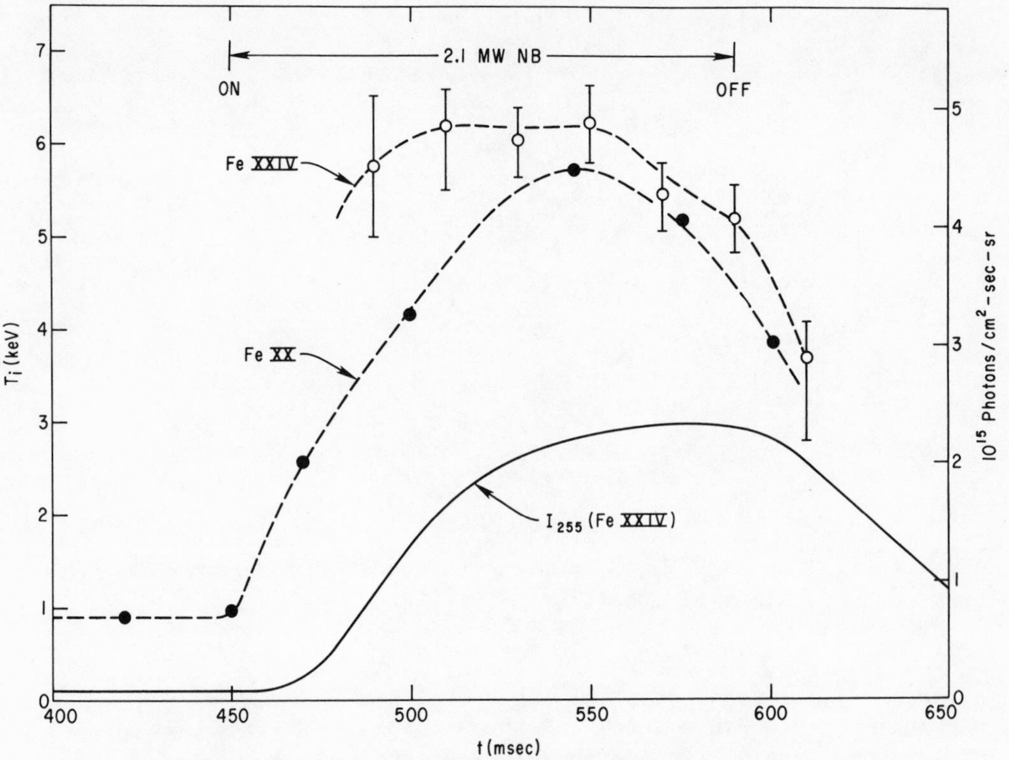

Fig. 10. Ion temperatures determined from the Doppler profiles of FeXX and FeXXIV lines, and the total intensity of the FeXXIV line, in the course of a NB injection in the PLT tokamak.

The FeXX ion radiates at somewhat larger radii (~ 10-15 cm) than FeXXIV. Figure 10 thus indicates a rapidly rising central ion temperature, which saturates in about 50 msec, while the $T_i(r)$ profile broadens for at least 50 msec longer. Both the saturation and the subsequent drop in ion temperature, while the beam is still on, are probably primarily caused by increasing plasma density, which changed rougly linearly from 1.7×10^{13} cm^{-3} at 430 msec to 5.3×10^{13} cm^{-3} at 600 msec.

An evident extension of the local Doppler temperature measurement is the determination of plasma rotations from the Doppler shift of the lines. Figure 11 shows the toroidal rotational velocities induced by unbalanced neutral beam injection in the PLT discharges.[30] The FeXX line again referes to the rotation near the center of the plasma, CV at about r = 30 cm , and HI near the periphery, r ≈ 40 cm. The latter, although emitted by a neutral atom, shows the shift because of the rapid charge-exchange collisions between H^0 and H^+. In principle, by using lines of ions of

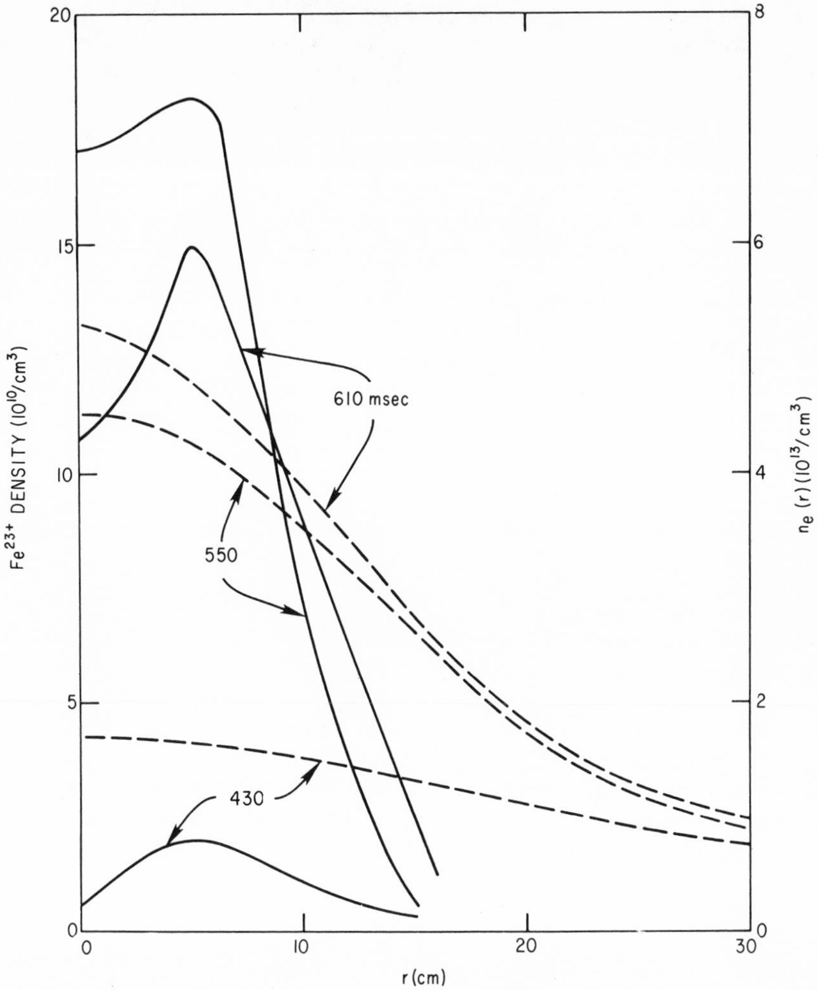

Fig. 12. Radial distributions of the lithiumlike FeXXIV ion (solid curves) and electron densities (dashed curves) during neutral beam injection. The times (msec) refer to Fig. 10.

lower ionization states, e.g. FeXX and FeXXII do not show appreciable increases, certainly nothing comparable to the FeXXIV behavior. Also, the rise of the FeXXIV light begins somewhat earlier near the center of the discharge, i.e. the density rise seems to spread outward from the center.

The probable cause of this phenomenon is enhanced recombination of the heliumlike FeXXV through charge-exchange with the neutral H^0 or D^0 atoms supplied by the beam. The near-central neutral hydrogen density is expected to rise to about 10^8 cm^{-3} or more during the four-beam injection,[4] and the charge-exchange cross-sections[43] appear to be sufficiently large to overbalance the ionization rates. Numerical modelling calculations[44] show that the magnitude of the observed effect can be explained by the charge-exchange process, although there may be some problems with detailed time-behavior. In any case, many of the potentially important aspects (e.g. the reflectivity of the hydrogen atoms on the walls, and toroidal variation of the neutral atom density) of the problem are still uncertain, and a considerable amount of detailed experimental data will be required for definitive answers. In particular, it should be very important to attempt to observe this phenomenon in different atoms, such as titanium, nickel and perhaps krypton in higher temperature (3-5 keV) plasmas, with the objective of changing the relationship between the relevant ionization and recombination rates, at the prevailing electron temperature.

Quite recently, the complementary part of the charge-exchange recombination of FeXXV has apparently been observed.[45] The iron K-line, observed with a high-resolution crystal spectrometer[39] shows a pronounced shift of the intensity of the components during the neutral beam injection, in the sense that the components due to heliumlike FeXXV drop and those of the lithiumlike FeXXIV increase. If quantitative analysis of the results substantiates this interpretation, a number of highly significant measurements will become possible. Among these are direct measurement of the FeXXV spatial distribution and its deviation from coronal equilibrium, as well as the radial profile of the H^0 density, and consequently the distribution of energy and momentum deposition by the neutral beam in the plasma.

The list of plasma diagnostics by means of atomic spectroscopy outlined in this paper is of course far from exhaustive. These experiments were chosen to illustrate various possibilities of measurements of important processes in high-temperature plasmas, and to point out some of the important atomic parameters required for quantitative interpretations of the measurements.

Acknowledgement

The experimental results presented in this paper are to a large extent due to the efforts of the scientist associated with the PLT tokamak and particularly to Dr. Szymon Suckewer.
The work has been supported by the United States Department of Energy under Contract No. EY-76-C-02-3073.

References

1. H. P. Furth, Nucl. Fusion 15, 487 (1975).
2. D. Grove et al., Plasma Physics and Controlled Nuclear Fusion

Research (Proc. 7th Intern. Conf., Innsbruck, 1978) IAEA, Vienna 1977, Vol I p. 21-33.
3. K. Bol et al., Plasma Physics and Controlled Nuclear Fusion Research (Proc. 7th Intern. Conf. Innsbruck, 1978) IAEA, Vienna 1979, Vol I, p. 1.
4. H. Eubank et al., Plasma Physics and Controlled Nuclear Fusion Research (Proc. 7th Intern. Conf. Innsbruck 1978) IAEA, Vienna 1979, Vol. I, p. 167.
5. E. Meservey, N. Bretz, D. Dimock, E. Hinnov, Nucl. Fusion 16, 593 (1976).
6. Equipe TFR, Nucl. Fusion 15, 1053 (1975).
7. E. Hinnov, et al., Plasma Physics 20, 723 (1978).
8. P. E. Stott, C. C. Daughney, R. A. Ellis, Jr., Nucl. Fusion 15, 431 (1975).
9. R. J. Hawryluk et al., PPPL-1534 (1979) to be published in Nuclear Fusion.
10. D. L. Moores, J. Phys. B 11 L403 (1978).
11. L. B. Golden and D. H. Sampson, J. Phys. B, 10, 2229 (1977).
12. D. H. Crandall, R. A. Phaneuf, B. E. Hasselquist, D. C. Gregory J. Phys. B 12 L249 (1979).
13. D. E. Post et al., Atomic Data and Nuclear Tables 20, 397 (1977).
14. L. Heroux, Proc. Phys. Soc. (London) 83, 121 (1964).
15. D. L. Dimock et al., Plasma Physics and Controlled Nuclear Fusion Research (Proc. 4th Intern. Conf., Madison 1971) IAEA, Vienna 1971 Vol I p. 451.
16. A. Merts et al., Report at Workshop on Atomic Processes in Controlled Thermonuclear Fusion Research, Los Alamos 1978.
17. O. Bely, Proc. Phys. Soc. 88, 587 (1966).
18. K. A. Berrington, P. G. Burke, P. L. Dufton, A. E. Kingston, J. Phys. B 10, 1465 (1977).
19. K. A. Berrington, P. G. Burke, P. L. Dufton, A. E. Kingston, A. L. Sinfailan, J. Phys. B 12, L275 (1979).
20. E. Hinnov, Astrop. J. 230 L197 (1979).
21. Y.-K. Kim and J. P. Desclaux, Phys. Rev. Lett. 36, 139 (1976).
22. L. Armstrong, Jr., W. R. Fiedler, Dong L. Lin, Phys. Rev. A 14, 1114 (1976).
23. J. Reader and J. Sugar, J. Phys. Chem, Ref. Data 4, 353 (1975).
24. K. G. Widing, Astrop. J. 222, 735 (1978).
25. H. Nussbaumer, Astrop. J. 166, 411 (1971).
26. S. O. Kastner, A. K. Bhatia, L. Cohen, Phys. Scr. 15, 259 (1977).
27. R. D. Cowan, "Spectra of Highly Ionized Atoms of Tokamak Interest" Los Alamos Scientific Laboratory Report LA-6679-MS (1977).
28. S. Suckewer and E. Hinnov, Phys. Rev. Lett. 41, 756 (1977).
29. S. Suckewer and E. Hinnov, Phys. Rev. A, 1979 (to be published).
30. S. Suckewer, H. P. Eubank, R. J. Goldston, E. Hinnov, N. Sauthoff, Phys. Rev. Lett. 43, 207 (1979).
31. S. Suckewer et al., PPPL Report 1563, to be published.

32. K. L. Aitken and M. F. A. Harrison, J. Phys. B 4, 1176 (1971).
33. G. H. Dunn, IEEE Transactions NS-23 929 (1976).
34. R. U. Datla, L. J. Nugent, H. R. Griem, Phys. Rev. A 14, 979 (1976).
35. A. Burgess, H. P. Summers, D. M. Cochrane, R. W. P. McWhirter, Mon. Not. R. Astron. Soc. 179 275 (1977).
36. E. Hinnov, Diagnostics for Fusion Experiments, Proc. Varenna Intern. School of Plasma Physics, 1978, E. Sindoni and C. Wharton, editors, Pergamon Press, Oxford, England, 1979.
37. E. Hinnov and F. W. Hofmann, J. Opt. Soc. Am. 53, 1259 (1963).
38. G. A. Doschek, U. Feldman, R. D. Cowan, L. Cohen, Astrop. J. 188 417 (1974).
39. M. Bitter et al., Phys. Rev. Lett. 42, 304, 1979.
40. K. Brau et al., Bull. Am. Phys. Soc. 23, 901 (1978).
41. E. Hinnov et al., Nucl. Fusion 18, 1305 (1978).
42. S. Suckewer and E. Hinnov, Nucl. Fusion 17, 945 (1977).
43. R. E. Olson et al., Phys. Rev. Lett. 41, 163 (1978).
44. D. E. Post, R. Hulse, E. Hinnov, S. Suckewer, (Proc. 11th Intern. Conf. of Physics of Electron and Atom Collisions) Kyoto, Japan, Sept. 1979.
45. M. Bitter, private communication, July 1979.

THERMALIZATION AND EXHAUST OF HELIUM IN A FUTURE THERMONUCLEAR REACTOR: PARTLY AN ATOMIC PHYSICS PROBLEM

H. W. Drawin

Association Euratom-CEA sur la Fusion

F-92260 Fontenay-aux-Roses, France

1. INTRODUCTION

The magnetically confined plasma of a continuously or quasi-continuously working thermonuclear fusion reactor must satisfy several severe conditions in order to allow operation in a self-sustained mode. Additional conditions should be fulfilled in order to get a high overall efficiency of the reactor: one concerns the slowing-down of the α-particles born in the D-T fusion process, the other is concerned with the exhaust of the helium.

The basic energy equation of a D-T fusion plasma is

$$^{2}D^{+} + {}^{3}T^{+} \rightarrow {}^{4}He^{++}\,(3.52\,\mathrm{MeV}) + {}^{1}n(14.1\,\mathrm{MeV});\; E^{total} = 17.6\,\mathrm{MeV} \quad (1)$$

The neutrons go directly to the walls where the energy of 14.1 MeV is transformed into heat (neglecting energy-producing or -consuming reactions in the T-production process within the blanket). The neutrons have no direct influence on the power balance of the plasma proper, although they determine indirectly the efficiency of the system. The dynamics of the plasma and especially its equilibrium state will to some extent depend on the behaviour of the He^{2+} ions (α-particles). The latter depends partly on the collision physics involved. In the following we will discuss some aspects of the α-particle problem in a future fusion reactor. Let us first recall some fundamental relations on which the concept of a D-T fusion reactor is based.

2. IGNITION, EQUILIBRIUM AND LAWSON CONDITION

2.1. Ignition Condition

Ignition of a fusion plasma is achieved when the power density $\dot{P}_\alpha$ given to the α-particles is equal to the loss rates of the radiation energy density, $\dot{P}_{rad}$, and of the thermal energy density, $\vec{\nabla}\cdot(E^{th}\,\vec{v})$. The ignition condition thus writes

$$\dot{P}_\alpha = \dot{P}_{rad} + \vec{\nabla}\cdot(E^{th}\,\vec{v}). \tag{2}$$

Denoting by σ_{DT} the cross-section for reaction (1), by n the particle density (with subscripts D, T and e for deuterons, tritons and electrons respectively) and by $Q_\alpha = 3.5$ MeV the energy gain by an α-particle per fusion event, we have for a pure D-T plasma (50% : 50% D-T mixture)

$$n_D + n_T = n_e \tag{3}$$

$$E^{th} = \frac{3}{2}(n_e + n_D + n_T)\,kT = 3n_e\,kT \tag{4}$$

$$\dot{P}_\alpha = \langle \sigma_{DT}\, v\rangle\, n_D n_T Q_\alpha = \frac{1}{4}\langle \sigma_{DT}\, v\rangle\, n_e^2\, Q_\alpha \tag{5}$$

In order to avoid the calculation of the energy flux density $E^{th}\,\vec{v}$ and its divergence, the diffusion term is simply replaced by

$$\vec{\nabla}\cdot(E^{th}\,\vec{v}) = \frac{E^{th}}{\tau} = \frac{3n_e\,kT}{\tau} \tag{6}$$

where τ is a characteristic diffusion time called the "mean confinement time". (In present large tokamaks τ lies between 25 and 120 ms). Assuming as the only radiation loss free-free bremsstrahlung one has

$$\dot{P}_{rad} = \dot{R}^{ff}(T)\,n_e^2 \tag{7}$$

where $\dot{R}^{ff}(T)$ is a function which depends only on electron temperature. With these relations the ignition condition can be put into the following form:

$$n_e\tau = \frac{3\,kT}{\frac{1}{4}\langle\sigma_{DT}\,v\rangle\,Q_\alpha - \dot{R}^{ff}(T)} \tag{8}$$

which - for given values of T and n_e - is a condition for the minimum confinement time τ. In slightly different form the ignition condition was first proposed by McNally[1,2]. The present form equals the one of Meade[3] who also considered the influence of impurities.

2.2. Equilibrium Condition

The equilibrium condition was first considered by Spitzer et al.[4]. We formulate it in a different manner.

For stationary operation, burned fuel must be replaced by incoming new fuel. This fuel might be "cold" or "hot". In the first case it must be ionised and heated to fusion temperatures by the fusion plasma itself. In the second case the neutral fuel might already have the "correct translational energy" of $(3/2)\,kT$ equivalent. In this case, the plasma has still to deliver the energy for ionising the neutrals and to heat the electrons to the plasma temperature. The plasma will receive energy from the injected fuel when the kinetic energy of the nuclei plus the one of the bound electrons is larger than the kinetic energy of the plasma particles. (This condition sets a lower limit of the energy in neutral beam injection heating.) In the following we assume that the fuel is initially "cold". Stationary operation requires that the following relations are fulfilled:

$$\frac{\partial}{\partial t}(n_D + n_T)\Big|_{\text{burned}} = \langle\sigma_{DT}\, v\rangle\, n_D\, n_T \qquad (9)$$

$$\dot{P}_\alpha = \dot{P}_{rad} + \vec{\nabla}\cdot(E^{th}\vec{v}) + \dot{P}_{fuel} \qquad (10)$$

where the first equation expresses the fact that the rate at which new fuel is fed into the system equals the rate at which it is burned. This ensures constancy of the plasma pressure provided T remains constant. $\dot{P}_{fuel}$ is given by the expression (with the ionisation energy neglected)

$$\dot{P}_{fuel} = 2\left[\frac{\partial}{\partial t}(n_D + n_T)\right]\frac{3}{2}kT = \frac{3}{4}\langle\sigma_{DT}\, v\rangle\, n_e^2\, kT . \qquad (11)$$

The factor two accounts for the fact that the electrons created in the ionisation process must be heated to the temperature T. The equation (10) leads thus to the following equilibrium condition

$$n_e\tau = \frac{3\,kT}{\frac{1}{4}\langle\sigma_{DT}\, v\rangle[Q_\alpha - 3\,kT] - \dot{R}^{ff}(T)} \qquad (12)$$

Since $3\,kT \ll Q_\alpha$ for all conditions of interest, the equilibrium condition for a self-sustained fusion plasma virtually equals the ignition condition (8).

2.3. Lawson Condition

The Lawson condition gives a lower limit for the product $n_e\tau$ in order to realise a reactor system with zero energy output. It was initially derived by Lawson[5] for a pulsed system and is always quoted in this form. For a pure D-T plasma the condition is (sign = for zero energy output, sign > for positive energy output)

$$n_e\tau \geqq \frac{3\,kT}{\frac{1}{4} < \sigma_{DT}\, v > Q_{DT}\, \frac{\eta}{1-\eta} - \dot{R}^{ff}(T)} \tag{13}$$

where η (with $0 < \eta < 1$) is the efficiency with which the total energy content of a duty cycle of duration τ can be reinjected into the plasma in order to repeat the cycle. $Q_{DT} = 17.6$ MeV is the total energy release per fusion event. It is interesting to see how the condition changes for a continuously operated system.

In the continuous regime, the plasma proper loses on the average per unit volume and unit time the energy $\vec{\nabla}\cdot(E^{th}\vec{v}) + \dot{P}_{rad}$. Fusion reactions yield neutrons of power density $\dot{P}_n = < \sigma_{DT}\, v > n_D\, n_T\, Q_n$, where $Q_n = 14.1$ MeV. Thus, the power density

$$\dot{P}_{wall} = \vec{\nabla}\cdot(E^{th}\vec{v}) + \dot{P}_{rad} + \dot{P}_n \tag{14}$$

is absorbed by the walls in form of heat and transformed into an energy form which can be reinjected into the plasma with an efficiency η in order to balance together with the α-particle power density $\dot{P}_\alpha$ the power loss rate $\vec{\nabla}\cdot(E^{th}\vec{v}) + P_{rad}$. Since D and T are burned refuelling is necessary in order to maintain the density and power output on a constant level. The fuel must be heated to fusion temperature. The power density required is $\dot{P}_{fuel}$ which must also be furnished by the system. The space average balance equation for the power density is therefore (sign = for zero energy output, sign > for positive energy output):

$$\eta[\vec{\nabla}\cdot(E^{th}\vec{v}) + \dot{P}_{rad} + \dot{P}_n] + \dot{P}_\alpha \geqq \vec{\nabla}\cdot(E^{th}\vec{v}) + \dot{P}_{rad} + \dot{P}_{fuel} \tag{15}$$

which can be put into the following form[6]

$$n_e\tau \geq \frac{3\,kT}{\frac{1}{4} < \sigma_{DT}\, v > \frac{1}{1-\eta}\,[\eta Q_n + Q_\alpha - 3\,kT] - \dot{R}^{ff}(T)} \tag{16}$$

The relations (13) and (16) are different. Since for all temperatures of interest $\eta Q_n + Q_\alpha - 3\,kT > \eta Q_{DT}$, the equation (16) yields for equal temperatures smaller values of $n_e\tau$ than equation (13). Although the difference is not very large it clearly shows that continuous operation is more advantageous than pulsed.

The general structure of all the above equations for $n_e\tau$ remains unchanged when impurity and synchrotron radiation are included. Only $\dot{R}^{ff}(T)$ has formally to be replaced by another expression. The radiation limit to tokamak operation has been discussed by Gibson[7]; space-dependent effects on the ignition and Lawson conditions have been treated by Kesner and Conn[8]. Curves for $n_e\tau$ with and without impurity radiation may be found in 4.

3. THERMALIZATION OF α-PARTICLES

According to reaction (1) the α-particles are born with an energy distribution function $f_\alpha = \delta(E - E_\alpha)$, peaked at $E = E_\alpha = 3.52$ MeV. In the above equations for $n_e\tau$ we always assumed equal temperature for all species. In a future reactor $kT \approx 10$ keV, i.e. $E_\alpha/kT \approx 350$. Hence, the α-particle energy must be reduced by more than two orders of magnitude until the α-particles are thermalized. This must be achieved in times $\tau_\alpha \ll \tau$, where τ is the confinement time defined in section 2; otherwise the plasma particles may have diffused to the walls before having interacted with the α-particles. It is thus necessary that the He-ions are well confined, as the fuel ions should be too.

At present there is no general agreement about the behaviour of the α-particles. There are predictions[9-12] that slowing-down and transport could be strongly influenced by a number of unstable modes; other theories predict simple collision-dominated behaviour with neo-classical diffusion.

For calculating the slowing-down of energetic test particles (α) in a heat bath filled with plasma particles (p) one starts with the kinetic (Boltzmann) equation

$$\frac{\partial f_\alpha}{\partial t} + \vec{v}_\alpha \cdot \frac{\partial f_\alpha}{\partial \vec{r}} + \frac{\vec{F}_\alpha}{m_\alpha} \cdot \frac{\partial f_\alpha}{\partial \vec{v}_\alpha} = \left[\frac{\partial f_\alpha}{\partial t}\right]_{\substack{\text{collision}\\ \text{radiation}}} \tag{17}$$

where $f_\alpha(\vec{r}, \vec{v}, t)$ is the velocity distribution function of the α's. The r.h.s. is the rate of change of f_α due to volume collision and radiation processes. Now, several assumptions are introduced:

1. The plasma is homogeneous : $\Rightarrow \frac{\partial f_\alpha}{\partial \vec{r}} = 0$.
2. The plasma is collision-dominated, i.e. force fields can be neglected : $\Rightarrow \vec{F}_\alpha = 0$.
3. The radiation field is so diluted that radiation processes can be neglected.

Thus, the equation (17) reduces to

$$\frac{\partial f_\alpha}{\partial t} = \left[\frac{\partial f_\alpha}{\partial t}\right]_{\text{collision}} \tag{18}$$

with the r.h.s. given by

$$\left[\frac{\partial f_\alpha}{\partial t}\right]_{\text{collision}} = n_p \int_{\vec{v}} d\vec{v}_p \left|\vec{v}_\alpha - \vec{v}_p\right| \int_\Omega d\Omega \frac{d\sigma_{\alpha p}}{d\Omega} \left[f'_p(\vec{v}'_p) f'_\alpha(\vec{v}'_\alpha) - f_p(\vec{v}_p) f_\alpha(\vec{v}_\alpha)\right] \tag{19}$$

where the primed quantities refer to post-collision velocities and distribution functions. The subscript p represents either deuterons, tritons or electrons: $d\sigma_{\alpha p}/d\Omega$ is the differential cross-section. The problem is thus reduced to a <u>pure collision problem</u> in velocity space.

4. It is finally assumed that the charged particles interact only through long-range Coulomb forces according to the classical cross-section formula

$$\frac{d\sigma_{\alpha p}}{d\Omega} = \frac{Z_\alpha^2 Z_p^2 e_o^4}{(4\pi\varepsilon_o)^2\, 4\, g_{\alpha p}^4\, \mu_{\alpha p}^2} \cdot \frac{1}{\sin^4(\Theta/2)} \tag{20}$$

where $d\Omega = \sin\Theta\, d\Theta\, d\psi$, $\vec{g}_{\alpha p} = \vec{v}_\alpha - \vec{v}_p$, $\mu_{\alpha p} = m_\alpha m_p/(m_\alpha + m_p)$. One further assumes that the Coulomb interactions lead to <u>elastic collisions</u>. (If this were true in general, free-free bremsstrahlung would not exist. The differences of particle momenta and energy before and after a Coulomb collision between electrons and ions go as bremsstrahlung to the photon field). For elastic collisions the following relations hold:

$$\vec{v}_\alpha = \vec{G}_{\alpha p} + \frac{m_p}{m_\alpha + m_p}\vec{g}_{\alpha p} ; \qquad \vec{v}_p = \vec{G}_{\alpha p} - \frac{m_\alpha}{m_\alpha + m_p}\vec{g}_{\alpha p}$$

$$\vec{v}'_\alpha = \vec{G}_{\alpha p} + \frac{m_p}{m_\alpha + m_p}\vec{g}'_{\alpha p} ; \qquad \vec{v}_p = \vec{G}_{\alpha p} - \frac{m_\alpha}{m_\alpha + m_p}\vec{g}'_{\alpha p} \tag{21}$$

$$\Delta\vec{v}_\alpha = \vec{v}'_\alpha - \vec{v}_\alpha = \frac{\mu_{\alpha p}}{m_\alpha}\Delta\vec{g}_{\alpha p} ; \qquad \Delta\vec{v}_p = \vec{v}'_p - \vec{v}_p = -\frac{\mu_{\alpha p}}{m_p}\Delta\vec{g}_{\alpha p} \tag{22}$$

where $\vec{G}_{\alpha p}$ is the constant center-of-mass velocity. From equation (22) follows

$$\Delta\vec{v}_p = -\frac{m_\alpha}{m_p}\Delta\vec{v}_\alpha \tag{23}$$

Due to the long-range interactions, many very very small deflections, assumed to be independent from each other, lead to a change $\Delta\vec{v}$ of $\vec{v}$. Also the $\Delta\vec{v}$ are still small quantities. Putting

$$\vec{v}'_\alpha = \vec{v}_\alpha + \Delta\vec{v}_\alpha , \qquad \vec{v}'_p = \vec{v}_p + \Delta\vec{v}_p \tag{24}$$

one can expand the post-collision distributions f' in powers of $\Delta\vec{v}$; for f'_α :

$$f'_\alpha(\vec{v}'_\alpha) = f_\alpha(\vec{v}_\alpha + \Delta\vec{v}_\alpha) = f_\alpha(\vec{v}_\alpha) + \frac{\partial f_\alpha}{\partial\vec{v}_\alpha}\cdot\Delta\vec{v}_\alpha + \frac{1}{2}\frac{\partial^2 f_\alpha}{\partial\vec{v}_\alpha\partial\vec{v}_\alpha} : \Delta\vec{v}_\alpha\,\Delta\vec{v}_\alpha + \cdots \tag{25}$$

and similarly for f'_p. Substituting the f' in equation (19) by the expanded form, with $\Delta\vec{v}_p$ substituted by equation (23), and retaining only terms through second order in $\Delta\vec{v}_\alpha$ yields the Boltzmann Fokker-Planck equation

$$\frac{\partial f_\alpha}{\partial t} = n_p \int_{\vec{v}_p} d\vec{v}_p\,\vec{g}_{\alpha p} \int_\Omega d\Omega \frac{d\sigma_{\alpha p}}{d\Omega}\Bigg[-\frac{m_\alpha}{m_p} f_\alpha\,\Delta\vec{v}_\alpha\cdot\frac{\partial f_p}{\partial\vec{v}_p} + f_p\,\Delta\vec{v}_\alpha\cdot\frac{\partial f_\alpha}{\partial\vec{v}_\alpha}$$

$$+\frac{1}{2}\left(\frac{m_\alpha}{m_p}\right)^2 f_\alpha\Delta\vec{v}_\alpha\Delta\vec{v}_\alpha : \frac{\partial^2 f_p}{\partial\vec{v}_p\,\partial\vec{v}_p} - \frac{m_\alpha}{m_p}\Delta\vec{v}_\alpha\Delta\vec{v}_\alpha : \frac{\partial f_\alpha}{\partial\vec{v}_\alpha}\frac{\partial f_p}{\partial\vec{v}_p} + \frac{1}{2}\Delta\vec{v}_\alpha\Delta\vec{v}_\alpha : \frac{\partial^2 f_\alpha}{\partial\vec{v}_\alpha\,\partial\vec{v}_\alpha}\Bigg] \tag{26}$$

In the present case, f_p represents the Maxwell distribution for the plasma particles, f_α the distribution function of the α-particles peaked at E_α = 3.52 MeV for t = 0. The $\Delta\vec{v}$ can be expressed by the collision coordinates, and integration over angles can directly be performed, see e.g. Montgomery and Tidman[13]. Terms proportional to $\partial f_\alpha/\partial \vec{v}_\alpha$ describe the slowing-down of the velocity along $\vec{v}_\alpha$, those proportional to $\partial^2 f_\alpha/\partial \vec{v}_\alpha \partial \vec{v}_\alpha$ a dispersion, i.e. the smearing-out of the initially peaked distribution function.

The solution of eq. (26) leads to the following general results (see also Spitzer[14] and Butler and Buckingham[15]): the high-energy α-particles heat preferentially the electrons. Only a few percent of the initial α-particle energy is given directly to the ions. Owing to collisions with the electrons the initial velocity $v_\alpha(0)$ decreases according to

$$v_\alpha \simeq v_\alpha(0)\, e^{-t/\tau_{\alpha e}} \tag{27}$$

with the relaxation time $\tau_{\alpha e}$ given by (in SI units)[14-16]

$$n_e \tau_{\alpha e} \cong \frac{3}{16}\left(\frac{8}{\pi}\right)^{1/2} \frac{m_\alpha}{m_e^{1/2}} \frac{(kT)^{3/2}}{\ln \Lambda_{\alpha e}} \left(\frac{4\pi \varepsilon_0}{Z_\alpha e_0^2}\right)^2 , \tag{28}$$

and numerically (n_e in cm^{-3}, T in ^oK, $\tau_{\alpha e}$ in s)

$$n_e \tau_{\alpha e} = \frac{0.508 \times 10^3\, T^{3/2}}{\ln(4.01 \times 10^5\, T/n_e^{1/2})} \tag{29}$$

In a time of order $\tau_{\alpha e}$ the α-particle energy has approximately decreased by a factor of ten. It is interesting to compare the values of $\tau_{\alpha e}$ with the confinement time τ. For $T = 10^8\ ^oK$ and $n_e = 5.10^{14}\ cm^{-3}$ the Lawson condition for a pure D-T plasma gives $\tau \approx 0.1$ s. Practically the same value is obtained from equation (29). That means that the slowing-down time $\tau_{\alpha e}$ of the α-particle energy is of the same order of magnitude as the confinement time.

The energy transfer from the electrons to the ions takes place also in times of the order of $\tau_{\alpha e}$.

From the Fokker-Planck equation it follows that the α-particles give their energy equally to the electrons and deuterons (tritons) when the actual α-particle energy E'_α is

$$E'_\alpha \approx 20\ kT \tag{30}$$

This relation defines the so-called characteristic α-particle velocity

$$v_{\alpha}^{char} = \left(\frac{2\,E'_{\alpha}}{m_{\alpha}}\right)^{\frac{1}{2}} \approx \left(\frac{40\ kT}{m_{\alpha}}\right)^{\frac{1}{2}} \tag{31}$$

When the α-particle energy has decreased below 20 kT the energy is given preferentially to the ions. As mentioned above, this direct transfer represents only a few percent of the total α-particle energy. Future experiments will show whether the α-particle behaviour is collision-dominated or not. Therefore a direct measurement of the distribution function f_{α} is essential in order to understand the physics of the thermalization process.

4. EXHAUST OF HELIUM

A D-T function reactor operating in a steady-state must continuously exhaust helium. From the energetic point of view the ideal case would be, of course, to exhaust helium only, with a rate equal to the incoming fuel. This is physically impossible due to the large diffusion fluxes to and from the walls of the plasma as a whole, in which helium must be considered as an impurity. This means that helium is exhausted with a large quantity of unburnt fuel.

Neo-classical impurity transport predicts an enrichment of higher Z-impurities in the central part of a tokamak plasma, see e.g. Papoular[17]. Consequently a high exhaust rate must be maintained in order to clean the plasma from the ash. Other theories predict that the α-particle transport is anomalously high due to instabilities, which would therefore favour continuous exhaust, see e.g. 12, but energy would be missing for α-particle heating.

Recently, Harbour and Harrison[18] have analysed theoretically the exhaust of a fusion reactor to a divertor target in the collisionless regime of a "pure" D-T plasma. They arrive at the following conclusion: "Thus it appears that the concept of a reactor exhausted by purely collisionless processes presents formidable practical problems ....., an alternative would be to exhaust in the collision-dominated mode whereby electrons lose energy by radiative collisions in a high-density exhaust plasma."

In the high-density regime a number of atomic processes may play a role which have not yet been discussed in the context of plasma exhaust from a fusion reactor. Figure 1 shows schematically the physical situation for a D-T plasma with helium as the only impurity. Electron and ion densities are maximum in the plasma centre. Since the temperature is also highest there, the α-particle production rate and, thus, the He^{2+} ion density has a maximum in the centre too. The local charge neutrality condition is $n_e = \Sigma z n_i^{(z)}$. The neutral particle density is highest in the

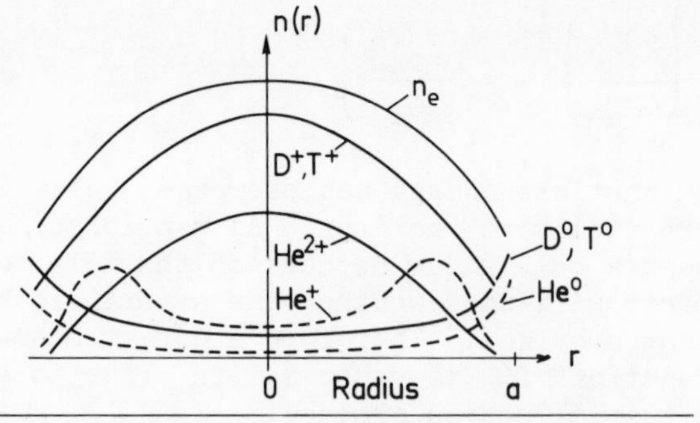

Fig. 1. Radial distribution of helium in a hypothetical D-T reactor. a is the plasma radius.

boundary region at $r \approx a$ and decreases rapidly towards the plasma centre. But the neutral particle density will never be negligible as far as concerns its effects on the plasma properties. In the smaller machines, the neutral hydrogen isotope density (D^0, T^0) will stay on a relatively high level[19,20] ($n_0 \approx 10^6 \ldots 10^8$ cm^{-3}) due to the common action of the divergence of the diffusion fluxes and the resonant charge exchange processes of type (H stands for the hydrogen isotopes)

$$H^+ + H^0 \rightarrow H^0 + H^+ \tag{32}$$

In future large machines and also in smaller machines operated at higher density (Alcator) the neutral particle density is simply determined by the local coronal ionisation-recombination equilibrium[21]. This will hold for both the hydrogen isotopes and helium until some critical distance λ from the plasma boundary. Within λ, the particle dynamics will then partly depend on other atomic processes. Since the dominant processes are different for hydrogen (isotopes) and helium, λ will be different for these two species. With regard to the exhaust problem in the collision-dominated regime it is therefore important to consider the various atomic processes in which helium is involved.

1. Central hot part of the plasma (1 keV $\leq$ kT $\leq$ 10 keV)

The He^{2+} ions created in the fusion process are subject to collisions with electrons, ions H^+ and neutral particles H^0.

Collisions with the electrons lead to He^+, mostly in the ground state, according to

$$He^{2+} + e \rightarrow He^+ + h\nu \tag{33}$$

the photon being lost as recombination radiation. Charge exchange processes with H^+ can be neglected. Collisions of He^{2+} with neutral hydrogen atoms according to

$$He^{2+} + H\ (1s) \rightarrow He^+ + H^+ \tag{34}$$

have cross-sections of the order of 10^{-17} to 10^{-15} cm^2, see also 22, 23.

We have now to consider the collisions of He^+ with the other plasma particles. The He^+ - H^0 charge exchange cross-sections are very small. Further, $n(H^0\ (1s)) \approx (10^{-8} \ldots 10^{-5})n_e$ in the central part of a future reactor. It thus follows that

$$n(He^+)\, n(H^0) < \sigma_{He^+,H^0}\, v > \ll n(He^+)\, n_e < \sigma_{He^+,e}\, v_e >$$

which means that the ratio $n(He^{2+})/n(H^+)$ will over a large part of the plasma cross-section be governed by the recombination reaction (33) and the ionisation reaction

$$He^+ + e \rightarrow He^{2+} + 2e\ . \tag{35}$$

As long as diffusion processes are negligible this ratio will correspond to the local coronal ionisation equilibrium values.

2. Boundary region (kT < 1 keV)

The situation changes in the boundary region where the density of the hydrogen neutrals is dominated by charge exchange processes (see Fig. 2). The density of hydrogen atoms in the ground state may reach values far above the coronal equilibrium values. Under such conditions also the excited state populations $n(H(i))$ are enhanced, since $n(H(i)) \propto n(H(1s))$. Table 1 gives some values of the ratio of excited hydrogen atoms with principal quantum number i to hydrogen ground state atoms at $kT \cong 10^2$ eV and 10^3 eV, after 24.

Table 1

Values of n(H(i))/n (H(1s) at $n_e = 10^{13}$ cm^{-3}.

kT [eV]	i = 2	i = 3	i = 5	i = 10
10^2	3.1×10^{-2}	2.7×10^{-3}	3.7×10^{-4}	2.7×10^{-5}
10^3	3.5×10^{-2}	2.2×10^{-3}	3.7×10^{-4}	3.1×10^{-5}

According to the measurements of Burniaux et al.[25] the charge exchange cross-sections of these excited atoms with He^{2+} ions are very large (see Fig. 3). Whether the excited neutral atoms will contribute to the recombination of He^{2+} or not will not only depend on the electron temperature (which determines the recombination coefficient of He^{2+} and the excitation coefficients of the hydrogen levels) but also on the ratio of n(H)/n_e in the boundary layer. If this ratio is larger than 10^{-3}, a non-negligible contribution to the recombination process can be expected to originate from charge exchange[26].

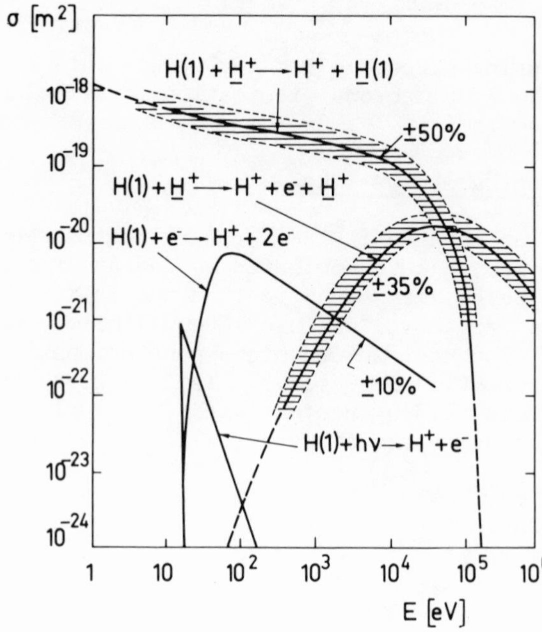

Fig. 2. Cross-sections for atomic hydrogen.

At present one does not know the neutral particle density and the electron density in the boundary region of tokamak plasmas, nor is the temperature known. All published values are order of magnitude extrapolations or assumptions. A value of $n(H^0)/n_e > 10^{-3}$ seems very probable.

The further recombination to neutral helium is expected to be governed by electronic collisions due to the small charge exchange cross-sections for He^+ - H collisions. It might be that the ultimate recombination to He^0 occurs only on the divertor plates, because of insufficient cooling of the plasma in the boundary layer.

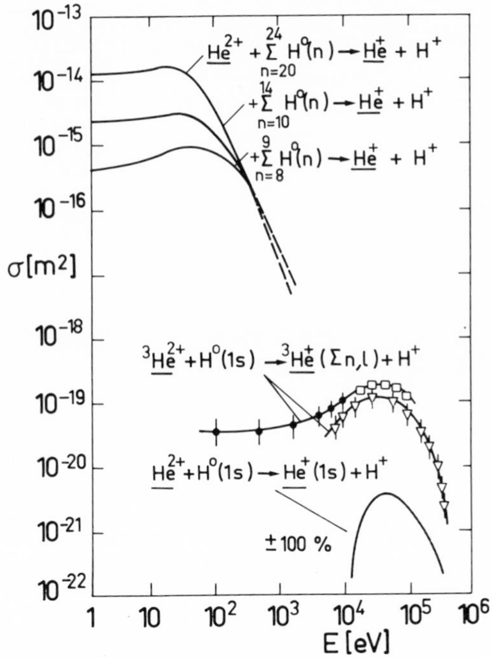

Fig. 3. Charge exchange cross-sections for He^{2+}, see also M. B. Shaw and H. B. Gilbody, J. Phys. B. 11, 121 (1978) Nutt et al. ref. 22, Burniaux et al. ref. 25.

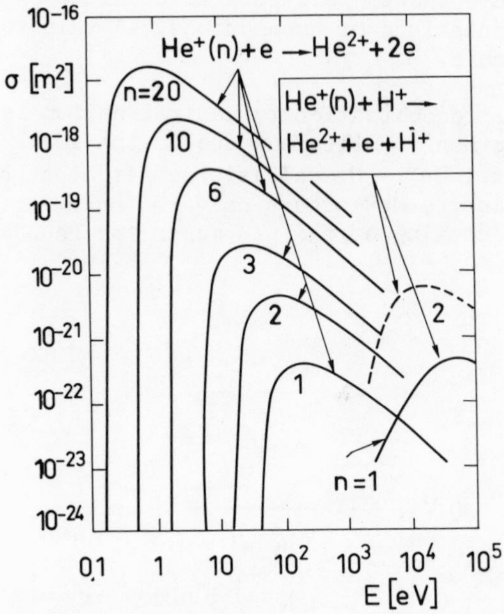

Fig. 4. Ionisation cross-sections for He^+ in different excited (n > 1) and unexcited states.

REFERENCES

1. J. Rand McNally, jr., in Nuclear Data in Science and Technology, Proc. Symp. Paris 1973, 2, I.A.E.A. Vienna (1973) 41.
2. J. Rand McNally, jr., Nuclear Fusion 17, 1273 (1977).
3. D. M. Meade, Nuclear Fusion 14, 289 (1974).
4. L. Spitzer et al., Report No. NYO-6047, USAEC, Washington, D.C. (1954).
5. J. D. Lawson, Proc. Phys. Soc. (London) B 70, 6 (1957).
6. H. W. Drawin, J. de Physique 40, C1, 73 (1979).
7. A. Gibson, Nuclear Fusion 16, 546 (1976).
8. J. Kesner, R. W. Conn, Nuclear Fusion 16, 397 (1976).
9. A. B. Mikhailovskii, Sov. Phys. J.E.T.P. 41, 890 (1976).
10. D. G. Lominadze, A. B. Mikhailovskii, W. M. Tang, Sov. J. Plasma Phys. 2, 286 (1976).
11. D. K. Bhadra, Report General Atomic No. A 14729, San Diego (1977).
12. D. J. Sigmar, H. C. Chan, Nuclear Fusion 18, 1569 (1978).

13. D. C. Montgomery, D. A. Tidman, Plasma Kinetic Theory, McGraw Hill 1964.
14. L. Spitzer, jr., Physics of Fully Ionized Gases, 2nd edit. John Wiley Interscience, New York (1967).
15. S. T. Butler, M. J. Buckingham, Phys. Rev. 126, 1 (1962).
16. O. Kofoed-Hansen, Risø report No. 385, Risø 1978, (On Alpha-Particle Heating of a Thermonuclear Reactor).
17. R. Papoular, Nuclear Fusion 16, 679 (1976).
18. P. J. Harbour, M. F. A. Harrison, Nuclear Fusion 19, 695 (1979).
19. D. F. Düchs, D. E. Post, P. H. Rutherford, Nuclear Fusion 17, 565 (1977).
20. J. G. Gilligan, S. L. Gralnick, W. G. Price, T. Kammash, Nuclear Fusion 18, 63 (1978).
21. Yu. N. Dnestrovskij, S. E. Lysensko, A. I. Kislyakov, Nuclear Fusion 19, 293 (1979).
22. W. L. Nutt, R. W. McCullough, K. Brady, M. B. Shaw, H. B. Gilbody, J. Phys. B. 11, 1457 (1978).
23. H. Ryufuku, T. Watanabe, Theoretical studies on one-electron charge transfer, in Contributions of the Research Group of Atoms and Molecules to Heavy Ions Science, edited by N. Oda, K. Hijikata, Institute for Nuclear Study, University of Tokyo, Tanashi-shi, Tokyo, Japan, March 1979.
24. H. W. Drawin, F. Emard, Physica 85 C, 333 (1977).
25. M. Burniaux, F. Brouillard, A. Jognaux, T. R. Govers, S. Szucs, J. Phys. B. 10, 2421 (1977).
26. H. W. Drawin, Physikal. Blätter 35, 119 (1979); 35, 150 (1979).

INDEX